New Insights in the Research of Bioactive Compounds from Plant Origin with Nutraceutical and Pharmaceutical Potential

New Insights in the Research of Bioactive Compounds from Plant Origin with Nutraceutical and Pharmaceutical Potential

Editors

Ivayla Dincheva
Ilian Badjakov
Bistra Galunska

MDPI • Basel • Beijing • Wuhan • Barcelona • Belgrade • Manchester • Tokyo • Cluj • Tianjin

Editors

Ivayla Dincheva
Agrobiotechologies
Agrobioinstitute, Agricultural Academy
Sofia
Bulgaria

Ilian Badjakov
Agrobiotechnologies
Agrobioinstitute, Agricultural Academy
Sofia
Bulgaria

Bistra Galunska
Biochemistry, Molecular Medicine and Nutrigenomics
Medical University
Varna
Bulgaria

Editorial Office
MDPI
St. Alban-Anlage 66
4052 Basel, Switzerland

This is a reprint of articles from the Special Issue published online in the open access journal *Plants* (ISSN 2223-7747) (available at: www.mdpi.com/journal/plants/special_issues/nutraceutical_pharmaceutical_potential).

For citation purposes, cite each article independently as indicated on the article page online and as indicated below:

LastName, A.A.; LastName, B.B.; LastName, C.C. Article Title. *Journal Name* **Year**, *Volume Number*, Page Range.

ISBN 978-3-0365-6510-1 (Hbk)
ISBN 978-3-0365-6509-5 (PDF)

Contents

About the Editors

Ivayla Dincheva

Assoc. Prof. Dr. Ivayla Dincheva is a leading scientist in the field of phytochemistry. Her research is concentrated in analyses of biologically active substances by LC/MS/MS and GC-MS, and the validation of analytical procedures according to ICH. She has actively participated in eight projects with international funding (EEA Grants, "America for Bulgaria" Foundation, Horizon 2020, etc.), and fifteen with national funding (BNSF) (five of which are ongoing). The results of her work are presented in 65 publications in journals with impact factors. She has been a Guest Editor of four Special Issues in the journal *Plants* (MDPI, Switzerland). Her research interests include metabolomics, natural product chemistry, bioactive compounds, and chromatographic methods.

Ilian Badjakov

Assoc. Prof. Dr. Ilian Badjakov is an established scientist in the field of genomics and metabolomics. The results of his work have been published in 50 articles (35 in journals with an impact factor). He has participated in numerous international and national projects and is a Guest Editor of four Special Issues in the journal *Plants* (MDPI, Switzerland). His research interests include natural products, bioactive compounds, GC-MS, LC-ESI-MS/MS, and biotechnology.

Bistra Galunska

Bistra Galunska is a Professor in Department of Biochemistry, Molecular Medicine and Nutrigenomics, Faculty of Pharmacy, Medical University "Prof. Dr. Paraskev Stoyanov" -Varna. Her research interests include molecular biomarkers, metabolic profiling, metabolomics and metabolome analysis, proteomics and proteomic analysis, clinical biochemistry and pathobiochemistry, pharmacobiochemistry, nutrient biochemistry, biochromatography, etc.

Preface to "New Insights in the Research of Bioactive Compounds from Plant Origin with Nutraceutical and Pharmaceutical Potential"

Focusing on the importance of bioactive compounds derived from plant origins for human health, this book presents solid scientific evidence on the latest research in this area. The development of high-throughput and efficient analytical approaches enabling the identification of plant-based compounds, methods for their extraction, isolation and structural characterization, as well as the establishment of new protocols for the evaluation of bioactivities, will contribute to discovering therapeutics for the prevention and treatment of various diseases.

This book is a contribution from a wide range of scientists from different research backgrounds, and its content aims to present a comprehensive overview on new insights into the research of bioactive compounds of plant origin with nutraceutical and pharmaceutical potential.

Scientists, researchers, and students will find recent and useful information to help them conduct their future research, analyze their key findings, and identify areas of differing perspectives where there are gaps in knowledge.

The production of this book involved the efforts of many people, especially the authors who contributed articles, and the reviewers and editors who have shown patience during the assembling and editing of all the contributions.

Ivayla Dincheva, Ilian Badjakov, and Bistra Galunska
Editors

plants MDPI

Editorial

New Insights into the Research of Bioactive Compounds from Plant Origins with Nutraceutical and Pharmaceutical Potential

Ivayla Dincheva [1,*], Ilian Badjakov [1] and Bistra Galunska [2]

1 Department of Agrobiotechnologies, Agrobioinstitute, Agricultural Academy, 8 Dragan Tsankov blvd., 1164 Sofia, Bulgaria
2 Department of Biochemistry, Molecular Medicine and Nutrigenomics, Faculties of Pharmacy, Medical University "Prof. Dr. Paraskev Stoyanov", 84 Tzar Osvoboditel Str., 9000 Varna, Bulgaria
* Correspondence: ivadincheva@yahoo.com

Plant bioactive compounds are essential for human health due to their multiple biological effects, such as antioxidant, anticarcinogenic, antiallergenic, anti-inflammatory, antimutagenic, and antimicrobial activities, which can have beneficial effects on various noncommunicable diseases, such as autoimmune, inflammatory, cardiovascular, cancer, metabolic, and neurodegenerative diseases. Identifying these components and establishing their beneficial health effects are extremely popular activities of scientific inquiry. The screening of natural sources for novel biologically active metabolites has been an essential part of several drug discovery programs.

The aim of this Special Issue is to present the recent developments in high-throughput and efficient analytical approaches, which have enabled the identification of plant-based compounds, the establishment of new protocols for the evaluation of bioactivities, and the development of methods for the extraction, isolation, and structural characterization of new bioactive components with nutraceutical and therapeutic potential.

Three papers in this Special Issue demonstrate the beneficial effects of bioactive compounds from natural sources such as *Pinus* spp., *Ranunculus* spp., and *Galphimia* spp. It has been proved that these components play an important role in the prevention of multiple pathologies. Some of them reduce the risk of numerous diseases, such as, for example, cancer, diabetes, and Alzheimer's dementia, due to their strong antioxidant activity, whereas others stimulate defense mechanisms, thereby enhancing the response to oxidative stress and preventing widespread damage. Pinosylvin has been shown to perform numerous and diverse actions through its ability to block, interfere, and/or stimulate the major cellular targets responsible for several disorders [1]. Antibacterial and antiprotozoal effects, immunomodulatory and anticarcinogenic properties, and the anti-inflammatory and analgesic actions of *Ranunculus* spp. plants used in traditional medicine applications have been confirmed [2]. *Galphimia* spp. is an important medicinal plant in Mexico, which has many applications in traditional medicine. The main investigations have focused on its anxiolytic and sedative effects due to the presence of the modified triterpenoids, galphimines. However, the genes encoding the enzymes of the galphimine synthesis pathway are still unknown. Hence, in this study, a comparative transcriptome analysis between two contrasting populations of *Galphimia* spp.—a galphimine producer and a non-galphimine producer—is performed using RNA-Seq with the Illumina NextSeq 550 platform to identify putative candidate genes that encode the enzymes of this metabolic pathway. Most of the transcripts were grouped at the molecular function level of gene ontology. In galphimine-producer plants, a larger number of highly expressed transcripts related to acyclic and polycyclic terpene synthesis were identified. As putative candidate genes involved in the galphimine synthesis pathway, P450 family members and enzymes with kinase activity were identified [3].

Citation: Dincheva, I.; Badjakov, I.; Galunska, B. New Insights into the Research of Bioactive Compounds from Plant Origins with Nutraceutical and Pharmaceutical Potential. *Plants* **2023**, *12*, 258. https://doi.org/10.3390/plants12020258

Received: 3 January 2023
Accepted: 3 January 2023
Published: 5 January 2023

Essential oils have been used for centuries by many cultures around the world for various purposes, including perfumery, aromatherapy, cosmetics, medicine, feed, bactericides, insecticides, fungicides, etc. Constant research in these areas represents a search for alternatives for more efficient drugs, as well as a way to obtain new products and supplies. Six papers provide the most recent perspectives on the impacts of essential oils on human health and disease.

One article presents the essential oil yield, composition, and bioactivity of *Sagebrush* spp. from the Bighorn Mountains [4]. There was significant variability in the chemical profile and antioxidant capacity between species and subspecies. However, none of the sagebrush EOs had significant antimicrobial, antimalarial, or antileishmanial activity, nor did they contain podophyllotoxin. The results obtained showed an immense chemical diversity that presents an opportunity for the selection of varieties with high concentrations of EOs with desirable compositions. These chemotypes can be selected and possibly introduced into a culture to be grown for the commercial production of these compounds to meet specific industry needs. Another two manuscripts describe the antioxidant and antimicrobial activities of *Thymus vulgaris* L. [5] and *Syzygium aromaticum* L. [6], which are essential oils produced in Slovakia. An in situ antifungal analysis of food models shows that the vapor phase of these oils can inhibit the growth of the microscopic filamentous fungi of the genus *Penicillium*. Additionally, the evaluation of the changes in the biofilm structure of microorganisms on glass and wood surfaces, such as *Salmonella enteritidis, Pseudomonas fluorescens, Pseudomonas expansum,* and *Pseudomonas chrysogenum*, were performed using MALDI-TOF MS. Due to the properties of the tested *T. vulgaris Syzygium aromaticum* L. essential oils, they can be used in the food industry as natural supplements to extend the shelf life of the foods. One study describes the metabolite profile of *Artemisia annua* L. [7], which was grown wild in Romania for the first time. Subsequently, a simple and inexpensive nanocarrier system was developed that capitalizes on the therapeutic properties of both the whole plant and magnetic Fe_3O_4 nanoparticles. Further studies are necessary to investigate the biological properties and the bioavailability of the new nanocarrier system. A recent investigation determined the composition of the essential oil and hydrolate of *Dracocephalum moldavica* L. [8], which was grown in the Republic of Serbia and distilled under semi-industrial conditions to determine the in vitro antioxidant and antimicrobial activities. The oil demonstrated better biological activities compared with the hydrolates against *Staphylococcus aureus, Escherichia coli, Salmonella Typhimurium,* and *Listeria monocytogenes,* which represent sources of food-borne illness at the global level. One paper compares the essential oil composition and antimicrobial activity of *Helichrysum arenarium* Moench and *Helichrysum italicum* G. Don, which already have an established international market [9]. Overall, the EO of *H. italicum* from all locations was more effective against *Staphylococcus aureus*. A moderate antimicrobial effect was found against *Candida krusei* and *Candida tropicalis.* The lowest antimicrobial activity was found against *Yersinia enterocolitica*. In general, the tested EOs were more effective against Gram-positive bacteria.

Eight manuscripts covered the topics of analytical strategies for bioactive compound identification and quantification, herbal preparations and natural medicines, and the in vivo and in vitro bioactivity of botanicals. One paper clearly demonstrated the immunostimulation, hematopoietic, and antiviral potential of *Sambucus ebulus* L.'s fruit aqueous extract (FAE), with attention given to its endoplasmic reticulum (ER) stress-reducing potential [10]. J774A.1 macrophages were treated with SE FAE alone or under the condition of lipopolysaccharide (LPS) stimulation. The phytochemical composition of the herbal extracts was analyzed by using GC-MS and LC–MS/MS. Transcription and protein levels were defined using qPCR and Western blot, respectively. Extracts exerted an immunostimulation potential by stimulating IL-6, TNFα, Ccl2, COX2, and iNOS transcription without inducing ER stress. The SE FAE suppressed the LPS-induced transcription of inflammation-related genes (IL-1β, IL-6, TNFα, Ccl2, Icam-1, Fabp4, COX2, iNOS, Noxo1, IL-1ra, and Sirt-1) and reduced the protein levels of iNOS, peIF2α, ATF6α, and CHOP. The effects were comparable to that of salicylic acid. SE suppresses LPS-stimulated inflammatory markers at the transcription and translation levels. The targeting of ER stress is possibly another mechanism underlying its anti-inflammatory potential. These findings reveal the potential of SE as being a beneficial therapeutic for inflammation and ER-stress-related pathological conditions.

Jacaranones are a small group of specific plant metabolites with promising biological activities. Moreover, they can serve as chemotaxonomic markers. The phytochemical investigation of *Crepis pulchra* L. (Asteraceae) resulted in three jacaranone derivatives (jacaranone; 2,3-dihydro-2-hydroxyjacaranone; and 2,3-dihydro-2-methoxyjacaranone) [11]. This is the first report on the isolation of jacaranones from a species belonging to the Cichorioideae subfamily of Asteraceae. Jacaranone derivatives were subjected to an in vitro antiproliferative assay against a panel of human cancer cell lines (MCF-7, MDA-MB-231, HeLa, and C33A), revealing high or moderate activities. Jacaranone (2) showed the highest antiproliferative activity against MDA-MB-231 (human breast cancer) and C33A (human cervical cancer) cells. The volatile compounds of eight peach varieties grown in Bulgaria (*Prunus persica* L.), including "Filina", "Gergana", "Ufo-4", "July lady", "Laskava", "Flat Queen", "Evmolpiya", and "Morsiani 90", were analyzed for the first time [12]. Gas chromatography–mass spectrometry (GC–MS) analysis and the HS-SPME technique revealed the presence of 65 volatile compounds. According to the provided principal component analysis (PCA) and hierarchical cluster analysis (HCA), the relative quantities of the identified volatile components depended on the studied peach variety. The results obtained could be successfully applied to the metabolic chemotaxonomy of peaches. Studies regarding the pharmaceutical potential of derivative products (polysaccharides and polyphenols) from plantain displayed prominent antioxidant, antifungal, antitumor, and prebiotic activities (particularly for the polysaccharide fraction) [13]. A comparative phytochemical investigation of *Aronia melanocarpa* L. fruit juices on the Bulgarian market clearly present with a high product quality due to their health-promoting compounds, such as phenolic acids, proanthocyanins, and anthocyanins, which define them as functional foods [14]. Promising new findings have been reported and discussed through the frame of the anti-inflammation, antiandrogen, and hair-growth-promoting potential of shallot (*Allium ascalonicum* L.) extract [15]. One publication clearly presents the antidiabetic activity of Mediterranean edible plant materials using the DIA-DB inverse virtual screening web server. Authors founded that the flavonoids are the most active phytochemicals as they modulate the function of 17 protein targets and present a high structural similarity to antidiabetic drugs. Their antidiabetic effects are linked with three mechanisms of action, namely: (i) the regulation of insulin secretion/sensitivity, (ii) the regulation of glucose metabolism, and (iii) the regulation of lipid metabolism [16]. Molecular docking studies reveal that the Swingle can be a potential source of natural products due to its antibiotic-potentiating activity and since it is anti-SARS-CoV-2 [17].

In recent decades, bioactive compounds from natural sources have attracted substantial attention and have been subjected to extensive research due to their antioxidant properties and their use potential in the promotion of health and the prevention of disease. Some of the most studied topics, in this regard, represent the focus of this Special Issue.

Author Contributions: All authors contributed to the design, writing and editing of this Editorial article. All authors have read and agreed to the published version of the manuscript.

Funding: This research received no external funding.

Conflicts of Interest: The authors declare no conflict of interest.

References

1. Bakrim, S.; Machate, H.; Benali, T.; Sahib, N.; Jaouadi, I.; Omari, N.E.; Aboulaghras, S.; Bangar, S.P.; Lorenzo, J.M.; Zengin, G.; et al. Natural sources and pharmacological properties of pinosylvin. *Plants* **2022**, *11*, 1541. [CrossRef] [PubMed]
2. Goo, Y.-K. Therapeutic potential of *Ranunculus* species (ranunculaceae): A literature review on traditional medicinal herbs. *Plants* **2022**, *11*, 1599. [CrossRef] [PubMed]

3. Iglesias, D.; Donato Capote, M.d.; Méndez Tenorio, A.; Valdivia, A.V.; Gutiérrez-García, C.; Paul, S.; Iqbal, H.M.N.; Villarreal, M.L.; Sharma, A. Identification of putative candidate genes from *Galphimia* spp. Encoding enzymes of the galphimines triterpenoids synthesis pathway with anxiolytic and sedative effects. *Plants* **2022**, *11*, 1879. [CrossRef]
4. Zheljazkov, V.D.; Cantrell, C.L.; Jeliazkova, E.A.; Astatkie, T.; Schlegel, V. Essential oil yield, composition, and bioactivity of sagebrush species in the bighorn mountains. *Plants* **2022**, *11*, 1228. [CrossRef]
5. Galovičová, L.; Borotová, P.; Valková, V.; Vukovic, N.L.; Vukic, M.; Štefániková, J.; Ďúranová, H.; Kowalczewski, P.Ł.; Čmiková, N.; Kačániová, M. *Thymus vulgaris* essential oil and its biological activity. *Plants* **2021**, *10*, 1959. [CrossRef]
6. Kačániová, M.; Galovičová, L.; Borotová, P.; Valková, V.; Ďúranová, H.; Kowalczewski, P.Ł.; Said-Al Ahl, H.A.H.; Hikal, W.M.; Vukic, M.; Savitskaya, T.; et al. Chemical composition, in vitro and in situ antimicrobial and antibiofilm activities of *Syzygium aromaticum* (clove) essential oil. *Plants* **2021**, *10*, 2185. [CrossRef]
7. Segneanu, A.-E.; Marin, C.N.; Ghirlea, I.O.-F.; Feier, C.V.I.; Muntean, C.; Grozescu, I. *Artemisia annua* growing wild in romania—A metabolite profile approach to target a drug delivery system based on magnetite nanoparticles. *Plants* **2021**, *10*, 2245. [CrossRef] [PubMed]
8. Aćimović, M.; Šovljanski, O.; Šeregelj, V.; Pezo, L.; Zheljazkov, V.D.; Ljujić, J.; Tomić, A.; Ćetković, G.; Čanadanović-Brunet, J.; Miljković, A.; et al. Chemical composition, antioxidant, and antimicrobial activity of *Dracocephalum moldavica* L. Essential oil and hydrolate. *Plants* **2022**, *11*, 941. [CrossRef] [PubMed]
9. Zheljazkov, V.D.; Semerdjieva, I.; Yankova-Tsvetkova, E.; Astatkie, T.; Stanev, S.; Dincheva, I.; Kačániová, M. Chemical profile and antimicrobial activity of the essential oils of *Helichrysum arenarium* (L.) moench. And *Helichrysum italicum* (Roth.) g. Don. *Plants* **2022**, *11*, 951. [CrossRef] [PubMed]
10. Tasinov, O.; Dincheva, I.; Badjakov, I.; Kiselova-Kaneva, Y.; Galunska, B.; Nogueiras, R.; Ivanova, D. Phytochemical composition, anti-inflammatory and er stress-reducing potential of *Sambucus ebulus* L. Fruit extract. *Plants* **2021**, *10*, 2446. [CrossRef] [PubMed]
11. Dávid, C.Z.; Kúsz, N.; Pinke, G.; Kulmány, Á.; Zupkó, I.; Hohmann, J.; Vasas, A. Jacaranone derivatives with antiproliferative activity from *Crepis pulchra* and relevance of this group of plant metabolites. *Plants* **2022**, *11*, 782. [CrossRef] [PubMed]
12. Mihaylova, D.; Popova, A.; Vrancheva, R.; Dincheva, I. HS-SPME-GC–MS volatile profile characterization of peach (*Prunus persica* L. Batsch) varieties grown in the eastern balkan peninsula. *Plants* **2022**, *11*, 166. [CrossRef] [PubMed]
13. Zaharie, M.-G.O.; Radu, N.; Pirvu, L.; Bostan, M.; Voicescu, M.; Begea, M.; Constantin, M.; Voaides, C.; Babeanu, N.; Roman, V. Studies regarding the pharmaceutical potential of derivative products from *Plantain*. *Plants* **2022**, *11*, 1827. [CrossRef] [PubMed]
14. Tasinov, O.; Dincheva, I.; Badjakov, I.; Grupcheva, C.; Galunska, B. Comparative phytochemical analysis of *Aronia melanocarpa* L. Fruit juices on bulgarian market. *Plants* **2022**, *11*, 1655. [CrossRef]
15. Ruksiriwanich, W.; Khantham, C.; Muangsanguan, A.; Chittasupho, C.; Rachtanapun, P.; Jantanasakulwong, K.; Phimolsiripol, Y.; Sommano, S.R.; Sringarm, K.; Ferrer, E.; et al. Phytochemical constitution, anti-inflammation, anti-androgen, and hair growth-promoting potential of shallot (*Allium ascalonicum* L.) extract. *Plants* **2022**, *11*, 1499. [CrossRef]
16. Goulas, V.; Banegas-Luna, A.J.; Constantinou, A.; Pérez-Sánchez, H.; Barbouti, A. Computation screening of multi-target antidiabetic properties of phytochemicals in common edible mediterranean plants. *Plants* **2022**, *11*, 1637. [CrossRef] [PubMed]
17. Zulkipli, M.; Mahbub, N.; Fatima, A.; Wan-Lin, S.L.; Khoo, T.-J.; Mahboob, T.; Rajagopal, M.; Samudi, C.; Kathirvalu, G.; Abdullah, N.H.; et al. Isolation and characterization of werneria chromene and dihydroxyacidissimol from *Burkillanthus malaccensis* (Ridl.) swingle. *Plants* **2022**, *11*, 1388. [CrossRef] [PubMed]

plants

MDPI

Review

Natural Sources and Pharmacological Properties of Pinosylvin

Saad Bakrim [1], Hamza Machate [2], Taoufiq Benali [3], Nargis Sahib [4], Imane Jaouadi [5], Nasreddine El Omari [6], Sara Aboulaghras [7], Sneh Punia Bangar [8], José Manuel Lorenzo [9,10], Gokhan Zengin [11], Domenico Montesano [12], Monica Gallo [13,*] and Abdelhakim Bouyahya [14,*]

1 Molecular Engineering, Valorization and Environment Team, Polydisciplinary Faculty of Taroudant, Ibn Zohr University, Agadir B.P. 32/S, Morocco; s.bakrim@hotmail.com
2 Laboratory of Biotechnology, Environment, Agri-Food and Health (LBEAS), Faculty of Sciences, University Sidi Mohamed Ben Abdellah (USMBA), Fez B.P. 1796, Morocco; hamza.mechchate@usmba.ac.ma
3 Environment and Health Team, Polydisciplinary Faculty of Safi, Cadi Ayyad University, Sidi Bouzid B.P. 4162, Morocco; benali.taoufiq@gmail.com
4 Laboratoire d'Amélioration des Productions Agricoles, Biotechnologie et Environnement (LAPABE), Faculté des Sciences, Mohammed Premier University, Oujda 60000, Morocco; n.sahib@ump.ac.ma
5 Laboratory of Organic Chemistry, Catalysis and Environment, Department of Chemistry, Faculty of Sciences, Ibn Tofail University, B.P.:133, Kenitra 14000, Morocco; jaouadi.imane@gmail.com
6 Laboratory of Histology, Embryology, and Cytogenetic, Faculty of Medicine and Pharmacy, Mohammed V University in Rabat, Rabat 10100, Morocco; nasrelomari@gmail.com
7 Physiology and Physiopathology Team, Faculty of Sciences, Genomic of Human Pathologies Research, Mohammed V University in Rabat, Rabat 10100, Morocco; sara.aboulghras@gmail.com
8 Department of Food, Nutrition and Packaging Sciences, Clemson University, Clemson, SC 29634, USA; snehpunia69@gmail.com
9 Centro Tecnológico de la Carne de Galicia, Rúa Galicia N° 4, Parque Tecnológico de Galicia, San Cibrao das Viñas, 32900 Ourense, Spain; jmlorenzo@ceteca.net
10 Facultade de Ciencias, Universidade de Vigo, Área de Tecnoloxía dos Alimentos, 32004 Ourense, Spain
11 Department of Biology, Science Faculty, Selcuk University, Konya 42130, Turkey; gokhanzengin@selcuk.edu.tr
12 Department of Pharmacy, University of Naples Federico II, Via D. Montesano 49, 80131 Naples, Italy; domenico.montesano@unina.it
13 Department of Molecular Medicine and Medical Biotechnology, University of Naples Federico II, Via Pansini, 5, 80131 Naples, Italy
14 Laboratory of Human Pathologies Biology, Department of Biology, Faculty of Sciences, Mohammed V University in Rabat, Rabat 10100, Morocco
* Correspondence: mongallo@unina.it (M.G.); boyahyaa-90@hotmail.fr (A.B.)

Citation: Bakrim, S.; Machate, H.; Benali, T.; Sahib, N.; Jaouadi, I.; Omari, N.E.; Aboulaghras, S.; Bangar, S.P.; Lorenzo, J.M.; Zengin, G.; et al. Natural Sources and Pharmacological Properties of Pinosylvin. *Plants* **2022**, *11*, 1541. https://doi.org/10.3390/plants11121541

Academic Editors: Ivayla Dincheva, Ilian Badjakov and Bistra Galunska

Received: 12 May 2022
Accepted: 7 June 2022
Published: 9 June 2022

Abstract: Pinosylvin (3,5-dihydroxy-*trans*-stilbene), a natural pre-infectious stilbenoid toxin, is a terpenoid polyphenol compound principally found in the Vitaceae family in the heartwood of *Pinus* spp. (e.g., *Pinus sylvestris*) and in pine leaf (*Pinus densiflora*). It provides defense mechanisms against pathogens and insects for many plants. Stilbenoids are mostly found in berries and fruits but can also be found in other types of plants, such as mosses and ferns. This review outlined prior research on pinosylvin, including its sources, the technologies used for its extraction, purification, identification, and characterization, its biological and pharmacological properties, and its toxicity. The collected data on pinosylvin was managed using different scientific research databases such as PubMed, SciFinder, SpringerLink, ScienceDirect, Wiley Online, Google Scholar, Web of Science, and Scopus. In this study, the findings focused on pinosylvin to understand its pharmacological and biological activities as well as its chemical characterization to explore its potential therapeutic approaches for the development of novel drugs. This analysis demonstrated that pinosylvin has beneficial effects for various therapeutic purposes such as antifungal, antibacterial, anticancer, anti-inflammatory, antioxidant, neuroprotective, anti-allergic, and other biological functions. It has shown numerous and diverse actions through its ability to block, interfere, and/or stimulate the major cellular targets responsible for several disorders.

Keywords: pinosylvin; pharmacological property; signaling pathway; antimicrobial; cancer

1. Introduction

Pinosylvin (3,5-dihydroxy-*trans*-stilbene), a natural pre-infectious stilbenoid toxin, is a stilbenoid polyphenol component mostly contained in the Pinaceae family, particularly in the heartwood of *Pinus* spp. (e.g., *Pinus sylvestris*) and in pine leaf (*Pinus densiflora*). Traditionally, different parts of pine trees have been used in East Asia for various purposes, such as treating liver toxicity, gastric disorders, and inflammation. In South Korea, pine needles were commonly consumed as tea and food [1]. The most stilbenes are contained in the Vitaceae family of plants containing, represented by the famous wine grape *Vitis vinifera* L., which is among the most abundant sources of new stilbenes currently known, along with other families, such as Fabaceae, Dipterocarpaceae, and Gnetaceae [2]. Pinosylvin is suggested as a functional compound responsible as a defense mechanism against pathogens and insects for a wide range of plants, especially pines [3]. They are also found in most berries or fruits, but can also be found in other types of plants, such as mosses and ferns [4]. Pinosylvin generates phytoalexins via a reaction between malonyl-CoA and cinnamoyl-CoA under the influence of different biotic and abiotic stresses like wounds, herbivores, fungi, ozone, and ultraviolet light [5]. Pinosylvin is widely explored for its relevance to plants and its characteristics favorable to human health, as it possesses several biological properties, including antimicrobial, anti-inflammatory, anticancer, antioxidant, neuroprotective, and antiallergic characteristics. Pharmacological examination of pinosylvin derivatives has shown a broad range of biological effects. In fact, they have antibacterial activity against different human pathogenic bacteria (Gram-positive and Gram-negative) such as *Staphylococcus aureus*, *Escherichia coli*, *Listeria monocytogenes*, *Lactobacillus plantarum*, *Salmonella infantis*, *Pseudomonas fluorescens*, *Campylobacter jejuni*, and *Campylobacter coli* [6–8]. Furthermore, pinosylvin exhibited significant antifungal effects against pathogenic fungi such as *Candida albicans*, *Saccharomyces cerevisiae*, *Trametes versicolor*, *Phanerochaete chrysosporium*, *Neolentinus lepideus*, *Gloeophyllum trabeum*, *Postia placenta*, *Rhizoctonia solani*, *Sclerotinia homoeocarpa*, etc. [9,10]. Also, pinosylvin, as a natural molecule, has been widely investigated in vitro and in vivo for its excellent anti-inflammatory potential, as evidenced in several studies [11–15]. The antioxidant activity of pinosylvin has been extensively studied by different investigators, not only in isolated case studies but also in association with several diseases such as rheumatoid arthritis, age-related diseases (age-related macular degeneration (AMD) and Alzheimer's disease), and oligoasthenospermia by reducing oxidative stress via the nuclear factor erythroid 2-related factor 2 (Nrf2)/antioxidant response element (ARE) pathway [16–19]. Pinosylvin is well recognized for its potential chemopreventive activity against cancer [20,21] even at low concentrations [22]. It showed anti-cancer activity against nasopharyngeal cancer [20], prostate cancer [23], fibrosarcoma [5], colorectal cancer [21], and oral cancer [20]. Indeed, in HCT116 colorectal cancer cells, pinosylvin was found to block the activation of proteins that play a role in the FAK/c-Src/ERK and PI3K/Akt/GSK-3b signaling pathways [21]. Additionally, in cultured HT1080 human fibrosarcoma cells, pinosylvin inhibited the production of matrix metalloproteinase (MMP)-2, MMP-9, and membrane type 1-MMP. The antimetastatic action of pinosylvin was associated with the downregulation of MMP-9 and cyclooxygenase-2 (COX-2) [5]. Despite the availability of certain investigations that have highlighted the different pharmacological functions of pinosylvin and its derivatives, to the best of our knowledge, no critical review has been carried out to provide suggestions for potential future clinical uses of this bioactive molecule. This synthesis article aims to provide a comprehensive review of the characteristics of this secondary metabolite, namely its sources, extraction technologies, purification, identification, characterization, and pharmacological and biological properties. We hope that this review will give a novel background for further investigations on this deterrent secondary metabolite and its pharmacological actions in order to explore new pharmaceutical opportunities for this natural molecule.

2. Sources of Pinosylvin

Pinosylvin, known as 3,5-dihydroxy-trans-stilbene, was first isolated by Erdtman in 1939 from extracts derived from *Pinus sylvestris*, hence the name pinosylvin [24]. It is a natural stilbenoid belonging to the phenolic group of compounds. Pinosylvin is found in a wide range of plant species, particularly in the leaves and wood of various *Pinus* species (Table 1).

Table 1. Sources of Pinosylvin.

Species	Extract/Essential Oil	References
Pinus sylvestris	Extract	[13,24–33]
Pinus resinosa	Extract	[9,28,31,34]
Pinus banksiana	Extract	[26,28,31]
Pinus nigra Arn.	Extract	[25,30]
Pinus densiflora	Extract	[35,36]
Pinus sibirica	Extract	[37]
Pinus contorta	Extract	[37]
Pinus strobus	Extract	[38–40]
Pinus taeda	Extract	[41]
Pinus cembra	Extract	[37]
Pinus pinaster	Extract	[42–44]
Hovenia dulcis Thunb.	Extract	[45]
Picea glauca	Extract	[9]
Nothofagus (Southern beeches)	Extract	[46,47]
Stemona cf. *pierrei*	Extract	[48]
Arachis hypogaea	Extract	[49]

3. Technology of Extraction and Purification

Stilbenes have attracted increasing interest in recent years for their health benefits in preventing disease. For these reasons, their extraction and subsequent purification are particularly interesting for the production of high-quality extracts. Due to its very low occurrence in plants, and the environmental and chemical hazards associated with multi-step isolation and purification processes, the extraction of pinosylvin from plants was difficult and unsustainable. These drawbacks have led scientists to introduce alternative production sources into the stilbene isolation process, such as the callus culture, plant culture, cell suspension culture, hairy root culture, genetically-modified plants, and the introduction of the stilbene biosynthetic genes into microbial hosts.

These different biotechnological approaches have been applied to avoid the formation of many undesirable biproducts of high medicinal risk in the extracts obtained by chemical synthesis. These biproducts subsequently necessitate a more complicated purification process, such as sequential flash chromatography in two repetitions in gradient mode during the mobile phase with cyclohexane (CX) and ethyl acetate (EtOAc) [50].

3.1. *Pinosylvin Production in Callus Cultures and Cell Suspension Cultures*

In 1961, Jorgensen showed that mechanical damage applied to the bark and cambium of red pine causes fungal penetration of the sapwood into the stems and roots. This fungal penetration affects the dying tissues and induces the formation of pinosylvin and its monomethyl ether, which are absent in the healthy sapwood [51]. At temperatures where cellular activity is possible, pinosylvin is formed by living cells in response to desiccation, causing slow tissue death in sections of living branches or callus tissues, thus producing the stilbenes.

In a study reported by Koo et al. [40], the authors have established a system of production of pinosylvin stilbene and its derivatives using the in vitro culture of *Pinus strobus* L. callus. From a culture of mature zygotic embryos in 1/2 Litvay medium with 1.0 mg/L

2.4-D and 0.5 mg/L BA, calli were obtained and the accumulation of pinosylvin significantly increased in prolonged callus cultures. In 1984, and for the first time, Schöppner and Kindl [49] described the purification of pinosylvin synthase from cell suspension cultures of peanuts using column chromatography. This study showed that from hypocotyls of 4-day-old seedlings, peanut cell cultures were initiated and propagated as callus cultures.

In a study reported by Lange et al. [52], the accumulation of the stilbenes pinosylvin and pinosylvin 3-O-methyl ether in methanolic cell extracts was induced by a cell suspension of treatment of *Pinus sylvestris* L. cultures with an elicitor preparation of the pine needle pathogen.

3.2. *Microbial Biosynthesis of Pinosylvin*

The bioproduction of pinosylvin was achieved by genetic engineering of the host strain to integrate the heterologous pathways of plants with the host strain. *E. coli* was the commonly used host for pinosylvin production. The bacterial hosts have a short life cycle, high growth rate, and easy genetic manipulation, and, therefore, significant overexpression of proteins and enzymes. At the same time, they severely lack the expression of large proteins and the post-translational modifications necessary for the correct folding and functional activity of recombinant proteins [53,54]. Therefore, pathway engineering is one of the pioneering methods in *E. coli* design (50% bioconversion rate) which facilitated the bioproduction of stilbenes in different micro-organisms to produce value-added bioactive pinosylvin [55].

Interestingly, another study described the development of an *E. coli* platform strain to produce the stilbene pinosylvin found in the heartwood of pines [56]. The authors of this study reported low pinosylvin concentrations (3 mg/L) after the optimization of gene expression and evaluation of different construction environments. To promote the production of pinosylvin stilbene, the authors added cerulenin to increase the intracellular reserves of malonlyl-CoA and subsequently obtained higher concentrations of pinosylvin of up to 70 mg/L from glycose and 91 mg/L by adding L-phenylalanine. Similar results were obtained in a study conducted by Xu et al. [57], who evaluated the biosynthetic pathway for pinosylvin production in engineered *E. coli*. It was shown in this study that the excessive accumulation of the precursor malonyl-CoA leading to malonylation of the biosynthetic enzymes decreases pinosylvin yield. In order to mitigate this decrease, several metabolic engineering techniques (PTM, PTM-ME) have been established to maintain an optimal level of intracellular acyl-CoA concentration, and thus increase the pinosylvin yield. Liang et al. [58] investigated an alternative approach to pinosylvin production using three bioengineering strategies to develop a simple and economical process for pinosylvin biosynthesis in *E. coli*. The authors were able to produce 47.49 mg/L of pinosylvin from glycerol, using these combinatory processes by promoting the expression of the pinosylvin pathway enzymes, increasing the level of the key precursor of pinosylvin bioproduction (malonyl-CoA) in the *E. coli* cell. The final step was to introduce phenylalanine super-producing *E. coli* to produce trans-cinnamic acid which is a precursor of pinosylvin. Other researchers have also established metabolic engineering techniques for *E. coli* for the biosynthesis of stilbene pinosylvin [59–61].

4. Technology of Identification and Characterization

Chromatographic analyses (GC-MS, LC-MS, GC-FID or HPLC) are frequently used to identify and characterize stilbenes using stilbene standards. In 1999, Holmgren et al. [62] used diffuse reflectance Fourier transform infrared spectroscopy (DRIFT) and FT-Raman near-infrared spectroscopy (NIR) to detect the presence of pinosylvin and its derivatives in the wood of *Pinus sylvestris* L. by a simple visual inspection of uniform wood blocks in disc form. Roupe et al. [63] developed a simple and novel high-performance liquid chromatography (HPLC) method to simultaneously determine pinosylvin and its metabolic products in rat serum and liver microsomes. The method consists of a preliminary precipitation of serum or microsomes with acetonitrile after adding an internal standard. The separation was then performed on a tris-3,5 dimethyl phenyl carbamate amylose column with UV detection at 308 nm. In another study performed by Ekeberg et al. [64], the quantitative identification of *P. sylvestris* L. heartwood extracts, including pinosylvin and its derivatives, was carried out using gas chromatography (GC) with flame ionization detection (FID). Similarly, a study on Scots pine/spruce wood residues from Norway conducted by Poljanšek et al. [50], describes the qualitative and quantitative analysis of the obtained extracts in terms of pinosylvin and pinosylvin monomethyl ether performed uisng gas chromatography with a flame ionization detector (GC-FID) and gas chromatography with mass spectrometry (GC-MS).

Somewhat removed from plant samples, Preusz et al. [65] conducted a study on organic residues in the form of black stains found at the sites of the ancient ports of Pyrgi and Castrum Novum on the Tyrrhenian coast, in which pinosylvin monomethyl ether was identified and confirmed for the first time in archaeological samples using GC-MS and HPLC with fluorimetric detection.

5. Biological and Pharmacological Properties

As evidenced in several investigations, pinosylvin was found to exhibit a wide range of biological and pharmacological properties, including antimicrobial [66], anti-inflammatory [11], antioxidant [17], anticancer [5,23], neuroprotective [67], and anti-allergic [45] effects (Figure 1).

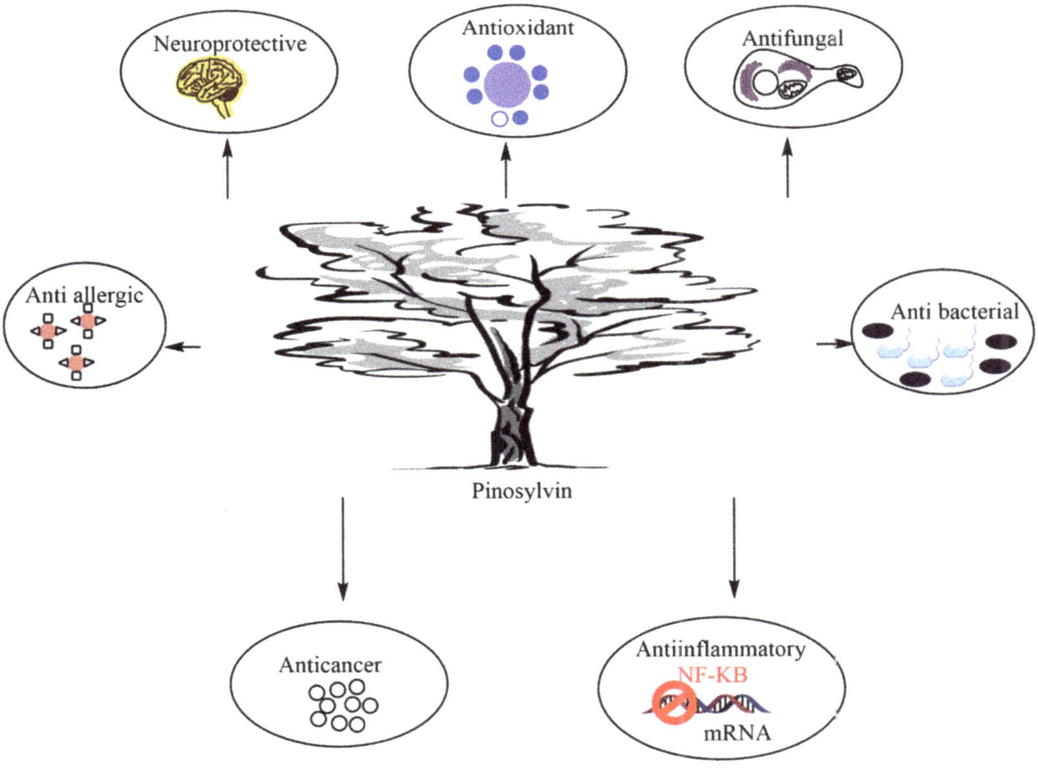

Figure 1. Major pharmacological properties of pinosylvin.

5.1. Antimicrobial Activity

With the increasing problems of the persistence and emergence of microbial resistance, much attention was given to the identification of new antimicrobial drugs derived from natural bioactive compounds [68–71]. Pinosylvin has been widely investigated for its health benefits and biological activities, including its antimicrobial effects. Lee et al. [66] showed the role of pinosylvin as an antimicrobial agent against various human pathogens, including Gram-positive (*S. aureus*) and Gram-negative (*E. coli*) bacteria, fungi (*C. albicans*), and yeasts (*S. cerevisiae*). *C. albicans* and *S. cerevisiae* appeared to be more sensitive to pinosylvin with minimal inhibitory concentration (MIC) values of 62.5 and 125 μg/mL, respectively, while the MIC for *E. coli* and *S. aureus* was 250 μg/mL. Moreover, pinosylvin extracted from the knot wood and bark of different *Pinus* species exhibits potent antimicrobial activity, effectively inhibiting the growth of a broad spectrum of pathogenic strains, including *Bacillus cereus*, *S. aureus*, *L. monocytogenes*, *L. plantarum*, *E. coli*, *S. infantis*, *P. fluorescens*, *C. albicans*, *S. cerevisiae*, *Aspergillus fumigatus*, and *Penicillium brevicompactum*, with inhibition diameters ranging from 19 ± 1 to 101 ± 6 mm. Sousa and collaborators [8] evaluated the potential interaction between pinosylvin and four antibiotics (tetracycline, chloramphenicol, erythromycin, and ciprofloxacin) against *Arcobacter butzleri* using checkerboard titration assays. Based on FICI values, no synergistic effects were observed for pinosylvin/four antibiotic combinations, while pinosylvin showed additive interactions on all the tested antibiotics, except ciprofloxacin. In addition, these researchers investigated the ability of pinosylvin to modulate the efflux pump activity using ethidium bromide (EtBr) accumulation assays. The results showed that pinosylvin causes a higher intracellular accumulation of EtBr, elucidating that it may attenuate the activity of efflux pumps (EPs) [8]. Overall, these findings shed light on the use of pinosylvin as a resistance modulator to control the decreasing susceptibility of *A. butzleri* to antibiotics and suggest the potential of pinosylvin as an efflux pump inhibitor. Furthermore, prenylation of stilbenes, including pinosylvin has been shown to enhance their antibacterial activity, which is explained by MIC values. In this regard, Bruijn et al. [72] demonstrated that prenylated pinosylvin derivatives isolated from *Rhizopus* extract exhibit potent antimicrobial activity against methicillin-resistant *Staphylococcus aureus* (MRSA), especially chiricanine A with a MIC value of 12.5 μg/mL.

As a natural compound, pinosylvin has potential applications in the development of antimicrobial food packaging systems due to its inherent antimicrobial activity, especially against Campylobacter spp. [6,7]. Indeed, it has been shown that pinosylvin or its inclusion complexes (ICs) with modified cyclodextrins (hydroxypropyl-b-cyclodextrin and hydroxypropyl-g-cyclodextrin) were able to inhibit the growth of *Campylobacter jejuni* and *Campylobacter coli* American type culture collection (ATTC) reference strains and clinical isolates [73]. The MIC values were between 25 to 50 mg/mL for the pure compound and between 16 and 64 mg/mL for the ICs. Furthermore, time-kill assays showed that pinosylvin ICs exhibit bactericidal action on both *Campylobacter* species at 37 °C and even at 4 or 20 °C [73]. Flow cytometric analysis shows that the mechanism behind this bactericidal action may mainly involve membrane damage mediated by the impairment of various cellular functions such as membrane polarization, permeability, and efflux activity [73]. These promising data make these pinosylvin ICs valuable lead compounds used in active food packaging to eradicate *Campylobacter* spp. in fresh poultry products. In this context, in their recent findings, the same authors demonstrated the role of coated pads containing pinosylvin ICs in controlling fresh chicken meat from *Campylobacter* contamination [7]. The above-mentioned compound exhibited effective in vitro bactericidal activity against *C. jejuni* with more than 99% colony count inhibition, even at the lowest concentrations (0.08 mg/cm^2). In vivo tests on chicken exudates and chicken fillets have also shown that these active pads exhibit promising anti-Campylobacter activity at 37 °C and 4 °C [7]. Additionally, coated pads-pinosylvin ICs are also effective against other major chicken foodborne bacteria, suggesting future uses of this coating as a new alternative to control the microbial growth in packaged chicken meat [7].

On the other hand, the antifungal potential of pinosylvin and pinosylvin monomethyl ether isolated from pine knot extract was assessed in vitro against *Plasmopara viticola*. This study showed that pinosylvin exhibits promising antimildew properties, inducing significant inhibition of zoospore mobility (IC_{50} = 34 μM) and mildew development (IC_{50} = 23 μM) [43]. These findings are corroborated by those described by other authors. Indeed, pinosylvin and pinosylvin monomethyl ether from *Pinus* trees have already demonstrated significant antifungal effects against white rot (*Trametes versicolor* and *Phanerochaete chrysosporium*) and brown-rot (*Neolentinus lepideus, Gloeophyllum trabeum*, and *Postia placenta*) [9] fungi. Furthermore, pinosylvin from the methylene chloride fraction of *Pinus densiflora* showed effective antifungal activity against plant pathogens such as *Rhizoctonia solani* AG1-1B, *R. solani* AG2-2IV, *R. cerealis*, and *S. homoeocarpa*. *S. homoeocarpa* showed the highest sensitivity with the lowest mean EC_{50} value (8.426 µg/mL), whereas among the *Rhizoctonia* pathogens, *R. cerealis* had the highest mean EC_{50} value (99.832 µg/mL). Pinosylvin could be a valuable lead compound for developing new effective and ecofriendly antifungal agents [10].

5.2. Anti-Inflammatory Activity

Over the past decades, several scientists have dedicated their efforts to developing novel anti-inflammatory drugs from natural molecules to overcome the serious and excessive side effects of current drugs [74,75]. As a natural molecule, pinosylvin has been extensively investigated (in vitro and in vivo) for its excellent potential anti-inflammatory effects (Table 2).

Table 2. Anti-inflammatory effects of pinosylvin.

Experimental Approaches	Key Results	References
Western blot analysis and reverse transcription-polymerase chain reaction (RT-PCR)	Inhibited COX-2, iNOS protein and gene expression	[76]
Murine adipocytes model, cytotoxicity assays, lipid staining, western blotting, and ELISA assays	Attenuated adipogenesis and inflammation through downregulation of the expression of PPARγ, C/EBP and TNF-a-stimulated IL-6 secretion	[14]
Cell viability and RNA interference analysis	Protected (10 µM) cell survival from oxidative damage by promoting HO-1 induction	[77]
LPS-induced mouse macrophage RAW 264.7 cells	Suppressed COX-2-mediated PGE_2 production (IC_{50} = 10.6 µM)	[78]
LPS-stimulated Macrophage cells Western blot analysis RT-PCR analysis	Inhibited LPS-induced iNOS protein and mRNA expression in dose-dependent manner (IC_{50} = 39.9 µM) Decreased the expression levels of interferon regulatory factor 3 (IRF-3) and interferon-E (IFN-E)	[79]
AITC-induced acute paw inflammation in mice model Fluo-3-AM assay and patch clamping	Reduced paw inflammation formation by inhibiting and attenuating IL-6 production at the site of inflammation	[15]
Adjuvant-induced arthritis in rats	Pinosylvin + MTX reduced oxidative stress by upregulating HO-1 expression in lungs and reducing plasma activity of thiobarbituricacid reactive substances (TBARS) and lipoxygenase (LOX) in the lungs	[16]
Primary cultures of human OA chondrocytes	Increased aggrecan expression Inhibited IL-6 production by attenuating NF-κB activity	[13]
AA in rats Chemiluminescence (CL) of the joint and myeloperoxidase (MPO) activity	Decreased HPV Reduced CL of the joint and MPO activity of the joint homogenate	[80]
Carrageenan-induced paw edema in male C57BL/6 mice	Reduced inflammatory response by downregulating the production of inflammatory cytokines IL6, MCP1, and NO	[11]
AA was induced in Lewis rats Fresh human blood neutrophils as model	Reduced the formation of oxidants, both extra- and intra-cellular Suppressed PKC activation induced by phorbol myristate acetate Reduced neutrophil countDecreased the amount of ROS (in vivo)	[81]

Table 2. *Cont.*

Experimental Approaches	Key Results	References
LPS-triggered apoptosis in the leukocyte	Enhanced apoptosis of LPS-preconditioned leukocytes via decreasing ALOX 15 expression mediated by ERK and JNK pathways	[82]
Humane monocytic THP-1 cell lines Western blotting analysis	Suppressed proinflammatory enzymes TNF-α and IL-8 by the inhibition of NF-κB activation	[66]
Murine and U937 Human macrophages model qRT-PCR and ELISA	Changed macrophage polarization from the proinflammatory M1 phenotype to the M2 phenotype Promoted resolution of inflammation and repair Enhanced PPAR-γ expression in IL-4 treated macrophages	[12]
LPS-induced mouse macrophage RAW 264.7 cells	Decreased inflammation on LPS-stimulated macrophages Inhibited PPARγ activity in vitro	[83]
Antigen-stimulated mast cell-like cell line rat basophilic leukemia (RBL)-2H3 and a passive cutaneous anaphylaxis (PCA) mouse model Degranulation assay RT-PCR, PCA Western blot analyses	Suppressed the release and expression of allergic and proinflammatory key enzymes (IL-4, TNF-α and PGE_2, COX-2, NFKB1, and NFKB2) in a dose-dependent manner	[45]

Research results by Park and colleagues (2005) [84] showed that pinosylvin downregulates the production of proinflammatory mediators such as prostaglandin E_2 (PGE_2) and nitric oxide (NO) in a dose-dependent manner. This effect was directly related to COX and inducible nitric oxide synthase (iNOS) inhibition. Moreover, pinosylvin significantly inhibited other key inflammatory enzymes, interleukin 6 (IL6) (IC_{50} = 32.1 μM) and monocyte chemotactic protein 1 (MCP1) (IC_{50} = 38.7 μM) [11]. Additionally, the in vivo investigation measuring carrageenan-induced paw edema in male C57BL/6 mice showed that pinosylvin at a dose of 30 mg/kg significantly reduced the inflammatory response by downregulating the production of inflammatory cytokines IL6, MCP1, and NO compared to an LY294002-treated group [11]. The similar anti-inflammatory effects of pinosylvin to those of the known commercial phosphatidylinositol-3 kinase (PI3K) inhibitor LY294002 suggest that these effects may be mediated by the inhibition of the PI3K/Akt pathway (Figure 2).

Furthermore, treatment with pinosylvin was shown to significantly inhibit stimulation-induced NO production of murine macrophages with lipopolysaccharide (LPS) in a dose-dependent manner, with an IC_{50} value of 39.9 μM compared to reference L-NMMA (IC_{50} = 30.7 μM) [79]. In addition, pinosylvin suppressed iNOS gene expression via downregulation of interferon regulatory factor 3 (IRF-3) and interferon-E (IFN-E) expression related to TIR-domain-containing adapter-inducing interferon-β (TRIF) mediated signaling pathway. These events were then associated with the suppression of JAK kinase phosphorylation, which decreased the phosphorylation of signal transducer and activator of transcription-1, one of the iNOS transcriptional activators [79].

Since the substitution patterns of the trans-stilbene have been shown to enhance various biological properties, Park et al. [78] assessed the substitutions of the dihydroxy group in pinosylvin with different lipophilic derivatives on LPS-induced RAW 264.7 cells. The results showed that the synthesized pinosylvin derivatives, especially 3,5-dimethoxy-trans-stilbene and 3-hydroxy-5-benzyloxy-trans-stilbene, significantly suppress COX-2 mRNA expression-mediated PGE_2 production. On the other hand, pinosylvin treatment at doses of 5 and 10 μM greatly enhanced human RPE cells from oxidative stress. The expression levels of heme oxygenase-1 (HO-1), an enzyme with anti-inflammatory and immunomodulatory activities, were upregulated by pinosylvin treatment and markedly correlated with cell survival [70]. These findings demonstrated the role of pinosylvin treatment in the protection against oxidative stress, induction of HO-1 expression in human

RPE cells and, therefore, potential health promotion against oxidative stress and aging-related diseases such as AMD and Alzheimer's disease [77].

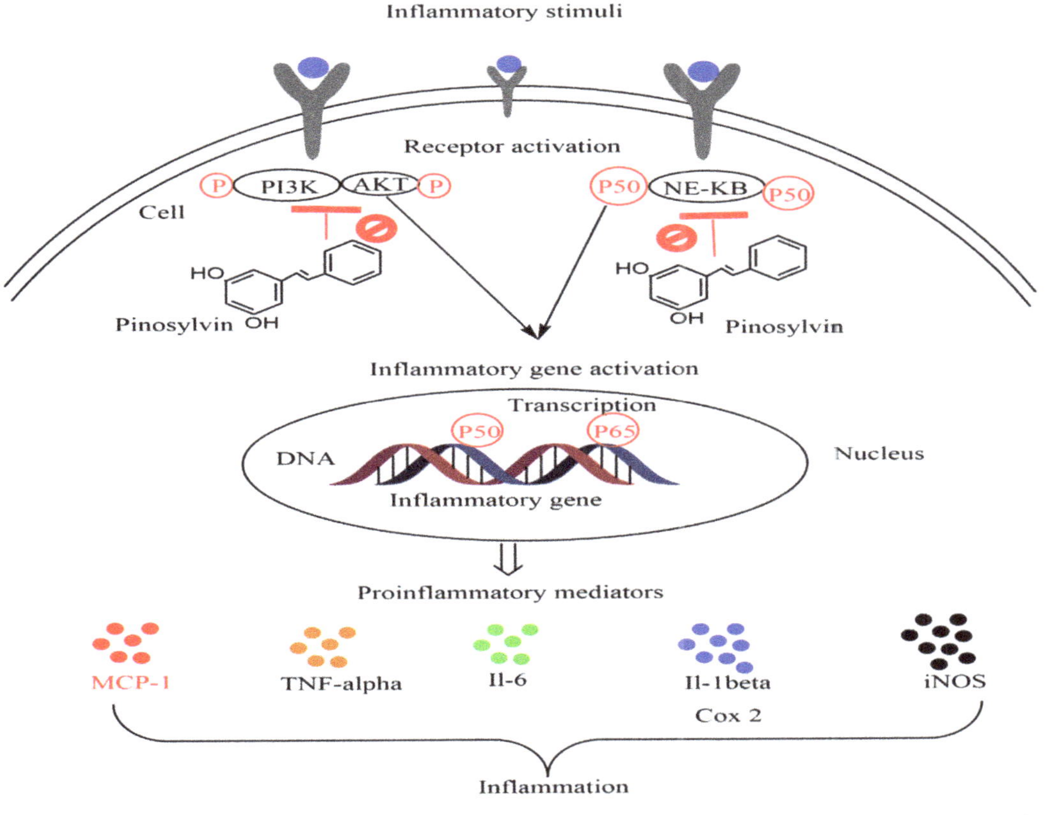

Figure 2. Anti-inflammatory effects of pinosylvin. This figure illustrates the ability of pinosylvin to reduce the expression of some proinflammatory cytokines and enzymes, probably via the inactivation of NF-κB and the PI3K/Akt pathway. Abbreviations: NF-κB, nuclear factor kappa B; NO, nitric oxide; COX-2, cyclooxygenase-2; iNOS, inducible nitric oxide synthase; IL6, interleukin 6; MCP1, monocyte chemotactic protein 1; TNF-α, tumor necrosis factor-α.

Interestingly, in addition to its ability to reduce the concentration of reactive oxygen and nitrogen species, pinosylvin has been shown to potentiate the therapeutic efficacy of methotrexate (MTX), an immunosuppressive drug in arthritis treatment. Indeed, Bauerova et al. [16] showed that the treatment of AA in rats with pinosylvin in combination with MTX (Orale doses of 50 mg/kg b.w. for PIN and 0.4 mg/kg b.w. for MTX) significantly reduced oxidative stress via upregulation of HO-1 expression in the lungs and reduction in plasmatic thiobarbituric acid reactive substances (TBARS) as well as markedly decreased lipoxygenase (LOX) activity in the lungs.

Ankyrin subtype 1 protein (TRPA1) has been involved in various inflammatory responses. Its suppression may provide promising targets for the treatment of many pathological conditions related to acute pain, inflammation, and hyperalgesia. In this respect, Moilanen and colleagues (2018) conducted their research to investigate the effect of pinosylvin on TRPA1 in vitro by measuring transient receptor potential ankyrin subtype 1 protein (TRPA1)-mediated Ca^{2+} influx and membrane currents. The findings reported a dose-dependent inhibitory effect of pinosylvin (IC_{50} = 16.7 μM) on AITC-induced

TRPA1-mediated responses. In vivo experiments using AITC-induced paw inflammation as a model demonstrated that pinosylvin treatment effectively reduced the formation of paw edema, attenuating the production of inflammatory cytokine IL-6 at the site of the inflammation [15].

5.3. Antioxidant Activity

The antioxidant activity of pinosylvin was extensively studied by different researchers, not only as an isolated study case (ORAC, $ABTS^+$, and FRAP in vitro assays), but in relation to many diseases such as rheumatoid arthritis, age-related diseases, and oligoasthenospermia [16–19] (Table 3).

Table 3. Pinosylvin antioxidant activity.

Origins	Cell Lines	Methods	Key Findings	References
Synthesized	Mouse model of oligoasthenospermia	Epididymal sperm concentration and motility evaluation Hormone level assessment Real-time PCR Western blot analysis Evaluation of the testicular levels of ROS and MDA	Decreased oxidative stress through glutathione peroxidase 3 drastically reduced oxidative stress (in vivo) by inhibiting the nuclear factor erythroid 2-related factor 2 (Nrf2)/antioxidant response element (ARE) pathway	[19]
Purchased	WT and NFE2L2 KO ($NFE2L2^{-/-}$) mice strain	ERG recording and processing of signals OCT imaging Antioxidant capacity analysis Immunohistochemical staining Confocal imaging	Retained retinal function Decreased accumulation of ubiquitin-tagged proteins Decreased chronic oxidative stress Preserved retinal function and morphology in the NFE2L2 KO disease model Reduced the risk of age-related macular degeneration (AMD) and halted its development	[85]
Purchased	In vitro non-enzymatic assays	ORAC-FL assay ABTS assay FRAP assay	Strong antioxidant and free radical scavenging properties	[18]
Synthetized	AA model induced in Lewis rats	Oral administration of pinosylvin to AA induced animals Monitoring of the hind paw volume Monitoring of the luminol-enhanced chemiluminescence (CL) of the joint Monitoring of myeloperoxidase (MPO) activity in hind paw joint	Reduced HPV (at days 14 and 28) Reduced joint CL and MPO activity in joint homogenate	[80]
Not reported	Human retinal pigment epithelial cells (ARPE-19)	Toxicity assessment Oxidative stress assessment MTT assay Real-time PCR Nrf2 and p62 RNA interference	Improved cell viability against oxidative stress (5 and 10 μM) Validated the importance of Nrf2 and HO-1 in pinosylvin-mediated protection against oxidative stress	[77]
Synthesized	Bovine aortic endothelial cells (BAECs)	Measurement of apoptosis Measurement of caspase-3 activity Cell proliferation Western blot analysis Cell migrationAdhesion of THP-1 to BAECs	Activated endothelial nitric oxide synthase Impacted cell proliferation in endothelial cells Stimulated cell migration and tube formation Avoided inflammatory cardiovascular disorders	[1]

Table 3. *Cont.*

Origins	Cell Lines	Methods	Key Findings	References
Synthesized	AA model induced in rats	Formation of reactive oxygen species Western blot analysis Measurement of ATP liberation Flow cytometry Effects of pinosylvin on arthritis	Reduced both extracellular and intracellular oxidant generation in isolated human neutrophils Inhibited PKC activation triggered by phorbol myristate acetate Increased the number of neutrophils in the blood of arthritic rats Improved whole blood chemiluminescence (both spontaneous and PMA-stimulated) Reduced the number of neutrophils and the number of reactive oxygen species in the blood	[81]
Not reported	AA model induced in rats	28 days of oral administration Changes in hind paw volume and arthrogram evaluation γ-glutamyltransferase (GGT) activity assessment Measurement of thiobarbituric acid reactive substances (TBARS)	Decreased the activity of GGT in the spleen Reduced the activity of GGT in joint tissue Exhibited moderate efficacy in preventing oxidative damage	[17]
Synthetized	AA model induced in rats	Assessment of hind paw volume Measurement of the C-reactive protein Monocyte chemotactic protein-1 (MCP-1) level measurement Plasma levels of thiobarbituric acid reactive substances (TBARS) and F2-isoprostanes measurement G-glutamyltransferase and lipoxygenase (LOX) activity evaluation	Increased NF-κB activation in the liver and lung, HO-1 expression and LOX activity in the lung, MCP-1 levels in plasma, and F2-isoprostane plasmatic levels Reduced the OS (an increase of HO-1 expression in the lung and reduction in plasmatic TBARS) Decreased the LOX activity in the lung	[16]

Bauerova et al. [16] assessed the impact of the treatment on selected parameters in AA (Alko, alcohol) rats when administered pinosylvin for 28 days as a monotherapy and in combination with methotrexate (MTX). The experiment included healthy controls, untreated AA, and AA given 50 mg/kg b.w. pinosylvin daily p.o. AA was monitored using hind paw volume, C-reactive protein, MCP-1 activity, TBARS, F_2-isoprostanes in plasma, g-glutamyltransferase activity in the spleen, lung LOX activity, HO-1 activity, and nuclear factor kappa B (NF-κB). Pinosylvin monotherapy enhanced NF-κB activation in the liver and lung, HO-1 expression and LOX activity in the lung, plasma MCP-1 levels (on the 14th day), and plasmatic levels of F_2-isoprostanes. The reduction in OS (an increase in HO-1 expression in the lungs and a reduction in plasmatic TBARS) and decrease in LOX activity in the lungs were substantial contributions of pinosylvin.

The pathophysiology of rheumatoid arthritis is strongly influenced by oxygen metabolism. Patients with rheumatoid arthritis have an altered antioxidant defense capacity barrier, which links oxidative stress, inflammation, and the immune system. Drafi et al. [17] investigated the impact of pinosylvin in monotherapy for the treatment of AA. Indeed, pinosylvin (30 mg/kg body mass daily per os) was provided in monotherapy to rats with AA for 28 days. In rats, parameters such as changes in hind paw volume and arthritis score were measured as indicators of destructive arthritis-related clinical changes, with determination of oxidative indicators, plasmatic levels of TBARS, and the latency of Fe^{2+}-induced lipid peroxidation (tau-FeLP) in plasma and the brain. CRP levels in the blood and glutamyltransferase (GGT) activity in the spleen and joints have been used as inflammatory indicators. Pinosylvin failed to significantly reduce the arthritic score in arthritic animals compared to untreated arthritic animals. Administration of pinosylvin somewhat reduced GGT activity in the spleen. Pinosylvin was less effective in reducing oxidative damage as determined by plasma TBARS levels.

Jančinová et al. [81] conducted their investigation to evaluate the effects of natural stilbenoid pinosylvin on neutrophil activity in vitro and experimental arthritis and to determine whether protein kinase C (PKC) activation functioned as an assumed target of pinosylvin action. The oxidative burst was assessed using enhanced chemiluminescence from neutrophils from fresh human blood. Flow cytometry was used to analyze neutrophil viability, and Western blotting was used to determine PKC phosphorylation. Adjuvant arthritis was produced in Lewis rats using heat-killed Mycobacterium butyricum, and the animals received pinosylvin (30 mg/kg, p.o.) daily for 21 days after arthritis was established. Pinosylvin (10 and 100 μmol/L) greatly reduced the generation of extracellular and intracellular oxidants and efficiently inhibited PKC activation triggered by phorbol myristate acetate (0.05 μmol/L) in isolated human neutrophils. However, inhibition did not occur due to neutrophil damage or increased apoptosis. Blood neutrophil counts were considerably elevated in arthritic rats, as was whole blood chemiluminescence (spontaneous and PMA-stimulated). The injection of pinosylvin reduced the number of neutrophils and considerably lowered the number of reactive oxygen species in blood. Pinosylvin is a potent inhibitor of neutrophil activity and has the potential to be beneficial as an adjunctive drug in conditions related to chronic inflammation. The observed results qualified pinosylvin as an efficient inhibitor of neutrophil activity, suggesting that it could be beneficial as a supplemental therapy in pathological situations related to chronic inflammation.

Koskela et al. [77] studied the capacity of pinosylvin to control oxidative stress in human RPE cells. ARPE-19 cells were treated with pinosylvin (5 μM) for 6 h, and mRNA was extracted at four timepoints (2 h, 6 h, 12 h, and 24 h) to determine changes in the expression of Nrf2, sequestosome 1 (p62/SQSTM1), HO-1, and glutathione S-transferase pi 1 (GSTP1). To further understand the molecular mechanism underlying pinosylvin-mediated protection, ARPE-19 cells were transfected with p62 and Nrf2 siRNAs for 24 h, and the roles of p62, Nrf2, and its target gene HO-1 in protection against oxidative stress were investigated using quantitative real-time PCR (qRT-PCR) and cell viability assay. At doses of 5 and 10 μM, pinosylvin dramatically improved cell survival against oxidative stress and increased the expression of HO-1, an enzyme with antioxidant, anti-inflammatory, and immunomodulatory capabilities, and was substantially linked to cell survival. However, pinosylvin treatment did not influence the expression of Nrf2 or its target genes, p62 or GSTP1, while having a strong effect on the expression of HO-1, another Nrf2-controlled gene. RNA interference study verified the importance of Nrf2 and HO-1 in PS-mediated protection against oxidative stress, whereas the contribution of p62 seemed minor at the levels of gene expression and cell viability. According to the findings of this research, pinosylvin therapy protects against oxidative stress by inducing HO-1 in human RPE cells.

In the study by Mačičková et al. [80], the research focused on the impact of pinosylvin on the development of adjuvant arthritis in rats. AA was developed in male Lewis rats using a single intradermal injection of *Mycobacterium butyricum* in inadequate Freund's adjuvant. Pinosylvin (30 mg/kg) was regularly given orally to arthritic animals. The therapy consisted of administering the chemicals examined from day 0 (day of immunization) to day 28 (experimental day), measuring several parameters, namely change in hind paw volume (HPV) at days 14, 21, and 28, joint chemiluminescence (CL), and myeloperoxidase (MPO) activity in hind paw joint homogenates (day 28). Arthritic animals treated with pinosylvin substantially reduced HPV at days 14 and 28. In contrast to untreated mice, pinosylvin lowered joint CL and joint homogenate MPO activity. This molecule demonstrated a favorable anti-inflammatory and antioxidant impact on oxidative stress-induced biochemical alterations in AA according to the three functional measures.

Rodríguez-Bonilla et al. [18] measured the antioxidant capacity of pinosylvin using a variety of analytical methodologies (ORAC, $ABTS^{+}$, or FRAP). Pinosylvin showed high antioxidant and free radical scavenging activity in all experiments due to phenolic hydroxy groups.

Considering that chronic oxidative stress eventually leads to protein aggregation in combination with impaired autophagy, as seen in AMD, Tamminen et al. [85] investigated the effects of commercial natural pinosylvin extract, Retinari™, on electroretinogram, optical coherence tomogram, autophagic activity, antioxidant capacity, and inflammation markers in their study. For 10 weeks before the experiments, wild-type and NFE2L2 knockout mice were given either ordinary or Retinari™ chow. Retinari™ therapy restored many retinal functions, with a- and b-wave amplitudes in electroretinogram responses being considerably preserved. Furthermore, this treatment reduced retinal thinning in NFE2L2 mutant animals that showed lower ubiquitin-tagged protein accumulation and local overexpression of complement factor H and the antioxidant enzymes superoxide dismutase 1 and catalase. Accordingly, in the NFE2L2 KO illness model, the therapy decreased chronic oxidative stress while maintaining retinal function and shape. The findings suggest that taking pinosylvin supplements may reduce the likelihood of developing age-related macular degeneration and halt its development.

Pinosylvin, a resveratrol analogue developed by Wang et al. [19], has been thoroughly studied in the treatment of oligoasthenospermia. They explored the molecular basis for improved sperm parameters in a mouse model of oligoasthenospermia produced using busulfan (BUS) therapy at 6 mg/kg b.w. Mice were given varying concentrations of pinosylvin daily for two weeks after receiving busulfan treatment. Then, epididymal sperm concentration and motility were evaluated and testicular histology was performed. Levels of serum hormones, including testosterone (T), luteinizing hormone (LH), and follicle-stimulating hormone (FSH), were tested using ELISA kits designed for each hormone. RNA sequencing was used to establish testicular mRNA expression profiles. Quantitative real-time PCR, Western blotting, and ELISA were used to confirm these results. After BUS therapy, pinosylvin improved epididymal sperm concentration and motility, increased testosterone levels, and facilitated morphological testicular recovery. The antioxidant glutathione peroxidase 3 dramatically decreased oxidative stress through the Nrf2/ARE-dependent antioxidant. Pinosylvin improved oligoasthenospermia in this mouse model by reducing oxidative stress via the Nrf2-ARE pathway.

5.4. Anticancer Activity

Pinosylvin is a functional compound in *Pinus* species known to exhibit potential cancer chemopreventive activity [20,21], even at low concentrations [22]. Based on that, the main concern of the researchers was to reveal its underlying molecular mechanisms [5] as well as its potential against resistant types of cancer [23] and metastasis [45]. (Table 4).

Table 4. Anticancer activities of pinosylvin.

Origins	Cell Lines	Methods	Key Findings	References
Not clear	THP1 and U937 monocytic cell lines	Trypan blue exclusion assay Cell sorting analysis RT-PCR Preparation of cell lysates Western blot analysis Detection of LC3 puncta DNA transfection	Increased (50–100 μmol/L) cell death Caused caspase-3 activation, phosphatidylserine flipping, LC3II accumulation, LC3 puncta, and p62 degradation Induced cell death Caused downregulation of AMP-activated protein kinase (AMPK) α1	[86]
Not reported	Bovine aortic endothelial cells	Apoptosis assay Western blot analysis Flow cytometry analysis Measurement of caspase-3 activity	Increased caspase-3 activity, phosphatidylserine flip-flop, and nuclear fragmentation Activated JNK and endothelial NO synthase	[87]

Table 4. *Cont.*

Origins	Cell Lines	Methods	Key Findings	References
Synthesized	Molt and Raji lymphoblastoid cell lines.	Growth inhibitory action Cell count and viability DNA and protein synthesis assessment	Inhibited cell proliferation Inhibited [3H] thymidine and leucine uptake	[88]
Pinus resinosa	A549, DLD-1, and WS1 cells	Cytotoxicity assay	$66 \pm 10 < IC_{50} < 75 \pm 14$ μM	[89]
Synthesized	HCT 116 colorectal cancer cells	Proliferation inhibitory potential testing Cell cycle distribution analysis Western blot analysis RT-PCR Identification of gene expression cDNA microarray Electrophoretic mobility shift assay	Slowed cell growth Slowed cell cycle transition from G_1 phase to S phase Decreased the levels of cyclin D1, cyclin E, CDK2, c-Myc, pRb, and p53 Stopped the activation of proteins involved in the FAK/c-Src/ERK signaling pathway and the PI3K/Akt/GSK-3b signaling pathway Inhibited b-nuclear catenin translocation	[21]
Synthesized	HT1080 human fibrosarcoma and Balb/c mice	RT-PCR Wound healing assay Colony dispersion assesement In vivo pulmonary metastasis method Gelatin zymography	Inhibited the production of matrix metalloproteinase (MMP)-2, MMP-9, and membrane type 1-MMP Reduced HT1080 cell migration Slowed tumor nodule growth and tumor weight in lung tissue Downregulated MMP-9 and cyclooxygenase-2 (COX-2) expression and ERK1/2 and Akt phosphorylation in lung carcinoma tissues	[5]
Not reported	ARPE-19 cells	Toxicity evaluation Oxidative stress assessment Cell viability RT-PCR Nrf2 and p62 RNA interference	Improved cell viability in the face of oxidative stress Increased HO-1 expression No effect on Nrf2 expression Protected against oxidative damage	[77]
Purchased	LNCaP-par and LNCaP-abl prostate cancer cells	High-throughput screening (HTS) Cell viability and apoptosis assays Gene expression analysisqPT-PCR	Inhibited androgen signaling and intracellular steroidogenesis in CRPC cells	[23]
Purchased	Nasal cavity cancer cells (RPMI 2650)	MTT assay Gap closure assay Cell migration assay Cell invasion assay Western blot analysis Proteome profiler human protease array	Suppressed migration and invasion of NPC039 and NPCBM cells at increasing doses Lowered the protein expression levels of MMP2 and MMP9 Decreased the enzyme activity of MMP2Reduced vimentin and N-cadherin expression (in NPC cells) Increased zonula occludens-1 and E-cadherin expression Inhibited NPC039 and NPCBM cell invasion and migration by modulating the p38, ERK1/2, and JNK1/2 pathways Inhibited NPC cell migration and invasion	[20]
Purchased	SCC-9 and HSC-3 cancer cells (tongue squamous)	MTT assay Wound closure assessment Gelatin zymography Cell migration and invasion evaluation Western blot analysis	Decreased the enzymatic activity of MMP-2 and lowered its protein level Raised the expression of TIMP-2 Stopped cancer cell growth in a wound-healing experiment Reduced ERK1/2 protein phosphorylation in SAS and SCC-9 cells	[20]
Not reported	Bovine aortic endothelial cells	Apoptosis experiment Western blot analysis Flow cytometry Measurement of caspase-3 activity Hoechst staining	Induced (100 μmol/L) cell death Boosted caspase-3 activation, nuclear condensation, and the "flip-flop" of phosphatidylserine (at high concentrations) Inhibited necrosis Promoted LC3 conversion from LC3-I to LC3-II and p62 degradation Stimulated AMP-activated protein kinase (AMPK) and an AMPK inhibitor Reversed the inhibitory impact of an AMPK inhibitor Induced autophagy via AMPK activation	[87]

The effects of pinosylvin on the migration and invasion of human oral cancer cells remain unknown, as do the underlying processes. Chen et al. [20] evaluated the effects of varying concentrations of pinosylvin (0–80 μM) on the metastatic and invasive capacities of SAS, SCC-9, and HSC-3 cells. Pinosylvin suppressed matrix metalloproteinase-2 (MMP-2) enzyme activity and lowered its protein level in Western blotting and gelatin zymography assays but enhanced the expression of tissue inhibitors of metalloproteinase-2 (TIMP-2). Pinosylvin also inhibited the migration of oral cancer cells (SAS, SCC-9, and HSC-3) in the wound healing experiment and using the transwell technique. Furthermore, this substance inhibited the phosphorylation of ERK1/2 protein expression in SAS and SCC-9 cells (Figure 3).

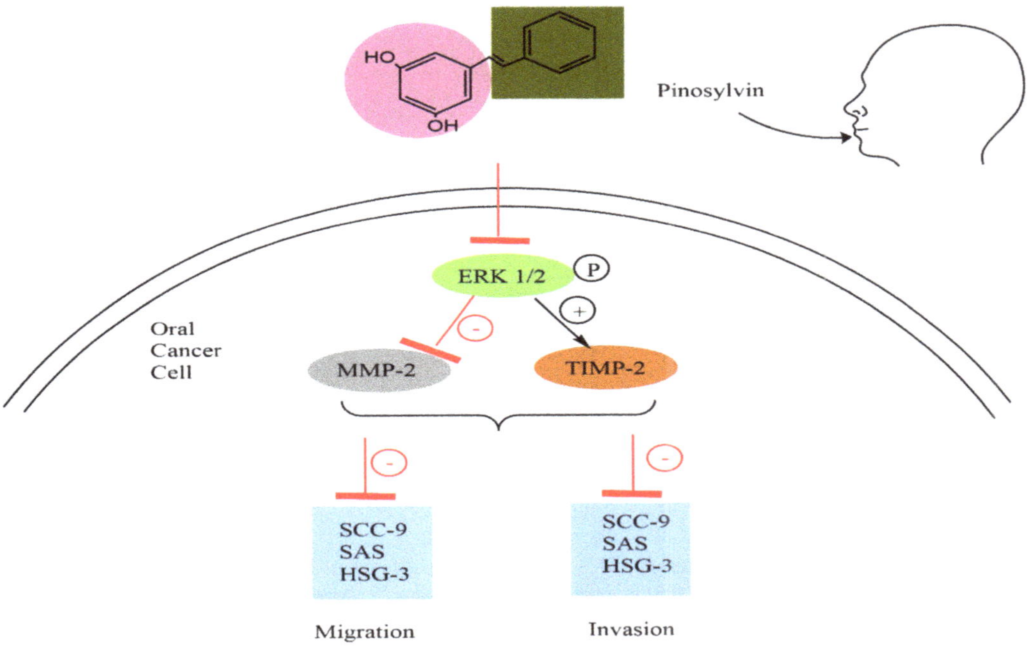

Figure 3. Anticancer activity of pinosylvin against oral cancer cells. Pinosylvin suppressed the invasion and migration of oral cancer cells by inhibiting the phosphorylation of ERK1/2 protein expression in SAS and SCC-9 cells. Abbreviations: MMP-2, matrix metalloproteinase-2; TIMP-2, tissue inhibitor of metalloproteinase-2; ERK, extracellular signal-regulated kinase.

These findings suggest that pinosylvin may be a promising anticancer drug to prevent oral cancer spread. Chuang et al. [90] aimed to examine the functional role of pinosylvin in nasopharyngeal carcinoma (NPC) cells (NPC039, NPCBM, and RPMI 2650). According to gap-closure and transwell assays, pinosylvin reduced the migration and invasion of NPC039 and NPCBM cells at increasing doses. It not only inhibited the activity of MMP2 enzymes, but also reduced the expression levels of MMP2 and MMP9 proteins. Pinosylvin inhibited the expression of vimentin and N-cadherin while dramatically increasing that of zonula occludens-1 and E-cadherin in NPC cells. It also inhibited the invasion and migration of NPC039 and NPCBM cells by modulating the p38, ERK1/2, and JNK1/2 pathways. According to the findings of this investigation, pinosylvin suppressed the migration and invasion of NPC cells.

There are currently few therapeutic options for castration-resistant prostate cancer (CRPC). A high-throughput screen of 4910 drugs and drug-like molecules was used in a study conducted by Ketola et al. [23] to detect antiproliferative substances on prostate cancer after androgen ablation therapy. The effects of compounds on cell survival were examined in androgen-ablated LNCaP prostate cancer cells, LNCaP cells cultured in androgens, and two non-malignant prostate epithelial cells (RWPE-1 and EP156T). Pinosylvin methyl ether (PSME) was a strong inhibitor of androgen-ablated LNCaP cell growth in cancer-specific antiproliferative drug validation assays. A genome-wide gene expression study in PSME-exposed cells was undertaken to obtain insight into growth inhibitory mechanisms in CRPC. In androgen-depleted LNCaP cells, pinosylvin affected the expression of genes involved in cell cycle, steroid, and cholesterol production. Reduced androgen-receptor expression and prostate-specific antigen in PSME exposed cells verified the decrease in androgen signaling. Taken together, our comprehensive screen revealed PSME as a new antiproliferative agent for CRPC. These findings provide a solid foundation for future preclinical and clinical investigations on CRPC treatment.

The capacity of pinosylvin to modify oxidative stress in human RPE cells was investigated by Koskela et al. [77]; by first evaluating the range of PS toxicity by exposing ARPE-19 cells to PS doses of 0.1–200 μM for 24 h, followed by a cell survival test. The ARPE-19 cells were then preincubated in pinosylvin for 24 h before being exposed to hydroquinone (HQ) without pinosylvin for another 24 h. Pinosylvin therapy at doses of 5 and 10 μM greatly improved cell survival against oxidative stress. Pinosylvin therapy elevated the production of HO-1, an enzyme with antioxidant, anti-inflammatory, and immunomodulatory abilities, which is positively associated with cell survival. Pinosylvin treatment did not influence the expression of Nrf2 or its target genes, p62 or GSTP1, while having a strong effect on the expression of HO-1, another Nrf2-controlled gene. RNA interference investigation verified the importance of Nrf2 and HO-1 in pinosylvin-mediated oxidative stress protection, whereas the contribution of p62 seemed minor at the gene expression and cell viability levels. The findings show that pinosylvin therapy protects against oxidative stress by inducing HO-1 in human RPE cells.

Pinosylvin is known to have an anti-inflammatory effect on endothelial cells. Hence, Kwon et al. [82] attempted to understand the exact process in their research. Pinosylvin was tested to determine if it increased COX or lipoxygenase (LOX) activity in THP-1 and U937 cells. Pinosylvin significantly increased LOX activity without affecting COX activity. Furthermore, it increased ALOX15 mRNA and protein levels, demonstrating that pinosylvin-induced LOX activity is due to increased ALOX15 expression. Pinosylvin appeared to enhance ERK and JNK phosphorylation in this cell signaling investigation. ERK and JNK inhibitors were observed to reduce ALOX15 expression and LPS-induced apoptosis produced by pinosylvin. Finally, pinosylvin promoted apoptosis in LPS-preconditioned leukocytes by increasing ALOX15 expression via ERK and JNK.

In cancer patients, metastases are a major cause of mortality [5]. Previous research revealed that pinosylvin has a potential cancer chemopreventive effect and suppresses the development of many human cancer cell lines by regulating cell cycle progression. In this study, the authors investigated the possible antimetastatic action of pinosylvin using in vitro and in vivo models. In cultured human fibrosarcoma HT1080 cells, pinosylvin inhibited the production of MMP-2, MMP-9, and membrane type 1-MMP. Pinosylvin has also been reported to interfere with HT1080 cell migration in colony dispersal and wound healing methods. Pinosylvin (10 mg/kg b.w., intraperitoneal treatment) effectively reduced tumor nodule growth and tumor weight in lung tissues in an in vivo model of spontaneous lung metastasis following injection of CT26 colon carcinoma into BALB/c mice. The study of tumors in lung tissue revealed that the antimetastatic impact of pinosylvin was associated with a decrease in the production of MMP-9 and COX-2 and the activation of ERK1/2 and Akt. These findings show that pinosylvin, via modulating MMPs, might be an effective inhibitor of tumor cell metastasis.

Park et al. [21] investigated the antiproliferative action of pinosylvin in human colorectal HCT-116 cancer cells to identify the underlying molecular processes. Pinosylvin inhibited HCT-116 cell proliferation by preventing the cell cycle from progressing from the G_1 to the S phase, as well as downregulating cyclin D1, cyclin E, cyclin A, cyclin-dependent kinase 2 (CDK2), CDK4, c-Myc, and retinoblastoma protein (pRb) and the upregulation of $p21^{WAF1/CIP1}$ and p53. Pinosylvin has also been shown to inhibit the activation of proteins involved in focal adhesion kinase and the phosphoinositide 3-kinase signaling system.

Pinosylvin, at high concentrations (100 µmol/L), was previously reported to promote cell death in bovine aortic endothelial cells. In the investigation conducted by Park et al. [22], it was attempted to reveal the role of pinosylvin in apoptosis, autophagy, and necrosis. Pinosylvin enhanced caspase-3 activation, nuclear condensation, and the "flip-flop" of phosphatidylserine at high concentrations, suggesting that pinosylvin triggers apoptosis. On the other hand, pinosylvin was found to suppress necrosis, a post-apoptotic process, based on flow cytometry data acquired using double-staining with annexin V and propidium iodide. Pinosylvin promoted LC3 conversion from LC3-I to LC3-II and p62 degradation, both of which are essential indications of autophagy. Furthermore, pinosylvin appeared to stimulate AMP-activated protein kinase (AMPK), and an AMPK inhibitor significantly reduced LC3 conversion. Pinosylvin reversed the inhibitory impact of an AMPK inhibitor. These findings imply that pinosylvin causes autophagy by activating AMPK. Additionally, an autophagy inhibitor was shown to enhance necrosis, which was later restored with pinosylvin, but the caspase-3 inhibitor had no impact on necrosis. These results show that pinosylvin-induced autophagy inhibits necrotic progression in endothelial cells.

In the study performed by Simard et al. [89], methanol extracts of *Pinus resinosa* wood containing pinosylvin were selectively cytotoxic against human lung cancer cells, A549 (IC_{50} = 41.6 µg/mL) and human colorectal adenocarcinoma cells, DLD-1 (IC_{50} = 47.4 µg/mL) compared to healthy cells, WS1 (IC_{50} = 130.11 µg/mL). Five known compounds were isolated and identified as: pinosylvin monomethyl ether (1), pinosylvin (2), pinosylvin dimethyl ether (3), pinobanksin (4), and (-)-norachelogenin using ^{1}H-, ^{13}C-NMR spectroscopy and HR-ESI-MS mass spectrometry (5). Compounds 1–5 were tested for their cytotoxicity against A549, DLD-1, and WS1. Compound 1 (pinosylvin monomethyl ether) had the highest cytotoxicity against both tumor and healthy cell lines, with IC_{50} values of 25.4, 20.1, and 34.3 µM for A549, DLD-1, and WS1, respectively

Skinnider and Stoessl [88] investigated the effects of phytoalexins lubimin, (-)-maackiain, pinosylvin, and related chemicals dehydroloroglossol and hordatine M on the development of the human lymphoblastoid cell lines Molt and Raji. The authors found that (-)-maackiain, pinosylvin, and dehydroloroglossol all significantly inhibited cell proliferation. The inhibition of [3H] thymidine and [3H] leucine absorption in pinosylvin and dehydroloroglossol was studied and shown to be effective. Phytoalexins and similar chemicals are abundant in plants and may serve as a source of antineoplastic drugs.

Several tests were carried out in the Song et al. [87] investigation to determine how high concentrations of pinosylvin (50 µM) promotes endothelial cell death. Pinosylvin, at high concentrations, was demonstrated to promote endothelial cell death by increasing caspase-3 activity, phosphatidylserine flip-flop, and nuclear fragmentation. They discovered that high concentrations of pinosylvin increased caspase-3 activity, which was amplified by serum deprivation or treatment with 100 µM etoposide. They also found that high concentrations of pinosylvin stimulated the activation of c-Jun N-terminal kinase (JNK) and endothelial nitric oxide synthase (eNOS). They then conducted a series of tests to determine which signaling molecule was important in pinosylvin-induced apoptosis. Finally, they found that SP-600125, a JNK inhibitor, inhibited pinosylvin-induced endothelial cell death, whereas L-NAME, an eNOS inhibitor, had no impact. These findings suggest that JNK is implicated in pinosylvin-induced apoptosis. At high concentrations, pinosylvin promotes cell death through JNK activation.

Resveratrol (pinosylvin analogue) has been shown to promote cell death in leukemia cells at high doses (50–100 µmol/L). Song et al. [86] foudn that cell death was significantly increased from 50 to 100 µmol/L pinosylvin in THP1 and U937 cells. Pinosylvin also induced caspase-3 activation, phosphatidylserine flipflop, LC3II accumulation, LC3 puncta, and p62 degradation in THP1 and U937 cells. These findings suggest that pinosylvin-induced cell death may occur through apoptosis and autophagy. Furthermore, we discovered that pinosylvin inhibits AMP-activated protein kinase 1 (AMPK1) in leukemia cells. As a result, a link was found between AMPK1 downregulation and leukemic cell death. Inhibition of AMPK1 reduces pinosylvin-induced apoptosis and autophagy in leukemia cells, indicating that AMPK is a crucial regulator of leukemia cell death. Moreover, when AMPK1-overexpressed leukemia cells were compared to vector-transfected cells, the progression of autophagy and apoptosis were inhibited by pinosylvin. Overexpression of AMPK1 increased cell death, but caspase-3 inhibitors or autophagy inhibitors significantly reduced pinosylvin-induced cell death. These findings imply that reducing AMPK1 by pinosylvin increases cell death by apoptosis and autophagy in leukemic cells.

5.5. *Neuroprotective Activity*

Based on the fact that neuroprotection is a typical technique to reduce the damage of cerebral ischemia, Xu et al. [67] set out to assess the neuroprotective efficacy of pinosylvin. Pinosylvin therapy reduced cell death in OGD/R-damaged PC12 cells and enhanced brain function in MCAO/R rats. Pinosylvin decreased the number of depolarized cells (low mitochondrial membrane potential) in OGD/R-damaged PC12 cells, implying a role in improving mitochondrial function. Further research revealed that pinosylvin triggers PINK1/Parkin-mediated protective mitophagy and activates the Nrf2 pathway, as shown by increased protein levels of LC3 II, Beclin1, PINK1, and Parkin, as well as Nrf2 translocation to the nucleus. Pinosylvin provided neuroprotection by triggering PINK1/Parkin-mediated mitophagy to eliminate damaged mitochondria and by activating the Nrf2 pathway to attenuate oxidative stress-induced mitochondrial dysfunction.

5.6. *Anti-Allergic Activity*

An extract of the branches of *H. dulcis* (containing pinosylvin) was tested for its anti-allergic potential using the rat basophilic leukemia (RBL)-2H3 cell line and the passive cutaneous anaphylaxis (PCA) mouse model using various assays [45]. The extract inhibited hexosaminidase secretion (indicating degranulation) and histamine release in antigen-stimulated RBL-2H3 cells, with decreased expression and production of the inflammatory mediators COX-2 and PGE_2, as well as the cytokines IL-4 and TNF-α, and suppression of NF-κB activation indicating the potential of the extract as a strong antiallergic agent.

6. Conclusions and Perspectives

Here, the main pharmacological characteristics and sources of pinosylvin have been documented and highlighted. Numerous published research has demonstrated that this natural molecule has exceptional biological properties, especially against tumor cell lines. Both molecular and cellular analyses revealed that pinosylvin blocks and inhibits the key pathways in nasopharyngeal cancer, prostate cancer, fibrosarcoma, colorectal cancer, and oral cancer with different target sites. This indicates that it may be a valuable anti-cancer drug component. Additionally, this molecule's antimicrobial, anti-inflammatory, antioxidant, and anti-allergic properties may qualify it as an effective bioactive ingredient in the treatment of cancer and neurodegenerative diseases. Nevertheless, a deeper insight into its pharmacokinetics and pharmacodynamics is required for its introduction as a chemotherapy drug. Furthermore, its safety requires validation by further toxicological studies.

Author Contributions: Conceptualization, A.B.; methodology, A.B., A.B., S.B., H.M. and T.B.; software, S.A. and I.J.; validation, A.B., G.Z., D.M. and M.G.; formal analysis, A.B. and D.M.; investigation, A.B., N.E.O. and G.Z.; resources, N.S., H.M. and S.B.; data curation, A.B.; writing—original draft preparation, I.J., S.B., H.M. and N.S.; writing—review and editing, S.P.B., J.M.L. and G.Z.; visualization, D.M. and M.G.; supervision, A.B.; project administration, A.B. All authors have read and agreed to the published version of the manuscript.

Funding: This research received no external funding.

Institutional Review Board Statement: Not applicable.

Informed Consent Statement: Not applicable.

Data Availability Statement: Not applicable.

Conflicts of Interest: The authors declare no conflict of interest.

Abbreviations

AMD	Age-related macular degeneration
MAPK	Mitogen-activated protein kinase
JNK	Jun amino-terminal kinase
NF-κB	Nuclear factor kappa B
Nrf2	Nuclear erythroid 2-related factor 2
ARE	Antioxidant response element
HCT	Colorectal cancer cell
FAK	Focal adhesion kinase
ERK	Extracellular signal-regulated kinase
GSK3	Glycogen synthase kinase-3
PI3K/AKT	Phosphatidylinositol 3-kinase/Protein kinase B
MMP	Matrix metalloproteinase
CX	Cyclohexane
EtOAc	Ethyl acetate
DRIFT	Diffuse reflectance fourier transform infrared spectroscopy
NIR	Near-infrared spectroscopy
FID	Flame ionization detection
GC	Gas chromatography
GC-MS	Gas chromatography-mass spectrometry
MIC	Minimal inhibitory concentration
EtBr	Ethidium bromide
Eps	Efflux pumps
MRSA	Methicillin-resistant *Staphylococcus aureus*;
ATTC	American type culture collection;
PGE2	Prostaglandin E2
NO	Nitric oxide
COX	Cyclooxygenase
iNOS	Inducible nitric oxide synthase
IL6	Interleukin 6
MCP1	Monocyte chemotactic protein 1
LPS	Lipopolysaccharide
IRF-3	Interferon regulatory factor 3
IFN-E	Interferon-E
TRIF	TIR-domain-containing adapter-inducing interferon-β
HO-1	Heme oxygenase-1
MTX	Methotrexate
TBARS	Plasmatic thiobarbituric acid-reactive substances
LOX	Lipoxygenase
TRPA1	Ankyrin subtype 1 protein
AA	Adjuvant arthritis
GGT	Glutamyltransferase
CRP	C-Reactive protein
PKC	Protein kinase C
GSTP1	Glutathione S-transferase pi 1
qRT-PCR	Quantitative reverse transcription polymerase chain reaction
HPV	Hind paw volume
CL	Chemiluminescence
MPO	Myeloperoxidase
LH	Luteinizing hormone
FSH	Follicle-stimulating hormone

ELISA	Enzyme-linked immunosorbent assay
NFE2L2	Nuclear factor erythroid 2 like 2
TIMP-2	Tissue inhibitor of metalloproteinase-2
NPC	Nasopharyngeal carcinoma
CRPC	Castration-resistant prostate cancer
PSME	Pinosylvin methyl ether
HQ	Hydroquinone
CDK2	Cyclin-dependent kinase 2
pRb	Retinoblastoma protein
AMPK	AMP-activated protein kinase
eNOS	Endothelial nitric oxide synthase
RBL	Rat basophilic leukemia
PCA	Passive cutaneous anaphylaxis
LC3-II	Microtubule associated protein 1 light chain 3-II

References

1. Jeong, E.; Lee, H.-R.; Pyee, J.; Park, H. Pinosylvin Induces Cell Survival, Migration and Anti-Adhesiveness of Endothelial Cells via Nitric Oxide Production: Pinosylvin is a vasoregulating compound. *Phytother. Res.* **2013**, *27*, 610–617. [CrossRef]
2. Riviere, C.; Pawlus, A.D.; Merillon, J.-M. Natural Stilbenoids: Distribution in the Plant Kingdom and Chemotaxonomic Interest in Vitaceae. *Nat. Prod. Rep.* **2012**, *29*, 1317–1333. [CrossRef] [PubMed]
3. Castelli, G.; Bruno, F.; Vitale, F.; Roberti, M.; Colomba, C.; Giacomini, E.; Guidotti, L.; Cascio, A.; Tolomeo, M. In Vitro Antileishmanial Activity of Trans-Stilbene and Terphenyl Compounds. *Exp. Parasitol.* **2016**, *166*, 1–9. [CrossRef] [PubMed]
4. Akinwumi, B.C.; Bordun, K.-A.M.; Anderson, H.D. Biological Activities of Stilbenoids. *Int. J. Mol. Sci.* **2018**, *19*, 792. [CrossRef] [PubMed]
5. Park, E.-J.; Park, H.J.; Chung, H.-J.; Shin, Y.; Min, H.-Y.; Hong, J.-Y.; Kang, Y.-J.; Ahn, Y.-H.; Pyee, J.-H.; Kook Lee, S. Antimetastatic Activity of Pinosylvin, a Natural Stilbenoid, Is Associated with the Suppression of Matrix Metalloproteinases. *J. Nutr. Biochem.* **2012**, *23*, 946–952. [CrossRef]
6. Plumed-Ferrer, C.; Väkeväinen, K.; Komulainen, H.; Rautiainen, M.; Smeds, A.; Raitanen, J.-E.; Eklund, P.; Willför, S.; Alakomi, H.-L.; Saarela, M. The Antimicrobial Effects of Wood-Associated Polyphenols on Food Pathogens and Spoilage Organisms. *Int. J. Food Microbiol.* **2013**, *164*, 99–107. [CrossRef]
7. Silva, F.; Domingues, F.C.; Nerín, C. Control Microbial Growth on Fresh Chicken Meat Using Pinosylvin Inclusion Complexes Based Packaging Absorbent Pads. *LWT* **2018**, *89*, 148–154. [CrossRef]
8. Sousa, V.; Luís, Â.; Oleastro, M.; Domingues, F.; Ferreira, S. Polyphenols as Resistance Modulators in Arcobacter Butzleri. *Folia Microbiol.* **2019**, *64*, 547–554. [CrossRef]
9. Celimene, C.C.; Micales, J.A.; Ferge, L.; Young, R.A. Efficacy of Pinosylvins against White-Rot and Brown-Rot Fungi. *Holzforschung* **1999**, *53*, 491–497. [CrossRef]
10. Lee, D.G.; Lee, S.J.; Rodriguez, J.P.; Kim, I.H.; Chang, T.; Lee, S. Antifungal Activity of Pinosylvin from Pinus Densiflora on Turfgrass Fungal Diseases. *J. Appl. Biol. Chem.* **2017**, *60*, 213–218. [CrossRef]
11. Eräsalo, H.; Hämäläinen, M.; Leppänen, T.; Mäki-Opas, I.; Laavola, M.; Haavikko, R.; Yli-Kauhaluoma, J.; Moilanen, E. Natural Stilbenoids Have Anti-Inflammatory Properties in Vivo and down-Regulate the Production of Inflammatory Mediators NO, IL6, and MCP1 Possibly in a PI3K/Akt-Dependent Manner. *J. Nat. Prod.* **2018**, *81*, 1131–1142. [CrossRef]
12. Kivimäki, K.; Leppänen, T.; Hämäläinen, M.; Vuolteenaho, K.; Moilanen, E. Pinosylvin Shifts Macrophage Polarization to Support Resolution of Inflammation. *Molecules* **2021**, *26*, 2772. [CrossRef]
13. Laavola, M.; Nieminen, R.; Leppänen, T.; Eckerman, C.; Holmbom, B.; Moilanen, E. Pinosylvin and Monomethylpinosylvin, Constituents of an Extract from the Knot of Pinus Sylvestris, Reduce Inflammatory Gene Expression and Inflammatory Responses in Vivo. *J. Agric. Food Chem.* **2015**, *63*, 3445–3453. [CrossRef]
14. Modi, S.; Yaluri, N.; Kokkola, T. Strigolactone GR24 and Pinosylvin Attenuate Adipogenesis and Inflammation of White Adipocytes. *Biochem. Biophys. Res. Commun.* **2018**, *499*, 164–169. [CrossRef]
15. Moilanen, L.J.; Hämäläinen, M.; Lehtimäki, L.; Nieminen, R.M.; Muraki, K.; Moilanen, E. Pinosylvin Inhibits TRPA 1-Induced Calcium Influx In Vitro and TRPA 1-Mediated Acute Paw Inflammation In Vivo. *Basic Clin. Pharmacol. Toxicol.* **2016**, *118*, 238–242. [CrossRef]
16. Bauerova, K.; Acquaviva, A.; Ponist, S.; Gardi, C.; Vecchio, D.; Drafi, F.; Arezzini, B.; Bezakova, L.; Kuncirova, V.; Mihalova, D.; et al. Markers of Inflammation and Oxidative Stress Studied in Adjuvant-Induced Arthritis in the Rat on Systemic and Local Level Affected by Pinosylvin and Methotrexate and Their Combination. *Autoimmunity* **2015**, *48*, 46–56. [CrossRef]
17. Drafi, F.; Bauerova, K.; Kuncirova, V.; Ponist, S.; Mihalova, D.; Fedorova, T.; Harmatha, J.; Nosal, R. Pharmacological Influence on Processes of Adjuvant Arthritis: Effect of the Combination of an Antioxidant Active Substance with Methotrexate. *Interdiscip. Toxicol.* **2012**, *5*, 84–91. [CrossRef]

18. Rodríguez-Bonilla, P.; Gandía-Herrero, F.; Matencio, A.; García-Carmona, F.; López-Nicolás, J.M. Comparative Study of the Antioxidant Capacity of Four Stilbenes Using ORAC, ABTS+, and FRAP Techniques. *Food Anal. Methods* **2017**, *10*, 2994–3000. [CrossRef]
19. Wang, C.; Sang, M.; Gong, S.; Yang, J.; Cheng, C.Y.; Sun, F. Two Resveratrol Analogs, Pinosylvin and 4,4′-Dihydroxystilbene, Improve Oligoasthenospermia in a Mouse Model by Attenuating Oxidative Stress via the Nrf2-ARE Pathway. *Bioorg. Chem.* **2020**, *104*, 104295. [CrossRef]
20. Chen, M.-K.; Liu, Y.-T.; Lin, J.-T.; Lin, C.-C.; Chuang, Y.-C.; Lo, Y.-S.; Hsi, Y.-T.; Hsieh, M.-J. Pinosylvin Reduced Migration and Invasion of Oral Cancer Carcinoma by Regulating Matrix Metalloproteinase-2 Expression and Extracellular Signal-Regulated Kinase Pathway. *Biomed. Pharmacother.* **2019**, *117*, 109160. [CrossRef]
21. Park, E.-J.; Chung, H.-J.; Park, H.J.; Kim, G.D.; Ahn, Y.-H.; Lee, S.K. Suppression of Src/ERK and GSK-3/β-Catenin Signaling by Pinosylvin Inhibits the Growth of Human Colorectal Cancer Cells. *Food Chem. Toxicol.* **2013**, *55*, 424–433. [CrossRef]
22. Park, J.; Pyee, J.; Park, H. Pinosylvin at a High Concentration Induces AMPK-Mediated Autophagy for Preventing Necrosis in Bovine Aortic Endothelial Cells. *Can. J. Physiol. Pharmacol.* **2014**, *92*, 993–999. [CrossRef]
23. Ketola, K.; Viitala, M.; Kohonen, P.; Fey, V.; Culig, Z.; Kallioniemi, O.; Iljin, K. High-Throughput Cell-Based Compound Screen Identifies Pinosylvin Methyl Ether and Tanshinone IIA as Inhibitors of Castration-Resistant Prostate Cancer. *J. Mol. Biochem.* **2016**, *5*, 12–22.
24. Chambers, V.H. British Bees and Wind-Borne Pollen. *Nature* **1945**, *155*, 145. [CrossRef]
25. Ioannidis, K.; Melliou, E.; Alizoti, P.; Magiatis, P. Identification of Black Pine (*Pinus nigra* Arn.) Heartwood as a Rich Source of Bioactive Stilbenes by QNMR. *J. Sci. Food Agric.* **2017**, *97*, 1708–1716. [CrossRef]
26. Rowe, J.W.; Bower, C.L.; Wagner, E.R. Extractives of Jack Pine Bark: Occurrence of Cis- and Trans-Pinosylvin Dimethyl Ether and Ferulic Acid Esters. *Phytochemistry* **1969**, *8*, 235–241. [CrossRef]
27. Seppänen, S.K.; Syrjälä, L.; Von Weissenberg, K.; Teeri, T.H.; Paajanen, L.; Pappinen, A. Antifungal Activity of Stilbenes in in Vitro Bioassays and in Transgenic Populus Expressing a Gene Encoding Pinosylvin Synthase. *Plant Cell Rep.* **2004**, *22*, 584–593. [CrossRef]
28. Pietarinen, S.P.; Willför, S.M.; Ahotupa, M.O.; Hemming, J.E.; Holmbom, B.R. Knotwood and Bark Extracts: Strong Antioxidants from Waste Materials. *J. Wood Sci.* **2006**, *52*, 436–444. [CrossRef]
29. Hovelstad, H.; Leirset, I.; Oyaas, K.; Fiksdahl, A. Screening Analyses of Pinosylvin Stilbenes, Resin Acids and Lignans in Norwegian Conifers. *Molecules* **2006**, *11*, 103–114. [CrossRef]
30. Vek, V.; Poljanšek, I.; Humar, M.; Willför, S.; Oven, P. In Vitro Inhibition of Extractives from Knotwood of Scots Pine (*Pinus sylvestris*) and Black Pine (*Pinus nigra*) on Growth of *Schizophyllum commune*, *Trametes versicolor*, *Gloeophyllum trabeum* and *Fibroporia Vaillantii*. *Wood Sci. Technol.* **2020**, *54*, 1645–1662. [CrossRef]
31. Lindberg, L.E.; Willför, S.M.; Holmbom, B.R. Antibacterial Effects of Knotwood Extractives on Paper Mill Bacteria. *J. Ind. Microbiol. Biotechnol.* **2004**, *31*, 137–147. [CrossRef]
32. Verkasalo, E.; Möttönen, V.; Roitto, M.; Vepsäläinen, J.; Kumar, A.; Ilvesniemi, H.; Siwale, W.; Julkunen-Tiitto, R.; Raatikainen, O.; Sikanen, L. Extractives of Stemwood and Sawmill Residues of Scots Pine (*Pinus sylvestris* L.) for Biorefining in Four Climatic Regions in Finland-Phenolic and Resin Acid Compounds. *Forests* **2021**, *12*, 192. [CrossRef]
33. Bergström, B. Chemical and Structural Changes during Heartwood Formation in *Pinus sylvestris*. *Forestry* **2003**, *76*, 45–53. [CrossRef]
34. François, S.; Jean, L.; Serge, L.; Vakhtang, M.; André, P. Inhibition of Cholinesterase and Amyloid-&bgr; Aggregation by Resveratrol Oligomers from *Vitis amurensis*. *Phytother. Res.* **2008**, *22*, 544–549. [CrossRef]
35. Kodan, A.; Kuroda, H.; Sakai, F. A Stilbene Synthase from Japanese Red Pine (*Pinus densiflora*): Implications for Phytoalexin Accumulation and down-Regulation of Flavonoid Biosynthesis. *Proc. Natl. Acad. Sci. USA* **2002**, *99*, 3335–3339. [CrossRef]
36. Dumas, M.T.; Hubbes, M.; Strunz, G.M. Identification of Some Compounds Associated with Resistance of *Pinus densiflora* to *Fomes annosus*. *Eur. J. For. Pathol.* **1983**, *13*, 151–160. [CrossRef]
37. Willför, S.M.; Ahotupa, M.O.; Hemming, J.E.; Reunanen, M.H.T.; Eklund, P.C.; Sjöholm, R.E.; Eckerman, C.S.E.; Pohjamo, S.P.; Holmbom, B.R. Antioxidant Activity of Knotwood Extractives and Phenolic Compounds of Selected Tree Species. *J. Agric. Food Chem.* **2003**, *51*, 7600–7606. [CrossRef]
38. Raiber, S.; Schröder, G.; Schröder, J. Molecular and Enzymatic Characterization of Two Stilbene Synthases from Eastern White Pine (*Pinus strobus*) A Single Arg/His Difference Determines the Activity and the PH Dependence of the Enzymes. *FEBS Lett.* **1995**, *361*, 299–302. [CrossRef]
39. Hwang, H.S.; Han, J.Y.; Choi, Y.E. Enhanced Accumulation of Pinosylvin Stilbenes and Related Gene Expression in *Pinus Strobus* after Infection of Pine Wood Nematode. *Tree Physiol.* **2021**, *41*, 1972–1987. [CrossRef]
40. Koo, H.B.; Hwang, H.S.; Han, J.Y.; Cheong, E.J.; Kwon, Y.S.; Choi, Y.E. Enhanced Production of Pinosylvin Stilbene with Aging of *Pinus strobus* Callus and Nematicidal Activity of Callus Extracts against Pinewood Nematodes. *Sci. Rep.* **2022**, *12*, 770. [CrossRef]
41. Hemingway, R.W.; Mcgraw, G.W.; Barras, S.J. Polyphenols in Ceratocystis Minor Infected *Pinus taeda*: Fungal Metabolites, Phloem and Xylem Phenols. *J. Agric. Food Chem.* **1977**, *25*, 717–722. [CrossRef]
42. Gabaston, J.; Leborgne, C.; Waffo-Téguo, P.; Pedrot, E.; Richard, T.; Mérillon, J.M.; Valls Fonayet. J. Separation and Isolation of Major Polyphenols from Maritime Pine (*Pinus pinaster*) Knots by Two-Step Centrifugal Partition Chromatography Monitored by LC-MS and NMR Spectroscopy. *J. Sep. Sci.* **2020**, *43*, 1080–1088. [CrossRef]

43. Gabaston, J.; Richard, T.; Cluzet, S.; Palos Pinto, A.; Dufour, M.C.; Corio-Costet, M.F.; Mérillon, J.M. *Pinus pinaster* Knot: A Source of Polyphenols against *Plasmopara viticola*. *J. Agric. Food Chem.* **2017**, *65*, 8884–8891. [CrossRef]
44. Conde, E.; Fang, W.; Hemming, J.; Willför, S.; Domínguez, H.; Parajó, J.C. Recovery of Bioactive Compounds from *Pinus pinaster* Wood by Consecutive Extraction Stages. *Wood Sci. Technol.* **2014**, *48*, 311–323. [CrossRef]
45. Lim, S.J.; Kim, M.; Randy, A.; Nho, C.W. Inhibitory Effect of the Branches of *Hovenia dulcis* Thunb. and Its Constituent Pinosylvin on the Activities of IgE-Mediated Mast Cells and Passive Cutaneous Anaphylaxis in Mice. *Food Funct.* **2015**, *6*, 1361–1370. [CrossRef]
46. Wollenweber, E.; Stevens, J.F.; Dörr, M.; Rozefelds, A.C. Taxonomic Significance of Flavonoid Variation in Temperate Species of Nothofagus. *Phytochemistry* **2003**, *62*, 1125–1131. [CrossRef]
47. Gyeltshen, T.; Jordan, G.J.; Smith, J.A.; Bissember, A.C. Natural Products Isolation Studies of the Paleoendemic Plant Species *Nothofagus gunnii* and *Nothofagus cunninghamii*. *Fitoterapia* **2022**, *156*, 105088. [CrossRef]
48. Kostecki, K.; Engelmeier, D.; Pacher, T.; Hofer, O.; Vajrodaya, S.; Greger, H. Dihydrophenanthrenes and Other Antifungal Stilbenoids from *Stemona* cf. Pierrei. *Phytochemistry* **2004**, *65*, 99–106. [CrossRef]
49. Schöppner, A.; Kindl, H. Purification and Properties of a Stilbene Synthase from Induced Cell Suspension Cultures of Peanut. *J. Biol. Chem.* **1984**, *259*, 6806–6811. [CrossRef]
50. Poljanšek, I.; Oven, P.; Vek, V.; Raitanen, J.E.; Hemming, J.; Willför, S. Isolation of Pure Pinosylvins from Industrial Knotwood Residue with Non-Chlorinated Solvents. *Holzforschung* **2019**, *73*, 475–484. [CrossRef]
51. Jorgensen, E.; Balsillie, D. Formation of Heartwood Phenols in Callus Tissue Cultures of Red Pine (*Pinus resinosa*). *Can. J. Bot.* **1969**, *47*, 1015–1016. [CrossRef]
52. Lange, B.M.; Trost, M.; Heller, W.; Langebartels, C.; Sandermann, H. Elicitor-Induced Formation of Free and Cell-Wall-Bound Stilbenes in Cell-Suspension Cultures of Scots Pine (*Pinus sylvestris* L.). *Planta* **1994**, *194*, 143–148. [CrossRef]
53. Makrides, S.C. Strategies for Achieving High-Level Expression of Genes in *Escherichia coli*. *Microbiol. Rev.* **1996**, *60*, 512–538. [CrossRef]
54. Wu, J.; Liu, P.; Fan, Y.; Bao, H.; Du, G.; Zhou, J.; Chen, J. Multivariate Modular Metabolic Engineering of *Escherichia coli* to Produce Resveratrol from L-Tyrosine. *J. Biotechnol.* **2013**, *167*, 404–411. [CrossRef]
55. Watts, K.T.; Lee, P.C.; Schmidt-Dannert, C. Biosynthesis of Plant-Specific Stilbene Polyketides in Metabolically Engineered *Escherichia coli*. *BMC Biotechnol.* **2006**, *6*, 22. [CrossRef]
56. Van Summeren-Wesenhagen, P.V.; Marienhagen, J. Metabolic Engineering of *Escherichia coli* for the Synthesis of the Plant Polyphenol Pinosylvin. *Appl. Environ. Microbiol.* **2015**, *81*, 840–849. [CrossRef]
57. Xu, J.Y.; Xu, Y.; Chu, X.; Tan, M.; Ye, B.C. Protein Acylation Affects the Artificial Biosynthetic Pathway for Pinosylvin Production in Engineered *E. coli*. *ACS Chem. Biol.* **2018**, *13*, 1200–1208. [CrossRef] [PubMed]
58. Liang, J.-L.; Guo, L.-Q.; Lin, J.-F.; He, Z.-Q.; Cai, F.-J.; Chen, J.-F. A Novel Process for Obtaining Pinosylvin Using Combinatorial Bioengineering in *Escherichia coli*. *World J. Microbiol. Biotechnol.* **2016**, *32*, 102. [CrossRef] [PubMed]
59. Salas-Navarrete, C.; Hernández-Chávez, G.; Flores, N.; Martínez, L.M.; Martinez, A.; Bolívar, F.; Barona-Gomez, F.; Gosset, G. Increasing Pinosylvin Production in *Escherichia coli* by Reducing the Expression Level of the Gene FabI-Encoded Enoyl-Acyl Carrier Protein Reductase. *Electron. J. Biotechnol.* **2018**, *33*, 11–16. [CrossRef]
60. Katsuyama, Y.; Funa, N.; Horinouchi, S. Precursor-Directed Biosynthesis of Stilbene Methyl Ethers in *Escherichia coli*. *Biotechnol. J.* **2007**, *2*, 1286–1293. [CrossRef] [PubMed]
61. Wang, S.; Zhang, S.; Xiao, A.; Rasmussen, M.; Skidmore, C.; Zhan, J. Metabolic Engineering of *Escherichia coli* for the Biosynthesis of Various Phenylpropanoid Derivatives. *Metab. Eng.* **2015**, *29*, 153–159. [CrossRef]
62. Holmgren, A.; Bergström, B.; Gref, R.; Ericsson, A. Detection of Pinosylvins in Solid Wood of Scots Pine Using Fourier Transform Raman and Infrared Spectroscopy. *J. Wood Chem. Technol.* **1999**, *19*, 139–150. [CrossRef]
63. Roupe, K.; Halls, S.; Davies, N.M. Determination and Assay Validation of Pinosylvin in Rat Serum: Application to Drug Metabolism and Pharmacokinetics. *J. Pharm. Biomed. Anal.* **2005**, *38*, 148–154. [CrossRef]
64. Ekeberg, D.; Flæte, P.O.; Eikenes, M.; Fongen, M.; Naess-Andresen, C.F. Qualitative and Quantitative Determination of Extractives in Heartwood of Scots Pine (*Pinus sylvestris* L.) by Gas Chromatography. *J. Chromatogr. A* **2006**, *1109*, 267–272. [CrossRef]
65. Preusz, M.; Tříska, J.; Vrchotová, N.; Vilímek, J.; Enei, F.; Preusz, K. Chemical Profile of Organic Residues from Ancient Amphoras Found in Pyrgi and Castrum Novum, Tyrrhenian Sea (Italy). *J. Archaeol. Sci. Rep.* **2019**, *24*, 565–573. [CrossRef]
66. Lee, S.K.; Lee, H.J.; Min, H.Y.; Park, E.J.; Lee, K.M.; Ahn, Y.H.; Cho, Y.J.; Pyee, J.H. Antibacterial and Antifungal Activity of Pinosylvin, a Constituent of Pine. *Fitoterapia* **2005**, *76*, 258–260. [CrossRef]
67. Xu, H.; Deng, R.; Li, E.T.S.; Shen, J.; Wang, M. Pinosylvin Provides Neuroprotection against Cerebral Ischemia and Reperfusion Injury through Enhancing PINK1/Parkin Mediated Mitophagy and Nrf2 Pathway. *J. Funct. Foods* **2020**, *71*, 104019. [CrossRef]
68. Sharifi-Rad, J.; Dey, A.; Koirala, N.; Shaheen, S.; El Omari, N.; Salehi, B.; Goloshvili, T.; Silva, N.C.C.; Bouyahya, A.; Vitalini, S. *Cinnamomum* Species: Bridging Phytochemistry Knowledge, Pharmacological Properties and Toxicological Safety for Health Benefits. *Front. Pharmacol.* **2021**, *12*, 600139. [CrossRef]
69. Bouyahya, A.; Chamkhi, I.; Benali, T.; Guaouguaou, F.-E.; Balahbib, A.; El Omari, N.; Taha, D.; Belmehdi, O.; Ghokhan, Z.; El Menyiy, N. Traditional Use, Phytochemistry, Toxicology, and Pharmacology of *Origanum majorana* L. *J. Ethnopharmacol.* **2021**, *265*, 113318. [CrossRef]

70. Bouyahya, A.; El Omari, N.; Elmenyiy, N.; Guaouguaou, F.-E.; Balahbib, A.; El-Shazly, M.; Chamkhi, I. Ethnomedicinal Use, Phytochemistry, Pharmacology, and Toxicology of *Ajuga iva* (L.) Schreb. *J. Ethnopharmacol.* **2020**, *258*, 112875. [CrossRef]
71. Marmouzi, I.; Bouyahya, A.; Ezzat, S.M.; El Jemli, M.; Kharbach, M. The Food Plant *Silybum marianum* (L.) Gaertn.: Phytochemistry, Ethnopharmacology and Clinical Evidence. *J. Ethnopharmacol.* **2021**, *265*, 113303. [CrossRef]
72. De Bruijn, W.J.; Araya-Cloutier, C.; Bijlsma, J.; de Swart, A.; Sanders, M.G.; de Waard, P.; Gruppen, H.; Vincken, J.-P. Antibacterial Prenylated Stilbenoids from Peanut (*Arachis hypogaea*). *Phytochem. Lett.* **2018**, *28*, 13–18. [CrossRef]
73. Silva, F.; Nerín, C.; Domingues, F.C. Stilbene Phytoallexins Inclusion Complexes: A Natural-Based Strategy to Control Foodborne Pathogen Campylobacter. *Food Control* **2015**, *54*, 66–73. [CrossRef]
74. Bouyahya, A.; Omari, N.E.; El Hachlafi, N.; Jemly, M.E.; Hakkour, M.; Balahbib, A.; El Menyiy, N.; Bakrim, S.; Naceiri Mrabti, H.; Khouchlaa, A. Chemical Compounds of Berry-Derived Polyphenols and Their Effects on Gut Microbiota, Inflammation, and Cancer. *Molecules* **2022**, *27*, 3286. [CrossRef]
75. Bouyahya, A.; Guaouguaou, F.-E.; El Omari, N.; El Menyiy, N.; Balahbib, A.; El-Shazly, M.; Bakri, Y. Anti-Inflammatory and Analgesic Properties of Moroccan Medicinal Plants: Phytochemistry, in Vitro and in Vivo Investigations, Mechanism Insights, Clinical Evidences and Perspectives. *J. Pharm. Anal.* **2021**, *12*, 35–37. [CrossRef]
76. Park, J.-H.; Choi, G.J.; Jang, K.S.; Lim, H.K.; Kim, H.T.; Cho, K.Y.; Kim, J.-C. Antifungal Activity against Plant Pathogenic Fungi of Chaetoviridins Isolated from *Chaetomium globosum*. *FEMS Microbiol. Lett.* **2005**, *252*, 309–313. [CrossRef]
77. Koskela, A.; Reinisalo, M.; Hyttinen, J.M.; Kaarniranta, K.; Karjalainen, R.O. Pinosylvin-Mediated Protection against Oxidative Stress in Human Retinal Pigment Epithelial Cells. *Mol. Vis.* **2014**, *20*, 760.
78. Park, E.-J.; Min, H.-Y.; Ahn, Y.-H.; Bae, C.-M.; Pyee, J.-H.; Lee, S.K. Synthesis and Inhibitory Effects of Pinosylvin Derivatives on Prostaglandin E2 Production in Lipopolysaccharide-Induced Mouse Macrophage Cells. *Bioorg. Med. Chem. Lett.* **2004**, *14*, 5895–5898. [CrossRef]
79. Park, E.-J.; Min, H.-Y.; Chung, H.-J.; Ahn, Y.-H.; Pyee, J.-H.; Lee, S.K. Pinosylvin Suppresses LPS-Stimulated Inducible Nitric Oxide Synthase Expression via the MyD88-Independent, but TRIF-Dependent Downregulation of IRF-3 Signaling Pathway in Mouse Macrophage Cells. *Cell. Physiol. Biochem.* **2011**, *27*, 353–362. [CrossRef]
80. Mačičková, T.; Drábiková, K.; Nosal, R.; Bauerová, K.; Mihálová, D.; Harmatha, J.; Pečivová, J. In Vivo Effect of Pinosylvin and Pterostilbene in the Animal Model of Adjuvant Arthritis. *Neuroendocrinol. Lett.* **2010**, *31*, 91.
81. Jančinová, V.; Perečko, T.; Nosál', R.; Harmatha, J.; Šmidrkal, J.; Drábiková, K. The Natural Stilbenoid Pinosylvin and Activated Neutrophils: Effects on Oxidative Burst, Protein Kinase C, Apoptosis and Efficiency in Adjuvant Arthritis. *Acta Pharmacol. Sin.* **2012**, *33*, 1285–1292. [CrossRef] [PubMed]
82. Kwon, O.; Seo, Y.; Park, H. Pinosylvin Exacerbates LPS-Induced Apoptosis via ALOX 15 Upregulation in Leukocytes. *BMB Rep.* **2018**, *51*, 302. [CrossRef] [PubMed]
83. Schuster, R.; Holzer, W.; Doerfler, H.; Weckwerth, W.; Viernstein, H.; Okonogi, S.; Mueller, M. Cajanus Cajan–a Source of PPARγ Activators Leading to Anti-Inflammatory and Cytotoxic Effects. *Food Funct.* **2016**, *7*, 3798–3806. [CrossRef] [PubMed]
84. Park, E.-J.; Ahn, Y.-H.; Pyee, J.-H.; Park, H.J.; Chung, H.-J.; Min, H.-Y.; Hong, J.-Y.; Kang, Y.-J.; Bae, I.-K.; Lee, S.K. Suppressive Effects of Pinosylvin, a Natural Stilbenoid, on Cyclooxygenase-2 and Inducible Nitric Oxide Synthase and the Growth Inhibition of Cancer Cells. *Cancer Res.* **2005**, *65*, 176.
85. Tamminen, T.; Koskela, A.; Toropainen, E.; Gurubaran, I.S.; Winiarczyk, M.; Liukkonen, M.; Paterno, J.J.; Lackman, P.; Sadeghi, A.; Viiri, J.; et al. Pinosylvin Extract Retinari™ Sustains Electrophysiological Function, Prevents Thinning of Retina, and Enhances Cellular Response to Oxidative Stress in NFE2L2 Knockout Mice. *Oxid. Med. Cell. Longev.* **2021**, *2021*, 8028427. [CrossRef]
86. Song, J.; Seo, Y.; Park, H. Pinosylvin Enhances Leukemia Cell Death via Down-Regulation of AMPKα Expression: Anti-Cancer Activity of Pinosylvin. *Phytother. Res.* **2018**, *32*, 2097–2104. [CrossRef]
87. Jina, S.; Jinsun, P.; Eunsil, J.; A-Young, S.; Jaeho, P.; Heonyong, P. Apoptotic Effect of Pinosylvin at a High Concentration Regulated by C-Jun N-Terminal Kinase in Bovine Aortic Endothelial Cells. *J. Life Sci.* **2015**, *25*, 416–424. [CrossRef]
88. Skinnider, L.; Stoessl, A. The Effect of the Phytoalexins, Lubimin, (−)-Maackiain, Pinosylvin, and the Related Compounds Dehydroloroglossol and Hordatine M on Human Lymphoblastoid Cell Lines. *Experientia* **1986**, *42*, 568–570. [CrossRef]
89. Simard, F.; Legault, J.; Lavoie, S.; Mshvildadze, V.; Pichette, A. Isolation and Identification of Cytotoxic Compounds from the Wood of *Pinus resinosa*. *Phytother. Res.* **2008**, *22*, 919–922. [CrossRef]
90. Chuang, Y.-C.; Hsieh, M.-C.; Lin, C.-C.; Lo, Y.-S.; Ho, H.-Y.; Hsieh, M.-J.; Lin, J.-T. Pinosylvin Inhibits Migration and Invasion of Nasopharyngeal Carcinoma Cancer Cells via Regulation of Epithelial-mesenchymal Transition and Inhibition of MMP-2. *Oncol. Rep.* **2021**, *46*, 143. [CrossRef]

plants

MDPI

Review

Therapeutic Potential of *Ranunculus* Species (Ranunculaceae): A Literature Review on Traditional Medicinal Herbs

Youn-Kyoung Goo

Department of Parasitology and Tropical Medicine, School of Medicine, Kyungpook National University, Daegu 41944, Korea; kuku1819@knu.ac.kr

Abstract: The genus *Ranunculus* includes approximately 600 species and is distributed worldwide. To date, several researchers have investigated the chemical and biological activities of *Ranunculus* species, and my research team has found them to have antimalarial effects. This review is based on the available information on the traditional uses and pharmacological studies of *Ranunculus* species. The present paper covers online literature, particularly from 2010 to 2021, and books on the ethnopharmacology and botany of *Ranunculus* species. Previous studies on the biological activity of crude or purified compounds from *Ranunculus* species, including *R. sceleratus* Linn., *R. japonicus* Thunb., *R. muricatus* Linn., *R. ternatus* Thunb., *R. arvensis* Linn., *R. diffusus* DC., *R. sardous* Crantz, *R. ficaria* Linn., *R. hyperboreus* Rotlb., and *R. pedatus* Waldst. & Kit., have provided new insights into their activities, such as antibacterial and antiprotozoal effects as well as antioxidant, immunomodulatory, and anticarcinogenic properties. In addition, the anti-inflammatory and analgesic effects of plants used in traditional medicine applications have been confirmed. Therefore, there is a need for more diverse studies on the chemical and pharmacological activities of highly purified molecules from *Ranunculus* species extracts to understand the mechanisms underlying their activities and identify novel drug candidates.

Keywords: ethnopharmacologic effect; herbal medicine; *Ranunculus* species

Citation: Goo, Y.-K. Therapeutic Potential of *Ranunculus* Species (Ranunculaceae): A Literature Review on Traditional Medicinal Herbs. *Plants* **2022**, *11*, 1599. https://doi.org/10.3390/plants11121599

Academic Editors: Ivayla Dincheva, Ilian Badjakov and Bistra Galunska

Received: 25 May 2022
Accepted: 16 June 2022
Published: 17 June 2022

1. Introduction

The genus *Ranunculus* includes approximately 600 species globally. Recent taxonomic reports suggest that this genus has a monophyletic origin and is divided into two subgenera and seventeen sections [1]. Owing to its wide distribution, the genus has high genetic diversity. Several *Ranunculus* species have been used in folk medicine to treat various diseases or symptoms, such as jaundice, nebula, edema, malaria, asthma, pain, gout, rheumatism, inflammatory skin disorders, cancer, and hypertension. In addition, researchers have reported that *Ranunculus* extracts possess antioxidant, anti-inflammatory, antimutagenic, antimalarial, antibacterial, antitumoral, cardioprotective, and wound-healing properties [2–7].

Over the last decade, various studies have investigated the chemical components and pharmacological activities of *Ranunculus* species [8,9]. However, no recent review has been published detailing the aspects of the plants that have been investigated, including their biology, traditional uses, phytoconstituents, therapeutic activities, and clinical applications, since a previous review article was reported in 2012 [10]. Thus, this article aims to provide an up-to-date survey of the advances in and prospects of the research on the phytochemicals and pharmacological potential of *Ranunculus* species.

2. Search Strategy

This review article is based on the information available on the phytochemical, toxicological, and pharmacological studies on the traditional uses of *Ranunculus* species. The present paper covers online literature (Google Scholar, PubMed, ScienceDirect, Scopus, SpringerLink, and Web of Science), particularly from 2010 to 2021, and books on the ethnopharmacology and botany of *Ranunculus* species. The following words were used as key search terms: ("*Ranunculus*" OR "*Ranunculus* species") AND ("herbal medicine" OR "herb medicine" OR "ethnopharmacological effects" OR "ethnopharmacological activity" OR "phytomedicine" OR "treatment" OR "drug". The range of the article publication year for the search (from 2010 to 2021) was selected because the previous review by Aslam et al. covered almost all literature data published by 2012 [10].

3. Taxonomy, Distribution, and Morphology

Ranunculaceae Juss., or the buttercup family, has a worldwide distribution, representing a large group comprising more than 2500 species belonging to 59 genera. Its family members live under a wide range of ecological conditions, especially in the Northern Hemisphere [1,11]. Among the family, *Ranunculus*, comprising 600 species, is distributed across all continents [11]. *Ranunculus* species are highly genetically diverse; therefore, their classification is challenging. As a result, generic delimitation and infrageneric classification of these species are still under consideration.

Initially, *Ranunculus* species were classified based on the descriptions of their achenes (e.g., the shape of their body and beak, pericarp structure, and indumentum), flowers (e.g., the number of sepals and honey-leaves, gloss and color of the petals, and shape of the nectaries), roots (e.g., whether they were uniform or dimorphic with fibrous and tuberous roots) [12], and fruit anatomy [13]. Later, Tamura classified the genera into seven subgenera based on the reassessment of the achene structure: *Pallasiantha, Coptidium, Ficaria, Batrachium, Crymodes, Gampsoceras,* and *Ranunculus* [1,11]. In this classification, the subgenera of *Ranunculus* were further subdivided into 20 sections [11].

Subsequently, DNA markers were utilized to delineate the phylogenetic relationships within Ranunculaceae [14–23]. The sequences of the internal transcribed spacer region of nuclear ribosomal DNA are mostly used as DNA barcode markers for phylogenetic studies at the generic/subgeneric level [24,25]. In combination with data from the chloroplast genome and other external data, this nuclear marker also offers insights into the reticulate patterns caused by hybridization [26,27]. Moreover, a complete study of the taxonomy of the genus using both DNA markers and morphological data suggested the separation of 226 species into two subgenera and 17 sections [20].

4. Phytochemical Investigations of *Ranunculus* Species

Ranunculus sceleratus Linn., commonly known as the celery-leaved buttercup, is a flowering plant species distributed over the Northern Hemisphere. The main constituents of *R. sceleratus* L. are flavonoids, steroids such as pyrogallol tannins, and the glycoside ranunculin [28]. Ranunculin is hydrolyzed after the leaves of *R. sceleratus* L. are dried or crushed and generates protoanemonin associated with the toxic properties of buttercups. Because of its instability, protoanemonin dimerizes to produce anemonin, a nonirritant form [29,30]. In addition, the 70% ethanolic extracts from the aerial parts of *R. sceleratus* L. have been found to be abundant in myristic acid [31], and sapigenin 4′-*O*-alpha-rhamnopyranoside, apigenin 7-*O*-beta-glucopyranosyl-4′-*O*-alpha-rhamnopyranoside, tricin 7-*O*-beta-glucopyranoside, tricin, and isoscopoletin have been identified as *R. sceleratus*-derived compounds in the extract [32].

Ranunculus ficaria Linn. is known as lesser celandine. The compositions found in *R. ficaria* L. were ranunculin and its enzymatic reaction products, flavonoids such as quercetin and rutoside, saponosides with hederagenin, oleanolic acid aglyca, macerate, and tinctures [33–35].

The components of *R. japonicus* Thunb. revealed by a Waters Acquity Ultra Performance liquid chromatography system were lactone glycosides, flavonoid glycosides, and aglycones including ranunculin, tricin, adonivernite, orientin, isorientin, vitexin, 6-C-β-D-glucosyl-8-C-α-L-arabinosylapigenin, and tricin-7-O-β-D-glucopyranoside [36].

Ranunculus muricatus Linn. is also known as spiny fruit buttercup. Phytochemical analysis of *R. muricatus* L. revealed the presence of saponins, tannins, phenols, flavonoids, alkaloids, cardiac glycosides, anthocyanins, carbohydrates, coumarins, and phytosterols [8,37,38]. The major constituents by HPLC were stigmast-4-ene-3,6-dione, stigmasterol, anemonin, β-sitosterol, protocatechuic aldehyde, protocatechuic acid, lutein, flavonoid glycosides, ranunculoside A, ranunculoside B, and ranunculone C, in addition to two potent antioxidants, caffeoyl-β-*D*-glucopyranoside, and 1,3-dihydroxy-2-tetracosanoylamino-4-(E)-nonadecene [9,39–41]. Moreover, four compounds, muriolide, muricazine, chalcone 4-benzyloxylonchocarpin, and new-to-nature anthraquinone muracatanes B, were recently isolated [42,43].

Phytochemical analyses of *R. ternatus* Thunb. reported that the plant contains flavonoids, glycosides, benzine, organic acids, sterols, esters, amino acids, and constant and trace elements [44]. Furthermore, *R. ternatus* ethyl acetate extract constitutes contain sternbin, methylparaben, 3-[(4-O-d-glucopyranosyl)-phenyl]-2-propenoic acid, linocaffein, β-d-glucose, robustaflavone-4′-methylether, kayaflavone, podocarpus flavone A, bilobetin, isoginkgetin, amentoflavone, ternatoside A, ternatoside B, and 4-*O*-d-glucopyranosyl-p-coumaric acid [45–47]. Furthermore, methyl (R)-3-[2-(3,4-dihydroxybenzoyl)-4,5-dihydroxyphenyl]-2-hydroxypropanoate was isolated from *R. ternatus* roots [48].

Ranunculus arvensis Linn. is commonly known as field buttercup. Phytochemical analysis indicated that *R. arvensis* L. possesses rutin, caffeic acids, and classes of flavonoids and phenolics, including flavonol glycosides of quercetin, kaempferol, isorhamnetin, and their aglycons [49].

In *Ranunculus* species, several bioactive compounds and *Ranunculus*-specific constituents have been identified, such as ranunculosides, muricazine, and muracatanes. Although many other species related to *Ranunculus* have also been studied to evaluate their pharmacological activities, the novel bioactive compounds found in *Ranunculus* species with high pharmacological effects show nutraceutical and pharmaceutical potential. Pharmacological properties and molecular formula of *Ranunculus* species compounds reported in articles published from 2010 to 2021 are summarized in Table 1.

Table 1. Pharmacological properties and molecular formula of *Ranunculus* species compounds reported in articles published from 2010 to 2021.

Ranunculus Species	Molecule	Molecular Formula	Pharmacological Activity	Ref
Ranunculus japonicus Thunb.	berberine yangonin	$C_{20}H_{18}NO_4^+$ $C_{15}H_{14}O_4$	inhibited the migration capacity of RA-FLSs in a dose-dependent manner	[50]
Ranunculus muricatus Linn.	muricazine	$C_{16}H_{10}N_2O_4$	antioxidant effect, lipoxygenase, and urease inhibitory activities.	[42]
	4-benzyloxylonchocarpin	$C_{27}H_{24}O_4$	acetylcholinesterase inhibitory effect	[43]
	muracatane B	$C_{14}H_8O_5$	alpha-glucosidase inhibitory effect	[43]
	4-methoxylonchocarpin	$C_{21}H_{20}O_4$	moderate cytotoxic effects towards ovarian carcinoma, colorectal adenocarcinoma, breast cancer, and thyroid carcinoma	[43]
	muriolide	$C_{15}H_{15}O_8$	antioxidant and lipoxygenase inhibitory activities	[42]
	caffeoyl-beta-D-glucopyranoside	$C_{14}H_{18}O_9$	antioxidant effect	[41]
	1,3-dihydroxy-2-tetracosanoylamino -4-(E)-nonadecene	$C_{43}H_{80}NO_3$	antioxidant effect	[41]
	ranuncoside	$C_{22}H_{11}O_7$	antioxidant effect	[42]
Ranunculus ternatus Thunb.	n-butyl-β-D-fructofuranoside	$C_{10}H_{20}O_6$	inhibitory effect of multidrug-resistant tuberculosis	[48]

5. Pharmacological Activities of *Ranunculus* Species

5.1. *Ranunculus sceleratus Linn.*

All parts of *Ranunculus sceleratus* Linn. are poisonous when fresh; however, the plant is used in folk medicine to treat various diseases after heating or drying [7]. In recent decades, ethnopharmacological effects have been experimentally proven by several studies (Table 2). The two ranunculins, protoanemonin and anemonin, have shown fungicidal, antimicrobial, antimutanenic, and antipyretic properties [29,30,51], and have been used for ethnopharmacological purposes in many countries [52,53]. Sharif et al. performed an in vivo study to evaluate the effects of hypertension treatment using normotensive and fructose-induced hypertensive rats, in which the aqueous fraction produced the most interesting effects. Furthermore, mechanistic studies with various pharmacological antagonists have demonstrated that the hypotensive response induced by *R. sceleratus* L. is caused by the involvement of a muscarinic receptor, angiotensin-converting enzyme inhibition, ganglionic block, and nitric oxide release [54]. In addition, the 70% ethanolic extracts from the aerial parts of *R. sceleratus* L. revealed that abundant myristic acid in the extract inhibited nitrite concentration in LPS-stimulated RAW 264.7 macrophage cell line [31]. Moreover, *R. sceleratus*-derived compounds, sapigenin 4′-O-alpha-rhamnopyranoside, apigenin 7-O-beta-glucopyranosyl-4′-O-alpha-rhamnopyranoside, tricin 7-O-beta-glucopyranoside, tricin, and isoscopoletin, showed inhibitory activity against the hepatitis B virus [32]. In addition to the treatment effect of R. sceleratus extract, fresh *R. sceleratus* for TianJiu therapy, which involves adding Chinese medicinal herbal paste on designated acupoints, showed good therapeutic effect on intrahepatic cholestasis in rats, although the fresh form of *R. sceleratus* L. is known as an irritant [55]. Specific mechanisms by which the extract induces irritant or nonirritant responses have not been revealed. To eluciate this phenomenon, a methanolic extract of *R. sceleratus* L. was used to demonstrate the mechanism of both irritant and non-irritant properties induced by the extract in topical inflammation. When arachidonic acid elicited the inflammatory process, the effect of the extract was generally proinflammatory or neutral. However, if the response was caused by the application of an irritant, such as etradecanoylphorbol acetate, the extract mainly resulted in anti-inflammatory effects. This effect was mentioned as a counter-irritant, and the extract itself could be an irritant in physiological conditions but could also counteract the action of previously applied irritants [7].

5.2. *Ranunculus ficaria Linn.*

Ranunculus ficaria Linn. is an herbal astringent commonly used to treat hemorrhoids internally or externally [67]. Various methods have been applied for ethnopharmacological use. Infusion or decoction of the leaves and roots of *R. ficaria* was known to have trophic and anti-inflammatory effects in varicose veins, hemorrhoids, and skin disorders in Romania. The macerate and tinctures obtained from this plant are used to treat hemorrhoids by stimulating blood circulation as a traditional medication [67]. The compositions found in *R. ficaria* could inhibit nitrite accumulation, and thus may be useful for preventing inflammatory diseases mediated by the excessive production of nitric oxide, according to an in vitro macrophage study. However, a previous report suggested that clinicians should consider using lesser celandine (pilewort, *R. ficaria*) as a causative agent owing to its hepatotoxicity [35,68].

5.3. *Ranunculus japonicus Thunb.*

Ranunculus japonicus Thunb. has been used to treat malaria, jaundice, migraines, stomachaches, arthralgia, crane-like arthropathy, ulcers, toothaches, and eye inflammation since Zhou Hou Bei Ji Fang was first recorded more than 1800 years ago [69]. Since then, studies have demonstrated various phytomedicinal activities, such as the protective effect on heart diseases including myocardial ischemic-reperfusion injury, hypertrophy in cardiomyocytes, and high blood pressure by alleviating chronic $[Ca^{2+}]$ *i* overload, as well as therapeutic effects on rheumatoid arthritis and decreasing intracellular $[Ca^{2+}]$ *i* in vascular smooth muscle cells [50,57,58]. In addition, *R. japonicus* extracts showed antimalarial effects in in vitro culture of *Plasmodium falciparum* and in vivo rodent malaria experimental systems of *Plasmodium berghei* [56].

Table 2. Therapeutic activity of *Ranunculus* species.

Ranunculus Species	Therapeutic Activity	Therapeutic Indications	Source	Ref.
Ranunculus sceleratus Linn.	Anti-inflammatory	Inhibits nitrite accumulation in macrophage	ethanolic extract of whole plant	[7,31]
	Hepatoprotective (treatment of cholestasis hepatitis)	Improves serum hepatic enzyme activity and hepatic pathologic changes in cholestatic rats	fresh *R. sceleratus* of whole plant	[55]
	Antihypertensive	Inhibits angiotensin converting enzyme (ACE) Involes muscarinic receptor, ganglionic block, and NO	aqueous fraction of aerial parts and roots	[28,54]
	Antiviral	Inhibits hepatitis B virus replication	isolated compounds of whole plant	[32]
Ranunculus japonicus Thunb.	Antirheumatoid arthritis	Inhibits migration capacity of rheumatoid arthritis fibroblast-like synoviocytes	methanolic extract of whole plant	[50]
	Antimalarial	Inhibits parasite growth in *Plasmodium falciparum* and *P. berghei* improve hepatic and renal parameters	ethanolic extract of whole plant	[56]
	Antihypertrophic	Suppresses elevated expression of the ANP, BNP, and beta-MHC inhibits up-regulation of $[Ca^{2+}]\,i$	total glycosides of whole plant	[57]
	Protective effect of myocardial ischemic-reperfusion injury	Improves heart function indexes Reduces the area of myocardial infarction	total glycosides of whole plant	[58]
	Antihypertensive	Decreases blood pressure and reduces calcium ions level in cells	total glycosides of whole plant	[36]
Ranunculus muricatus Linn.	Antioxidant	Scavenges the DPPH free radical Inhibits lipoxygenase and urease enzyme activity	methanolic extract of whole plant ethyl acetate fraction of whole plant	[41,42]
	Anticarcinogenic	Shows cytotoxic activity to cancer cells Inhibits acetylcholinesterase and alpha glucosidase	ethanolic extract of whole plant	[43]
	Anti-inflammatory Analgesic	Inhibits paw edema, paw licking and abdominal constrictions/stretching of hind limbs	methanolic extract of whole plant	[59]
	Cardiotonic (cardiovascular)	Increases perfusion pression and force of contraction Increases heart rate	methanolic extract of whole plant	[60]
Ranunculus ternatus Thunb.	Anticarcinogenic	Induces cell death depending on caspase-7	ethyl acetate extract of whole plant	[61]
	Antibacterial	Shows inhibitory activity against *Mycobacterium tuberculosis* Inhibits multidrug-resistant tuberculosis	ethanolic extract of roots	[48]
Ranunculus arvensis Linn.	Antioxidant	Shows antioxidant activity in DPPH free radical scavenging assay	methanolic extract of whole plant	[49]
	Anticarcinogenic	Induces cell death	aqueous and methanolic extract of whole plant	[62]
Ranunculus diffusus DC.	Anti-inflammatory	Suppresses NF-kB signaling targeting Src and Syk	methanolic extract of aerial parts	[63,64]
Ranunculus sardous Crantz.	Anti-inflammatory	Inhibits nitrite accumulation in macrophage	ethanolic extract from aerial and root parts	[35]
Ranunculus ficaria Linn.	Anti-inflammatory	Inhibits nitrite accumulation in macrophage	ethanolic extract from aerial and root parts	[33,35]
Ranunculus hyperboreus Rotlb.	Anti-inflammatory	Decreases the elevated nitrate amount Regulates the expression and protein levels of inflammation-related enzymes, iNOS and COX-2, and proinflammatory cytokines, TNF-α, IL-1β, and IL-6 Suppresses activation of MAPK pathway	aqueous and methanolic extract of whole plant	[65]
Ranunculus pedatus Waldst. & Kitt.	Anti-inflammatory	Inhibits increased capillary permeability induced by acetic-acid	methanolic extract of whole plant	[66]
	Wound healing	Shows fast dermal remodeling and re-epithelization in epidermis Enhances hydroxyproline content	methanolic and aqueous extract of whole plant	
Ranunculus constantinapolitanus (DC.) d'Urv	Anti-inflammatory	Inhibits increased capillary permeability induced by acetic-acid	methanolic extract of whole plant	[66]
	Wound healing	Enhances hydroxyproline content	methanolic extract of whole plant	

5.4. *Ranunculus muricatus Linn.*

Ranunculus muricatus Linn. has tremendous medicinal potentials [70]. It is used by the local population as a folk medicine for cough, asthma, heart disease, jaundice, diarrhea, dysentery, urinary infection, eczema, lymphatic tuberculosis, dental diseases, ringworm infection, and leprosy [71,72]. In addition, it exhibits antioxidant, anti-inflammatory, antibacterial, antifungal, analgesic, and cytotoxic activities [37,59,60,73]. Therefore, among *Ranunculus* species, *R. muricatus* L. is the most extensively studied. Several constituents identified in *R. muricatus* L. exhibit phytochemical activities. For example, the major isolated constituents are stigmast-4-ene-3,6-dione, stigmasterol, anemonin, β-sitosterol, protocatechuic aldehyde, protocatechuic acid, lutein, flavonoid glycosides, ranunculoside A, ranunculoside B, and ranunculone C, in addition to the two potent antioxidants, caffeoyl-β-*D*-glucopyranoside and 1,3-dihydroxy-2-tetracosanoylamino-4-(E)-nonadecene [9,39–41]. Moreover, two recently isolated compounds, muriolide, a new lactone, and muricazine, a new hydrazine derivative, exhibited robust free radical scavenging properties and exerted an inhibitory effect on lipoxygenase [42]. Finally, chalcone 4-benzyloxylonchocarpin, which inhibits AcheE, and the new-to-nature anthraquinone muracatanes B, which inhibits α-glucosidase, were isolated from *R. muricatus* L. [43].

5.5. *Ranunculus diffusus DC.*

The phytomedicinal effects of *Ranunculus diffusus* DC. have recently been reported. The methanol extract of *R. diffusus* showed photoaging protective effects on ultraviolet B radiation-induced skin by inhibiting the p38-AP-1 signal cascade. In addition, the extract exerted anti-inflammatory effects without toxicity by suppressing Src and Syk, which are targets of NF-κB signaling [63,64].

5.6. *Ranunculus ternatus Thunb.*

Ranunculus ternatus Thunb. has been used in traditional Chinese medicine [74] because of its effects on malignant lymphoma, leukemia, pulmonary tuberculosis, breast tumors, goiters, esophageal tumors, lung disease, gastric problems, and other health conditions [75–77]. Constituents of *R. ternatus*, such as amentoflavone and podocarpus flavone A, induce apoptosis [78,79]; however, their mechanisms have not been evaluated. Furthermore, n-butyl-β-D-fructofuranoside, isolated from *R. ternatus* roots, demonstrated significant therapeutic activity against tuberculosis [48]. Finally, the ethyl acetate extract of *R. ternatus* exerts caspase-7-dependent apoptosis in a cancer model [61].

5.7. *Ranunculus arvensis Linn.*

Ranunculus arvensis Linn. has been widely used to treat arthritis, asthma, hay fever, rheumatism, psoriasis, gut diseases, and rheumatic diseases [49]. Moreover, *R. arvensis* extracts showed antioxidant and anticarcinogenic activities [49,62]. However, topical use of the plant may cause contact dermatitis, such as skin inflammation, skin burns, and injury of mucous membranes [80–82].

5.8. *Ranunculus hyperboreus Rotlb.*

Ranunculus hyperboreus Rotlb. is a subarctic and subalpine plant that lives in extreme environmental conditions. *R. hyperboreus* extract induces anti-inflammatory activity by regulating the gene expression and protein levels of inflammation-related enzymes, such as iNOS and COX-2, and proinflammatory cytokines, such as TNF-α, IL-1β, and IL-6 [65].

5.9. *Ranunculus pedatus Waldst. & Kit.*

The wound healing activity of *Ranunculus pedatus* Waldst. & Kitt. was evaluated using its methanolic extract and was found to exert significant effects on wound healing with robust anti-inflammatory activity in both incision and excision wound animal models [66].

6. Conclusions

The chemical and biological activities of *Ranunculus* species have been investigated using plant extracts. Contemporary research on the biological activity of the extracts of the species mentioned above has uncovered many activities, including antibacterial, antiviral, and antiprotozoal effects, as well as antioxidant and anticarcinogenic properties. In addition, these studies have demonstrated that herbal extracts exert hepatoprotective, hypoglycemic, and thyroid regulatory effects. Moreover, the anti-inflammatory and analgesic effects of the plants, known from the application of traditional medicine, have been confirmed. Furthermore, the molecules isolated from *Ranunculus* species showed promising pharmacological activity. Therefore, it is expected that effective purified molecules could be discovered from *Ranunculus* species to develop novel drugs through intensive research.

Funding: This research was financially supported by the Basic Science Research Program (NRF-2019R1C1C1002170) through the National Research Foundation of Korea (NRF), funded by the Ministry of Science, ICT and Future Planning.

Institutional Review Board Statement: Not applicable.

Informed Consent Statement: Not applicable.

Data Availability Statement: Not applicable.

Conflicts of Interest: The author declares no conflict of interest.

References

1. Tamura, M. Ranunculaceae. In *Flowering Plants Dicotyledons. The Families and Genera of Vascular Plants*; Kubitzki, K., Rohwer, J.G., Bittrich, V., Eds.; Springer: Berlin/Heidelberg, Germany, 1993; Volume 2.
2. Gürhan, G.; Ezer, N. Plants used in the treatment of hemorrhoids in folk medicine-1. *J. Hacetterpe Univ. Fac. Pharm.* **2004**, *24*, 37–55.
3. Zou, Y.P.; Tan, C.H.; Wang, B.D.; Jiang, S.H.; Zhu, D.Y. Flavonoid glycosides from *Ranunculus chinensis* Bge. *Helv. Chim. Acta* **2007**, *90*, 1940–1945. [CrossRef]
4. Sezik, E.; Yeşilada, E.; Honda, G.; Takaishi, Y.; Takeda, Y.; Tanaka, T. Traditional medicine in Turkey, X. Folk medicine in Central Anatolia. *J. Ethnopharmacol.* **2001**, *75*, 95–115. [CrossRef]
5. Newall, D.R.; Beedles, K.E. The stem-cell test: An in vitro assay for teratogenic potential. Results of a blind trial with 25 compounds. *Toxicol. In Vitro* **1996**, *10*, 229–240. [CrossRef]
6. Barbour, E.K.; Al Sharif, M.; Sagherian, V.K.; Habre, A.N.; Talhouk, R.S.; Talhouk, S.N. Screening of selected indigenous plants of Lebanon for antimicrobial activity. *J. Ethnopharmacol.* **2004**, *93*, 1–7. [CrossRef]
7. Prieto, J.M.; Recio, M.C.; Giner, R.M.; Máñez, S.; Ríos, J.L. Pharmacological approach to the pro- and anti-inflammatory effects of *Ranunculus sceleratus* L. *J. Ethnopharmacol.* **2003**, *89*, 131–137. [CrossRef]
8. Khan, F.A.; Zahoor, M.; Khan, E. Chemical and biological evaluation of *Ranunculus muricatus*. *Pak. J. Pharm. Sci.* **2016**, *29*, 503–510.
9. Wu, B.L.; Zou, H.L.; Qin, F.M.; Li, H.Y.; Zhou, G.X. New ent-kaurane-type diterpene glycosides and benzophenone from *Ranunculus muricatus* Linn. *Molecules* **2015**, *20*, 22445–22453. [CrossRef]
10. Aslam, M.S.; Choudhary, B.A.; Uzair, M.; Ijaz, A.S. The genus *Ranunculus*: A phytochemical and ethnopharmacological review. *Int. J. Pharm. Pharm. Sci.* **2012**, *4*, 15–22.
11. Tamura, M. Angiospermae. Ordnung Ranunculales. Fam. Ranunculaceae. II. Systematic Part. In *Natürliche Pflanzenfamilien*, 2nd ed.; Hiepko, P., Ed.; Duncker & Humblot: Berlin, Germany, 1995; pp. 223–519.
12. Candolle, A.P.D. *Prodromus Systematis Naturalis Regni Vegetabilis*; Treuttel & Wurz: Paris, France, 1838.
13. Prantl, K. *Beiträge zur Morphologie und Systematik der Ranunculaceen*; Engelmann: Leipzig, Germany, 1887; Volume 9, pp. 225–273.
14. Johansson, J.T. Chloroplast DNA restriction site mapping and the phylogeny of Ranunculus (Ranunculaceae). *Plant Syst. Evol.* **1998**, *213*, 1–19. [CrossRef]
15. Hörandl, E.; Paun, O.; Johansson, J.T.; Lehnebach, C.; Armstrong, T.; Chen, L.; Lockhart, P. Phylogenetic relationships and evolutionary traits in *Ranunculus* s.l. (Ranunculaceae) inferred from ITS sequence analysis. *Mol. Phylogenet. Evol.* **2005**, *36*, 305–327. [CrossRef]
16. Paun, O.; Greilhuber, J.; Temsch, E.M.; Hörandl, E. Patterns, sources and ecological implications of clonal diversity in apomictic *Ranunculus carpaticola* (*Ranunculus auricomus* complex, Ranunculaceae). *Mol. Ecol.* **2006**, *15*, 897–910. [CrossRef]
17. Hoffmann, M.H.; von Hagen, K.B.; Hörandl, E.; Röser, M.; Tkach, N.V. Sources of the arctic flora: Origins of arctic species in Ranunculus and related genera. *Int. J. Plant Sci.* **2010**, *171*, 90–106. [CrossRef] [PubMed]
18. Emadzade, K.; Lehnebach, C.; Lockhart, P.; Hörandl, E. A molecular phylogeny, morphology and classification of genera of Ranunculeae (Ranunculaceae). *Taxon* **2010**, *59*, 809–828. [CrossRef]

19. Emadzade, K.; Gehrke, B.; Linder, H.P.; Hörandl, E. The biogeographical history of the cosmopolitan genus *Ranunculus* L. (Ranunculaceae) in the temperate to meridional zones. *Mol. Phylogenet. Evol.* **2011**, *58*, 4–21. [CrossRef]
20. Hörandl, E.; Emadzade, K. Evolutionary classification: A case study on the diverse plant genus *Ranunculus* L. (Ranunculaceae), Perspectives in Plant Ecology. *Evol. Syst.* **2012**, *14*, 310–324.
21. Emadzade, K.; Lebmann, M.J.; Hoffmann, M.H.; Tkach, N.; Lone, F.A.; Hörandl, E. Phylogenetic Relationships and Evolution of High Mountain Buttercups (Ranunculus) in North America and Central Asia, Perspectives in Plant Ecology. *Evol. Syst.* **2015**, *17*, 131–141.
22. Li, Z.; Yang, L.; Lu, W.; Guo, W.; Gong, X.; Xu, J.; Yu, D. Spatial patterns of leaf carbon, nitrogen stoichiometry and stable carbon isotope composition of Ranunculus natans C.A. Mey. (Ranunculaceae) in the arid zone of northwest China. *Ecol. Eng.* **2015**, *77*, 9–17. [CrossRef]
23. Almerekova, S.; Shchegoleva, N.; Abugalieva, S.; Turuspekov, Y. The molecular taxonomy of three endemic Central Asian species of Ranunculus (Ranunculaceae). *PLoS ONE* **2020**, *15*, e0240121. [CrossRef]
24. Lockhart, P.J.; McLenachan, P.A.; Havell, D.; Glenny, D.; Huson, D.; Jensen, U. Phylogeny, radiation, and transoceanic dispersal of New Zealand alpine buttercups: Molecular evidence under split decomposition. *Ann. Mo. Bot. Gard.* **2001**, *88*, 458–477. [CrossRef]
25. Winkworth, R.C.; Wagstaff, S.J.; Glenny, D.; Lockhart, P.J. Evolution of the New Zealand mountain flora: Origins, diversification and dispersal. *Org. Divers. Evol.* **2005**, *5*, 237–247. [CrossRef]
26. Suh, Y.J.; Lee, S.; Lee, S.H.; Lee, N.S. Molecular evidence for the taxonomic identity of Korean Adonis (Ranunculaceae). *J. Plant Res.* **2002**, *115*, 217–223. [CrossRef]
27. Li, T.; Fu, X.; Deng, H.; Han, X.; Wen, F.; Xu, L. The complete chloroplast genome of *Ranunculus* Cantoniensis. *Mitochondrial DNA B* **2019**, *4*, 1095–1096. [CrossRef]
28. Mahran, G.H.; Saber, A.H.; el-Alfy, T. Spectrophotometric determination of protoanemonin, anemonin and ranunculin in *Ranunculus sceleratus* L. *Planta Med.* **1968**, *16*, 323–328. [CrossRef]
29. Martín, M.L.; San Román, L.; Domínguez, A. In vitro activity of protoanemonin, an antifungal agent. *Planta Med.* **1990**, *56*, 66–69. [CrossRef]
30. Minakata, H.; Komura, H.; Nakanishi, K.; Kada, T. Protoanemonin, an antimutagen isolated from plants. *Mutat. Res.* **1983**, *116*, 317–322. [CrossRef]
31. Marrelli, M.; De Marco, C.T.; Statti, G.; Neag, T.A.; Toma, C.C.; Conforti, F. *Ranunculus* species suppress nitric oxide production in LPS-stimulated RAW 264.7 macrophages. *Nat. Prod. Res.* **2021**, *6*, 1–5. [CrossRef]
32. Li, H.; Zhou, C.; Pan, Y.; Gao, X.; Wu, X.; Bai, H.; Zhou, L.; Chen, Z.; Zhang, S.; Shi, S.; et al. Evaluation of antiviral activity of compounds isolated from *Ranunculus sieboldii* and *Ranunculus sceleratus*. *Planta Med.* **2005**, *71*, 1128–1133. [CrossRef]
33. Bonora, A.; Botta, B.; Menziani-Andreoli, E.; Bruni, A. Organ-specific distribution and accumulation of protoanemonin in *Ranunculus fiscaria* L. *Biochem. Physiol. Pflanz.* **1988**, *183*, 443–447. [CrossRef]
34. Tomczyk, M.; Gudej, J.; Sochacki, M. Flavonoids from Ficaria verna Huds. *Z. Nat. C* **2002**, *57*, 440–444. [CrossRef]
35. Neag, T.; Olah, N.K.; Hanganu, D.; Benedec, D.; Pripon, F.F.; Ardelean, A.; Toma, C.C. The anemonin content of four different *Ranunculus* species. *Pak. J. Pharm. Sci.* **2018**, *31*, 2027–2032.
36. Rui, W.; Chen, H.; Tan, Y.; Zhong, Y.; Feng, Y. Rapid analysis of the main components of the total glycosides of *Ranunculus japonicus* by UPLC/Q-TOF-MS. *Nat. Prod. Commun.* **2010**, *5*, 783–788. [CrossRef]
37. Ibrar, M.; Samreen, U. Phytochemical screening and evaluation of cytotoxic and phytotoxic effects of *Ranunculus muricatus* L. *Pak. J. Plant. Sci* **2012**, *18*, 35–45.
38. Aslam, M.S.; Choudhary, B.A.; Uzair, M.; Ijaz, A.S. Phytochemical study of Ariel parts of *Ranunculus muricatus* for the pharmacological active compounds. *J. Appl. Pharm.* **2013**, *5*, 827–832.
39. Wang, L.J.; Gao, X.Z. Studies on the chemical constituents in *Ranunculus muricatus* L. *Chin. JMAP* **2009**, *26*, 460–462.
40. Wu, B.; Qin, F.; Zhou, G. Studies on chemical constituents of *Ranunculus muricatus* Linn. *Nat. Prod. Res. Dev.* **2013**, *25*, 736–741.
41. Azam, F.; Chaudhry, B.A.; Ijaz, H.; Qadir, M.I. Caffeoyl-β-d-glucopyranoside and 1,3-dihydroxy-2-tetracosanoylamino-4-(E)-nonadecene isolated from *Ranunculus muricatus* exhibit antioxidant activity. *Sci. Rep.* **2019**, *9*, 15613. [CrossRef]
42. Raziq, N.; Saeed, M.; Ali, M.S.; Zafar, S.; Shahid, M.; Lateef, M. A new glycosidic antioxidant from *Ranunculus muricatus* L. (Ranunculaceae) exhibited lipoxygenasae and xanthine oxidase inhibition properties. *Nat. Prod. Res.* **2017**, *31*, 1251–1257. [CrossRef]
43. Hussain, H.; Ali, I.; Wang, D.; Mamadalieva, N.Z.; Hussain, W.; Csuk, R.; Loesche, A.; Fischer, L.; Staerk, D.; Anam, S.; et al. 4-Benzyloxylonchocarpin and Muracatanes A-C from *Ranunculus muricatus* L. and Their Biological Effects. *Biomolecules* **2020**, *10*, 1562. [CrossRef]
44. Miao, Y.D.; Li, X.J.; Jia, Y.J. Research progress on chemical constituents of *Ranunculi Ternati* Radix and their pharmacological effects. *Chin. Tradit. Herb. Drugs* **2014**, *45*, 1651–1654.
45. Zhang, X.G.; Tian, J.K. Studies on chemical constituents of *Ranunculus ternatus* (III) *Chin. Pharmacol. J.* **2006**, *41*, 1460–1461.
46. Xiong, Y.; Deng, K.Z.; Guo, Y.Q.; Gao, W.Y. Studies on Toxiological chemical constituents of flavonoids and glycosides in *Ranunculus ternatus*. *Chin. Tradit. Herb. Drugs* **2008**, *39*, 1449–1452.
47. Tian, J.K.; Sun, F.; Cheng, Y.Y. Chemical constituents from the roots of *Ranunculus ternatus*. *J. Asian Nat. Prod. Res.* **2006**, *8*, 35–39. [CrossRef]

48. Deng, K.Z.; Xiong, Y.; Zhou, B.; Guan, Y.M.; Luo, Y.M. Chemical constituents from the roots of *Ranunculus ternatus* and their inhibitory effects on *Mycobacterium tuberculosis*. *Molecules* **2013**, *18*, 11859–11865. [CrossRef]
49. Bhatti, M.Z.; Ali, A.; Saeed, A.; Saeed, A.; Malik, S.A. Antimicrobial, antitumor and brine shrimp lethality assay of *Ranunculus arvensis* L. extracts. *Pak. J. Pharm. Sci.* **2015**, *28*, 945–949.
50. Wang, Z.Y.; Chu, F.H.; Gu, N.N.; Wang, Y.; Feng, D.; Zhao, X.; Meng, X.D.; Zhang, W.T.; Li, C.F.; Chen, Y.; et al. Integrated strategy of LC-MS and network pharmacology for predicting active constituents and pharmacological mechanisms of *Ranunculus japonicus* Thunb. for treating rheumatoid arthritis. *J. Ethnopharmacol.* **2021**, *271*, 113818. [CrossRef]
51. Misra, S.B.; Dixit, S.N. Antifungal principle of *Ranunculus sceleratus*. *Econ. Bot.* **1980**, *34*, 362–367. [CrossRef]
52. Cappelletti, E.M.; Trevisan, R.; Caniato, R. External antirheumatic and antineuralgic herbal remedies in the traditional medicine of north-eastern Italy. *J. Ethnopharmacol.* **1982**, *6*, 161–190. [CrossRef]
53. Turner, N.J. Counter-irritant and other medicinal uses of plants in Ranunculaceae by native peoples in British Columbia and neighbouring areas. *J. Ethnopharmacol.* **1984**, *11*, 181–201. [CrossRef]
54. Sharif, A.; Saleem, M.; Alotaibi, N.H.; Alharbi, K.S.; Bukhari, S.N.A.; Irfan, H.M.; Younis, W. Blood pressure lowering effects of *Ranunculus scleratus* Linn. in normal and fructose induced hypertensive rats and estimation of underlying mechanisms. *Pak. J. Pharm. Sci.* **2020**, *33*, 2243–2247.
55. Zhang, Z.; Miao, Y.; Xu, M.; Cheng, W.; Yang, C.; She, X.; Geng, Q.; Zhang, Q. TianJiu therapy for α-naphthyl isothiocyanate-induced intrahepatic cholestasis in rats treated with fresh *Ranunculus sceleratus* L. *J. Ethnopharmacol.* **2020**, *248*, 112310. [CrossRef]
56. Yun, H.S.; Dinzouna-Boutamba, S.D.; Lee, S.; Moon, Z.; Kwak, D.; Rhee, M.H.; Chung, D.I.; Hong, Y.; Goo, Y.K. Antimalarial Effect of the Total Glycosides of the Medicinal Plant, *Ranunculus japonicus*. *Pathogens* **2021**, *10*, 532. [CrossRef]
57. Dai, H.L.; Jia, G.Z.; Zhao, S. Total glycosides of *Ranunculus japonicus* prevent hypertrophy in cardiomyocytes via alleviating chronic Ca(2+) overload. *Chin. Med. Sci. J.* **2015**, *30*, 37–43. [CrossRef]
58. Gao, X.W.; Liu, Y.; Yang, Z.C.; Tan, Y.Z. Protective effect of total glycosides of *Ranunculus japonicus* on myocardial ischemic-reperfusion injury in isolated rat hearts. *Zhong Yao Cai* **2014**, *37*, 1429–1433.
59. Nasreen, P.; Uttra, A.M.; Asif, H.; Younis, W.; Hasan, U.H.; Irfan, H.M.; Sharif, A. Evaluation of anti-inflammatory and analgesic activities of aqueous methanolic extract of *Ranunculus muricatus* in albino mice. *Pak. J. Pharm. Sci.* **2020**, *33*, 1121–1126.
60. Khan, A.Q.; Ahmad, T.; Mushtaq, M.N.; Malik, M.N.H.; Naz, H.; Ahsan, H.; Asif, H.; Noor, N.; Rahman, M.S.U.; Dar, U.; et al. Phytochemical analysis and cardiotonic activity of methanolic extract of *Ranunculus muricatus* Linn. in isolated rabbit heart. *Acta Pol. Pharm.* **2016**, *73*, 949–954.
61. Fang, M.; Shinomiya, T.; Nagahara, Y. Cell death induction by *Ranunculus ternatus* extract is independent of mitochondria and dependent on Caspase-7. *3 Biotech* **2020**, *10*, 123. [CrossRef]
62. Bhatti, M.Z.; Ali, A.; Ahmad, A.; Saeed, A.; Malik, S.A. Antioxidant and phytochemical analysis of *Ranunculus arvensis* L. extracts. *BMC Res. Notes* **2015**, *8*, 279. [CrossRef]
63. Hong, Y.H.; Kim, J.H.; Cho, J.Y. Photoaging Protective Effects of *Ranunculus bulumei* Methanol Extract. *Evid.-Based Complement. Alternat. Med.* **2020**, *2020*, 1761785. [CrossRef]
64. Hong, Y.H.; Kim, J.H.; Cho, J.Y. *Ranunculus bulumei* Methanol Extract Exerts Anti-Inflammatory Activity by Targeting Src/Syk in NF-κB Signaling. *Biomolecules* **2020**, *10*, 546. [CrossRef]
65. Kong, C.S.; Lee, J.I.; Karadeniz, F.; Kim, H.; Seo, Y. Effect of the Arctic terrestrial plant *Ranunculus hyperboreus* on LPS-induced inflammatory response via MAPK pathways. *Z. Nat. C J. Biosci.* **2018**, *73*, 273–279. [CrossRef] [PubMed]
66. Akkol, E.K.; Süntar, I.; Erdoğan, T.F.; Keleş, H.; Gonenç, T.M.; Kıvçak, B. Wound healing and anti-inflammatory properties of *Ranunculus pedatus* and *Ranunculus constantinapolitanus*: A comparative study. *J. Ethnopharmacol.* **2012**, *139*, 478–484. [CrossRef] [PubMed]
67. Tita, I.; Mongosanu, G.D.; Tita, M.G. Ethnobotanical inventory of medicinal plants from the South-West of Romania. *Farmacia* **2009**, *57*, 141–156.
68. Yilmaz, B.; Yilmaz, B.; Aktaş, B.; Unlu, O.; Roach, E.C. Lesser celandine (pilewort) induced acute toxic liver injury: The first case report worldwide. *World J. Hepatol.* **2015**, *7*, 285–288. [CrossRef]
69. Fawen, K. *A comprehensive Chinese-Latin-English dictionary of the names of Chinese Herbal Medicines*; World Publishing Corporation: Shanghai, China, 1998; pp. 671–682.
70. Ullah, M.; Khan, M.U.; Mahmood, A.; Malik, R.N.; Hussain, M.; Wazir, S.M.; Daud, M.; Shinwari, Z.K. An ethnobotanical survey of indigenous medicinal plants in Wana district South Waziristan agency. *Pak. J. Ethnopharmacol.* **2013**, *150*, 918–924. [CrossRef]
71. Iqbal, H.; Sher, Z.; Khan, Z.U. Medicinal plants from salt range Pind Dadan Khan, district Jhelum, Punjab, Pakistan. *J. Med. Plant Res.* **2011**, *5*, 2157–2168.
72. Rahman, I.U.; Ijaz, F.; Iqbal, Z.; Afzal, A.; Ali, N.; Afzal, M.; Khan, M.A.; Muhammad, S.; Qadir, G.; Asif, M. A novel survey of the ethno medicinal knowledge of dental problems in Manoor Valley (Northern Himalaya). *Pak. J. Ethnopharmacol.* **2016**, *194*, 877–894. [CrossRef]
73. Nazir, S.; Tahir, K.; Naz, R.; Khan, Z.; Khan, A.; Islam, R.; Rehman, A.U. In vitro screening of *Ranunculus muricatus* for potential cytotoxic and antimicrobial activities. *J. Pharmacol.* **2014**, *8*, 427–431.
74. Pan, Z.H.; Sun, Y.X. Observation on the antibacterial effect of *Artemisia annua* and *Ranunculus ternatus* on mycobacterium tuberculosis drug resistant. *Inf. Tradit. Chin. Med.* **1986**, *5*, 28.
75. Zhang, J.H.; Wan, M.R. Toxic effect of Rannuculin on leukemic cells in vitro. *Chin. J. Clin. Oncol.* **1993**, *12*, 941–943.

76. Chen, B.C.; Hang, Y.Y.; Chen, B.R. Advances in medicinal plant *Ranunculus ternatus*. *Chin. Wild Plant. Res.* **2002**, *1*, 7–9.
77. Tong, Y.L.; Yang, F.; Dai, G.H.; Ren, Z.M.; Wang, B.B. Study on activity in vitro of radix *Ranunculus ternati* saponins on cell A549 of non-small cell lung cancer. *Chin. Arch. Tradit. Chin. Med.* **2013**, *31*, 2181–2184.
78. Pei, J.S.; Liu, C.C.; Hsu, Y.N.; Lin, L.L.; Wang, S.C.; Chung, J.G.; Bau, D.T.; Lin, S.S. Amentoflavone induces cell-cycle arrest and apoptosis in MCF-7 human breast cancer cells via mitochondria-dependent pathway. *In Vivo* **2012**, *26*, 963–970.
79. Yeh, P.H.; Shieh, Y.D.; Hsu, L.C.; Kuo, L.M.; Lin, J.H.; Liaw, C.C.; Kuo, Y.H. Naturally occurring cytotoxic [3′→8″]-biflavonoids from *Podocarpus nakaii*. *J. Tradit. Med.* **2012**, *2*, 220–226. [CrossRef]
80. An, I.; Ucmak, D.; Esen, M.; Gevher, O.D. Phytocontact dermatitis due to *Ranunculus arvensis*: Report of three cases. *North. Clin. Istanb.* **2018**, *6*, 81–84. [CrossRef] [PubMed]
81. Kocak, A.O.; Saritemur, M.; Atac, K.; Guclu, S.; Ozlu, I. A rare chemical burn due to *Ranunculus arvensis*: Three case reports. *Ann. Saudi Med.* **2016**, *36*, 89–91. [CrossRef]
82. Polat, M. A case of phytodermatitis due to *Ranunculus arvensis* used as an herbal remedy. *Int. J. Dermatol.* **2016**, *55*, e37–e38. [CrossRef]

plants

MDPI

Article

Identification of Putative Candidate Genes from *Galphimia* spp. Encoding Enzymes of the Galphimines Triterpenoids Synthesis Pathway with Anxiolytic and Sedative Effects

Dianella Iglesias [1,2], Marcos de Donato Capote [2], Alfonso Méndez Tenorio [3], Ana Victoria Valdivia [2], Claudia Gutiérrez-García [2], Sujay Paul [2], Hafiz M. N. Iqbal [4], María Luisa Villarreal [1,*] and Ashutosh Sharma [2,*]

1 Centro de Investigación en Biotecnología, Universidad Autónoma del Estado de Morelos, Cuernavaca 62210, CP, Mexico; dianella.iglesiasrod@uaem.edu.mx

2 Centre of Bioengineering, NatProLab, School of Engineering and Sciences, Tecnologico de Monterrey, Querétaro 76130, CP, Mexico; mdedonate@tec.mx (M.d.D.C.); victoria.valdivia@tec.mx (A.V.V.); a00887520@itesm.mx (C.G.-G.); spaul@tec.mx (S.P.)

3 Escuela Nacional de Ciencias Biológicas, Instituto Politécnico Nacional, Ciudad de Mexico 56565, CP, Mexico; amendezt@ipn.mx

4 School of Engineering and Sciences, Tecnologico de Monterrey, Monterrey 64849, CP, Mexico; hafiz.iqbal@tec.mx

* Correspondence: luisav@uaem.mx (M.L.V.); asharma@tec.mx (A.S.)

Citation: Iglesias, D.; Donato Capote, M.d.; Méndez Tenorio, A.; Valdivia, A.V.; Gutiérrez-García, C.; Paul, S.; Iqbal, H.M.N.; Villarreal, M.L.; Sharma, A. Identification of Putative Candidate Genes from *Galphimia* spp. Encoding Enzymes of the Galphimines Triterpenoids Synthesis Pathway with Anxiolytic and Sedative Effects. *Plants* **2022**, *11*, 1879. https://doi.org/10.3390/plants11141879

Academic Editors: Ivayla Dincheva, Ilian Badjakov and Bistra Galunska

Received: 9 June 2022
Accepted: 5 July 2022
Published: 20 July 2022

Abstract: *Galphimia* spp. is popularly used in Mexican traditional medicine. Some populations of *Galphimia* exert anxiolytic and sedative effects due to the presence of the modified triterpenoids galphimines. However, the galphimine synthesis pathway has not yet been elucidated. Hence, in this study, a comparative transcriptome analysis between two contrasting populations of *Galphimia* spp., a galphimine-producer, and a non-galphimine-producer, is performed using RNA-Seq in the Illumina Next Seq 550 platform to identify putative candidates genes that encode enzymes of this metabolic pathway. Transcriptome functional annotation was performed using the Blast2GO in levels of gene ontology. For differential expression analysis, edgeR, pheatmap, and Genie3 library were used. To validate transcriptome data, qPCR was conducted. In producer and non-producer plants of both populations of *Galphimia* spp., most of the transcripts were grouped in the Molecular Function level of gene ontology. A total of 680 differentially expressed transcripts between producer and non-producer plants were detected. In galphimine-producer plants, a larger number of highly expressed transcripts related to acyclic and polycyclic terpene synthesis were identified. As putative candidate genes involved in the galphimine synthesis pathway, P450 family members and enzymes with kinase activity were identified.

Keywords: functional annotation; *Galphimia* spp.; transcriptome

1. Introduction

Galphimia spp. is a medicinal plant widely used in Mexico to treat various ailments, including fever, labor pains [1], diarrhea [2], and anxiety, due to its anti-inflammatory and anxiolytic properties [3,4]. The anxiolytic and sedative effects of the extract have made this plant the target of pharmacological [5–7], phytochemicals [3,8–10], biotechnological [6,11,12], metabolomic [13,14] and genetics studies [15,16]. Nevertheless, the effects of some populations of the plant, are due to the presence of the modified triterpenoid galphimines [17].

Different metabolomic studies have already been carried out to detect the presence of galphimines as well as their concentration among six Mexican populations of *Galphimia* spp. [13,14], from which only two (located in Jalpan, (Querétaro) and Dr. Mora

(Guanajuato)) were identified as galphimines producers. Moreover, to identify the individual populations of *Galphimia* a DNA barcode study was conducted by Sharma et al. [15], and the clades obtained showed clear differences between the galphimine-producer populations (Doctor Mora and Jalpan) and the other analyzed populations. In another study, conducted by Gesto-Borroto et al. [16], among nine populations of *Galphimia* spp. evaluated by DNA barcode, four galphimine-producing populations were identified, including Dr. Mora (Guanajuato) and Jalpan (Querétaro), corroborating the previous studies.

In several plants, stages of triterpene synthesis have been studied [18–20]; however, the galphimine-producing pathway in *Galphimia* spp., is still elusive. Nevertheless, a biogenetic proposal for its production has been published from the oxidation and decarboxylation of the Taraxasteryl cation [21], although it has not yet been demonstrated.

Regarding the biological properties of the galphimines, the hypothesis to be tested was the following: „If a transcriptomic study and a comparison are carried out in two populations of *Galphimia* spp.; a gaphimine producer population (GP) with a non-galphimine producer population (NPG), the putative candidates genes, that encode the enzymes involved in the production pathway of these triterpenoids, can be identified. Hence, this paper aimed to analyze transcriptomes from a GP and NPG populations of *Galphimia* spp., to identify putative candidates genes that encode enzymes for the synthesis of these triterpenoids.

2. Results

2.1. Plant Material

Plants with phenotypic characteristics similar to those described for the *Galphimia* genus [22], were obtained from both populations GP and NPG. They were characterized by having several adventitious roots, 5–6 cm long, with abundant adsorbent hairs. They are 1.4 m tall shrubs, with approximately 15 lateral branches in the GP plants and 10 in the NGP plants. The leaves were elliptical, with prominent secondary veins and their area was 13.75 cm^2. Yellow inflorescences were observed arranged in terminal racemes.

2.2. Transcriptome Sequencing

RNA samples had a high quality with RNA integrity number (RIN) $\geq$8. Around 56.6 Gb of clean data were obtained from transcriptome sequencing. Q30 percentages of clean data for all samples were 97.2%, and the Guanine/Cytosine (GC) were rated between 50 to 51%. The total bases per sample obtained were in a range between 2,704,225 and 675,148.

After FastQC analysis, it was observed that the quality of bases per sequence was optimal since all readings were located in the green area of the graphic. The fast QC analysis showed that the filtered reads had quality valius >Q30, but the reads by sample GC content and duplication levels differed. After trimming the reads with Trimmomatic, these metrics reached the standard values. Post-assembly, around 9000 transcripts from leaf samples and 5000 transcripts from roots samples were obtained (Table 1).

Table 1. Transcripts were obtained after de novo assembly of transcriptomes.

Sample and Replica	Number of Transcripts
Leaves of galphimine producing-population	9023
Leaves of galphimine producing population, replica	8915
Leaves of non-galphimine producing-population	8667
Leaves of non-galphimine producing-population, replica	8896
Roots of galphimine producing-population	5004
Roots of galphimine producing population, replica	4667

2.3. Transcriptome Functional Annotation

The gene ontology analysis showed a different number of annotated genes: 5900 in GP plants, 3600 in the NGP ones, and 1500 in the controls. In GP plants and controls, most of the transcripts were grouped in gene ontology categories MF and CC. In the NGP plants, most of the transcripts were grouped mostly in the MF and BP categories. At the level of gene ontology BP, the largest number of transcripts were associated with cellular processes, biological processes, and biological regulation (Figure 1), while at the MF level, most of the transcripts annotated were related to catalysis, binding, and transporter. In the CC, transcripts of three samples are located as a cellular anatomical entity and protein-containing complex.

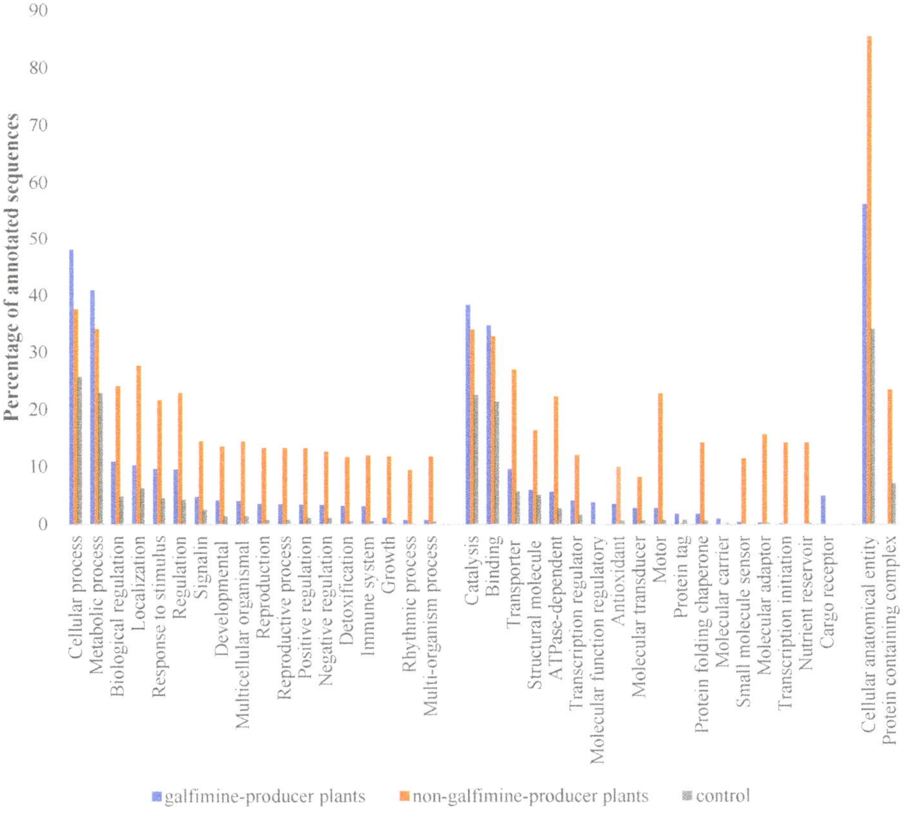

Figure 1. Percentage of annotated sequences of galphimine-producer plants (leaves samples, non-galphimine producer plants, (leaves samples) and control (roots samples) in gene ontology levels Biological Process (BP), Molecular Function (MF), and Cellular Component (CC).

According to the enzyme distribution code in the GP population, several transcripts were identified (Figure 2), such as 1300 with transferase activity, 1020 with hydrolase activity, and 810 with oxidoreductase activity; while in the NGP population transcripts were observed as follows: 2250 with oxidoreductase activity, 2220 with transferase activity, and 2050 with translocase activity. In control, 270 transcripts with hydrolase activity, 260 with transferase activity, and 250 with translocase activity were observed.

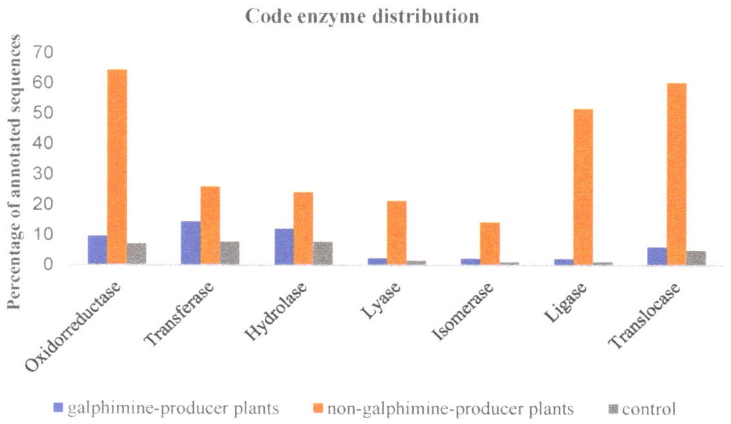

Figure 2. Enzyme distribution code of annotated sequences (percentage) in galphimine producer plants (leaves samples), non-galphimine producer plants (leaves samples), and control (roots samples).

2.4. Differential Expression Analysis

When comparing the GP population with the NGP population, a total of 680 transcripts with differential expressions were found. Of these, 313 were positively regulated and 367 negatively regulated. The volcano plot shows that both positively regulated (represented by red dots) and negatively regulated transcripts (represented by green dots) have a large-magnitude logFC due to the significant separation from the center, and a high level of statistical significance since the location of dots over 100 of log10 (PValue). The greatest statistical significance was observed in positive regulated transcripts. (Figure 3a).

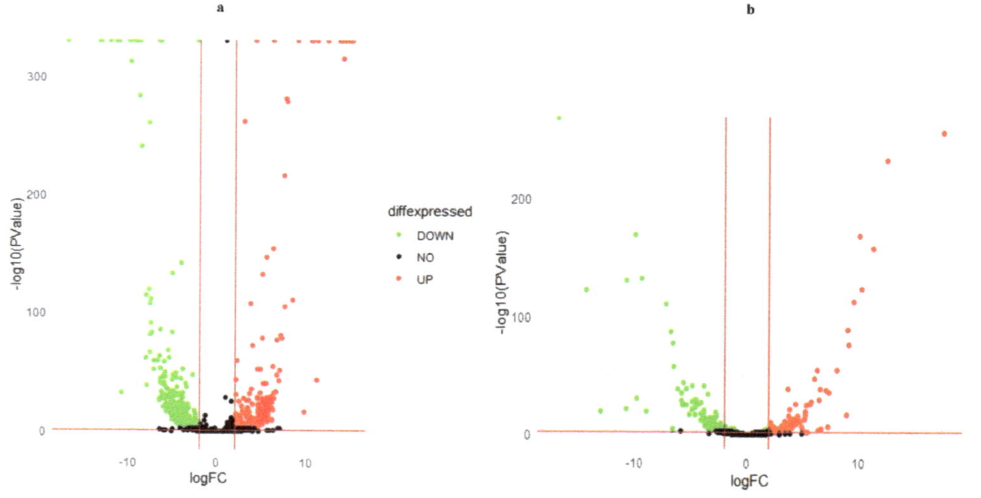

Figure 3. Volcano plot of differentially expressed genes (**a**) galphimine producer plants (leaves samples) versus non-galphimine producer plants (leaves samples) (**b**) galphimine producer plants (leaves samples) versus control (roots samples). X-axis and y-axis represent log2 fold-change differences between the compared samples and statistical significance as the negative log of differentially expressed genes p-values, respectively. The significantly up-regulated and down-regulated genes are indicated with red and green dots, respectively, while non-significant genes are shown as black dots.

A total of 203 transcripts were differentially regulated between GP plants and control. The number of negatively regulated transcripts (logFC < 2) was 116, higher than positively regulated transcripts (logFC > 2), which was 87 in number. The positively regulated transcripts presented a higher statistical significance (Figure 3b). In addition, there were changes in the logFC and high values of -log10 (PValue), indicators of the statistical significance of some transcripts.

In the GP plants, a greater number of differentially expressed transcripts with positive regulation were observed as compared to the NGP and the control (Figure 4), suggesting a high metabolic activity in these plants. Moreover, the transcripts with the highest expression values in the GP plants were mostly hypothetical proteins, ribosomal proteins, transferases, endonucleases, and carboxypeptidases. However, in NGP plants, the transcripts with the highest expression encoded mitochondrial receptors, ribosomal proteins, enzymes with synthase activity, and polygalacturonases.

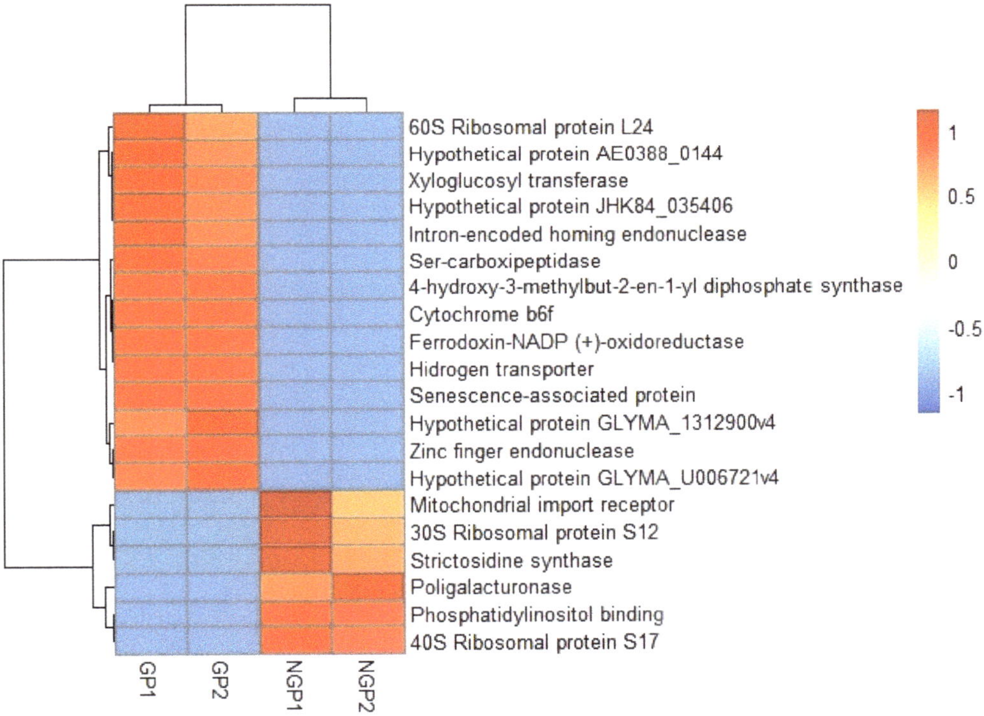

Figure 4. Comparison of transcripts with higher differential expression values between two populations of *Galphimia* spp. GP1, GP2 (leaves of galphimine producer plants). NGP1, NPG2 (leaves of non-galphimine producer plants). High values are represented in red color, the lowest values are represented in blue.

In the comparison between GP plants and controls, the transcripts with the highest expression values in the GP plants were ribosomal proteins, proteins associated with senescence, and enzymes with endonuclease, polymerase, and dehydrogenase activities (Figure 5). In controls, the transcripts with higher expression values code for receptors, aquaporins, chaperones, endonucleases, and methyltransferases.

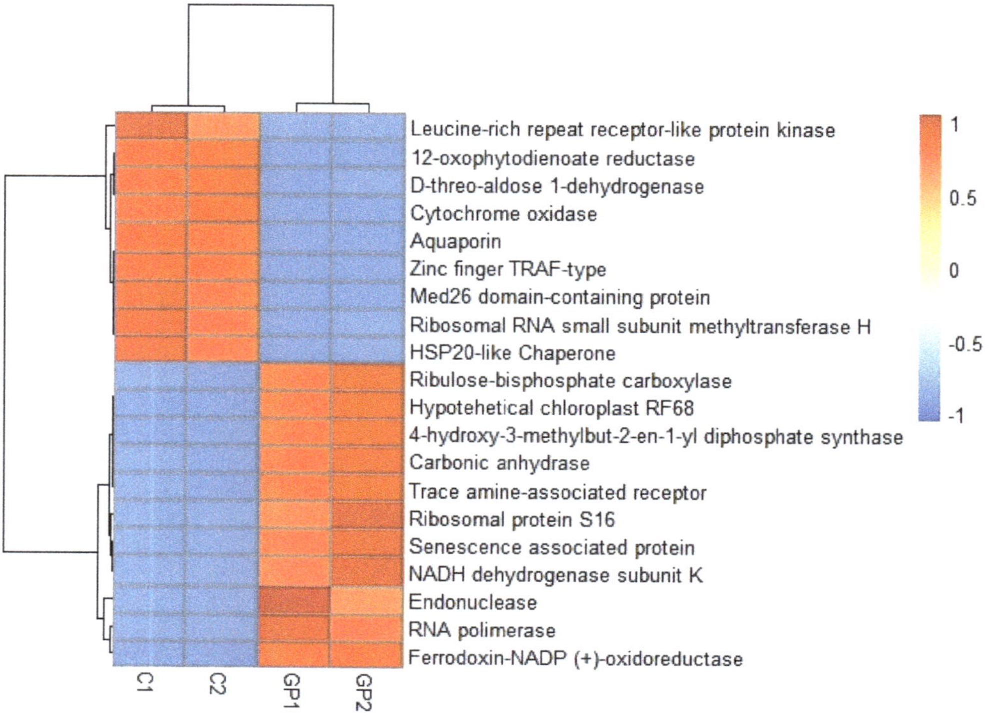

Figure 5. Comparison of transcripts with higher differential expression values between leaves and roots of galphimine producer plants. C1, C2 (roots). GP1, GP2 (leaves). High values are represented in red color, the lowest values are represented in blue.

2.5. Identification of Putative Candidate Genes That Encode Enzymes of the Galphimine Synthesis Pathway

Based on the results of functional annotation, 44 transcripts associated with the terpenes synthesis pathway were identified in GP plants (Table 2). Out of those, two transcripts encode for 3-hydroxy-3-methylglutaryl-CoA reductase (a regulatory enzyme of the mevalonic acid pathway), and 13 transcripts encode enzymes of the methylerythritol phosphate pathway. Among other transcripts, three encode for enzymes associated with intermediates in the isoprenoid biosynthesis, one related to terpenes synthesis, and 25 with triterpenes.

In the NGP plants, 35 transcripts were found associated with terpene synthesis, among which one encodes the regulatory enzyme of the mevalonic acid pathway, 10 associated with methylerythritol phosphate pathway, five with the intermediary in the isoprenoid biosynthesis, three transcripts related to terpenes synthesis and 16 with triterpenes (Table 2). In the control, the following transcripts were observed: one that encodes for 3-hydroxy-3-methylglutaryl-CoA reductase, one for P450, and five for terpene synthase.

The GP plants presented a greater number and diversity of transcripts related to the synthesis of terpenes with high values of gen counts compared to the NGP plants. This suggests a higher metabolic activity in the terpene synthesis pathway in these plants. P450 TBP, P450, and terpene synthase were the enzymes with the highest gen count. In the NGP plants, the enzymes with the highest gen counts were terpene synthase and CYP82G1.

Table 2. Annotation and gene counts of transcripts related to terpene synthesis of two populations of *Galphimia* spp.

	Galphimine-Producer Population		Non-Galphimine-Producer Population		Control	
Annotated Transcripts	**Number of Annotations**	**Gen Counts**	**Number of Annotations**	**Gen Counts**	**Number of Annotations**	**Gen Counts**
3-hydroxy-3-methylglutaryl-CoA reductase	2	13	1	8	1	6
1-deoxy-D-xylulose-5-phosphate synthase	4	15	4	29	-	-
4-diphosphocytidyl-2-C-methyl-D-erythritol kinase	4	18	1	10	-	-
4-hydroxy-3-methylbut-2-en-1-yl diphosphate synthase	5	19	5	16	-	-
isopentenyl diphosphate isomerase	1	52	1	17	-	-
farnesyl transferase	1	6	3	4	-	-
geranyl diphosphate synthase	1	14	1	4	-	-
terpene synthase	1	347	3	253	5	78
squalene synthase	1	7	-	-	-	-
squalene epoxidase	1	16	-	-	-	-
beta-amyrin synthase	2	23	4	14	-	-
lanosterol synthase	7	18	6	23	-	-
P450	2	304.444	-	-	1	4
P450 TBP	19	417.183	-	-	-	-
CYP71A24	1	5	2	3	-	-
CYP71A26	-	-	1	14	-	-
CYP71B10	-	-	1	12	-	-
CYP71B34	1	9	-	-	-	-
CYP71D11	-	-	-	-	1	15
CYP72A	2	10	-	-	-	-
CYP72A15	-	-	1	7	-	-
CYP81D1	1	18	-	-	-	-
CYP82C4	1	231	-	-	-	-
CYP82G1	1	41	1	129	-	-
PLAC8	2	12	-	-	-	-
5′-AMP-activated protein kinase	1	30	-	-	-	-
serine/threonine-protein kinase AtPK2/AtPK19	1	6	-	-	-	-

In the co-expression network, 22 nodes were observed (Figure 6), among which, HMGCR (3-hydroxy-3-methylglutaryl-CoA reductase), IPPI (isopentenyl diphosphate isomerase), and FT (farnesyl transferase) showed the highest number of interactions with other nodes. The expression of the majority of P450 family members was closely related.

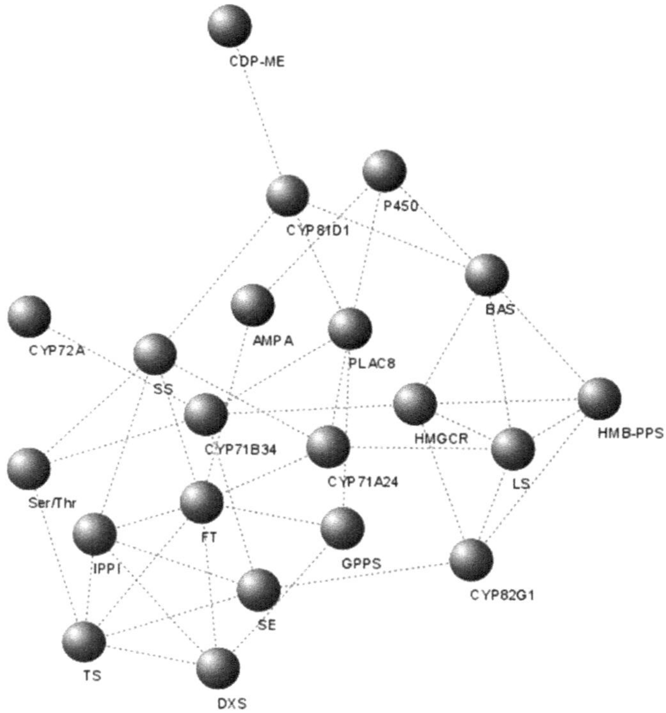

Figure 6. Co-expression network of transcripts related to the synthesis pathway of triterpenes and members of the cytochrome P450 family in leaves of the galphimine producer plants and non-galphimine producer plants. Transcripts are represented by spheres and dotted lines are their interaction. HMGCR (3-hydroxy-3-methylglutaryl-CoA reductase), DXS (1-deoxy-D-xylulose-5-phosphate synthase), CDP-ME (4-diphosphocytidyl-2-C-methyl-D-erythritol kinase), HMB-PPS (4-hydroxy-3-methylbut-2-en-1-yl diphosphate synthase), IPP isomerase (isopentenyl diphosphate isomerase), GPPS (geranyl diphosphate synthase), FT (farnesyl transferase), TS (terpene synthase), SS (squalene synthase), SE (squalene epoxidase), BAS (beta-amyrin synthase), LS (lanosterol synthase).

2.6. Validation of Transcriptome Data by qPCR

By qPCR analysis, the results of the transcriptomic study were validated. The expression levels of putative candidate genes related to galphimine synthesis were found to be similar in both differential gene expression analysis (Table 2) and qPCR (Figure 7). The expression was higher in GP plants because CT values were lower than in NGP plants (Figure 8) and significant differences between plants of both populations in fold change value were observed (Figure 6). In GP plants the genes with a higher level of relative expression were P450 TBP, 5′-AMP-activated protein kinase, and P450. Overall, the relative expression of these genes in NGP plants was low.

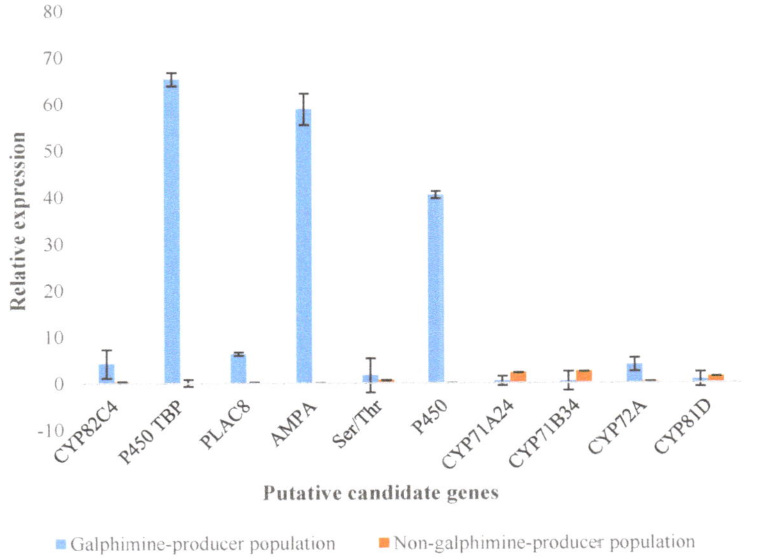

Figure 7. Relative expression of putative candidate genes in leaves of galphimine producer plants (GP) and non-galphimine producer plants (NGP) was calculated by the $2^{-\Delta\Delta Ct}$ method in quantitative PCR analysis. Y-axis represents the fold change value between GP and NGP. AMPA (5′-AMP-activated protein kinase), Ser/Thr (serine/threonine-protein kinase AtPK2/AtPK19).

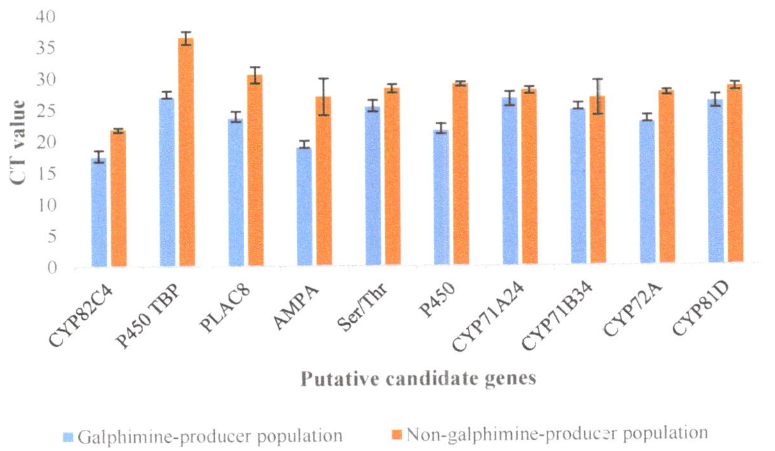

Figure 8. CT values of leaves of galphimine producer plants (GP) and non-galphimine producer plants (NGP) obtained by quantitative PCR analysis. AMPA (5′-AMP-activated protein kinase), Ser/Thr (serine/threonine-protein kinase AtPK2/AtPK19).

3. Discussion

Galphimia spp, is an important medicinal plant in Mexico, with several applications in traditional medicine; thus, the main investigations with this plant are based on its pharmacological effects, especially for the modified triterpenoids galphimines. However, genes that encode enzymes of the galphimines synthesis pathway are unknown. Nevertheless, identifying new genes is a complex and great important process, especially when the genome is not available. The least complex and economical strategy is the use of omics tools such as transcriptome and digital gene expression profile analysis. These strategies allowed to explore and analyze genes with a differential expression which increases the chance of discovering new genes as well as the characterization of their structure. In this study, the Illumina Next seq 550 platform has been used to investigate the transcriptomes of *Galphimia* spp, which allowed the successful identification of several galphimine synthesis-related genes.

The GO annotation provides a preliminary indication of the nature of a gene product. In this study, most of the transcripts have been recorded at the MF and CC gene ontology levels, which provided an approximation to the elementary functions in which the transcripts participate as well as the parts of the cells or extracellular environment they're located in. Given that galphimines are triterpenoids of polycyclic structure and that the synthesis of these metabolites includes oxidation and transfer of functional groups, it is predicted that most of the transcripts identified are associated with transferase and oxidoreductase activities.

Triterpenes are one of the most important groups of secondary metabolites in plants. These are generally produced from the acyclic 30-carbon precursors 2,3-oxidosqualene in eukaryotes. The conversion of 2,3-oxidosqualene to various cyclic triterpenes by squalene cyclase is the first step in the diversification of the biosynthetic pathway of the triterpenes. Subsequently, the generated intermediaries undergo a large number of modifications regio and stereospecific structures catalyzed by cytochrome P450 (CYP), acetylases, and UDP glucosyltransferases that catalyze oxidation, methylation, acetylation, malonylation, and glycosylation, [23]. Plant CYPs transfer two electrons from cofactors NADPH and atmospheric oxygen to their substrates. Generally, plants require a CYP reductase to support electron transfer [24]. The oleanane, derived from β-amyrin, is one of the largest representatives of triterpene scaffolds pentacyclic. Approximately 55 CYPs act on scaffolds of pentacyclic triterpenes of plants. Most of them belong to members of the CYP716 family, although other families such as CYP51, CYP71, CYP72, CYP87, CYP88, and CYP93 also participate in modifications of pentacyclic triterpenes [23].

Members of the cytochrome P450 family are involved in the synthesis of triterpenes. Geisler et al. performed a functional analysis of CYP51 in *Nicotiana benthamiana* leaves and demonstrated that this enzyme can catalyze the hydroxylation and the epoxidation of simple triterpene β-amyrin to give 12,13β-epoxy-3β, 16β-dihydroxy-oleanane (12, 13β-epoxy-16β-hydroxy-β-amyrin) [25].

Also, three oxidosqualene cyclases (OSC) TwOSC1, TwOSC2, and TwOSC3 were isolated and characterized from *Tripterygium wilfordii*. TwOSC1 and TwOSC3 were multiproduct friedelin synthases and were involved in celastrol biosynthesis, while TwOSC2 was a β-amirin synthase [26]. In that work, the authors proposed a biogenetic pathway for celastrol in which hydroxyfriedelanes synthesis was mediated by P450 [26]. On the other hand, in *Aralia elata*, CYP716A295 and CYP716A296 were identified as the candidate genes most likely associated with oleanolic acid synthesis [27].

Galphimines are modified triterpenoids, classified as Nor-secofriedelanes. These are derived from oleanane and later from friedelin. The arrangements in the intermediates between friedelin and norsecofriedelanes involve oxidation-reduction reactions for opening a ring and the C3-C4 bond break, resulting in the two functional groups required for the formation of the seven-membered lactone characteristic of galphimines. Considering the functions that have been reported for the P450 family in the synthesis of triterpenes and the gene count of the transcripts encoding members of the P450 family in the GP population (Table 2), as well as the absence of many of these in the NGP population, these enzymes could be considered potential candidates for galphimines synthesis, as suggested in Figure 9.

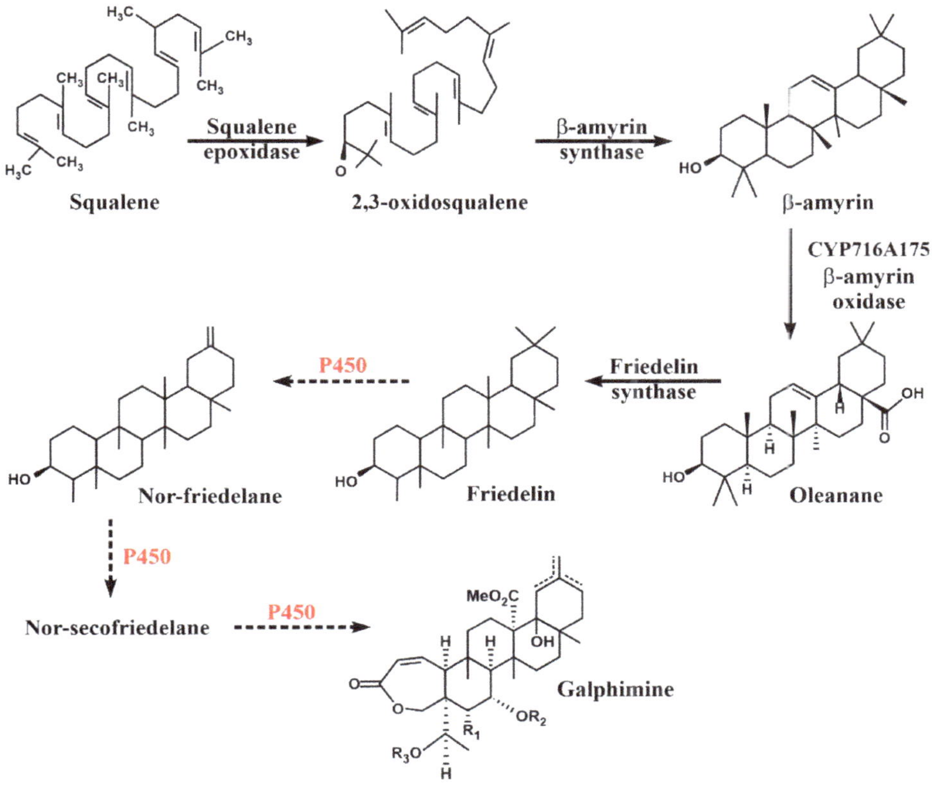

Figure 9. Proposed galphimines synthesis pathway mediated by cytochrome P450 in *Galphimia* spp. plants. Solid arrows indicate pathway reactions identified in previous work [18,28]. Dotted arrows indicate reactions proposed in this study.

In differential expression analysis, the GP plants presented a greater number of genes related to the synthesis of terpenes as compared to the NGP plants. In both analyzed populations, some transcripts encode the mevalonic acid and methylerythritol phosphate pathways as well as precursors of the biosynthesis of isoprenoids, terpenes, and triterpenes. However, in the GP plants, the number and diversity of identified enzymes were higher. Since terpenes are a group of secondary metabolites that are highly represented in plants [29], they are present in both studied populations. However, the GP plants presented a higher expression of the enzymes of this pathway and, as a consequence, a higher content of these metabolites. This corroborates with the results of the study by Sharma et al. where the chemical profile of the *Galphimia* spp., the populations used in this study was analyzed by thin-layer chromatography (TLC), suggesting the presence of galphimines in the GP population, while other terpenes were in the NGP population [15].

Among the enzymes with the highest levels of expression in both populations were terpene synthase and CYP82G1. This is expected because terpene synthase diversifies isoprenoid precursors into a large number of mono and sesquiterpenes [30]. CYP82G1 participates in the synthesis of volatile homoterpenes. These homoterpenes are considered among the most common plant compounds that act as part of defense mechanisms against herbivores [31].

Co-expression networks are a new resource for understanding convergent pathways and their relations. This resource has been used to address various biological questions [32].

In the present study, the enzymes selected as possible candidates for the galphimines synthesis pathway interacted closely with each other, as well as with other enzymes involved in stages before the triterpenes synthesis. It has shown the genes that participate in the same metabolic pathway or those are under the same regulatory mechanism. It is worth remarking on the interaction of 3-hydroxy-3-methylglutaryl-CoA reductase, with a considerable number of the identified enzymes that are associated with the synthesis of polycyclic triterpenes. This enzyme is a regulator of the mevalonic acid pathway, which seems to be the proper route for synthesizing triterpenes [33]. The 3-hydroxy-3-methylglutaryl-CoA reductase seems to control the core of the terpene's metabolism in the cytosol, regulating the production of the compound obtained, as well as the activity of the gene products involved in these reactions.

In this study, relative expression determined by qPCR corroborated the results from the transcriptomic analysis. Our results allow us to support that the genes selected as candidates involved in galphimine biosynthesis are present in high concentrations in the GP plants and are in low concentrations or absent in the NGP plants.

Given the limited availability of reference genomes in plants, the transcriptomic analysis of bioactive metabolites in producer and nonproducer plants is a valuable strategy to detect the genes involved in the synthesis pathways of the compounds of interest.

4. Material and Methods

4.1. Plant Material

The collection of plants and seeds was carried out manually under the permission of Secretaría de Medio Ambiente y Recursos Naturales (SEMARNAT) in a GP population, Doctor Mora (Guanajuato) (W 100° 19.22; N 21° 08.7; 2120 m altitude) and in a NGP population, Tepoztlán (Morelos) (W 99° 06.97 N 18° 59.35, 1700 m altitude). The samples were deposited in the HUMO Herbarium of the Sierra de Huatla Environmental Education and Research Center of the Universidad Autónoma del Estado de Morelos (UAEM) to be properly identified, voucher numbers 15189 and 15485, respectively. From 10 seeds collected from each population, *Galphimia* spp. seedlings were obtained and cultivated at 20 °C of temperature and 60% of humidity for 2 months and then in natural conditions for 15 months. The experiment was performed from September 2019 to January 2021 in the Tissue Culture Lab at the Bioengineering Center of Instituto Tecnológico y de Estudios Superiores de Monterrey, Querétaro, México. During the cultivation period of the plants, parameters related to their growth and development were evaluated, such as height, number of lateral branches, leaf area by ellipse equation, and appearance of inflorescences.

4.2. RNA Extraction, Library Preparation, and Transcriptome Sequencing

Total RNA was extracted using RNeasy Plant Mini Kit (QIAGEN, Hilden, Germany) from the leaves of plants originating from both populations as well as from the roots of the plants of the producer population as a control. The quality and integrity of the total RNA were evaluated using a Qsep 100 Advance, Bioptic. RNA concentration was measured using a NanoDrop 2000 spectrophotometer (NanoDrop Technologies, Technologies, Wilmington, DE, USA) and a Qubit 4^{TD} fluorometer (Invitrogen, Thermo Fisher, Waltham, MA, USA).

Six RNA-Seq libraries (leaves samples from a GP population plant, leaves samples from the NGP population, root samples from the GP population, and one replica of each sample) were constructed using the Illumina TruSeq Stranded Total RNA protocol. Roots were used as a control since the production of galphimines is negligible in this organ. The quantity and quality of enriched dscDNA were assessed using a Qubit fluorometer (Thermo Fisher, Waltham, MA, USA) and Agilent 2100 Bioanalyzer (Agilent Technologies, Santa Clara, CA, USA). The libraries were sequenced using the Illumina Next Seq 550 platform; for 2 × 76 sequencing, cycles to obtain paired-end readings.

Reads Quality Control and De Novo Transcriptome Assembly

The quality of the reads obtained by sequencing was analyzed using the FastQC software (v.0.11.8) and Trimmomatic (v.0.38) tool was used to trim the sequence reads. Afterward, de novo transcriptomes assembly was carried out, for each sample, by Trinity (v.2.9.1) using the default settings mode.

4.3. *Transcriptome Functional Annotation*

Transcriptome functional annotation was performed using the Blast2GO (v.4.1.9). The assembled transcripts were aligned by a tblastx with the NCBI database. A BLAST expectation value of 1.0×10^{-3} was applied, and for each sequence, 20 alignments were made. A sequence mapping was conducted, followed by functional annotation in levels of gene ontology Biological Process (BP), Molecular Function (MF), and Cellular Component (CC). Sequences that were not annotated by gene ontology were identified by InterPro, UniProt, and TAIR databases.

4.4. *Differential Expression Analysis*

To identify corresponding gene pairs between individuals in the two populations of *Galphimia* spp., Corset programs (v.1.09) (clustering de novo assembled transcripts) was used. After transcripts were compared and grouped, results were also verified by functional annotation. An abundance quantification of the transcripts assembled in each transcriptome was conducted using the Kallisto program (v.0.46.0.4). Then, the counting matrix was built in R (v.4.0.4) for Windows, with the use of the plyr, dplyr, and stats libraries. Genes with low library counts were filtered using the CPM Filter on a per-million count (CPM) basis. For the analysis of differential expression, the edgeR library was used. Samples in the counting matrix were normalized on a transcript-per-million (TPM) basis. Transcripts expression in the leaves of the GP plants was compared with transcripts of the NGP and with the control. The differentially expressed transcripts were considered to be those with a false positive rate (FDR) <0.01 and classified as positively regulated when log fold Change (logFC) > 2 and negative when logFC < −2. The transcripts with the highest differential expression values in each sample were visualized on a heat map constructed with the pheatmap library.

4.5. *Identification of Putative Genes Related to the Galphimines Synthetic Pathway*

Data obtained from the functional annotation of transcriptomes and the results of the differential expression analysis were used to identify genes involved in the synthesis of terpenes and putative candidate genes for the synthesis of galphimines. To identify the interactions between the genes related to the synthesis of terpenes and the possible candidate genes for the synthesis of galphimines, a co-expression network was also built, using the libraries of R, GENIE3, igraph, RCy3, and Rgraphviz. The co-expression network was calculated in the form of a weighted adjacency matrix, using ensembles of regression trees. Candidate regulatory genes were not used. The tree method was Random Forests. The number of candidate regulators randomly selected at each tree node was the square root of the total number of candidate regulators. The number of trees in each target was 1000. The co-expression network was visualized by Cytoscape (v.3.8.2).

4.6. *Validation of Transcriptome Data by qPCR*

cDNA of GP plants and NGP plants was obtained using the PrimerScript RT Reagent Kit with gDNA Eraser (TaKaRa, San Jose, CA, USA). The concentration and purity of cDNA were evaluated by Nano-Drop 2000 spectrophotometer (NanoDrop Technologies, Technologies, Wilmington, DE). According to transcriptome data of GP plants, nucleotide sequences of ten putative candidate genes related to galphimines synthetic pathway (CYP82C4, P450 TBP, PLAC8, 5'-AMP-activated protein kinase, serine/threonine-protein kinase AtPK2/AtPK19, P450, CYP71A24, CYP71B34, CYP72A, CYP81D), were used for primers designing (Table 3) by Primer3 plus program. The quantitative PCR was conducted using the TB Green® Advantage® qPCR Premix (Takara Bio USA, Inc., San Jose, CA, USA) following the manufacturer's protocol in a StepOne™ Real-Time PCR System (Applied Biosystems, Waltham, MA, USA). The amplification conditions used were as follows: activation/denaturation at 95 °C for 10 s, 45 cycles of denaturation at 95 °C for 5 s, and annealing/extension at 60 °C for 30 sec followed by a melting curve analysis ranging from 56–95 °C. Three biological replicates with three technical replicates were used for each sample. The relative expression of the RNA was quantified by the $2^{-\Delta\Delta Ct}$ method using U6 as the reference gene [34].

Table 3. Sequences of synthesized primers for qPCR.

Gene Description	Gene Symbol	Primer Sequence 5′-3′	PCR Product Length (bp)
CYP82C4	CYP82C4	F 5′TCATTTGGTGGGCTAAAAGC 3′ R 5′ACGAGTCCGTTAATGGTTGC 3′	191
P450 TBP	P450 TBP	F 5′AGCTTCGTCGCAAGTGAAAT 3′ R 5′TAGGACGCCTGCGTTATCTT 3′	189
PLAC8	PLAC8	F 5′ACTGTTGATAACCCCGATGG 3′ R 5′TCTCAGGAAGTCCGAACTGG 3′	170
5′-AMP-activated protein kinase	AMPA	F 5′CCCAAGGAAAAGGTTTCACA 3′ R 5′GGCAAAAGCAATGGCTAAGA 3′	229
serine/threonine-protein kinase AtPK2/AtPK19	Ser/Thr	F 5′TTTCAACTGGACCACAAAGG 3′ R 5′CGTTCGGGAAAGTCTACCAA 3′	164
P450	P450	F 5′CCTTGAGGTCATGGCTGAAT 3′ R 5′GCTTCTCTCCAAAGGCACAC 3′	232
CYP71A24	CYP71A24	F 5′CAAACCGGCCTAAATCAAAG 3′ R 5′GCTCTTGTTCATAACTTTCTCAATC 3′	200
CYP71B34	CYP71B34	F 5′CCGTGAGAGAGGCCATTAAC 3′ R 5′GTGCGAAAGATTTGCGTTCT 3′	160
CYP72A	CYP72A	F 5′TTGTTGGCTTTTCGAGGAAT 3′ R 5′AAGCATCAGGAGTGGCAAAC 3′	167
CYP81D1	CYP81D1	F 5′TCGGAGGATTGGACTACGAC 3′ R 5′TTCCGCCATAACATTTCTCC 3′	219

5. Conclusions

Transcriptome analysis of two populations of *Galphimia* spp. allowed the detection of differences between a galphimine producer GP plant and a non-galphimine producer NGP plant. The higher expression of P450 family members and kinases in GP plants and their well-known role in triterpene synthesis make them putative candidates to encode enzymes involved in the galphimine synthesis pathway.

Author Contributions: Conceptualization, A.S. and M.L.V.; Methodology, D.I., A.V.V., M.d.D.C. and C.G.-G.; Software, A.M.T. and D.I.; Validation, A.S., M.L.V., M.d.D.C., S.P., H.M.N.I. and D.I.; Formal Analysis, A.S., M.L.V., M.d.D.C., A.M.T. and D.I.; Investigation, D.I., Resources, A.S., Data Curation, A.S., M.L.V., M.d.D.C., A.M.T. and D.I.; Writing-Review & Editing, A.S., M.L.V., S.P., M.d.D.C. and D.I.; Visualization, D.I., Supervision, A.S. and M.L.V.; Project Administration, A.S.; Funding Acquisition, A.S. All authors have read and agreed to the published version of the manuscript.

Funding: The authors want to thank the Regional Department of Bioengineering, South Central Region of the Tecnológico de Monterrey, for the logistics and equipment to support the conduction of this work.

Institutional Review Board Statement: Not applicable.

Informed Consent Statement: Not applicable.

Data Availability Statement: Not applicable.

Acknowledgments: Thanks to CONACYT, Mexico, for the scholarship granted to carry out Dianella Iglesias' Ph.D. thesis (946795).

Conflicts of Interest: The authors declare no conflict of interest.

References

1. Tortoriello, J.; Ortega, A.; Herrera-Ruíz, M.; Trujillo, J.; Reyes-Vázquez, C. Galphimine-B modifies electrical activity of ven-tral tegmental area neurons in rats. *Planta Med.* **1998**, *64*, 309–313. [CrossRef]
2. Camacho, M.D.R.; Phillipson, J.D.; Croft, S.L.; Marley, D.; Kirby, G.C.; Warhurst, D.C. Assessment of the Antiprotozoal Activity of Galphimia glauca and the Isolation of New Nor-secofriedelanes and Nor-friedelanes. *J. Nat. Prod.* **2002**, *65*, 1457–1461. [CrossRef]
3. González-Cortazar, M.; Herrera-Ruiz, M.; Zamilpa, A.; Jiménez-Ferrer, E.; Marquina, S.; Álvarez, L.; Tortoriello, J. An-ti-Inflammatory activity and chemical profile of Galphimia glauca. *Planta Med.* **2014**, *80*, 90–96. [PubMed]
4. Campos, M.G.; Toxqui, E.; Tortoriello, J.; Oropeza, M.V.; Ponce, H.; Vargas, M.H.; Montaño, L.M. Galphimia glauca organic fraction antagonizes LTD4-induced contraction in guinea pig airways. *J. Ethnopharmacol.* **2001**, *74*, 7–15. [CrossRef]
5. Dorsch, W.; Wagner, H. New antiasthmatic drugs from traditional medicine? *Int. Arch. Allergy Immunol.* **1991**, *94*, 262–265. [CrossRef] [PubMed]
6. Osuna, L.; Pereda-Miranda, R.; Tortoriello, J.; Villarreal, M.L. Production of the sedative triterpene Galphimine B in Gal-phimia glauca tissue culture. *Planta Med.* **1999**, *65*, 149–152. [CrossRef]
7. Romero-Cerecero, O.; Islas-Garduño, A.L.; Zamilpa, A.; Pérez-García, M.D.; Tortoriello, J. Therapeutic effectiveness of Gal-phimia glauca in young people with social anxiety disorder: A pilot study. *Evid. Based Complement. Altern. Med.* **2018**. [CrossRef]
8. Taketa, A.T.C.; Lozada-Lechuga, J.; Fragoso-Serrano, M.; Villarreal, M.L.; Pereda-Miranda, R. Isolation of Nor-secofriedelanes from the Sedative Extracts of Galphimia glauca. *J. Nat. Prod.* **2004**, *67*, 644–649. [CrossRef]
9. González-Cortazar, M.; Tortoriello, J.; Alvarez, L. Nor-secofriedelanes as spasmolytics, advances of structure-activity relationships. *Planta Med.* **2005**, *71*, 711–716. [CrossRef]
10. Ortega, A.; Pastor-Palacios, G.; Ortiz-Pastrana, N.; Ávila-Cabezas, E.; Toscano, R.A.; Joseph-Nathan, P.; Morales-Jiménez, J.; Bautista, E. Further galphimines from a new population of Galphimia glauca. *Phytochemistry* **2019**, *169*, 112180. [CrossRef]
11. Náder, B.L.; Cardoso-Taketa, A.T.; Iturriaga, G.; Pereda-Miranda, R.; Villarreal, M.L. Genetic transformation of Galphimia glauca by Agrobacterium rhizogenes and the production of Norfriedelanes. *Planta Med.* **2004**, *70*, 1174–1179. [CrossRef] [PubMed]
12. Ortíz, A.; Cardoso-Taketa, A.; Rodríguez, M.; Arellano, J.; Hernández, G.; Villarreal, M.L. Transformed cell suspension cul-ture of Galphimia glauca producing sedative Nor-friedelanes. *Planta Med.* **2010**, *76*, 386–392. [CrossRef] [PubMed]
13. Cardoso-Taketa, A.T.; Pereda-Miranda, R.; Choi, Y.H.; Verpoorte, R.; Villarreal, M.L. Metabolic profiling of the Mexican anxiolytic and sedative plant Galphimia glauca using Nuclear Magnetic Resonance spectroscopy and multivariate data analy-sis. *Planta Med.* **2008**, *74*, 1295–1301. [CrossRef] [PubMed]
14. Sharma, A.; Cardoso-Taketa, A.; Choi, Y.H.; Verpoorte, R.; Villarreal, M.L. A comparison on the metabolic profiling of the Mexican anxiolytic and sedative plant Galphimia glauca four years later. *J. Ethnopharmacol.* **2012**, *141*, 964–974. [CrossRef] [PubMed]
15. Sharma, A.; Folch, J.L.; Cardoso-Taketa, A.; Lorence, A.; Villarreal, M.L. DNA barcoding of the Mexican sedative and anxio-lytic plant Galphimia glauca. *J. Ethnopharmacol.* **2012**, *144*, 371–378. [CrossRef] [PubMed]
16. Gesto-Borroto, R.; Cardoso-Taketa, A.; Yactayo-Chang, J.P.; Medina-Jiménez, K.; Hornung-Leoni, C.; Lorence, A.; Villarreal, M.L. DNA barcoding and TLC as tools to properly identify natural populations of the Mexican medicinal species Galphimia glauca Cav. *PLoS ONE* **2019**, *14*, e0217313. [CrossRef]
17. Toscano, R.A.; Maldonado, E.; Lozoya, X.; Tortoriello, J.; Ortega, A.; Gaviño, R. Structure of galphimine B. *Acta Crystallogr. Sect. C Cryst. Struct. Commun.* **1993**, *49*, 774–776. [CrossRef]
18. Ma, C.H.; Gao, Z.J.; Zhang, J.J.; Zhang, W.; Shao, J.H.; Hai, M.R.; Chen, J.W.; Yang, S.C.; Zhang, G.H. Candidate genes in-volved in the biosynthesis of triterpenoid saponins in Platycodon grandiflorum identified by transcriptome analysis. *Front. Plant Sci.* **2016**, *7*, 1–14. [CrossRef]
19. Srivastava, G.; Sandeep; Garg, A.; Misra, R.C.; Chanotiya, C.S.; Ghosh, S. Transcriptome analysis and functional characterization of oxidosqualene cyclases of the arjuna triterpene saponin pathway. *Plant Sci.* **2019**, *292*, 110382. [CrossRef]
20. Lertphadungkit, P.; Qiao, X.; Sirikantaramas, S.; Satitpatipan, V.; Ye, M.; Bunsupa, S. De novo transcriptome analysis and identification of candidate genes associated with triterpenoid biosynthesis in *Trichosanthes cucumerina* L. *Plant Cell Rep.* **2021**, *40*, 1845–1858. [CrossRef]
21. Rios, M.Y.; Ortega, A.; Domínguez, B.; Déciga, M.; de la Rosa, V. Glaucacetalin E and galphimidin B from Galphimia glauca and their anxiolytic activity. *J. Ethnopharmacol.* **2020**, *259*, 112939. [CrossRef] [PubMed]
22. Anderson, C. Revision of Galphimia (Malpighiaceae). *Contrib. Univ. Mich. Herb.* **2007**, *25*, 1–82.
23. Ghosh, S. Triterpene Structural Diversification by Plant Cytochrome P450 Enzymes. Front. *Plant Sci.* **2017**, *8*, 1886. [CrossRef]
24. Bathe, U.; Tissier, A. Cytochrome P450 enzymes. *Phytochemistry* **2019**, *161*, 149–162. [CrossRef]
25. Geisler, K.; Hughes, R.K.; Sainsbury, F.; Lomonossoff, G.P.; Rejzek, M.; Fairhurst, S.; Olsen, C.E.; Motawia, M.S.; Melton, R.E.; Hemmings, A.M.; et al. Biochemical analysis of a multifunctional cytochrome P450 (CYP51) enzyme required for synthesis of antimicrobial triterpenes in plants. *Proc. Natl. Acad. Sci. USA* **2013**, *110*, 3360–3367. [CrossRef]

26. Zhou, J.; Hu, T.; Gao, L.; Su, P.; Zhang, Y.; Zhao, Y.; Chen, S.; Tu, L.; Song, Y.; Wang, X.; et al. Friedelane-type triterpene cyclase in celastrol biosynthesis from Tripterygium wilfordii and its application for triterpenes biosynthesis in yeast. *New Phytol.* **2019**, *223*, 722–735. [CrossRef]
27. Cheng, Y.; Liu, H.; Tong, X.; Liu, Z.; Zhang, X.; Li, D.; Jiang, X.; Yu, X. Identification and analysis of CYP450 and UGT super-gene family members from the transcriptome of Aralia elata (Miq.) seem to reveal candidate genes for triterpenoid saponin biosynthesis. *BMC Plant Biol.* **2020**, *20*, 1–17. [CrossRef]
28. Souza-Moreira, T.M.; Alves, T.B.; Pinheiro, K.; Felippe, L.G.; De Lima, G.M.A.; Watanabe, T.F.; Barbosa, C.C.; Santos, V.A.F.F.M.; Lopes, N.P.; Valentini, S.R.; et al. Friedelin Synthase from Maytenus ilicifolia: Leucine 482 Plays an Essential Role in the Production of the Most Rearranged Pentacyclic Triterpene. *Sci. Rep.* **2016**, *6*, 1–3. [CrossRef]
29. Tholl, D. Biosynthesis and biological functions of terpenoids in plants. In *Biotechnology of Isoprenoids. Advances in Biochemical Engineering/Biotechnology*, 1st ed.; Chrader, J., Bohlmann, J., Eds.; Springer: Luxemburgo, 2015; Volume 148, pp. 63–106.
30. Booth, J.K.; Page, J.E.; Bohlmann, J. Terpene synthases from Cannabis sativa. *PLoS ONE* **2017**, *12*, e0173911. [CrossRef]
31. Lee, S.; Badieyan, S.; Bevan, D.R.; Herde, M.; Gatz, C.; Tholl, D. Herbivore-induced, and floral homoterpene volatiles are biosynthesized by a single P450 enzyme (CYP82G1) in Arabidopsis. *Proc. Natl. Acad. Sci. USA* **2010**, *107*, 21205–21210. [CrossRef]
32. Hartl, C.L.; Ramaswami, G.; Pembroke, W.G.; Muller, S.; Pintacuda, G.; Saha, A.; Parsana, P.; Battle, A.; Lage, K.; Geschwind, D.H. Coexpression network architecture reveals the brain-wide and multiregional basis of disease susceptibility. *Nat. Neurosci.* **2021**, *24*, 1313–1323. [CrossRef] [PubMed]
33. Itoh, D.; Kawano, A.K.; Nabeta, K. Biosynthesis of Chloroplastidic and Extrachloroplastidic Terpenoids in Liverwort Cultured Cells: 13C Serine as a Probe of Terpene Biosynthesis via Mevalonate and Non-mevalonate Pathways. *J. Nat. Prod.* **2003**, *66*, 332–336. [CrossRef] [PubMed]
34. Tang, F.; Chu, L.; Shu, W.; He, X.; Wang, L.; Lu, M. Selection and validation of reference genes for quantitative expression analysis of miRNAs and mRNAs in Poplar. *Plant Methods* **2019**, *15*, 35. [CrossRef] [PubMed]

plants

MDPI

Article

Essential Oil Yield, Composition, and Bioactivity of Sagebrush Species in the Bighorn Mountains

Valtcho D. Zheljazkov [1,*], Charles L. Cantrell [2], Ekaterina A. Jeliazkova [1], Tess Astatkie [3] and Vicki Schlegel [4]

1 Crop and Soil Science Department, Oregon State University, 3050 SW Campus Way, Corvallis, OR 97331, USA; ekaterina.jeliazkova@oregonstate.edu
2 Natural Products Utilization Research, USDA-Agricultural Research Service, University of Mississippi, University, MS 38677, USA; charles.cantrell@usda.gov
3 Faculty of Agriculture, Dalhousie University, Truro, NS B2N 5E3, Canada; astatkie@dal.ca
4 Department of Food Science and Technology, University of Nebraska-Lincoln, 326 Food Technology Complex, Lincoln, NE 68583, USA; vschlegel3@unl.edu
* Correspondence: valtcho.jeliazkov@oregonstate.edu

Abstract: Sagebrush (*Artemisia* spp.) are dominant wild plants in large areas of the U.S., Canada and Mexico, and they include several species and subspecies. The aim was to determine if there are significant differences in essential oil (EO) yield, composition, and biological activity of sagebrush within the Bighorn Mountains, U.S. The EO yield in fresh herbage varied from 0.15 to 1.69% for all species, including 0.25–1.69% in *A. tridentata* var. *vaseyana*, 0.64–1.44% in *A. tridentata* var. *tridentata*, 1% in *A. tridentata* var. *wyomingensis*, 0.8–1.2% in *A. longifolia*, 0.8–1% in *A. cana*, and 0.16% in *A. ludoviciana*. There was significant variability in the EO profile between species, and subspecies. Some EO constituents, such as α-pinene (0–35.5%), camphene (0–21.5%), eucalyptol (0–30.8%), and camphor (0–45.5%), were found in most species and varied with species and subspecies. The antioxidant capacity of the EOs varied between the species and subspecies. None of the sagebrush EOs had significant antimicrobial, antimalarial, antileishmanial activity, or contained podophyllotoxin. Some accessions yielded EO with significant concentrations of compounds including camphor, eucalyptol, cis-thujone, α-pinene, α-necrodol-acetate, fragranol, grandisol, para-cymene, and arthole. Therefore, chemotypes can be selected and possibly introduced into culture and be grown for commercial production of these compounds to meet specific industry needs.

Keywords: *Artemisia tridentata*; *A. longifolia*; *A. ludoviciana*; *A. cana*; camphor; antioxidant; eucalyptol; cis-thujone; α-pinene; α-necrodol-acetate

Citation: Zheljazkov, V.D.; Cantrell, C.L.; Jeliazkova, E.A.; Astatkie, T.; Schlegel, V. Essential Oil Yield, Composition, and Bioactivity of Sagebrush Species in the Bighorn Mountains. *Plants* **2022**, *11*, 1228. https://doi.org/10.3390/plants11091228

Academic Editor: William N. Setzer

Received: 1 April 2022
Accepted: 28 April 2022
Published: 1 May 2022

1. Introduction

The genus *Artemisia*, distributed mostly in the Northern hemisphere, comprises small herbs and shrubs of over 500 species occurring in North America, Europe, and Asia [1]. Sagebrush (*Artemisia* spp.) are woody shrub species of the genus and the dominant wild species in large areas of Western United States, Canada, and Mexico. They occupy around 69 million ha in the Western United States alone [2,3]. The woody sagebrush include several species such as big sagebrush (*Artemisia tridentata* Nutt.), black sagebrush (*Artemisia nova* A. Nelson), silver sagebrush (*Artemisia cana* Pursh), several subspecies and hybrids (crosses) between these [4], distributed in various states and provinces in North America [5]. Sagebrush communities contribute to the survival of various wildlife species, especially the greater sage-grouse (*Centrocercus urophasianus*) and the Gunnison sage-grouse (*Centrocercus minimus*) [3,6–9]. Pronghorn antelope (*Antilocapra americana*) is thought to have evolved with sagebrush and is now perhaps the only animal species to browse it extensively, whereas sage-grouse is endemic for sagebrush ranges. Other bigger animals, such as mule deer (*Odocoileus hemionus*), elk (*Cervus canadensis*), and bighorn sheep (*Ovis canadensis*), browse sagebrush. It may also be browsed by domestic sheep and cattle especially during the winter months, when the ground is covered with snow and there is limited other forage available [10,11]. The grazing and browsing of big and small sagebrush by wild ungulates and cattle have been shown to modify density of sagebrush species and to shift the species composition [12]. Sagebrush density habitat changes have been shown to affect mortality of pronghorn [13]. The importance of sagebrush rangelands for many animal and bird species has been outlined in the literature; sagebrush affords habitats for obligate species (the ones that live only in sagebrush ecosystems) and facultative species (the animals that may use sagebrush ecosystems and other ecosystems) [3].

In addition to providing cover for wildlife, sagebrush is extensively used in the reclamation of disturbed lands, especially in oil and natural gas exploration and development areas, which started in the 1920s and peaked in 2010–2013. Oil and gas exploration, renewable energy such as wind and solar developments, and intensification of recreational activities have all been affecting sagebrush and associated species, such as sage-grouse, habitat [14,15]. Fragmentation of habitat by anthropogenic activities such as mining and energy developments has generally had a negative effect on the sagebrush and sage-grouse population in Wyoming, where 37% of sage-grouse population resides [14,15].

Sagebrush species have been used extensively by native peoples in North America as medicinal plants [16]. A comprehensive review on sagebrush species research conducted in North America provided insights into the evolution, botany and taxonomy, phytochemical complexity, and also pharmacological findings [17]. Sagebrush EO and extracts have shown antifungal activity [16,18,19]. A study with extracts from 100 medicinally active plants reported that extracts of aerial parts of *A. ludoviciana* and *A. tridentata* (among other species) had significant fungal inhibitory activity [16].

Sagebrush plant species contain substantial amounts of EO with a strong specific aroma; however, the EO is toxic [20]. The EO of aboveground plant parts of sagebrush species, including *A. tridentata*, has been investigated in the past [21,22]. However, the EO yield, composition, and antimicrobial activity of different sagebrush species are yet to be fully characterized. The hypothesis of this study was that there would be significant between-species (interspecies) and within-species (intraspecies) variations of EO yield, composition, and antimicrobial activity. The objective of this study was to assess sagebrush species in the Bighorn Mountains area of Wyoming, and their EO yield, composition, and bioactivity.

2. Results and Discussion

There were two separate collections of sagebrush (in 2011 and in 2014) that were treated as two different studies. Response variables were compared within each collection.

2.1. The 2011 Collections

In the 2011 collection, 13 accessions were identified as *A. tridentata* Natt. var. *vaseyana*, and were collected from elevations ranging from 1907 to 2980 m; 3 accessions were *A. longifolia*, collected at elevations ranging from 1147 to 1166 m; 2 accessions were *A. cana* var. *cana*, collected at elevations of 1292 and 1333 m; another 2 accessions were *A. tridentata* var. *tridentata*, collected at 2141 and 2299 m; 1 accession was identified as *A. ludoviciana* ssp. *ludoviciana*, collected at 1643 m; 1 accession identified as *A. tridentata* var. *wyomingensis*, collected at 1453 m (Table 1).

2.1.1. *Artemisia tridentata* var. *vaseyana*

The chemical constituents found in the oil of the *A. tridentata* var. *vaseyana* accessions ranged from 13 to 48 per accession (Table S1). However, the amassed number of identified constituents amounted to 116 different constituents, indicating significant variability in the EO profile within the *A. tridentata* var. *vaseyana* subspecies. The major constituents present in the EO of *A. tridentata* var. *vaseyana* were α-pinene at 35.5% in accession #210, camphor at 45.4% for accession #212, eucalyptol at 17.9% for accession #211, chrysanthenone at 17.9%

for accession #201, fragranol at 20.3% and grandisol at 36.2% for accession #203 (Table S1). A previous study found 42 constituents in the EO obtained by steam distillation from leaves and branches of *A. tridentata*, with camphor, camphene, and 1,8-cineole (eucalyptol) accounting to 28.6, 16.9, and 13.2% of the total oil [23].

Table 1. Latin name, collection date, elevation and GPS coordinates for sagebrush accessions collected from Bighorn Mountains in Wyoming in fall of 2011 and fall of 2014.

Accession	Latin Name	Collection Date	Elevation, m	GPS Coordinates
201	*A. tridentata* Nutt. var. *vaseyana* (Rydb.) Boivin	24 August 2011	2383	–
202	*A. tridentata* Nutt. var. *vaseyana* (Rydb.) Boivin	24 August 2011	2406	44.8048, −107.5413
203	*A. tridentata* Nutt. var. *vaseyana* (Rydb.) Boivin	24 August 2011	2559	44.8126, −107.6095
204	*A. tridentata* Nutt. var. *vaseyana* (Rydb.) Boivin	24 August 2011	2377	44.7888, −107.9297
205	*A. tridentata* Nutt. var. *vaseyana* (Rydb.) Boivin	26 August 2011	2980	44.749, −107.7471
206	*A. tridentata* Nutt. var. *vaseyana* (Rydb.) Boivin	26 August 2011	2825	44.7588, −107.7556
207	*A. tridentata* Nutt. var. *vaseyana* (Rydb.) Boivin	26 August 2011	1907	44.7822, −107.968
209	*A. tridentata* Nutt. var. *vaseyana* (Rydb.) Boivin	23 September 2011	2253	44.3156, −106.9416
210	*A. tridentata* Nutt. var. *vaseyana* (Rydb.) Boivin	23 September 2011	2396	44.2512, −106.9562
211	*A. tridentata* Nutt. var. *vaseyana* (Rydb.) Boivin	23 September 2011	2423	44.1571, −107.2517
212	*A. tridentata* Nutt. var. *vaseyana* (Rydb.) Boivin	23 September 2011	2221	44.1325, −107.2526
213	*A. ludoviciana* Nutt. ssp. *ludoviciana*	26 September 2011	1643	44.6327, −107.0786
214	*A. tridentata* Nutt. var. *vaseyana* (Rydb.) Boivin	26 September 2011	2157	44.619, −107.1014
215	*A. tridentata* Nutt. var. *vaseyana* (Rydb.) Boivin	26 September 2011	2164	44.6186, −107.1102
216	*A. longifolia* Nutt.	27 September 2011	1166	44.8117, −106.9116
217	*A. longifolia* Nutt.	27 September 2011	1147	44.8354, −106.8735
218	*A. longifolia* Nutt.	27 September 2011	1149	44.8376, −106.8404
219	*A. cana* Pursh var. *cana*	27 September 2011	1292	44.8233, −107.2269
220	*A. cana* Pursh var. *cana*	27 September 2011	1333	44.826, −107.236
221	*A. tridentata* Nutt. var. *tridentata*	30 September 2011	2299	44.569, −107.5337
222	*A. tridentata* Nutt. var. *tridentata*	30 September 2011	2141	44.5742, −107.5666
223	*A. tridentata* Nutt. var. *wyomingensis* (Beetle and Young) Welsh	30 September 2011	1453	44.0279, −107.5646
250	*Artemisia* ssp.	28 October 2014	1166	44.8318, −106.8314
251	*A. tridentata* Nutt. var. *wyomingensis* (Beetle and Young) Welsh	28 October 2014	1192	44.8318, −106.8338
252	*A. cana* Pursh var. *cana*	28 October 2014	1128	44.8376, −106.8403
253	*A. cana* Pursh var. *cana*	28 October 2014	1126	44.8433, −106.8407
254	*A. cana* Pursh var. *cana*	28 October 2014	1136	44.8462, −106.8395
255	*A. cana* Pursh var. *cana*	29 October 2014	1118	44.8925, −107.0287
256	*A. cana* Pursh var. *cana*	29 October 2014	1119	44.8886, −107.0293
257	*A. cana* Pursh var. *cana*	29 October 2014	1130	44.8887, −107.0327
258	*A. cana* Pursh var. *cana*	29 October 2014	1105	44.8967, −107.0289
259	*A. tridentata* Nutt. var. *vaseyana* (Rydb.) Boivin	30 October 2014	2172	44.618, −107.1121
260	*A. tridentata* Nutt. var. *wyomingensis* (Beetle and Young) Welsh	30 October 2014	2524	44.7161, −107.4587
261	*A. tridentata* Nutt. var. *vaseyana* (Rydb.) Boivin	30 October 2014	2300	44.5691, −107.5338
262	*A. tridentata* Nutt. var. *vaseyana* (Rydb.) Boivin	30 October 2014	2895	44.6653, −107.5451

2.1.2. *Artemisia tridentata* var. *tridentata*

The chemical constituents found in the EO of *A. tridentata* var. *tridentata* are listed in Table S2 and were 24 and 27 for accession #221 and #222, respectively. Camphor, eucalyptol, chrysanthenone, and camphene were the major EO constituents for accession #221. For accession #222, the major EO constituents were camphor, camphene, eucalyptol, and α-pinene (Table S2).

2.1.3. *Artemisia tridentata* var. *wyomingensis*

The 18 EO constituents found in the oil of *A. tridentata* var. *wyomingensis* accession are listed in Table S3. Camphor, arthole, α-santoline alcohol, and eucalyptol were the major constituents at 20.6, 17.4, 14.8, and 13.0%, respectively.

2.1.4. *Artemisia cana*

Among the 30 identified constituents in the EO of *A. cana*, camphor, eucalyptol, α-santoline alcohol, and arthole were the major constituents in accession #219 at 35.3%, 15.3%, 11.3%, and 9.0%, respectively (Table S4). The major constituents in the EO of accession #220 were camphor, eucalyptol, camphene, and arthole at 40.6%, 20.5%, 6.6%, and 6.2% of the total oil, respectively.

2.1.5. *Artemisia longifolia*

The chemical constituents found in the EO of *A. longifolia* are listed in Table S5. Among the 30 identified constituents, 11, 12, and 9 constituents were present above 1% in the oil of accessions #216, 217, and 218 accounting for 90.9%, 90.9%, and 91.4%, respectively. Eucalyptol at 30.8% was the major constituent in the EO of accession #216, while camphor was the major constituent at 27.7% and 43.4% in the EO of accessions #217 and #218, respectively.

2.1.6. *Artemisia ludoviciana* ssp. *ludoviciana*

Among the 33 constituents found in the EO of *A. ludoviciana* ssp. *ludoviciana* accession, 9 were present at concentrations of above 1% and accounted for 90% of the oil (Table S6). Among these 9 constituents, camphor, eucalyptol, camphene, and borneol were present at concentrations of 46.2%, 17.9%, 13.3%, and 4.6%, respectively.

2.2. *The 2014 Collections*

In the 2014 collection, 3 accessions were identified as *A. tridentata* var. *vaseyana*, collected at elevations of 2172, 2300, and 2895 m; 2 accessions were *A. tridentata* var. *wyomingensis*, collected at elevations of 1192 and 2524 m; and 7 accessions as *A. cana* var. *cana*, collected at elevations ranging from 1105 to 1136 m (Table 1).

2.2.1. *Artemisia tridentata* var. *vaseyana*

The number of identified chemical constituents found in the EO of *A. tridentata* var. *vaseyana* leaves obtained by hydro-distillation were 19, 28, and 9 for accessions #259, 261, and 262, respectively (Table S7). Artemiseole at 42.8%, camphor at 53.6%, and cyclooctadiene at 38.3% were the major constituents in the oils of accessions #259, 261, and 262, respectively.

2.2.2. *Artemisia tridentata* var. *wyomingensis*

There were 27 identified constituents in the EO obtained by hydro-distillation of *A. tridentata* var. *wyomingensis* inflorescences. However, the identified constituents in the EO obtained by hydro-distillation of leaves were 13 for accession #251 and 27 for #260, with 7 out of these 13 constituents and 14 out of the 27 constituents unique to each accession (Table S8). Artemiseole at 32.6% and 26.3% was the major constituent in the EO of the leaves or inflorescences, respectively, for accession #251. Cis-thujone at 71% was the major constituent in the oil obtained from the leaves of accession #260.

2.2.3. *Artemisia cana* var. *cana*

The 28 identified constituents in the EO of either leaves or inflorescences of *A. cana* var. *cana* accessions obtained by hydro-distillation are presented in Table S9. Camphor and eucalyptol were the major EO constituents for both leaves and inflorescences of *A. cana* Pursh var. *cana* accessions, except for accession #253. The major oil constituents for the leaves and inflorescences of accession #253 were ortho-cymene/para-cymene and α-phellandrene. Eucalyptol and camphor were the two major EO constituents in *A. cana* flowers, leaves, and stalks, collected from the Central Alberta Prairies, western Canada, although in different concentrations [24].

Overall, the results from this study demonstrated significant variation in EO content and composition in sagebrush species and subspecies collected from the Bighorn Mountains in Wyoming. Previous research from Oregon reported differences in EO composition between the three main subspecies of *A. tridentata*: *wyomingensis*, *tridentata*, and *vaseyna* [21]. The major constituents were methacrolein, camphene, artemiseole, eugenol, and artemisia ketone, with wide variations between and within the subspecies [21]. The wide variation in the concentration of individual chemical constituents between and within a subspecies is generally consistent with the results from this study. However, the chemical profile of the three *Artemisia tridentata* subspecies presented in the previous research from Oregon [21] is different from the ones in this study, which underlines the possibility for the existence of even greater chemical diversity among *A. tridentata* subspecies.

Earlier report on *A. tridentata* identified camphor (40–45%) and eucalyptol (1,8-Cineole), α-pinene, β-pinene, camphene, thujone, and α-phellandrene being the major constituents; however, the exact subspecies was not identified [25].

The EO of *A. ludoviciana* var. *latiloba* from South Dakota was characterized with high concentration of oxygenated monoterpenes, such as camphor (20%), borneol, (around 15%), and eugenol (around 10%); however, it was obtained from a single accession [26]. Eugenol and camphor were also among the major constituents of the EO of *A. longifolia* and *A. ludoviciana*, and the latter contained significant amount of davanone [19].

2.3. Essential Oil Yield

The EO yield of the species and subspecies collected in 2011 varied widely, from 0.15% to 1.69% in fresh herbage within *Artemisia* spp. (Figure 1). Within *A. tridentata* var. *vaseyana* alone, the EO yield of fresh herbage varied from 0.25 to 1.69%. The EO yield within *A. longifolia* varied from 0.8 to 1.2%, whereas the EO yield in *A. cana* was 0.8–1% (not significantly different within the species), while the EO yield of *A. ludoviciana* was only 0.16%, in *A. tridentata* var. *tridentata* EO the yield was 0.64–1.44% and was 1% in *A. tridentata* var. *wyomingensis* (Figure 1). The EO yield of the species and subspecies collected in 2014 varied within and between species (Table S10). EO obtained by hydrodistillation of *A. tridentata* var. *vaseyana* leaves was 0.23% on average and varied from 0.06 to 0.42% (Table S10). Essential oil obtained by hydrodistillation of *A. tridentata* var. *wyomingensis* leaves or inflorescences was 0.89 and 1.15%, respectively (Table S10). The EO obtained from *A. cana* var. *cana* leaves was 0.47% on average and varied from 0.04 to 0.96%, while the average EO content of inflorescences was 0.61% and varied from 0.17 to 1.10% (Table S10).

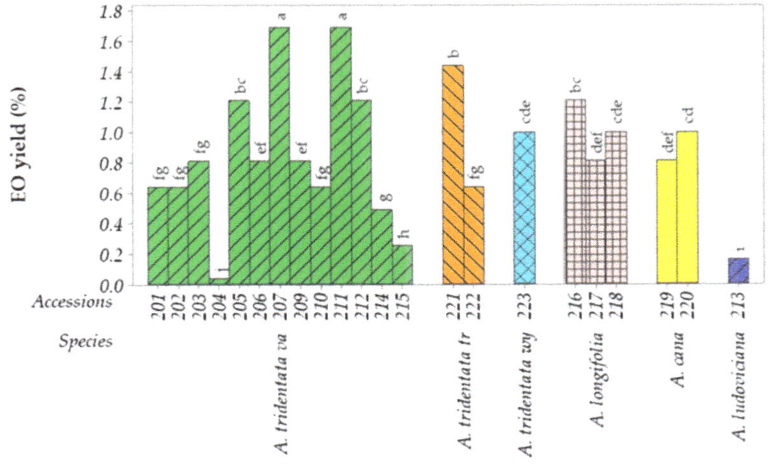

Figure 1. Mean EO yield (%) from six species and their accessions collected in the fall of 2011 from the Bighorn Mountains in Wyoming. Means sharing the same letter are not significantly different at the 5% level. *A. tridentata va* = *A. tridentata* Nutt. var. *vaseyana*; *A. tridentata tr* = *A. tridentata* Nutt. var. *tridentata*; *A. tridentata wy* = *A. tridentata* Nutt. var. *wyomingensis*.

2.4. The Essential Oil Constituents Found in All Artemisia Species

Alpha-pinene was found in the EO of most species and subspecies with the exception of two accessions of *A. tridentata* var. *vaseyana*, #205 and 206, collected at elevations of 2980 and 2825 m, respectively. The concentration of α-pinene in the oil varied from undetectable amounts to 35.5% in *A. tridentata* var. *vaseyana*; the concentration of this monoterpene in the oils of the other Artemisia species did not vary significantly within a species (Table 2). Alpha-pinene, an alkene ($C_{10}H_{16}$), is a major constituent in orange peel EO, but also in the EO of many other species ranging from *Cannabis sativa*, *Rosemarinus officinalis* to *Pinus* ssp. [27–29]. Mercier et al. (2009) [28] have reviewed the many biological effects of α- and β-pinenes (in turpentine and its fractions). Table 3 shows the square root of Mean Squares Error (MSE) of the studied variables that estimates the common standard deviation (σ).

Table 2. Mean composition of alpha-pinene, camphene, eucalyptol and camphor (all in %) from six species and corresponding accessions. Blank cells indicate the constituents were not obtained from these accessions. *A. tridentata va* = *A. tridentata* Nutt. var. *vaseyana*; *A. tridentata tr* = *A. tridentata* Nutt. var. *tridentata*; *A. tridentata wy* = *A. tridentata* Nutt. var. *wyomingensis*.

Species	Accession	Alpha-Pinene	Camphene	Eucalyptol	Camphor
A. tridentata va	201	3.8 def	8.5 de	1.1 i	31.9 bcde
A. tridentata va	202	1.9 ef	6.7 e	2.4 i	27.3 de
A. tridentata va	203	0.9 f	0.1 g	0.8 i	0.3 h
A. tridentata va	204	2.9 def			
A. tridentata va	207	1.6 ef	8.4 de	14.8 e	29.3 cde
A. tridentata va	209	29.0 b	2.9 fg	12.3 g	1.8 gh
A. tridentata va	210	35.5 a	2.2 g	12.8 fg	0.6 h
A. tridentata va	211	5.3 d	13.8 b	17.9 d	44.8 a
A. tridentata va	212	5.6 d	13.6 bc	14.1 ef	45.4 a
A. tridentata va	214	1.3 ef	6.5 e	9.7 h	13.5 fg
A. tridentata va	215	23.5 c	2.1 g	2.5 i	3.2 gh
A. tridentata tr	221	3.2 def	10.5 cd	13.6 efg	43.2 ab
A. tridentata tr	222	4.3 de	21.5 a	21.2 c	41.3 ab
A. tridentata wy	223	1.1 ef	6.6 e	12.5 fg	21.6 ef
A. tridentata wy	216	2.8 def	5.8 ef	30.8 a	24.2 def
A. longifolia	217	2.8 def	7.0 e	24.6 b	27.7 de
A. longifolia	218	3.2 def	6.9 e	25.5 b	43.4 ab
A. longifolia	219	2.7 def	7.4 de	15.3 e	35.2 abcd
A. cana	220	3.2 def	6.6 e	20.5 c	40.6 abc
A. cana	213	1.9 ef	13.3 bc	17.9 d	46.2 a

Within each column, means followed by the same letter are not significantly different at the 5% level using Tukey's method.

Similar to α-pinene, camphene (another monoterpene) was found in all *Artemisia* species and subspecies with the exception of the same two accessions of *A. tridentata* var. *vaseyana* (#205 and 206). Camphene concentrations varied from undetectable amounts to 13.8% within *A. tridentata* var. *vaseyana*, from 10.5 to 21.5% in *A. tridentata* var. *tridentata*, and was 6.6% in *A. tridentata* var. *wyimingensis*. Camphene concentration was 5.8–7% in *A. longifolia*, 6.6–7.4% in *A. cana*, and 13.3% in *A. ludoviciana*, and did not vary significantly within these species.

Similar to α-pinene and camphene, eucalyptol was found in all *Artemisia* species and subspecies with the exception of the same two accessions #205 and 206 of *A. tridentata*. The concentration of eucalyptol varied from undetectable amounts to 30.8%; its concentration in *A. tridentata* was 0–21.2%, 24.6–30.8% in *A. longifolia*, 15.3–20.5% in *A. cana*, and 17.9% in *A. ludoviciana* (Table 2). The monoterpenoid eucalyptol (1,8 cineole), is a common EO constituent in other plant species and especially in eucalyptus (*Eucalyptus globulus*), *Artemisia vestita*, bay laurel (*Laurus nobilis*), but also in a number of other plants such as ginger (*Zingiber officinale*) and even a chemotype of lavender (*Lavandula stoechas*) [27,30–34]. Eucalyptol has wide applications as flavoring agent in consumer products such as mouthwashes, in perfumery and cosmetics, in cigarettes, and also in food products at low concentrations. Research has shown that eucalyptol has anti-inflammatory and anti-depressive effects [27].

Table 3. Square root of mean squares error (MSE) that estimates the common standard deviation (σ).

Variable	$\sqrt{MSE}$
EO yield	0.073
antioxidant activity	6.71
α-pinene	0.949
camphene	1.02
eucalyptol	0.601
camphor	3.66
trans-α-necrodol-acetate	2.06
fragranol	2.43
grandisol	3.10
piperitenone/citronellyl acetate	1.04
borneol	0.154
trans-pinocarveol	1.14
cis-arbusculone	0.158
pinocarvone	0.420
4-terpineol	0.048
Myrtenol	0.123
santolina triene	0.304
borneol/lavandulol	0.070
arthole	1.09
gamma-terpinene	0.063
α-santoline alcohol	0.319
chrysanthemyl alcohol	1.06
beta-pinene	0.956
chrysanthenone	4.17

Similar to α-pinene, camphene, and eucalyptol, camphor was found in all *Artemisia* species and subspecies with the exception of the same two accessions #205 and #206 of *A. tridentata*. Overall, the concentration of camphor varied from undetectable amounts to 45.5% in the oil of *A. tridentata* var. *vasseyana*, from 24.2 to 43.4% in the oil of *A. longifolia*, 35.2–40.6% in the oil of *A. cana* (Table 2). The waxy aromatic solid, camphor, a terpenoid ($C_{10}H_{16}O$), is isolated from wood of camphor tree, *Cinnamomum camphora*, and is also found in EO from other species such as *Cedrus libani*, *Ocimum kilimandscharicum*, *A. annua*, *A. vestita*, *Piper angustifolium*, *Sassafras albidum*, and *Rosmarinus officinalis* among others. Camphor has been known for centuries as an aromatic substance, and it has been used in ancient China and Japan, in other Asian countries and in Europe medicinally, but also in culinary applications and cosmetics [35]. It is a toxic substance with various biological activities ranging from antimicrobial, antiviral, insecticidal, and antitussive to anticancer activities [35,36].

Trans-α-necrodol-acetate, fragranol, and grandisol were found in the EOs of the collected accessions #201, 202, 203, 205, and 206 of *A. tridentata* var. *vasseyana* (Table 4). Additionally, trans-α-necrodol-acetate was found in the collected accessions #214 and 215 of *A. tridentata* var. *vasseyana*. The concentration of trans-α-necrodol-acetate varied from undetectable amounts (in most accessions) to 45.1% in *A. tridentata* var. *vasseyana*; this compound was not found in the oils of the other species (Table 4). Alpha-necrodol acetate ($C_{12}H_{20}O_2$) is a rather rare constituent in plants and has been reported in the EO of two other plant species, *Lavandula suisieri* [37,38] and *Evolvulus alsinoides* (slender dwarf morning glory) [39].

The concentration of fragranol varied from 5.7 to 20.3% in the above-mentioned collected accessions of *A. tridentata* var. *vasseyana*, whereas the concentration of grandisol varied from 6.5 to 36.2% in the same accessions (Table 4). The monoterpene grandisol ($C_{10}H_{18}O$) is used as pheromone (sex attractant) for agricultural pests such as cotton boll weevil (*Anthonomus grandis*) and other weevils [40]. It has been reported in the EO of other *Artemisia* species such as *A. vestita* [32] and also in *Achillea falcata* [41].

Table 4. Mean trans-α-necrodol-acetate, fragranol, grandisol, cis-arbusculone, trans-pinocarveol and chrysanthenone (all in %) from 10 collections of the *A. tridentata* Nutt. var. *vaseyana* (Rydb.) Boivin. Blank cells indicate the constituents were not obtained from these accessions. *A. tridentata va* = *A. tridentata* Nutt. var. *vaseyana*.

Species	Accession Number	Trans-α-Necrodol-Acetate	Fragranol	Grandisol	Cis-Arbusculone	Trans-Pinocarveol	Chrysanthenone
A. tridentata va	201	5.7 e	5.7 c	6.5 d			21.61 a
A. tridentata va	202	13.5 d	12.4 bc	17.6 c			1.21 b
A. tridentata va	203	25.9 c	20.3 a	36.2 a			
A. tridentata va	205	34.8 b	14.9 ab	31.1 ab			
A. tridentata va	206	45.1 a	15.0 ab	26.2 bc			
A. tridentata va	207						11.86 ab
A. tridentata va	209				3.1 a	24.8 a	
A. tridentata va	210				2.3 b	20.9 b	
A. tridentata va	214	8.2 de					
A. tridentata va	215	4.8 e				6.0 c	

Within each column, means followed by the same letter are not significantly different at the 5% level using Tukey's method.

Trans-arbusculone was found in the oils of only two collected accessions (#209 and 210) of *A. tridentata* var. *vasseyana*, and its concentration was 2.3–3.1% of the oil. Trans-pinocarveol was found in four accessions (#207, 209, 210, and 215) of *A. tridentata* var *vasseyana* and its concentration varied from 2 to 24.8% of the oil.

Chrysanthenone was found in the oils of three accessions (#201, 202, and 207) of *A. tridentata* var. *vasseyana*, and its concentration ranged from 1.2 to 21.6% of the oil (Table 4).

Borneol was not found in *A. tridentata*, it was only present in the oils of *A. longifolia*, *A. cana*, and *A. ludoviciana*, and its concentration was 2.3–2.6% in the oils of *A. longifolia* and *A. cana* and 4.6% in the oil of *A. ludoviciana* (Table 5). Pinocarvone was found in six accessions of *A. tridentata* var. *vasseyana* only and its concentration varied from 1.2 to 6.9% of the oil. The monoterpene 4-terpineol was found in four accessions of *A. tridentata* var *vasseyana* and also in *A. longifolia* and *A. cana*. The concentration of 4-terpineole varied from 0.9 to 1.4% in the oil of *A. tridentata* var. *vasseyana*, from 1.7 to 2.5% in the oil of *A. longifolia*, and 1.5–2.1% in the oil of *A. cana*. Santolina triene was found in two accessions of *A. tridentata* var. *vasseyana* (#214, 215), and also in the oils of *A. tridentata* var. *tridentata* *A. logifolia*, and *A. cana*. The concentration of santolina triene was 1.7–2.5% in the oil of *A. tridentata*, 3.5–3.7% in *A. cana*, and varied from 2.7 to 6.8% in the oil of *A. longifolia*.

Arthole was found in the oils of the same two accession of *A. tridentata vasseyana*, in *A. tridentata* var. *tridentata*, in *A. tridentata* var. *wyomingensis*, and in the oils of *A. longifolia* and *A. cana*. The concentration of arthole was 10.3 and 13.0% in the oils of *A. tridentata* var *vasseyana*, 3.8% in the oil of *A. tridentata* var. *tridentata*, 20.1% in the oil of *A. tridentata* var *wyomingensis*, from 3.2 to 9.7% in the oil of *A. longifolia*, and 6.2 and 9.0% in the oil of *A. cana* (Table 5).

Gama-terpinene was found in the oil of two accessions of *A. tridentata* var. *vasseyana* (at 1.0 and 2.2%), in *A. longifolia* (1.5–2.5% range) and in *A. cana* (1.7–1.8%) (Table 6). Alpha santoline alcohol was found in the same two accessions of *A. tridentata* var. *vasseyana* (at 3.7 and 5.9%), in two accessions of *A. longifolia* (at 6.0 and 7.6%), and in one accession of *A. cana* (at 11.3%). Chrysanthemyl alcohol was found only in the same two accessions of *A. tridentata* var. *vasseyana* (at 4.4 and 11.4%). Beta-pinene was only found in the oil of one accession of *A. tridentata* var. *tridentata* at 2.2% concentration (Table 6).

Table 5. Mean borneol, pinocarvone, 4-terpineol, santolina triene and arthole (all in %) from six species and their corresponding collection number. Blank cells indicate the constituents were not obtained from these accessions. *A. tridentata va* = *A. tridentata* Nutt. var. *vaseyana; A. tridentata tr* = *A. tridentata* Nutt. var. *tridentata; A. tridentata wy* = *A. tridentata* Nutt. var. *wyomingensis.*

Species	Accession Number	Borneol	Pinocarvone	4-Terpineol	Santolina Triene	Arthole
A. tridentata va	209		5.3 b	1.3 e		
A. tridentata va	210		6.9 a	1.4 de		
A. tridentata va	211		1.0 c	1.1 f		
A. tridentata va	212		1.2 c	0.9 g		
A. tridentata va	214		8.1 a		2.5 cde	13.0 b
A. tridentata va	215		1.9 c		1.8 de	10.3 c
A. tridentata tr	221				1.7 e	3.8 f
A. tridentata wy	223					20.1 a
A. longifolia	216	2.3 c		2.5 a	2.7 cd	9.7 cd
A. longifolia	217	2.4 bc		1.7 c	6.8 a	6.8 e
A. longifolia	218	2.4 bc		2.2 b	3.1 bc	3.2 f
A. cana	219	2.3 c		1.5 d	3.5 b	9.0 d
A. cana	220	2.6 b		2.1 b	3.7 b	6.2 e
A. ludoviciana	213	4.6 a		1.3 e	2.2 de	

Within each column, means followed by the same letter are not significantly different at the 5% level using Tukey's method.

Table 6. Mean gamma-terpinene, alpha-santoline alcohol, chrysanthemyl alcohol and beta-pinene (all in %) from four species and their subspecies. Blank cells indicate the constituents were not obtained from these accessions. *A. tridentata va* = *A. tridentata* Nutt. var. *vaseyana; A. tridentata tr* = *A. tridentata* Nutt. var. *tridentata.*

Species	Accession Number	Gamma-Terpinene	Alpha-Santoline Alcohol	Chrysanthemyl Alcohol	Beta-Pinene
A. tridentata va	204				7.23 a
A. tridentata va	214	1.0 e	3.7 d	11.37 a	
A. tridentata va	215	2.2 b	5.9 c	4.44 b	
A. tridentata tr	222				2.21 b
A. longifolia	216	1.5 d	7.6 b		
A. longifolia	217	2.5 a	6.0 c		
A. longifolia	218	1.5 d			
A. cana	219	1.8 c	11.3 a		
A. cana	220	1.7 cd			

Within each column, means followed by the same letter are not significantly different at the 5% level using Tukey's method.

The following more significant oil constituents were only found in one or two accessions of *A. tridentata* var. *vasseyana* and did not vary within the accessions: periritenone (average concentration of 4.1%) in three accessions, artemisyl acetate (2.9%) in one accession, myrtenol (at 1.2%) in two accessions, borneol/lavandulol (1.9%) in two accessions, *trans*-arbusculone (1.9%) in one accession, piperitenone/citronellyl acetate (at 7.3%) in two accessions (Table 7).

Table 7. Piperitenone/citronellyl acetate, artemisyl acetae, myrtenol, borneol/lavandulol, trans-rbusculone, sabinene, alpha-phellendrene, para-cymene and *trans*-ocimene that were obtained from only one accession (shown in bracket in Column 1) or without significant difference among accessions.

Species (Accession Numbers)	Constituent	Mean (%)
A. tridentata Nutt. var. *vaseyana* (201, 202, 203)	Piperitenone/citronellyl acetate	4.13
A. tridentata Nutt. var. *vaseyana* (209)	Artemisyl acetae	2.88
A. tridentata Nutt. var. *vaseyana* (209, 210)	Myrtenol	1.2
A. tridentata Nutt. var. *vaseyana* (211, 212)	Borneol/Lavandulol	1.9
A. tridentata Nutt. var. *vaseyana* (215)	Trans-arbusculone	1.88
A. tridentata Nutt. var. *vaseyana* (214, 215)	Piperitenone/Citronellyl acetate	7.26
A. tridentata Nutt. var. *vaseyana* (204)	Sabinene	12.32
A. tridentata Nutt. var. *vaseyana* (204)	Alpha-phellendrene	16.86
A. tridentata Nutt. var. *vaseyana* (204)	Para-cymene	31.83
A. tridentata Nutt. var. *vaseyana* (204)	Trans-ocimene	0.94

2.5. Antioxidant Capacity of Artemisia Species and Subspecies

The antioxidant capacity of the EOs varied between the species and subspecies, with the highest being 80.5 μmol/g for *A. tridentata* var. *wyomingensis* and 60.5 μmol/g in *A. tridentata* var. *tridentata*; the antioxidant capacity of the rest of the oils was not different from the antioxidant capacity of the oils from the above two species (Table 8). This is the first report on antioxidant activity of the EO of the sagebrush species tested in this study; however, there is a previous report on *A. tridentata* var. *wyomingensis* leaf extract (70% ethanol extraction for 24 h) [42], but not on sagebrush EO.

Table 8. Mean antioxidant activity (μM Trolox equivalents/g) of the essential oils from the five species.

Species	Accession Number	Antioxidant Activity
A. tridentata Nutt. var. *vaseyana*	207	63.1 ab
A. tridentata Nutt. var. *tridentata*	221	60.5 b
A. tridentata Nutt. var. *wyomingensis*	223	80.5 a
A. longifolia	216	73.4 ab
A. cana	220	71.5 ab

Means followed by the same letter are not significantly different at the 5% level using Tukey's method.

None of the accessions contained any podophyllotoxin, indicating that *Artemisia* species and subspecies collected in this study did not contain podophyllotoxin.

2.6. Antileishmanial Evaluations

Compounds from some other *Artemisia* species (e.g., the endoperoxide artemisinin from *A. annua*) were reported to possess leishmanicidal activity [43]. Compounds extracted from *A. aucheri* have shown antileishmanial effect [44]. Ethanol extract from *A. absinthium* was found to be effective against *Leishmania major* L. in vitro [45]. The essential oil of *A. absinthium* was also reported to have promise as active compounds source against *Leishmania* [46].

In this study, representative essential oil samples from each of the species *A. cana* Pursh var. *cana*, *A. longifolia*, *A. tridentata* Nutt. var. *tridentata*, *A. tridentata* Nutt. var. *vaseyana*, and *A. tridentata* Nutt. var. *wyomingensis* were each evaluated against *Leishmania donovani* promastigotes and none demonstrated activity above 50% inhibition when evaluated at 80 μg/mL (Table S11). The bioassays used in this study are suited for the discovery of new therapeutic agents and hence the concentrations used in the analysis were much lower than those in previous studies.

2.7. Antiplasmodial Evaluations

Representative essential oil samples from each of the species *A. cana* Pursh var. *cana*, *A. longifolia*, *A. tridentata* Nutt. var. *tridentata*, *A. tridentata* Nutt. var. *vaseyana*, and *A. tridentata* Nutt. var. *wyomingensis* were evaluated against *Plasmodium falciparum* D6 at 15,867 ng/mL and none of them demonstrated activity above 50% inhibition to warrant secondary evaluations for LC_{50} determinations (Table S11).

2.8. Antimicrobial Evaluations

Representative essential oil samples from each of the species *A. cana* Pursh var. *cana*, *A. longifolia*, *A. tridentata* Nutt. var. *tridentata*, *A. tridentata* Nutt. var. *vaseyana*, and *A. tridentata* Nutt. var. *wyomingensis* were evaluated against *Candida albicans*, *Candida glabrata*, *Candida krusei*, *Aspergillus fumigatus*, *Cryptococcus neoformans*, *Staphylococcus aureus*, Methicillin-resistant *S. aureus*, *Escherichia coli*, *Pseudomonas aeruginosa*, and *Mycobacterium intracellulare*, but none of them demonstrated activity above 50% inhibition when evaluated at 50 μg/mL (Table S12).

3. Materials and Methods

3.1. Collection of the Plant Material

Two separate collections of sagebrush species were conducted and included in this study. In 2011, we conducted a comprehensive study in the Bighorn Mountains area to identify the species of sagebrush and analyze their EO content, profile, and bioactivity. Accessions were collected from a total of 22 sites at various elevations (1150–2988 m a.s.l.) (Table 1). In a separate study conducted in 2014, additional accessions were collected from 12 sites at elevations ranging from 1108 to 2902 m a.s.l. (Table 1). The sagebrush aboveground herbage was collected from each collection site, and the GPS coordinates were recorded (Table 1). Plant material from each collection site was identified by Ms. Bonnie Heidel at the Rocky Mountain Herbarium, at the University of Wyoming (http://www.uwyo.edu/wyndd/about-wyndd/staff/bonnie-heidel.html), accessed on 28 April 2022.

The majority of the collected accessions belonged to the big sagebrush subspecies: *A. tridentata* Natt. var. *vaseyana* (Rydb.) Beetle (mountain big sagebrush), *A. tridentata* Natt. var. *wyomingensis* Beetle and Young (Wyoming big sagebrush), and *A. tridentata* Natt. var. *tridentata* (basin big sagebrush). The most frequent and widespread species was *A. tridentata* Natt. var. *vaseyana* (Rydb.). In addition, we collected and identified the following species in the area: *Artemisia cana* Pursh (silver sagebrush), *A. ludoviciana* Nutt. (white sagebrush), and *A. longifolia* Nutt. (longleaf wormwood).

3.2. Essential Oil Extraction

3.2.1. Steam Distillation

Subsamples of *Artemisia* species and subspecies collected in 2011 were subjected to steam distillation for extraction of the EO. Representative samples of 500 g fresh material that included all aboveground plant parts, leaves, inflorescences, and annual stems not thicker than 2 mm were cut into approximately 2.5 cm long pieces. Then, each sample was immediately placed into the bioflask of 2 L steam distillation units (HeartMagic, Rancho Santa Fe, CA, USA) and steam distilled for 60 min as described previously for spearmint and peppermint [47,48].

3.2.2. Hydro-Distillation

Essential oil from the accession samples collected in 2014 was extracted via hydro-distillation. A 100 g fresh plant biomass sample consisting of stems and leaves only or inflorescences only, cut into approximately 2.5 cm long pieces, was placed into a 2 L boiling flask along with 1.5 L water and distilled for 60 min.

Distillation time was measured from when the first drop of EO appeared in the glass Florentine vessel (the separator), and at the end of the 60 min period, the power was shut down and the distillation discontinued. After the distillation of each sample, the water was drained from the Florentine vessel and the oil was collected in a glass vial and placed in a freezer. The collected EOs were separated from the remaining water, measured on an analytical scale, and the oil content (oil yield) was calculated as grams of oil per 100 g of fresh aboveground sagebrush biomass. Afterward, the EO samples were stored in at −5 °C in a freezer until the oils were analyzed for chemical profile on gas chromatography-mass spectrometry (GC-MS).

3.3. Gas Chromatography (GC) Mass Spectroscopy (MS) of the Sagebrush Species Essential Oil (EO)

The GC-MS analysis of the sagebrush EO samples was conducted as described previously [49]. Briefly, the sagebrush EO samples (all samples in three replicates) were analyzed on a GC-MS instrument (Hewlett Packard Model 6890; Hewlett-Packard, Palo Alto, CA, USA). The carrier gas was helium at a mean speed of 40 cm/s^{-1}, at 11.7 psi (60 °C), and a constant flow rate at 2.5 mL/min^{-1}. The injection was split 60:1, 0.5 μL, with 220 °C injector temperature. The GC oven temperature program was as described previously: 60 °C for 1 min and 10 °C/min to 250 °C. The column was HP-INNOWAX (crosslinked polyethylene glycol; 30 m × 0.32 mm × 0.5 mm), and the flame ionization detector temperature was 275 °C. The sagebrush EO constituents were expressed as percentage of all the constituents in the EO. The identification of individual constituent peaks was completed using standard compounds, through retention index values and through MS spectra comparison.

3.4. Podophyllotoxin Extraction and Measurements

The *Artemisia* samples collected in 2011 were also subjected to chemical analyses for podophyllotoxin. The podophyllotoxin extraction and purification followed a previously described procedures [50] and as described in a podophyllotoxin analyses of *Juniperus* paper [51].

3.5. Antimicrobial, Antimalarial, and Antileishmanial Activity and Cytotoxicity

The EO from the accessions collected in 2011 were submitted for analyses of antimicrobial, antimalarial, and antileishmanial activity and cytotoxicity. The assays were conducted at The University of Mississippi, National Center for Natural Products Research using methods developed at the Center and as described previously [51].

3.6. Antioxidant Activity of the EOs of Artemisia Species and Subspecies from This Study

The antioxidative capacity of the *Artemisia* oils collected in 2011 was determined by the oxygen radical absorbance capacity (ORAC) [52,53] and as described previously [54]. All samples were taken in triplicate.

3.7. Statistical Analyses of the Data

Analysis of variance was completed for EO yield, antioxidant activity, and the concentrations of α-pinene, camphene, eucalyptol, camphor, trans-α-necrodol-acetate, fragranol, grandisol, piperitenone/citronellyl acetate, borneol, trans-pinocarveol, cis-arbusculone, pinocarvone, 4-terpineol, myrtenol, santolina triene, borneol/lavandulol, arthole, gamma-terpinene, α-santoline alcohol, chrysanthemyl alcohol, beta-pinene, and chrysanthenone using either a nested design or a completely randomized design model, and both designs used 3 replications. A nested design (species, and subspecies nested within species effects in the model) was used for EO yield, and the constituents obtained from more than one species. However, for the constituents obtained only within a species, a completely randomized design was used to compare the subspecies. The analysis was completed using the Mixed Procedure of SAS [54], and for each response, the validity of model assumptions was verified by examining the residuals as described in [55]. Since the effects of species (where applicable) and subspecies were significant (with the exception of piperitenone/citronellyl acetate, myrtenol, and borneol/lavandulol where the subspecies were not significantly different), multiple means comparison of the subspecies was completed using Tukey's multiple means comparison method at the 5% level of significance.

4. Conclusions

This study investigated the EO composition and bioactivity of several sagebrush species in the Western United States, and the findings revealed wide variation in the EO yield and composition between the sagebrush species, as well as within a species or subspecies, partially confirming the hypothesis. These findings suggest the presence of chemotypes within some species of sagebrush.

The outcomes from this study refuted part of our hypothesis that the essential oil obtained from the sagebrush species would have antileishmanial, antimalarial, and antimicrobial activities. The hypothesis was based on reports of such activities for extracts and derivatives from other *Artemisia* species, such as *A. annua*, *A. aucheri*, *A. absinthium*. Synthetic 1,2-dioxanes were shown to possess leishmanicidal activity, whilst the natural endoperoxide artemisinin was not effective against *L. donovani* [43]. Compounds extracted from *A. aucheri* have shown antileishmanial effects against *L. major* [44]. Ethanol extract from *A. absinthium* was found to be effective against *Leishmania major* L. in vitro [45]. The EO of *A. absinthium* was also reported to have promise as active compounds against *Leishmania* [46]. It is worth noting that the bioassays used in this study are suited to the discovery of new therapeutic agents and the concentrations used in the analysis are perhaps much lower than the concentrations used in previous reports. In addition, we used whole oils from different *Artemisia* species.

The EOs of the sagebrush species in this study did not show significant antimicrobial activities, which contradicts other reports [18]. Most probably, the differences between the antimicrobial activity of the tested EO in this study and literature reports were due to the following: (1) higher concentrations used in the previous reports; (2) different assays; (3) differences in chemical constituents of the EOs in this study vs. literature reports. Indeed, a report [18] found differences in antifungal activity between different populations of *A. tridentata*. The authors of the report explained these differences with compositional dissimilarities of secondary metabolites between the different populations with respect to antifungal activity.

Overall, wild grown sagebrush species and subspecies seem to be a largely untapped resource for EO with interesting and possibly desirable composition. Some of the accessions have yielded EO with significant concentrations of compounds such as camphor, eucalyptol, cis-thujone, α-pinene, α-necrodol-acetate, fragranol, grandisol, para-cymene, and arthole among others. Therefore, chemotypes can be selected and possibly introduced into culture and be grown for commercial production of these compounds.

The results of our studies suggest immense chemical diversity exists that presents an opportunity for the selection of chemotypes/varieties with high concentration of EO with desirable composition.

Supplementary Materials: The following are available online at https://www.mdpi.com/article/10.3390/plants11091228/s1, Table S1: Mean (using n = 3) content (%) of constituents in essential oil obtained by steam distillation of *A. tridentata* Nutt. var. *vaseyana* (Rydb.) Boivin accessions biomass collected from Bighorn Mountains in Wyoming; Table S2: Mean (using n = 3) content (%) of constituents in essential oil obtained by steam distillation of *A. tridentata* Nutt. var. *tridentata* accessions biomass collected from Bighorn Mountains in Wyoming; Table S3: Mean (using n = 3) content (%) of constituents in essential oil obtained by steam distillation *A. tridentata* Nutt. var. *wyomingensis* (Beetle and Young) Welsh accessions biomass collected from Bighorn Mountains in Wyoming; Table S4: Mean (using n = 3) content (%) of constituents in essential oil obtained by steam distillation of *A. cana* Pursh var. *cana* accessions biomass collected from Bighorn Mountains in Wyoming; Table S5: Mean (using n = 3) content (%) of constituents in essential oil obtained by steam distillation of *A. longifolia* Nutt. accessions biomass collected from Bighorn Mountains in Wyoming; Table S6: Mean (using n = 3) content (%) of constituents in essential oil obtained by steam distillation of *A. ludoviciana* Nutt. ssp. *ludoviciana* accessions biomass collected from Bighorn Mountains in Wyoming; Table S7: Mean (using n = 3) content (%) of constituents in essential oil obtained by hydrodistillation of *A. tridentata* var. *vaseyana* accessions biomass (leaves) collected from Bighorn Mountains in Wyoming; Table S8: Mean (using n = 3) content (%) of constituents in essential oil obtained by hydrodistillation of *A. tridentata* var. *wyomingensis* accessions biomass (leaves or inflorescences) collected from Bighorn Mountains in Wyoming; Table S9: Mean (using $n = 3$) content (%) of constituents in essential oil obtained by hydrodistillation of *A. cana* var. *cana* accessions biomass (leaves or inflorescences) collected from Bighorn Mountains in Wyoming; Table S10: Mean (using $n = 3$) content (%) and range of EO obtained by hydrodistillation for accessions from Bighorn Mountains in Wyoming in the Fall of 2014; Table S11. Evaluation of select essential oils against *Leishmania donovani* and *Plasmodium falciparum* D6. Data are percent inhibition. Primary evaluations performed at 80 ug/mL for *L. donovani* and 15,867 ng/mL for *P. falciparum* D6; Table S12. Evaluation of select essential oils against opportunistic infection pathogens. Primary evaluations performed at 50 ug/mL and data are percent inhibition.

Author Contributions: Conceptualization, V.D.Z. and E.A.J.; methodology, V.D.Z., C.L.C., E.A.J. and V.S.; software, T.A. and C.L.C.; validation, V.D.Z., C.L.C., E.A.J. and V.S.; formal analysis, T.A.; investigation, V.D.Z., C.L.C., E.A.J. and V.S.; resources, V.D.Z., C.L.C., E.A.J. and V.S.; data curation, V.D.Z.; writing—original draft preparation, V.D.Z.; writing—review and editing, C.L.C., E.A.J., T.A. and V.S.; visualization, T.A.; supervision, V.D.Z. and E.A.J.; project administration, V.D.Z. and E.A.J.; funding acquisition, V.D.Z. All authors have read and agreed to the published version of the manuscript.

Funding: The research was supported by funded project "Landscape Restoration Through Science Based Reclamation and Education" by the Bureau of Land Management WY-State, Grant number L14AC00099. This study was also partially supported by the University of Wyoming and Oregon State University startup funding awarded to Dr. Valtcho D. Zheljazkov (Jeliazkov).

Institutional Review Board Statement: Not applicable.

Informed Consent Statement: Not applicable.

Data Availability Statement: Data are contained within the article.

Acknowledgments: We thank the Forest Service of the U.S. Department of Agriculture for issuing us a permit to sample sagebrush in the Bighorn National Forest. We thank Travis Fack, Natural Resource Specialist, and Clarke McClung, Tongue District Ranger, for helping with the permit. We are also thankful to Lyn Ciampa, Santosh Shiwakoti, and Shital Poudyal for their help with the collections of sagebrush accessions in the Big Horn Mountains.

Conflicts of Interest: The authors declare no conflict of interest.

References

1. Bora, K.S.; Sharma, A. The Genus *Artemisia*: A Comprehensive Review. *Pharm. Biol.* **2011**, *49*, 101–109. [CrossRef] [PubMed]
2. Boyle, S.A.; Reeder, D.R. Colorado Sagebrush: A Conservation Assessment and Strategy. Grand Junction: Colorado Division of Wildlife. 2005. Available online: https://cpw.state.co.us/Documents/WildlifeSpecies/Sagebrush/CHAPTER0contentsfrontmatter.pdf (accessed on 10 October 2020).
3. Dumroese, R.K. Sagebrush rangelands and greater sage-grouse in Northeastern California. In *Northeastern California Plateaus Bioregion Science Synthesis*; Dumroese, R.K., Moser, W.K., Eds.; Gen. Tech. Rep. RMRS-GTR-409; U.S. Department of Agriculture, Forest Service, Rocky Mountain Research Station: Fort Collins, CO, USA, 2020; pp. 112–130.
4. Byrd, D.W.; McArthur, E.D.; Wang, H.; Graham, J.H.; Freeman, D.C. Narrow hybrid zone between two subspecies of big sagebrush, *Artemisia tridentata* (Asteraceae). VIII. Spatial and temporal pattern of terpenes. *Biochem. Syst. Ecol.* **1999**, *27*, 11–25. [CrossRef]
5. Shultz, L.M. Artemisia. In *Flora of North America: North of Mexico*; Flora of North America Editorial Committee, Ed.; Oxford University Press: New York, NY, USA, 2006; Volume 19, pp. 503–534.
6. Davies, K.W.; Bates, J.D.; Johnson, D.D.; Nafus, A.M. Influence of mowing *Artemisia tridentata* ssp. wyomingensis on winter habitat for wildlife. *Environ. Manag.* **2009**, *44*, 84–92. [CrossRef] [PubMed]
7. Davis, A.J.; Phillips, M.L.; Doherty, P.F., Jr. Nest success of Gunnison sage-grouse in Colorado, USA. *PLoS ONE* **2015**, *10*, e0136310. [CrossRef]
8. Dziba, L.E.; Provenza, F.D.; Villalba, J.J.; Atwood, S.B. Supplemental energy and protein increase use of sagebrush by sheep. *Small Rumin. Res.* **2007**, *69*, 203–207. [CrossRef]
9. Welch, B.L.; Pederson, J.C.; Rodrigues, R.L. Selection of big sagebrush by sage grouse. *Great Basin Nat.* **1988**, *48*, 274–279.
10. Oldemeyer, J.L.; Barmore, W.J.; Gilbert, D.L. Winter ecology of Bighorn sheep in Yellowstone National Park. *J. Wildl. Manag.* **1971**, *35*, 257–269. [CrossRef]
11. Sheehy, D.P.; Winward, A.H. Artemisia taxa to mule deer and sheep. *J. Range Manag.* **1981**, *34*, 397–399. [CrossRef]

12. Veblen, K.E.; Nehring, K.C.; McGlone, C.M.; Ritchie, M.E. Contrasting effects of different mammalian herbivores on sagebrush plant communities. *PLoS ONE* **2015**, *10*, e0118016. [CrossRef]
13. Jacques, C.N.; Jenks, J.A.; Grovenburg, T.W.; Klaver, R.W. Influence of habitat and intrinsic characteristics on survival of neonatal pronghorn. *PLoS ONE* **2015**, *10*, e0144026. [CrossRef]
14. Copeland, H.E.; Pocewicz, A.; Naugle, D.E.; Griffiths, T.; Keinath, D.; Evans, J.; Platt, J. Measuring the effectiveness of conservation: A novel framework to quantify the benefits of sage-grouse conservation policy and easements in Wyoming. *PLoS ONE* **2013**, *8*, e67261. [CrossRef]
15. Doherty, K.; Tack, J.D.; Evans, J.S.; Naugle, D.E. Mapping breeding densities of greater sage-grouse: A tool for range-wide conservation planning. In *BLM Completion Report: Interagency Agreement # L10PG00911*; Bureau of Land Management: Washington, DC, USA, 2010; pp. 1–29.
16. McCutcheon, A.R.; Ellis, S.M.; Hancock, R.E.W.; Towers, G.H.N. Antifungal screening of medicinal plants of British Columbian native peoples. *J. Ethnopharmacol.* **1994**, *44*, 157–169. [CrossRef]
17. Turi, C.E.; Shipley, P.R.; Murch, S.J. North American *Artemisia* species from the subgenus Tridentatae (*Sagebrush*): A phytochemical, botanical and pharmacological review. *Phytochemistry* **2014**, *98*, 9–26. [CrossRef]
18. Talley, S.M.; Coley, P.D.; Kursar, T.A. Antifungal leaf-surface metabolites correlate with fungal abundance in sagebrush populations. *J. Chem. Ecol.* **2002**, *28*, 2141–2168. [CrossRef]
19. Lopes-Lutz, L.; Alviano, D.S.; Alviano, C.S.; Kolodziejczyk, P.P. Screening of chemical composition, antimicrobial and antioxidant activities of *Artemisia essential* oils. *Phytochemistry* **2008**, *69*, 1732–1738. [CrossRef]
20. Jassbi, A.R.; Zamanizadehnajari, S.; Baldwin, I.T. Phytotoxic volatiles in the roots and shoots of *Artemisia tridentata* as detected by headspace solid-phase microextraction and gas chromatographic-mass spectrometry analysis. *J. Chem. Ecol.* **2010**, *36*, 1398–1407. [CrossRef]
21. Kelsey, R.G.; Wright, W.E.; Sneva, F.; Winward, A.; Britton, C. The concentration and composition of big sagebrush essential oils from Oregon. *Biochem. Syst. Ecol.* **1983**, *11*, 353–360. [CrossRef]
22. Epstein, W.W.; Gaudioso, L.A.; Brewster, G.B. Essential oil constituents of *Artemisia tridentata rothrockii*. The isolation and characterization of two new irregular monoterpenes. *J. Org. Chem.* **1984**, *49*, 2748–2754. [CrossRef]
23. Borek, T.T.; Hochrein, J.M.; Irwin, A.N. *Composition of the Essential Oils from Rocky Mountain Juniper (Juniperus scopulorum), Big Sagebrush (Artemisia tridentata), and White Sage (Salvia apiana)*; SAND2003-3081; Sandia National Laboratories: Albuquerque, NM, USA, 2003; pp. 1–19.
24. Lopes-Lutz, D.; Mckay, T.; Kolodziejczyk, P.P. Distribution of volatiles in *Artemisia cana*. *Pharm. Biol.* **2008**, *46*, 373–376. [CrossRef]
25. Buttkus, H.A.; Bose, R.J.; Shearer, D.A. Terpenes in the essential oil of sagebrush (*Artemisia tridentata*). *J. Agric. Food Chem.* **1977**, *25*, 288–291. [CrossRef]
26. Collin, G.; St-Gelais, A.; Turcotte, M.; Gagnon, H. Composition of the essential oil and of some extracts of the aerial parts of *Artemisia ludoviciana* var. *latiloba* Nutt. *Am. J. Essent. Oils* **2017**, *5*, 28–38.
27. Borges, R.S.; Ortiz, B.L.S.; Pereira, A.C.; Keita, H.; Carvalho, J.C.T. *Rosmarinus officinalis* essential oil: A review of its phytochemistry, anti-inflammatory activity, and mechanisms of action involved. *J. Ethnopharmacol.* **2019**, *229*, 29–45. [CrossRef] [PubMed]
28. Mercier, B.; Prost, J.; Prost, M. The essential oil of turpentine and its major volatile fraction α- and beta-pinenes. A review. *Int. J. Occup. Med. Environ. Health* **2009**, *22*, 331–342. [CrossRef]
29. Zheljazkov, V.D.; Sikora, V.; Dincheva, I.; Kačániová, M.; Astatkie, T.; Semerdjieva, I.B.; Latkovic, D. Industrial, CBD, and wild hemp: How different are their essential oil profile and antimicrobial activity? *Molecules* **2020**, *25*, 4631. [CrossRef]
30. Boukhatem, M.N.; Sudha, T.; Darwish, N.H.E.; Chader, H.; Belkadi, A.; Rajabi, M.; Houche, A.; Benkebailli, F.; Oudjida, F.; Mousa, S.A. A new eucalyptol-rich lavender (*Lavandula stoechas* L.) essential oil: Emerging potential for therapy against inflammation and cancer. *Molecules* **2020**, *25*, 3671. [CrossRef] [PubMed]
31. Boukhatem, M.N.; Boumaiza, A.; Nada, H.G.; Rajabi, M.; Mousa, S.A. *Eucalyptus globulus* essential oil as a natural food preservative: Antioxidant, antibacterial and antifungal properties in vitro and in a real food matrix (*Orangina fruit* juice). *Appl. Sci.* **2020**, *10*, 5581. [CrossRef]
32. Chu, S.C.; Liu, Q.R.; Liu, Z.L. Insecticidal activity and chemical composition of the essential oil of *Artemisia vestita* from China against *Sitophilus zeamais*. *Biochem. Syst. Ecol.* **2010**, *38*, 489–492. [CrossRef]
33. Fidan, H.; Stefanova, G.; Kostova, I.; Stankov, S.; Damyanova, S.; Stoyanova, A.; Zheljazkov, V.D. Chemical composition and antimicrobial activity of *Laurus nobilis* L. essential oils from Bulgaria. *Molecules* **2019**, *24*, 804. [CrossRef]
34. Stefanova, G.; Girova, T.; Gochev, V.; Stoyanova, M.; Petkova, Z.; Stoyanova, A.; Zheljazkov, V.D. Comparative study on the chemical composition of laurel (*Laurus nobilis* L.) leaves from Greece and Georgia and the antibacterial activity of their essential oil. *Heliyon* **2020**, *6*, e05491. [CrossRef]
35. Chen, W.; Vermaak, I.; Viljoen, A. Camphor-a fumigant during the Black Death and a coveted fragrant wood in ancient Egypt and Babylon-a review. *Molecules* **2013**, *18*, 5434–5454. [CrossRef]
36. Manoharan, R.K.; Lee, J.H.; Lee, J. Antibiofilm and antihyphal activities of cedar leaf essential oil, camphor, and fenchone derivatives against *Candida albicans*. *Front. Microbiol.* **2017**, *8*, 1476. [CrossRef]
37. Garcia-Vallejo, M.C.M.I.; Sanz, J.; Bernabe, M.; Velasco-Negueruela, A. Necrodane (1, 2, 2, 3, 4-pentamethylcyclopentane) derivatives in Lavandula luisieri, new compounds to the plant kingdom. *Phytochemistry* **1994**, *36*, 43–45. [CrossRef]

38. Pombal, S.; Rodrigues, C.F.; Araújo, J.P.; Rocha, P.M.; Rodilla, J.M.; Diez, D.; Granja, Á.P.; Gomes, A.C.; Silva, L.A. Antibacterial and antioxidant activity of Portuguese *Lavandula luisieri* (Rozeira) Rivas-Martinez and its relation with their chemical composition. *Springerplus* **2016**, *5*, 1711. [CrossRef]
39. Kashima, Y.; Miyazawa, M. Chemical composition and aroma evaluation of essential oils from *Evolvulus alsinoides* L. *Chem. Biodivers.* **2014**, *11*, 396–407. [CrossRef]
40. Phillips, T.W.; West, J.R.; Foltz, J.L.; Silverstein, R.M.; Lanier, G.N. Aggregation pheromone of the deodar weevil, *Pissodes nemorensis* (Coleoptera: Curculionidae): Isolation and activity of grandisol and grandisal. *J. Chem. Ecol.* **1984**, *10*, 1417–1423. [CrossRef]
41. Senatore, F.; Napolitano, F.; Arnold, N.A.; Bruno, M.; Herz, V. Composition and antimicrobial activity of the essential oil of *Achillea falcata* L. (Asteraceae). *Flavour Fragr. J.* **2005**, *20*, 291–294. [CrossRef]
42. Pu, X.Z.; Lam, L.; Gehlken, K.; Ulappa, A.C.; Rachlow, J.L.; Forbey, J.S. Antioxidant capacity of Wyoming big sagebrush (*Artemisia tridentata* ssp. *wyomingensis*) varies spatially and is not related to the presence of a sagebrush dietary specialist. *West. N. Am. Nat.* **2015**, *75*, 78–87. [CrossRef]
43. Ortalli, M.; Varani, S.; Rosso, C.; Quintavalla, A.; Lombardo, M.; Trombini, C. Evaluation of synthetic substituted 1, 2-dioxanes as novel agents against human leishmaniasis. *Eur. J. Med. Chem.* **2019**, *170*, 126–140. [CrossRef]
44. Saryazdi, A.K.P.; Ghaffarifar, F.; Dalimi, A.; Dayer, M.S. In-vitro and in-vivo comparative effects of the spring and autumn-harvested *Artemisia aucheri* Bioss extracts on *Leishmania major*. *J. Enthopharmacol.* **2020**, *257*, 112910. [CrossRef]
45. Azizi, K.; Shahidi-Hakak, F.; Asgari, Q.; Hatam, G.R.; Fakoorziba, M.R.; Miri, R.; Moemenbellah-Fard, M.D. In vitro efficacy of ethanolic extract of *Artemisia absinthium* (Asteraceae) against *Leishmania major* L. using cell sensitivity and flow cytometry assays. *J. Parasit. Dis.* **2016**, *40*, 735–740. [CrossRef]
46. Monzote, L.; Piñón, A.; Scull, R.; Setzer, W.N. Chemistry and leishmanicidal activity of the essential oil from *Artemisia absinthium* from Cuba. *Nat. Prod. Commun.* **2014**, *9*, 1799–1804. [CrossRef] [PubMed]
47. Cannon, J.B.; Cantrell, C.L.; Astatkie, T.; Zheljazkov, V.D. Modification of yield and composition of essential oils by distillation time. *Ind. Crops Prod.* **2013**, *41*, 214–220. [CrossRef]
48. Zheljazkov, V.D.; Cantrell, C.L.; Astatkie, T.; Hristov, A. Yield, content, and composition of peppermint and spearmints as a function of harvesting time and drying. *J. Agric. Food Chem.* **2010**, *58*, 11400–11407. [CrossRef] [PubMed]
49. Zheljazkov, V.D.; Astatkie, T.; O'Brocki, B.; Jeliazkova, E. Essential oil composition and yield of anise from different distillation times. *HortScience* **2013**, *48*, 1393–1396. [CrossRef]
50. Canel, C.; Dayan, F.E.; Ganzera, M.; Khan, I.A.; Rimando, A.; Burandt, C.L., Jr.; Moraes, R.M. High yield of podophyllotoxin from leaves of *Podophyllum peltatum* by in situ conversion of podophyllotoxin 4-O-β-D-glucopyranoside. *Planta Med.* **2001**, *67*, 97–99. [CrossRef]
51. Cantrell, C.L.; Zheljazkov, V.D.; Osbrink, W.A.; Castro, A.; Maddox, V.; Craker, L.E.; Astatkie, T. Podophyllotoxin and essential oil profile of *Juniperus* and related species. *Ind. Crops Prod.* **2013**, *43*, 668–676. [CrossRef]
52. Huang, D.; Ou, B.; Hampsch-Woodill, M.; Flanagan, J.; Demmer, E.K. Development and validation of oxygen radical absorbance capacity assay for lipophilic antioxidants using randomly methylate B-cylodextrin as the solubility enhancer. *J. Agric. Food Chem.* **2002**, *50*, 1815–1821. [CrossRef]
53. Huang, D.; Ou, B.; Hampsch-Woodill, M.; Flanagan, J.; Prior, R. High-throughput assay of oxygen radical absorbance capacity (ORAC) using a multichannel liquid handling system coupled with a microplate fluorescence reader in 96-well format. *J. Agric. Food Chem.* **2002**, *50*, 4437–4444. [CrossRef]
54. SAS Institute Inc. *SAS/STAT®9.4 User's Guide*; SAS Institute Inc.: Cary, NC, USA, 2014.
55. Montgomery, D.C. *Design and Analysis of Experiments*, 10th ed.; Wiley: New York, NY, USA, 2020.

plants

MDPI

Article

Thymus vulgaris Essential Oil and Its Biological Activity

Lucia Galovičová [1,*], Petra Borotová [2,3], Veronika Valková [1,3], Nenad L. Vukovic [4], Milena Vukic [4], Jana Štefániková [3], Hana Ďúranová [3], Przemysław Łukasz Kowalczewski [5], Natália Čmiková [1] and Miroslava Kačániová [1,6,*]

1 Institute of Horticulture, Faculty of Horticulture and Landscape Engineering, Slovak University of Agriculture, Tr. A. Hlinku 2, 94976 Nitra, Slovakia; veronika.valkova@uniag.sk (V.V.); xcmikova@uniag.sk (N.Č.)
2 Institute of Applied Biology, Faculty of Biotechnology and Food Sciences, Slovak University of Agriculture, Tr. A. Hlinku 2, 94976 Nitra, Slovakia; petra.borotova@uniag.sk
3 AgroBioTech Research Centre, Slovak University of Agriculture, Tr. A. Hlinku 2, 94976 Nitra, Slovakia; jana.stefanikova@uniag.sk (J.Š.); hana.duranova@uniag.sk (H.Ď.)
4 Department of Chemistry, Faculty of Science, University of Kragujevac, 34000 Kragujevac, Serbia; nvchem@yahoo.com (N.L.V.); milena.vukic@pmf.kg.ac.rs (M.V.)
5 Department of Food Technology of Plant Origin, Poznań University of Life Sciences, 31 Wojska Polskiego St., 60-624 Poznań, Poland; przemyslaw.kowalczewski@up.poznan.pl
6 Department of Bioenergy, Food Technology and Microbiology, Institute of Food Technology and Nutrition, University of Rzeszow, 4 Zelwerowicza St, 35-601 Rzeszow, Poland
* Correspondence: l.galovicova95@gmail.com (L.G.); miroslava.kacaniova@gmail.com (M.K.)

Citation: Galovičová, L.; Borotová, P.; Valková, V.; Vukovic, N.L.; Vukic, M.; Štefániková, J.; Ďúranová, H.; Kowalczewski, P.Ł.; Čmiková, N.; Kačániová, M. *Thymus vulgaris* Essential Oil and Its Biological Activity. *Plants* **2021**, *10*, 1959. https://doi.org/10.3390/plants10091959

Academic Editors: Bistra Galunska, Ilian Badjakov and Ivayla N. Dincheva

Received: 30 August 2021
Accepted: 17 September 2021
Published: 19 September 2021

Abstract: *Thymus vulgaris* essential oil has potential good biological activity. The aim of the research was to evaluate the biological activity of the *T. vulgaris* essential oil from the Slovak company. The main components of *T. vulgaris* essential oil were thymol (48.1%), *p*-cymene (11.7%), 1,8-cineole (6.7), γ-terpinene (6.1%), and carvacrol (5.5%). The antioxidant activity was 85.2 ± 0.2%, which corresponds to 479.34 ± 1.1 TEAC. The antimicrobial activity was moderate or very strong with inhibition zones from 9.89 to 22.44 mm. The lowest values of MIC were determined against *B. subtilis*, *E. faecalis*, and *S. aureus*. In situ antifungal analysis on bread shows that the vapor phase of *T. vulgaris* essential oil can inhibit the growth of the microscopic filamentous fungi of the genus *Penicillium*. The antimicrobial activity against *S. marcescens* showed 46.78–87.80% inhibition at concentrations 62.5–500 μL/mL. The MALDI TOF MS analyses suggest changes in the protein profile of biofilm forming bacteria *P. fluorescens* and *S. enteritidis* after the fifth and the ninth day, respectively. Due to the properties of the *T. vulgaris* essential oil, it can be used in the food industry as a natural supplement to extend the shelf life of the foods.

Keywords: *Thymus vulgaris*; biofilm; DPPH; *P. fluorescens*; *S. enteritidis*

1. Introduction

Thymus vulgaris is a flowering plant of the Lamiaceae family. The origin of this plant is in southern Europe. It is an evergreen shrub with small, strongly aromatic, gray-green leaves and with purple or pink bunches of flowers [1]. Thyme was used in folk medicine for centuries for its significant antimicrobial and anti-inflammatory effects. It is often used in cooking as a seasoning [2]. *T. vulgaris* is characterized by chemical polymorphism according to the main volatile component. There are six known chemotypes of *T. vulgaris* essential oils: geraniol, linalool, α-terpineol, tujanol-4, thymol, and carvacrol. In the vast majority of plants and subsequent essential oils, one chemotype occurs, but in some, we can find two to three different chemotypes [3,4]. Due to the antimicrobial features of the main constituents, *T. vulgaris* essential oils are effective not only against bacteria and yeasts but also show inhibition activity against microscopic filamentous fungi [5,6].

Bacteria are very rarely found in planktonic lifeform as they are often exposed to an adverse environment. For higher resistance, bacteria begin to form a biofilm which is a

3D structure surrounded by an extracellular matrix of polysaccharides [7]. These bacterial forms are difficult to control and are highly resistant to antibiotics [8].

Salmonella enteritidis is a gram-negative bacterium (G^-) that causes gastrointestinal problems due to contamination of predominantly animal products such as meat and eggs [9]. Its cells can form biofilms and adhere to various surfaces that are in contact with food. The majority of the biofilms occur in food processing industries, on work surfaces such as stainless steel, glass, rubber, and polyurethane [10]. As a result, cross-contamination of products may occur. Due to biofilm formation, the bacteria are more resistant to commonly used disinfectants which increase the risk of the contamination [11].

Pseudomonas fluorescens is a gram-negative foodborne pathogen that causes the spoilage of foods with a high-water content. Compared to the other pathogens, it can grow even at low temperatures [12]. In the dairy industry, it is a source of contamination for milk and dairy products. *P. fluorescens* is a suitable and recognized model organism for the study of biofilms [13].

The most commonly used techniques for studying biofilms are microscopy-based methods, which are very limited. Mass spectrometry is a suitable option for the study of biofilms [14]. The MALDI-TOF MS Biotyper is a suitable and new method for the analysis of phenotypic differences in biofilm progression. Additionally, the changes in the properties of the growth related to the different surfaces was detectable [15]. To this date, only a few authors have followed the development and structure of biofilms using MALDI-TOF MS Biotyper [15–17].

Essential oils from medicinal plants were used since ancient times to treat certain diseases due to their antimicrobial effects. Currently, there is a trend of increasing antibiotic resistance of many microorganisms, and it is necessary to look for alternative compounds to eliminate them. Due to their antimicrobial effects, essential oils represent a promising potential as a natural alternative to antibiotics [18].

This study's aim was to analyze the biological activity of the essential oil from *T. vulgaris*. To determine the antioxidant, antimicrobial, antibiofilm activity, as well as the chemical composition, of the essential oil. Moreover, the evaluation of the changes in biofilm structure on glass and wood surfaces on microorganisms *Salmonella enteritidis* and *Pseudomonas fluorescens* using MALDI-TOF MS Biotyper was performed. We also focused on the antifungal effect of the vapor phase of essential oil on a food model.

2. Results

2.1. Chemical Composition of *T. vulgaris* Essential Oil

Using gas chromatography/mass spectrometry (GC/MS) and gas chromatography (GC-FID), we detected the main components of *T. vulgaris* essential oil is thymol 48.1%, *p*-cymene 11.7%, 1,8-cineole 6.7%, γ-terpinene 6.1%, and carvacrol 5.5% (Table 1). Based on the chemical composition, we classify the essential oil *T. vulgaris* into the thymol chemotype.

Table 1. Chemical composition of essential oil of *T. vulgaris*.

	ERI [a]	LRI [b]	Compound [c]	% [d]
1	926	930	α-thujene	0.5
2	938	939	α-pinene	2.1
3	948	954	camphene	0.7
4	977	975	sabinene	0.2
5	980	979	β-pinene	0.9
6	992	990	β-myrcene	1.0
7	993	991	3-octanol	tr
8	1004	1002	α-phellandrene	0.1
9	1009	1011	δ-3-carene	tr
10	1016	1017	α-terpinene	0.8

Table 1. *Cont.*

	ERI [a]	LRI [b]	Compound [c]	% [d]
11	1023	1024	*p*-cymene	11.7
12	1028	1029	α-limonene	1.3
13	1033	1031	1,8-cineole	6.7
14	1047	1050	(*E*)-β-ocimene	tr
15	1060	1059	γ-terpinene	6.1
16	1088	1088	α-terpinolene	0.3
17	1089	1086	trans-linalool oxide	tr
18	1098	1096	linalool	4.4
19	1148	1146	camphor	1.3
20	1151	1152	menthone	0.2
21	1170	1169	borneol	2.2
22	1178	1177	4-terpinenol	1.9
26	1189	1188	α-terpineol	0.5
27	1245	1244	carvacrol methyl ether	0.3
28	1255	1257	linalool acetate	tr
29	1256	1252	geraniol	tr
30	1286	1285	bornyl acetate	tr
31	1290	1290	thymol	48.1
32	1302	1299	carvacrol	5.5
33	1422	1419	(*E*)-caryophyllene	2.3
34	1507	1505	β-bisabolene	0.4
35	1525	1523	δ-cadinene	0.1
36	1583	1583	caryophyllene oxide	0.3
	total			99.7

[a] Experimental values of retention indices on HP-5MS column; [b] literature values of retention indices; [c] identified compounds; [d] the percentage of the identified compound; tr = compounds identified in amounts less than 0.1%.

2.2. Antioxidant, Antimicrobial Activity, and Minimal Inhibition Concentration (MIC)

The antioxidant activity of *T. vulgaris* essential oil was determined to be 85.2 ± 0.2% using the DPPH radical and expressed as the equivalent of the standard substance Trolox 479.34 ± 1.1 TEAC. The antimicrobial activity was evaluated by the disk diffusion test evaluating the effect against gram-positive bacteria (*Bacillus subtilis* CCM 1999, *Enterococcus faecalis* CCM 4224, *Staphylococcus aureus*,), gram-negative bacteria (*Pseudomonas aeruginosa* CCM 3955, *Yersinia enterocolitica* CCM 7204, *Salmonella enterica* subsp. *enterica* ser. Enteritidis CCM 4420, *Serratia marcescens* CCM 8588), yeast (*Candida krusei* CCM 8271, *Candida albicans* CCM 8261, *Candida tropicalis* CCM 8223, *Candida glabrata* CCM 8270), and biofilm-forming bacteria (*Salmonella enteritidis*, *Pseudomonas fluorescens*). In all tested microorganisms, including biofilm-forming bacteria, we observed moderate and very strong inhibitory activity except in *C. tropicalis*, which showed moderate inhibitory activity slightly below the threshold value of strong activity. MIC 50 and MIC 90 were determined by analysis of the minimum inhibitory concentrations. Low values of MIC 50 (12.12–16.56 µL/mL) and MIC 90 (16.43–19.26 µL/mL) were found in *B. subtilis*, *E. faecalis*, and *S. aureus*. The highest MIC 50 and MIC 90 values were determined for *S. enteritidis* biofilm. Moderate MIC 50 and MIC 90 values were determined for *S. enterica* subs. *enterica* ser. Enteritidis, *P. aeruginosa*, *Y. enterocolitica*, *C. albicans*, *C. krusei*, *C. tropicalis*, *C. glabrata*, *S. marcescens*, and *P. fluorescens* biofilm. Details of the results of antimicrobial activity and minimum inhibitory concentrations are given in Table 2.

Table 2. Antimicrobial activity of essential oil of *T. vulgaris*.

Microorganism	Zone Inhibition (mm)	Activity of EO	MIC 50 (μL/mL)	MIC 90 (μL/mL)	ATB
S. enterica subsp. *enterica* ser. *enteritidis*	17.00 ± 0.87	***	86.35	121.31	28.00 ± 0.06
P. aeruginosa	10.67 ± 0.87	**	103.28	169.19	25.00 ± 0.03
Y. enterocolitica	10.56 ± 1.67	**	64.49	71.59	24.00 ± 0.08
S. aureus	10.67 ± 1.00	**	16.56	19.26	24.00 ± 0.08
B. subtilis	11.33 ± 1.53	**	12.12	16.56	26.00 ± 0.05
E. faecalis	10.22 ± 1.30	**	13.85	16.43	25.00 ± 0.08
C. albicans	10.56 ± 1.13	**	121.56	159.26	26.00 ± 0.08
C. krusei	11.56 ± 1.67	**	165.46	183.21	24.00 ± 0.09
C. tropicalis	9.89 ± 1.27	*	135.38	164.43	25.00 ± 0.02
C. glabrata	12.22 ± 1.20	**	146.82	169.34	28.00 ± 0.04
S. marcescens	17,11 ± 1,27	***	84.27	136.41	22.00 ± 0.04
S. enteritidis biofilm	16.67 ± 1.22	***	274.37	311.56	25.00 ± 0.02
P. fluorescens biofilm	22.44 ± 1.33	***	97.78	108.82	24.00 ± 0.01

* Weak antimicrobial activity (zone 5–10 mm). ** Moderate inhibitory activity (zone 5–10 mm). *** Very strong inhibitory activity (zone > 15 mm), ATB—antibiotics, positive control (cefoxitin for G^-, gentamicin for G^+, fluconazole for yeast).

2.3. Analysis of Biofilm Developmental Phases and Evaluation of Molecular Differences on Different Surfaces Using MALDI-TOF MS Biotyper

The effect of *T. vulgaris* essential oil on the molecular structure and growth inhibition of *S. enteritidis* and *P. fluorescens* biofilms was evaluated using MALDI TOF MS Biotyper. The spectra of biofilms and planktonic cells in the control group developed identically and, therefore, the spectra of planktonic cells were used for greater clarity than the control spectrum. For each day, two experimental spectra from different surfaces (glass, wood) and a planktonic spectrum representing the development of the control group are shown.

Figure 1 shows the mass spectra of *S. enteritidis* biofilm during the individual days of the experimental evaluation.

The mass spectra obtained on the third day of the experiment represent very small differences in peaks between the experimental and control planktonic spectra (Figure 1). From day 5 of the experiment, we noticed differences between the two experimental spectra and the control spectrum. During 5 and 7 days, we noticed a more significant difference in the experimental group on wood. From day 9 to day 14, the differences were significant in both experimental groups compared to the control. Due to the influence of the essential oil *T. vulgaris*, we were able to observe changes in the protein spectrum of the biofilm. This finding suggests that the essential oil disrupts biofilm homeostasis leading to the degradation of this form of microorganism.

To visualize the similarity of mass spectra, a dendrogram was constructed based on MSP distances (Figure 2). The control groups and spectra of young biofilms (3 days) had the shortest distance together with planktonic cells. From day 5, we could see an increase in the distance of MSP experimental groups. During days 5 and 7, it can be seen that the experimental group on glass has a much shorter MSP distance than the experimental group on wood. This trend was maintained with a smaller difference throughout the duration of the experiment. The distance MSP of control groups from all tested days was significantly shorter than in the experimental groups. The increasing length of the MSP in experimental groups suggests changes in the protein profile of the bacterial biofilm of *S. enteritidis*.

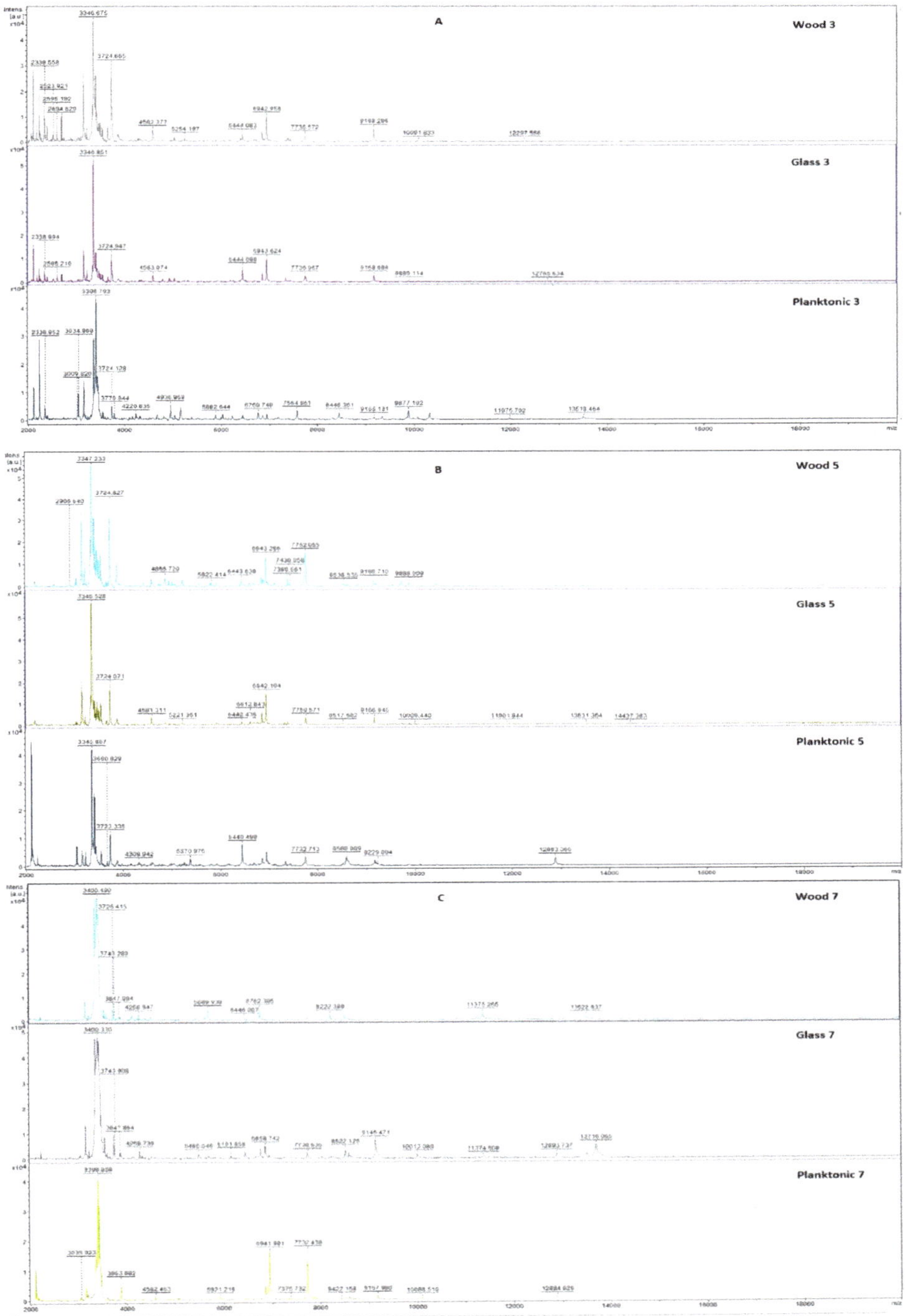

Figure 1. *Cont.*

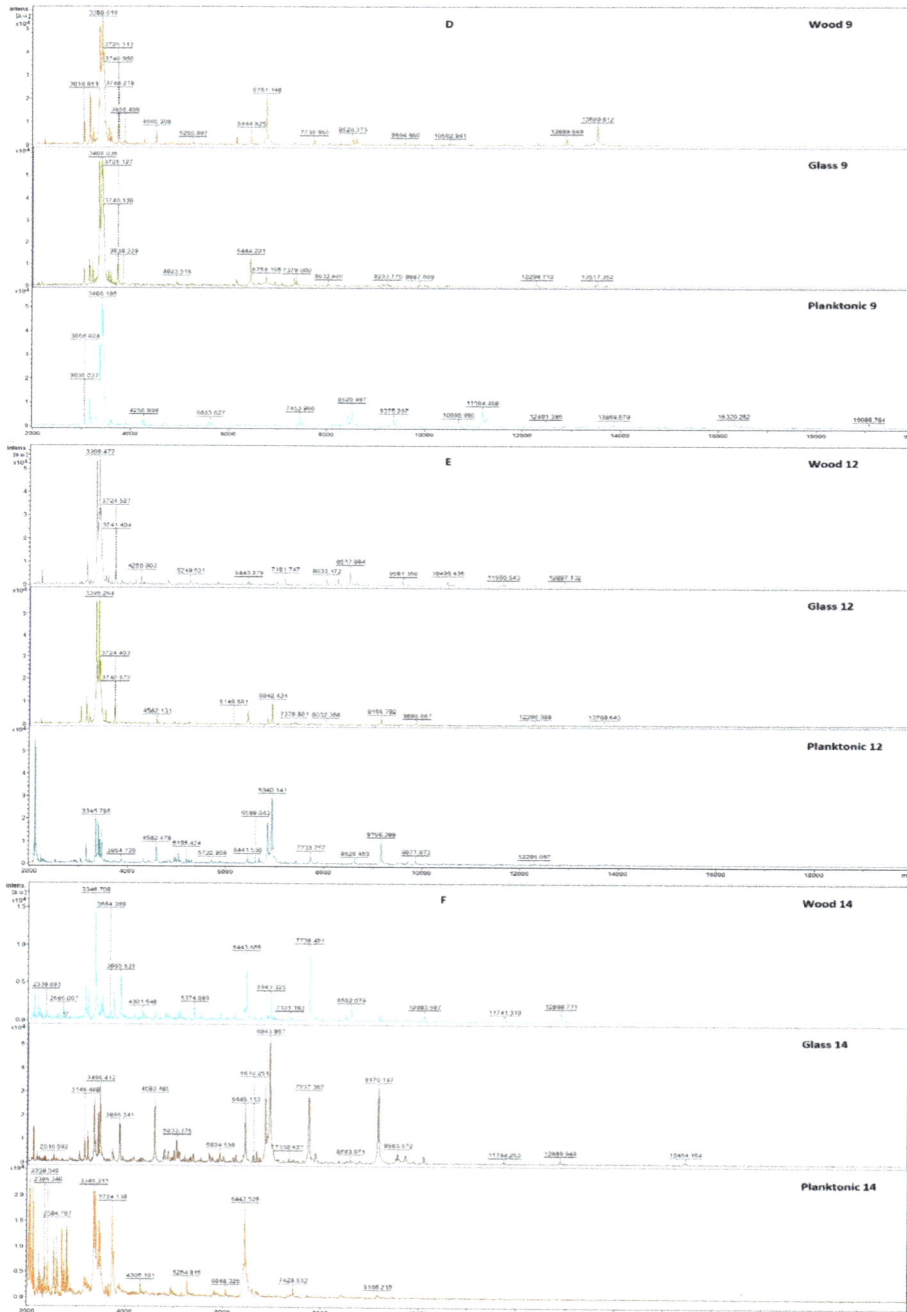

Figure 1. MALDI-TOF mass spectra of *S. enteritidis* biofilm during development after the addition of *T. vulgaris* EO: (**A**) 3rd day, (**B**) 5th day, (**C**) 7th day, (**D**) 9th day, (**E**) 12th day, and (**F**) 14th day.

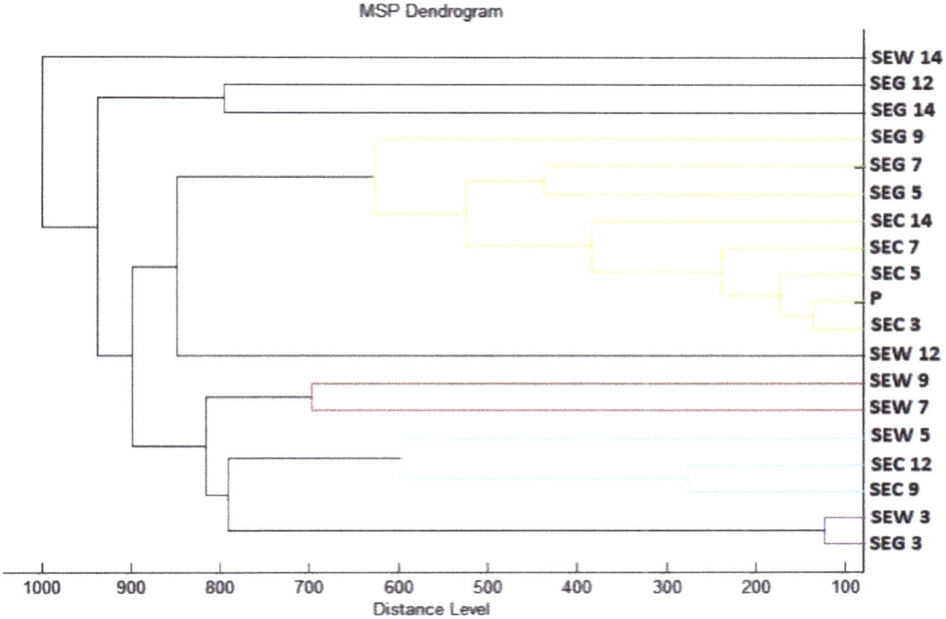

Figure 2. Dendrogram of *S. enteritidis* generated using MSPs of the planktonic cells and the control. SE, *S. enteritidis*; C, control; G, glass; W, wood; and P, planktonic cells.

Figure 3 shows the mass spectra of *P. fluorescens* biofilm during the individual days of the experimental evaluation.

In the first days of the experiment, the experimental groups of *P. fluorescens* biofilm with the addition of *T. vulgaris* essential oil had similar mass spectra as the control group (Figure 3). From the third to the seventh day of the experiment, there was no change in the protein spectrum of the experimental group due to the action of *T. vulgaris* essential oil. From day 7 of the experiment, we recorded lower peaks of the experimental group compared to the control group, which indicates the influence of the essential oil in the experimental group. From the ninth day, the appearance of the spectra of the experimental group changed more significantly. Due to the change in the protein spectrum of the experimental group compared to the control spectrum, we evaluate the effect of the essential oil as positive for disrupting the viability of the biofilm structure of *P. fluorescens* during the longer exposure.

To visualize the similarity of mass spectra, a dendrogram was constructed based on MSP distances. The control groups and spectra of biofilms of the third, fifth, and seventh day had the shortest distance together with planktonic cells (Figure 4). From day 9, we observed an increase in the distance of MSP experimental groups. During days 12 and 14, the MSP distance of the experimental group was the longest. These findings are evidence of the effect of *T. vulgaris* essential oil on altering the molecular structure of biofilms. Based on the constructed dendrogram, it can be stated that the MSP distance of the control groups was significantly shorter than the distance of the experimental groups from day 9, which confirms the inhibitory effect on the development of *P. fluorescens* biofilm at a longer exposure.

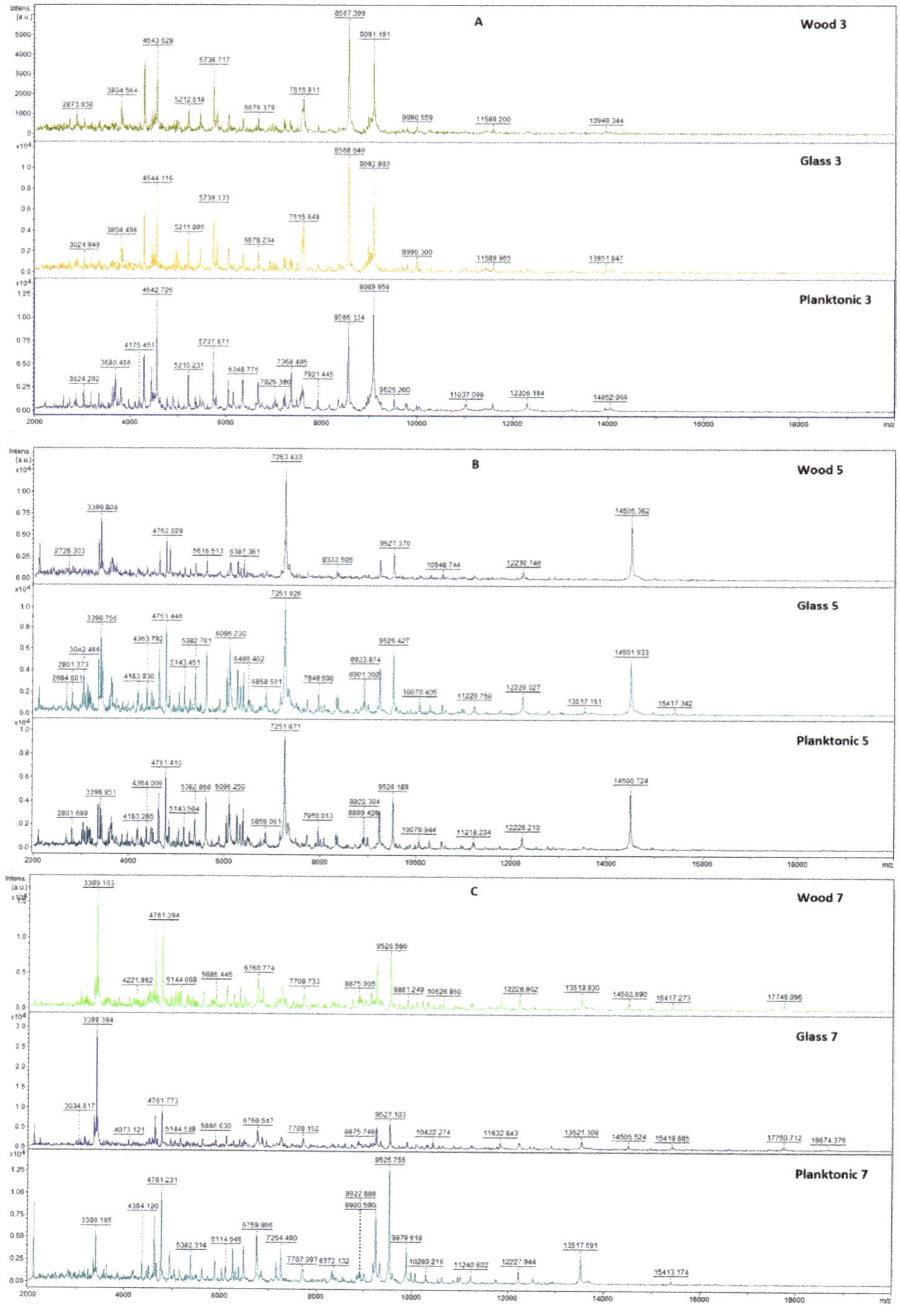

Figure 3. *Cont.*

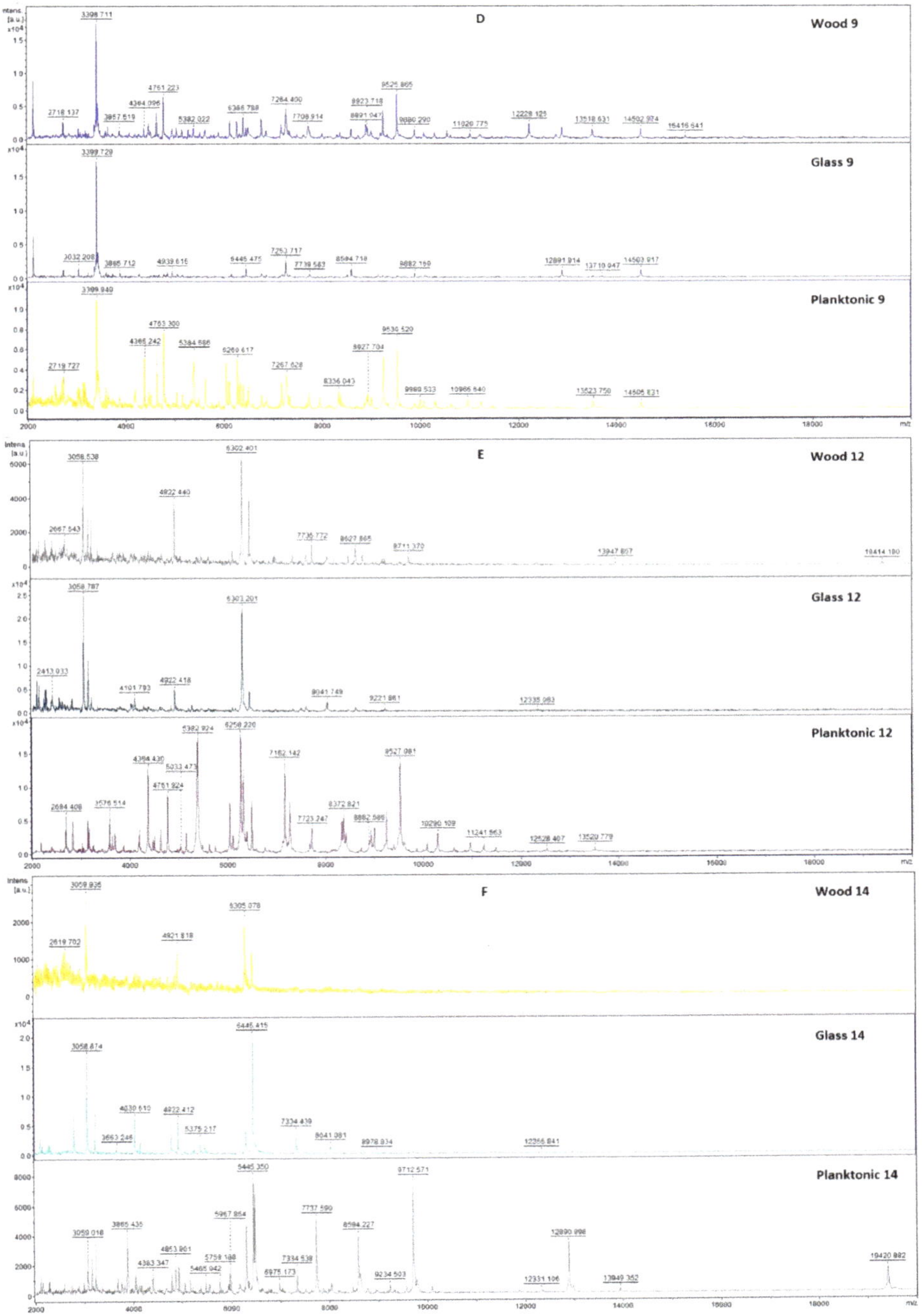

Figure 3. Representative MALDI-TOF mass spectra of *P. fluorescens*: (**A**) 3rd day, (**B**) 5th day, (**C**) 7th day, (**D**) 9th day, (**E**) 12th day, and (**F**) 14th day.

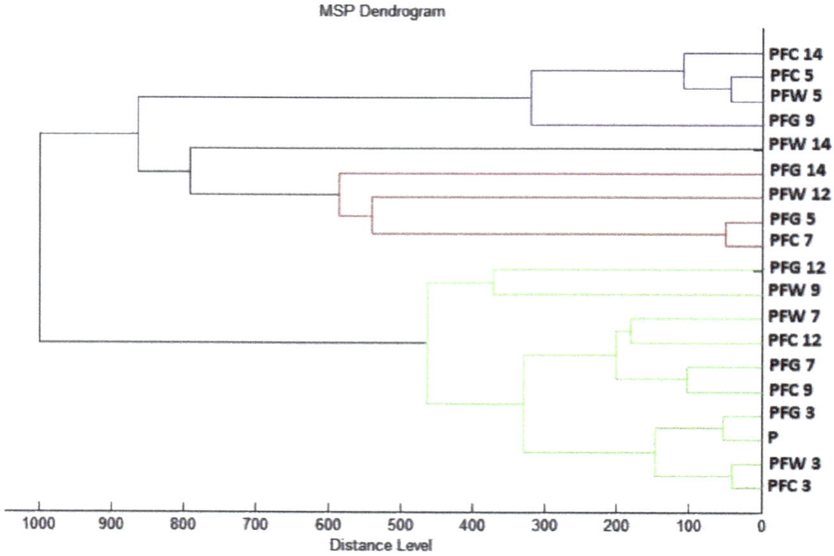

Figure 4. Dendrogram of *P. fluorescens* generated using MSPs of the planktonic cells and the control. PF, *P. fulorescens*; C, control; G, glass; W, wood; and P, planktonic cells.

2.4. Antimicrobial Analysis of Bread In Situ

In situ antimicrobial analysis on bread shows that microscopic filamentous fungi of the genus *Penicillium* were inhibited by all tested concentrations (Table 3). The highest percentage of inhibition at 62.5 μL/L was 82.17% for *P. commune*. The lowest rate of inhibition at this concentration was observed for *P. chrysogenum*. At the concentration of 125 μL/L, the inhibition rate was 80–89% for all microorganisms. At the concentration of 250 μL/L, the inhibition was 92–96%. At the highest tested concentration, the inhibition was 98–100%. For *P. glabrum*, a significance was noted between concentration 62.5 μL/L compared to 250 μL/L and 500 μL/L. For *P. chrysogenum*, differences were visible between 62.5 μL/L and the remaining concentrations. For *P. expansum*, no significant differences were observed between the tested concentrations. For *P. commune*, a significant difference was observed between concentrations of 62.5 μL/L and 500 μL/L. These findings suggest the inhibitory effects of *T. vulgaris* against the potentially pathogenic fungi.

Table 3. In situ analysis of the antibacterial activity of the vapor phase of *T. vulgaris* essential oil in bread.

	Mycelial Growth Inhibition [%]			
Concentration of EO / **Microorganisms**	**62.5 μL/L**	**125 μL/L**	**250 μL/L**	**500 μL/L**
P. glabrum	62.83 ± 10.81 [a]	80.65 ± 7.22 [ab]	94.05 ± 7.75 [b]	99.48 ± 0.74 [b]
P. chrysogenum	22.93 ± 15.53 [a]	86.07 ± 4.30 [b]	92.18 ± 8.82 [b]	98.25 ± 2.48 [b]
P. expansum	75.51 ± 14.95	80.30 ± 3.21	96.38 ± 2.82	100.00 ± 0.00
P. commune	82.17 ± 1.37 [a]	89.87 ± 4.51 [ab]	93.67 ± 5.58 [ab]	99.43 ± 0.11 [b]

Means ± standard deviation. Values followed by different superscript within the same row are significantly different ($p < 0.05$). The statistical differences ($p < 0.05$) between individual EO concentrations were as follows: *P. glabrum*: 62.5 μL/L vs. 250 μL/L and 500 μL/L; *P. chrysogenum*: 62.5 μL/L vs. 125 μL/L, 250 μL/L and 500 μL/L; *P. commune*: 62.5 μL/L vs. 500 μL/L.

2.5. In Situ Antimicrobial Analysis on Carrots

From in situ analysis on carrots, the effect of the vapor phase of *T. serpyllum* essential oil against *S. marcescens* was recorded at all tested concentrations (Table 4). The highest inhibition was observed at a concentration of 500 µL/L by 87.8%. The lowest inhibition rate was at a concentration of 125 µL/L by 46.78%. The vapor phase of EO from *T. vulgaris* has the inhibitory effect on the growth of the bacteria on the food model.

Table 4. Results of in situ analysis of antibacterial activity of the vapor phase of *T. vulgaris* essential oil on carrots.

	Bacterial Growth Inhibition [%]			
Concentration of EO / **Microorganisms**	**62.5 µL/L**	**125 µL/L**	**250 µL/L**	**500 µL/L**
S. marcescens	57.35 ± 1.43 [a]	46.78 ± 2.59 [b]	69.82 ± 3.23 [c]	87.80 ± 1.41 [d]

Means ± standard deviation. Values followed by different superscript within the same row are significantly different ($p < 0.05$). The statistical differences ($p < 0.05$) between individual EO concentrations were as follows: 62.5 µL/L vs. 125 µL/L, 250 µL/L and 500 µL/L; 125 µL/L vs. 250 µL/L and 500 µL/L; 250 µL/L vs. 500 µL/L.

3. Discussion

Plants, and subsequently essential oils of *T. vulgaris*, can have different chemotypes. Using chemical composition analysis, we verified that the essential oil we tested belongs to the thymol chemotype. Al-Asmari et al. [19] analyzed the essential oil of *T. vulgaris* according to its chemical composition was also the thymol type but as other main components they determined furan 12.19% and *p*-cymene 2.78%. Micucci et al. [20] determined essential oil as thymol-type, and as other major components carvacrol and *p*-cymene was observed. Thymol and carvacrol are terpenoids that commonly occur as major components of essential oils [21]. They are recognized as safe in food by the Food and Drug Administration [22]. Thymol and carvacrol are very effective against foodborne pathogens such as *Salmonella* spp. and *Staphylococcus aureus* that can produce biofilms by which they adhere to various surfaces and thus endanger food operations [23,24].

For the analysis of antioxidant activity of essential oil *T. vulgaris*, we used DPPH radical dissolved in methanol. The percentage inhibition of DPPH free radical for *T. vulgaris* essential oil was determined to be 85.2 ± 0.2%, which corresponds to 479.34 ± 1.1 TEAC. Bistgani et al. [25] determined the antioxidant activity of the methanol extract of *T. vulgaris* 69.7 ± 4.8%. Punya et al. [26] set the DPPH radical scavenging activity at 78.73%. Kulisic et al. [27] set the percentage inhibition at 91.30 ± 0.30%. Although the method of DPPH free radical scavenging is not standardized and each of these authors performed it with modifications, our results agree that *T. vulgaris* shows strong antioxidant activity.

Using the disk diffusion method, we found moderate to very strong antimicrobial activity in all tested microorganisms, including biofilm-forming bacteria. For *C. tropicalis*, we found a borderline moderate inhibitory activity. Al Maqtari [28] evaluated the effect of *T. vulgaris* against *B. subtilis*, *S. aureus*, *P. aeruginosa*, and *C. albicans* by disk diffusion method and recorded a zone of inhibition more than 20 mm in all tested microorganisms, which is considered to be a very strong inhibition. Borugă et al. [29] evaluated the effect against *S. aureus*, *E. faecalis*, *C. albicans*, *S. typhimurium*, and *P. aeruginosa* and recorded zones of inhibition more than 13 mm when 10 µL per disc was applied. They rated the effect of the essential oil as very strong. Rota et al. [30] recorded zones of inhibition more than 20 mm in the tested microorganisms and evaluated the essential oil of *T. vulgaris* as very effective. Boukhatem et al. [31] reported a marked inhibition of the genus *Candida* by *T. vulgaris* essential oil. Our findings on the strong inhibitory activity of *T. vulgaris* essential oil are consistent with the authors work above.

Nikolić et al. [32] evaluated the effect of *T. vulgaris* against genus *Candida* strains where they found MIC 80–160 µL/mL and MBC 160–320 µL/mL. Jafri and Ahmad [33] evaluated the effect of *T. vulgaris* essential oil on the elimination of biofilm produced by the genus *Candida*. They found that the essential oil has lower MIC values than the tested antibiotics.

Al-Shuneigat et al. [34] found in their study that *T. vulgaris* essential oil showed strong antimicrobial and antibiofilm effects MIC values were 0.0625–2% *v*/*v*.

Myszka et al. [35] evaluated the effect of *T. vulgaris* against *P. fluorescens* biofilm and found that the essential oil can degrade the biofilm formed on stainless steel. Čabarkapa et al. [36] evaluated the effect of several essential oils, including *T. vulgaris*, against *S. enteritidis* biofilm. All tested essential oils showed inhibition of *S. enteritidis* biofilm formation at subminimum concentrations of essential oils. Al-Shuneigat et al. [34] evaluated the effect of *T. vulgaris* against *P. aeruginosa* biofilm and found that a very small amount is sufficient to eliminate both biofilms and planktonic cells. De Oliveira et al. [37] stated that *T. vulgaris* essential oil has the potential to be used as a biofilm control agent. MALDI-TOF MS Biotyper was used in only a small number of publications. Li et al. [12] used this method to analyze *B. subtilis* biofilm and to determine the spatial distribution of specific peptides and lipopeptides that are produced in biofilms. Kubesová et al. [38] used MALDI-TOF MS to analyze the biofilm produced by the genus *Candida*. Rams et al. [39] found that the phenotypic identification of culturable *P. gingivalis* biofilms is 100% accurate using MALDI-TOF MS because of the differences in the protein profile. The changes in the mass spectra profile were demonstrated by MALDI-TOF in Kačániová et al. [40,41] where the inhibitory effects on the biofilm of *Coriandrum sativum* and *Citrus aurantium* EOs was detected. Use of MALDI-TOF can be a fast and easy method for the assessment of the biofilm growth and degradation due to the structural and molecular changes. Kloucek et al. [42] found out that vapor phase essential oils represent a suitable alternative to antimicrobials in the food industry due to the need of lower concentrations of EO than during use of contact effect of the liquid phase. Reyes-Jurado et al. [43] evaluated the effect of the vapor phase of *T. vulgaris* essential oil against filamentous microscopic fungi. Based on the antimicrobial activity of the vapor phase, they found that the essential oil has the potential to be used to protect packaged foods. Mani López et al. [44] found that vapor phase essential oils inhibited the growth of microscopic filamentous fungi of the genus *Penicillium* on bread and it would be appropriate to verify the effect on sensory properties.

4. Materials and Methods

4.1. Essential Oil

T. vulgaris EO of thymol chemotype was obtained from Hanus, s.r.o. (Nitra, Slovakia). The EO was prepared by steam distillation of the partially dried stalk.

4.2. Chemical Characterization of *T. vulgaris* EO by Gas Chromatography/Mass Spectrometry (GC/MS) and Gas Chromatography (GC-FID)

GC/MS analysis of *T. vulgaris* EO was performed using the Agilent 6890N gas chromatograph (Agilent Technologies, Santa Clara, CA, USA) coupled to a quadrupole mass spectrometer 5975B (Agilent Technologies, Santa Clara, CA, USA). A HP-5MS capillary column (30 m × 0.25 mm × 0.25 µm). The temperature program was set from 60 °C to 150 °C (increasing rate 3 °C/min) and from 150 °C to 280 °C (increasing rate 5 °C/min). The total run time of the program was 60 min. Helium 5.0 was used as the carrier gas with a flow rate of 1 mL/min. The injection volume was 1 µL (EO sample was diluted in pentane), while the split/splitless injector temperature was set at 280 °C. The investigated sample was injected in the split mode with split ratio at 40.8:1. Electron-impact mass spectrometric data (EI-MS; 70 eV) were acquired in scan mode over the m/z range 35–550. MS ion source and MS quadrupole temperatures were 230 °C and 150 °C, respectively. Acquisition of data started after 3 min of solvent delay time. GC-FID analyses were performed on the Agilent 6890N gas chromatograph coupled to a FID detector. Column (HP-5MS) and chromatographic conditions were the same as for GC-MS. The temperature of the FID detector was set at 300 °C.

The individual volatile constituents of *T. vulgaris* EO sample were identified according to their retention indices [45] and they were compared with the reference spectra (Wiley and NIST databases). The retention indices were determined experimentally by a standard method that included retention times of n-alkanes (C6-C34) injected under the same chromatographic conditions [46]. The percentages of the identified compounds (amounts higher than 0.1%) were derived from their GC peak areas.

4.3. Determination of Antioxidant Activity

The radical scavenging of 2,2-diphenyl-1-picrylhydrazyl (DPPH, Sigma Aldrich, Germany) was used to measure the antioxidant activity of *T. vulgaris* EO. The solution of DPPH (0.025 g/L dissolved in methanol) was adjusted to absorbance 0.7 at wavelength 515 nm. 5 μL of EO sample was added to 195 μL DPPH solution in 96 well microplate and the reaction solution was incubated for 30 min in the dark with continuous shaking at 1000 rpm. The antioxidant activity was expressed as the percentage of DPPH inhibition and was calculated according to the formula $(A0 - AA)/A0 \times 100$, where A0 was absorbance of DPPH and AA was absorbance of the sample.

Antioxidant activity was calculated in relation to standard reference Trolox (Sigma Aldrich, Schnelldorf, Germany) dissolved in methanol (Uvasol® for spectroscopy, Merck, Darmstadt, Germany) to concentration range 0–100 μg/mL. Total antioxidant activity was expressed according to calibration curve as 1 μg of Trolox to 1 mL of the EO sample (TEAC).

4.4. Microorganisms

Gram-positive bacteria (*Bacillus subtilis* CCM 1999, *Staphylococcus aureus* subsp. aureus CCM 8223, *Enterococcus faecalis* CCM 4224,), gram-negative bacteria (*Pseudomonas aeruginosa* CCM 3955, *Yersinia enterocolitica* CCM 7204, *Salmonella enterica* subsp. *enterica* ser. Enteritidis CCM 4420, *Serratia marcescens* CCM 8588), and yeasts (*Candida krusei* CCM 8271, *Candida albicans* CCM 8261, *Candida tropicalis* CCM 8223, *Candida glabrata* CCM 8270) were obtained from the Czech collection of microorganisms (Brno, Czech Republic). The biofilm-forming bacterial strain *Pseudomonas fluorescens* was obtained from the fish and *Salmonella enteritidis* was obtained from the sample of meat. Bacteria were identified with 16S rRNA sequencing and MALDI-TOF MS Biotyper. A fungi *P. glabrum*, *P. chrysogenum*, *P. expansum*, and *P. commune* were obtained from grape samples, and were identified by 16S rRNA sequencing and MALDI-TOF MS Biotyper.

4.5. Determination of Antimicrobial Activity

Antimicrobial activity of *T. vulgaris* EO was determined by the disc diffusion method. The microbial inoculum was cultivated for 24 h on tryptone soya agar (TSA, Oxoid, Basingstoke, UK) at 37 °C for bacteria and sabouraud dextrose agar (SDA, Oxoid, Basingstoke, UK) at 25 °C for yeasts. The inoculum was adjusted to optical density 0.5 McFarland standard (1.5×10^8 CFU/mL) and 100 μL was added on plates with Mueller Hinton agar (MHA, Oxoid, Basingstoke, UK). Sterile 6 mm discs were saturated with 10 μL of *T. vulgaris* EO and placed on the layer of agar with microbial suspension. Samples were incubated for 24 h at 37 °C for bacteria and 25 °C for yeasts. Two antibiotics (cefoxitin, gentamicin; Oxoid, Basingstoke, UK), and one antifungal (fluconazole; Oxoid, Basingstoke, UK) were used as positive controls for gram-negative, gram-positive bacteria and yeasts, respectively. Disks impregnated with 0.1% DMSO (dimethylsulfoxid, Centralchem, Bratislava, SK) served as the negative control.

An inhibition zone above 15 mm was determined as very strong antimicrobial activity, the inhibition zone above 10 mm was determined as moderate activity, and inhibition zone above 5 mm was determined as weak activity. Antimicrobial activity was measured in triplicate.

4.6. Minimum Inhibitory Concentration (MIC)

Microbial inoculum was cultivated for 24 h in Mueller Hinton broth (MHB, Oxoid, Basingstoke, UK) at 37 °C for bacteria and sabouraud dextrose broth (SDB, Oxoid, Basingstoke, UK) at 25 °C for yeasts. A total of 50 μL of inoculum with optical density 0.5 of the McFarland standard was added to a 96-well microtiter plate. Subsequently, the *T.*

vulgaris EO was prepared by serial dilution to the concentration range of 400 μL/mL to 0.2 μL/mL in MHB/SDB and 100 μL of suspension was thoroughly mixed with bacterial inoculum in wells. Bacterial samples were incubated for 24 h at 37 °C. Yeast samples were incubated for 24h at 25 °C. MHB/SDB with EO was used as a negative control and MHB/SDB with inoculum was used as positive control of the maximal growth.

For non-adherent microorganisms, the absorbance was measured after the incubation period at 570 nm by Glomax spectrophotometer (Promega Inc., Madison, WI, USA). The MIC of biofilm-forming bacteria was measured with the use of crystal violet. The suspension with non-attached cells was discarded, the wells were washed with distilled water three times, and left to dry at room temperature. A total of 200 μL of 0.1% (*w*/*v*) crystal violet was added to the wells and samples were incubated for 15 min. Subsequently, the wells were repeatedly washed and dried. Stained biofilms were resolubilized with 200 μL of 33% acetic acid [47]. Absorbance was measured at 570 nm. The concentration of EO which absorbance was lower than the absorbance of the maximal growth control was determined as the minimum inhibitory concentration. The test was prepared in triplicate.

4.7. Analysis of Differences in Biofilm Development with MALDI-TOF MS Biotyper

The changes of protein spectra during biofilm development after *T. vulgaris* EO addition were evaluated by MALDI-TOF MS Biotyper. The biofilm forming bacteria were added to 50 mL polypropylene tubes with 20 mL of MHB; subsequently, a wooden toothpick and a glass slide were added as a model of different surfaces. The experimental groups were treated with 0.1% *T. vulgaris* EO, and control group samples were left untreated. The samples were incubated at 37 °C on shaker with 170 rpm.

The samples were analyzed after 3, 5, 7, 9, 12, and 14 days. The biofilm samples were taken from a glass slide and wooden toothpick with a sterile cotton swab and were imprinted onto a MALDI-TOF metal target plate. The planktonic cells were obtained from 300 μL of culture medium, cells were centrifuged for 1 min at 12,000 rpm, and the supernatant was discarded. The pellet was three times resuspended in 30 μL of ultrapure water and the suspension was centrifuged for 1 min at 12,000 rpm. In the last step, 1 μL of planktonic cells suspension was applied to a target plate.

The target plate was dried and 1 μL of α-Cyano-4-hydroxycinnamic acid matrix (10 mg/mL) was applied. The samples were processed with MALDI-TOF MicroFlex (Bruker Daltonics) linear and positive mode for the range of m/z 200–2000 after crystallization. The spectra were obtained by an automatic analysis and the same sample similarities were used to generate the standard global spectrum (MSP) and the 19 MSP was generated from the spectra by MALDI Biotyper 3.0 and were grouped into dendrograms using Euclidean distance [40].

4.8. Antimicrobial Analysis In Situ on a Food Model

The antifungal effect of the *T. vulgaris* EO vapor phase was evaluated in 0.5 L sterile glass jars (Bormioli Rocco, Italy) on a bread used as a food model. The fungi of *Penicillium* genus were cultivated for 5 days on sabouraud dextrose agar (SDA; Oxoid, Basingstoke, UK) at 25 °C. The cultures were applied to the bread slices (15 × 15 × 1.5 cm) by three stabs. A 6 cm sterile filter paper was placed to the jar lid and 100 μL of *T. vulgaris* EO (62.5, 125, 250, and 500 μL/L diluted in ethyl acetate) were applied. The control group was left untreated. The jars were hermetically sealed and were incubated in the dark for 14 days at 25 °C ± 1 °C.

In situ antibacterial analysis in the vapor phase was tested on *S. marcescens*. Warm MHA was poured into 60 mm Petri dishes (PD) and the lid. Sliced carrots (0.5 mm) were placed on agar. Then, an inoculum was prepared as previously described. *T. vulgaris* EO was diluted twice in ethyl acetate to 500, 250, 125, and 62.5 μL/L and used for sterile filter paper inoculation. The filter paper was placed in for 1 min to evaporate the remaining ethyl acetate, sealed, and incubated at 37 °C for 7 days.

An inhibition of the fungal growth was evaluated by stereological methods. A volume density (Vv) of fungi was estimated using ImageJ software. The stereological grid points of the colonies (P) and substrate (p) were counted. The density of fungal growth was calculated in % according to the formula Vv = P/p × 100. The antifungal activity of EO was expressed as mycelial growth inhibition in % (MGI): MGI = [(C − T)/C] × 100, where C was the density of the fungal growth in the control group and T was the density of the fungal growth in the treatment group [48,49].

In situ bacterial growth was determined using stereological methods. In this concept, the volume density (Vv) of bacterial colonies was firstly estimated using ImageJ software and counting the points of the stereological grid hitting the colonies (P), and those (p) falling to the reference space (growth substrate used). The volume density of bacterial colonies was consequently calculated as follows: Vv (%) = P/p. The antibacterial activity of EO was defined as the percentage of bacterial growth inhibition (BGI) BGI = [(C − T)/C] × 100, where C and T were bacterial growth (expressed as Vv) in the control group and the treatment group, respectively. The negative results represented the growth stimulation.

4.9. Statistical Data Evaluation

There was SAS® software version 8 used for data processing. The results of the MIC value (concentration that caused 50% and 90% inhibition in bacterial growth) were determined by logit analysis.

5. Conclusions

The main components of *T. vulgaris* essential oil were thymol 48.1%, *p*-cymene 11.7%, 1,8-cineole 6.7%, γ-terpinene 6.1%, and carvacrol 5.5%. The antioxidant activity of the essential oil was 85.2 ± 0.2%, which corresponds to 479.34 ± 1.1 TEAC. We rate this antioxidant activity as high. *T. vulgaris* essential oil had very good antimicrobial effects as well as antibiofilm effects, as observed on various surfaces and detected by MALDI-TOF MS Biotyper. In the test of the antimicrobial activity of the vapor phase of the essential oil, very good antifungal effects against the genus *Penicillium* were observed on the model food (bread). Good effects on inhibition of carrots were also observed in the vapor phase test against *S. marcescens*. There is a possibility of future use of *T. vulgaris* essential oil in extending the shelf life of bakery products and it could find application in the storage of root vegetables. In the case of essential oils with a dominant proportion of volatile components, a stronger effect of the vapor phase is often observed compared to contact application. The vapor phase has a weaker effect on sensory properties than the contact phase. In the future, it would be necessary to evaluate the influence of essential oil in the vapor phase on the sensory properties of model foods.

Author Contributions: Conceptualization, L.G. and M.K.; Data curation, L.G., P.B., N.L.V., M.V., J.Š., H.Ď., V.V., N.Č., and M.K.; Methodology, L.G., P.B., N.L.V., M.V., J.Š., H.Ď., V.V., and M.K.; Supervision, and M.K.; Writing—original draft, L.G., P.B., P.Ł.K., and M.K. All authors have read and agreed to the published version of the manuscript.

Funding: This work was supported by the grant APVV-20-0058 "The potential of the essential oils from aromatic plants for medical use and food preservation."

Institutional Review Board Statement: Not applicable.

Informed Consent Statement: Not applicable.

Data Availability Statement: Data is contained within the article.

Acknowledgments: This work has been supported by the grants of the VEGA no. 1/0180/20.

Conflicts of Interest: The authors declare no conflict of interest.

References

1. Hosseinzadeh, S.; Jafarikukhdan, A.; Hosseini, A.; Armand, R. The Application of Medicinal Plants in Traditional and Modern Medicine: A Review of *Thymus vulgaris*. *Int. J. Clin. Med.* **2015**, *06*, 635. [CrossRef]
2. Kuete, V. Chapter 28—*Thymus vulgaris*. In *Medicinal Spices and Vegetables from Africa*; Kuete, V., Ed.; Academic Press: Cambridge, MA, USA, 2017; pp. 599–609. ISBN 978-0-12-809286-6.
3. György, Z.; Incze, N.; Pluhár, Z. Differentiating *Thymus vulgaris* Chemotypes with ISSR Molecular Markers. *Biochem. Syst. Ecol.* **2020**, *92*, 104118. [CrossRef]
4. Thompson, J.D.; Chalchat, J.-C.; Michet, A.; Linhart, Y.B.; Ehlers, B. Qualitative and Quantitative Variation in Monoterpene Co-Occurrence and Composition in the Essential Oil of *Thymus vulgaris* Chemotypes. *J. Chem. Ecol.* **2003**, *29*, 859–880. [CrossRef]
5. Mahboubi, M.; Heidarytabar, R.; Mahdizadeh, E.; Hosseini, H. Antimicrobial Activity and Chemical Composition of *Thymus* Species and Zataria Multiflora Essential Oils. *Agric. Nat. Resour.* **2017**, *51*, 395–401. [CrossRef]
6. Segvić Klarić, M.; Kosalec, I.; Mastelić, J.; Piecková, E.; Pepeljnak, S. Antifungal Activity of Thyme (*Thymus vulgaris* L.) Essential Oil and Thymol against Moulds from Damp Dwellings. *Lett. Appl. Microbiol.* **2007**, *44*, 36–42. [CrossRef]
7. Miquel, S.; Lagrafeuille, R.; Souweine, B.; Forestier, C. Anti-Biofilm Activity as a Health Issue. *Front. Microbiol.* **2016**, *7*, 592. [CrossRef] [PubMed]
8. Koo, H.; Allan, R.N.; Howlin, R.P.; Stoodley, P.; Hall-Stoodley, L. Targeting Microbial Biofilms: Current and Prospective Therapeutic Strategies. *Nat. Rev. Microbiol.* **2017**, *15*, 740–755. [CrossRef] [PubMed]
9. Galanis, E.; Wong, D.M.A.L.F.; Patrick, M.E.; Binsztein, N.; Cieslik, A.; Chalermchaikit, T.; Aidara-Kane, A.; Ellis, A.; Angulo, F.J.; Wegener, H.C. Web-Based Surveillance and Global *Salmonella* Distribution, 2000–2002. *Emerg. Infect. Dis.* **2006**, *12*, 381–388. [CrossRef] [PubMed]
10. Veluz, G.A.; Pitchiah, S.; Alvarado, C.Z. Attachment Of *Salmonella* Serovars And *Listeria monocytogenes* to Stainless Steel and Plastic Conveyor Belts. *Poult. Sci. J.* **2012**, *91*, 2004–2010. [CrossRef] [PubMed]
11. Joseph, B.; Otta, S.K.; Karunasagar, I.; Karunasagar, I. Biofilm Formation by *Salmonella* Spp. on Food Contact Surfaces and Their Sensitivity to Sanitizers. *J. Food Microbiol.* **2001**, *64*, 367–372. [CrossRef]
12. Li, T.; Wang, D.; Liu, N.; Ma, Y.; Ding, T.; Mei, Y.; Li, J. Inhibition of Quorum Sensing-Controlled Virulence Factors and Biofilm Formation in *Pseudomonas Fluorescens* by Cinnamaldehyde. *Int. J. Food Microbiol.* **2018**, *269*, 98–106. [CrossRef]
13. Sillankorva, S.; Neubauer, P.; Azeredo, J. *Pseudomonas Fluorescens* Biofilms Subjected to Phage PhiIBB-PF7A. *BMC Biotechnol.* **2008**, *8*, 79. [CrossRef]
14. Floyd, K.A.; Moore, J.L.; Eberly, A.R.; Good, J.A.D.; Shaffer, C.L.; Zaver, H.; Almqvist, F.; Skaar, E.P.; Caprioli, R.M.; Hadjifrangiskou, M. Adhesive Fiber Stratification in Uropathogenic *Escherichia coli* Biofilms Unveils Oxygen-Mediated Control of Type 1 Pili. *PLoS Pathog.* **2015**, *11*, e1004697. [CrossRef]
15. Pereira, F.D.E.S.; Bonatto, C.C.; Lopes, C.A.P.; Pereira, A.L.; Silva, L.P. Use of MALDI-TOF Mass Spectrometry to Analyze the Molecular Profile of *Pseudomonas aeruginosa* Biofilms Grown on Glass and Plastic Surfaces. *Microb. Pathog.* **2015**, *86*, 32–37. [CrossRef]
16. Benagli, C.; Rossi, V.; Dolina, M.; Tonolla, M.; Petrini, O. Matrix-Assisted Laser Desorption Ionization-Time of Flight Mass Spectrometry for the Identification of Clinically Relevant Bacteria. *PLoS ONE* **2011**, *6*, e16424. [CrossRef] [PubMed]
17. Ferreira, L.; Sánchez-Juanes, F.; García-Fraile, P.; Rivas, R.; Mateos, P.F.; Martínez-Molina, E.; González-Buitrago, J.M.; Velázquez, E. MALDI-TOF Mass Spectrometry Is a Fast and Reliable Platform for Identification and Ecological Studies of Species from Family Rhizobiaceae. *PLoS ONE* **2011**, *6*, e20223. [CrossRef] [PubMed]
18. Wińska, K.; Mączka, W.; Łyczko, J.; Grabarczyk, M.; Czubaszek, A.; Szumny, A. Essential Oils as Antimicrobial Agents—Myth or Real Alternative? *Molecules* **2019**, *24*, 2130. [CrossRef] [PubMed]
19. Al-Asmari, A.K.; Athar, M.T.; Al-Faraidy, A.A.; Almuhaiza, M.S. Chemical Composition of Essential Oil of *Thymus vulgaris* Collected from Saudi Arabian Market. *Asian Pac. J. Trop. Biomed.* **2017**, *7*, 147–150. [CrossRef]
20. Micucci, M.; Protti, M.; Aldini, R.; Frosini, M.; Corazza, I.; Marzetti, C.; Mattioli, L.B.; Tocci, G.; Chiarini, A.; Mercolini, L.; et al. *Thymus vulgaris* L. Essential Oil Solid Formulation: Chemical Profile and Spasmolytic and Antimicrobial Effects. *Biomolecules* **2020**, *10*, 860. [CrossRef]
21. Hyldgaard, M.; Mygind, T.; Meyer, R.L. Essential Oils in Food Preservation: Mode of Action, Synergies, and Interactions with Food Matrix Components. *Front. Microbiol.* **2012**, *3*, 12. [CrossRef]
22. Igoe, R.S.; Hui, Y.H. Substances for use in foods: Listing under Title 21 of the Code of Federal Regulations. In *Dictionary of Food Ingredients*; Igoe, R.S., Hui, Y.H., Eds.; Springer US: Boston, MA, USA, 1996; pp. 159–185. ISBN 978-1-4615-6838-4.
23. Casarin, L.S.; de Casarin, F.O.; Brandelli, A.; Novello, J.; Ferreira, S.O.; Tondo, E.C. Influence of Free Energy on the Attachment of *Salmonella enteritidis* and *Listeria monocytogenes* on Stainless Steels AISI 304 and AISI 316. *LWT-Food Sci. Technol.* **2016**, *69*, 131–138. [CrossRef]
24. Kostaki, M.; Chorianopoulos, N.; Braxou, E.; Nychas, G.-J.; Giaouris, E. Differential Biofilm Formation and Chemical Disinfection Resistance of Sessile Cells of *Listeria monocytogenes* Strains under Monospecies and Dual-Species (with *Salmonella Enterica*) Conditions. *Appl. Environ. Microbiol.* **2012**, *78*, 2586–2595. [CrossRef] [PubMed]
25. Bistgani, Z.E.; Hashemi, M.; DaCosta, M.; Craker, L.; Maggi, F.; Morshedloo, M.R. Effect of Salinity Stress on the Physiological Characteristics, Phenolic Compounds and Antioxidant Activity of *Thymus vulgaris* L. and *Thymus daenensis* Celak. *Ind. Crops. Prod.* **2019**, *135*, 311–320. [CrossRef]

26. Punya, H.N.; Mehta, N.; Chatli, M.K.; Wagh, R.; Panwar, H. In-Vitro Evaluation of Antimicrobial and Antioxidant Efficacy of Thyme (*Thymus vulgaris* L.) Essential Oil. *J. Anim. Res.* **2019**, *9*, 443–449.
27. Kulisic, T.; Radonić, A.; Milos, M. Antioxidant Properties of Thyme (*Thymus vulgaris* L.) and Wild Thyme (*Thymus serpyllum* L.) Essential Oils. *Ital. J. Food Saf.* **2005**, *17*, 315–324.
28. Al Maqtari, M. Chemical Composition and Antimicrobial Activity of Essential Oil of *Thymus vulgaris* from Yemen. *Turk. J. Biochem.* **2011**, *36*, 342–349.
29. Borugă, O.; Jianu, C.; Mişcă, C.; Goleţ, I.; Gruia, A.; Horhat, F. *Thymus vulgaris* Essential Oil: Chemical Composition and Antimicrobial Activity. *J. Med. Life* **2014**, *7*, 56–60.
30. Rota, M.C.; Herrera, A.; Martínez, R.M.; Sotomayor, J.A.; Jordán, M.J. Antimicrobial Activity and Chemical Composition of *Thymus vulgaris, Thymus zygis* and *Thymus hyemalis* Essential Oils. *Food Control* **2008**, *19*, 681–687. [CrossRef]
31. Boukhatem, M.N.; Darwish, N.H.E.; Sudha, T.; Bahlouli, S.; Kellou, D.; Benelmouffok, A.B.; Chader, H.; Rajabi, M.; Benali, Y.; Mousa, S.A. In Vitro Antifungal and Topical Anti-Inflammatory Properties of Essential Oil from Wild-Growing *Thymus vulgaris* (Lamiaceae) Used for Medicinal Purposes in Algeria: A New Source of Carvacrol. *Sci. Pharm.* **2020**, *88*, 33. [CrossRef]
32. Nikolić, M.; Glamočlija, J.; Ferreira, I.C.F.R.; Calhelha, R.C.; Fernandes, Â.; Marković, T.; Marković, D.; Giweli, A.; Soković, M. Chemical Composition, Antimicrobial, Antioxidant and Antitumor Activity of *Thymus serpyllum* L., *Thymus algeriensis* Boiss. and Reut and *Thymus vulgaris* L. Essential Oils. *Ind. Crop. Prod.* **2014**, *52*, 183–190. [CrossRef]
33. Jafri, H.; Ahmad, I. *Thymus vulgaris* Essential Oil and Thymol Inhibit Biofilms and Interact Synergistically with Antifungal Drugs against Drug Resistant Strains of *Candida Albicans* and *Candida Tropicalis*. *J. Mycol. Med.* **2020**, *30*, 100911. [CrossRef] [PubMed]
34. Al-Shuneigat, J.; Al-Sarayreh, S.; Al-saraireh, Y.; Al-Qudah, M.; Al-Tarawneh, I. Effects of Wild *Thymus vulgaris* Essential Oil on Clinical Isolates Biofilm-Forming bacteria. *IOSR J. Dent. Med. Sci.* **2014**, *13*, 62–66. [CrossRef]
35. Myszka, K.; Schmidt, M.T.; Majcher, M.; Juzwa, W.; Olkowicz, M.; Czaczyk, K. Inhibition of Quorum Sensing-Related Biofilm of *Pseudomonas Fluorescens* KM121 by *Thymus Vulgare* Essential Oil and Its Major Bioactive Compounds. *Int. Biodeterior. Biodegrad.* **2016**, *114*, 252–259. [CrossRef]
36. Čabarkapa, I.; Čolović, R.; Đuragić, O.; Popović, S.; Kokić, B.; Milanov, D.; Pezo, L. Anti-Biofilm Activities of Essential Oils Rich in Carvacrol and Thymol against *Salmonella* Enteritidis. *Biofouling* **2019**, *35*, 361–375. [CrossRef]
37. de Oliveira, M.A.; da C Vegian, M.R.; Brighenti, F.L.; Salvador, M.J.; Koga-Ito, C.Y. Antibiofilm Effects of *Thymus vulgaris* and *Hyptis Spicigera* Essential Oils on Cariogenic Bacteria. *Future Microbiol.* **2021**, *16*, 241–255. [CrossRef] [PubMed]
38. Kubesová, A.; Šalplachta, J.; Horká, M.; Růžička, F.; Šlais, K. *Candida "Psilosis"*—Electromigration Techniques and MALDI-TOF Mass Spectrometry for Phenotypical Discrimination. *Analyst* **2012**, *137*, 1937–1943. [CrossRef] [PubMed]
39. Rams, T.E.; Sautter, J.D.; Getreu, A.; van Winkelhoff, A.J. Phenotypic Identification of Porphyromonas Gingivalis Validated with Matrix-Assisted Laser Desorption/Ionization Time-of-Flight Mass Spectrometry. *Microb. Pathog.* **2016**, *94*, 112–116. [CrossRef]
40. Kačániová, M.; Terentjeva, M.; Galovičová, L.; Ivanišová, E.; Štefániková, J.; Valková, V.; Borotová, P.; Kowalczewski, P.Ł.; Kunová, S.; Felšöciová, S.; et al. Biological Activity and Antibiofilm Molecular Profile of *Citrus Aurantium* Essential Oil and Its Application in a Food Model. *Molecules* **2020**, *25*, 3956. [CrossRef] [PubMed]
41. Kačániová, M.; Galovičová, L.; Ivanišová, E.; Vukovic, N.L.; Štefániková, J.; Valková, V.; Borotová, P.; Žiarovská, J.; Terentjeva, M.; Felšöciová, S.; et al. Antioxidant, Antimicrobial and Antibiofilm Activity of Coriander (*Coriandrum sativum* L.) Essential Oil for Its Application in Foods. *Foods* **2020**, *9*, 282. [CrossRef]
42. Kloucek, P.; Smid, J.; Frankova, A.; Kokoska, L.; Valterova, I.; Pavela, R. Fast Screening Method for Assessment of Antimicrobial Activity of Essential Oils in Vapor Phase. *Int. Food Res. J.* **2012**, *47*, 161–165. [CrossRef]
43. Reyes-Jurado, F.; Cervantes-Rincón, T.; Bach, H.; López-Malo, A.; Palou, E. Antimicrobial Activity of Mexican Oregano (*Lippia berlandieri*), Thyme (*Thymus vulgaris*), and Mustard (*Brassica nigra*) Essential Oils in Gaseous Phase. *Ind. Crops. Prod.* **2019**, *131*, 90–95. [CrossRef]
44. Mani López, E.; Valle Vargas, G.P.; Palou, E.; López Malo, A. *Penicillium Expansum* Inhibition on Bread by Lemongrass Essential Oil in Vapor Phase. *J. Food Prot.* **2018**, *81*, 467–471. [CrossRef] [PubMed]
45. Adams, R.P. *Identification of Essential Oil Components by Gas Chromatography/Mass Spectrometry*; Allured Publishing Corporation: Carol Stream, IL, USA, 2007.
46. van Den Dool, H.; Kratz, D.P. A Generalization of the Retention Index System Including Linear Temperature Programmed Gas—Liquid Partition Chromatography. *J. Chromatogr. A* **1963**, *11*, 463–471. [CrossRef]
47. Dao, T.; Bensoussan, M.; Gervais, P.; Dantigny, P. Inactivation of Conidia of *Penicillium chrysogenum, P. Digitatum* and *P. Italicum* by Ethanol Solutions and Vapours. *Int. J. Food Microbiol.* **2008**, *122*, 68–73. [CrossRef]
48. Talibi, I.; Askarne, L.; Boubaker, H.; Boudyach, E.H.; Msanda, F.; Saadi, B.; Aoumar, A.A.B. Antifungal activity of some Moroccan plants against *Geotrichum candidum*, the causal agent of postharvest citrus sour rot. *Crop. Prot.* **2013**, *35*, 41–46. [CrossRef]
49. Aman, M.; Rai, V.R. Antifungal activity of fungicides and plant extracts against yellow sigatoka disease *causing Mycosphaerella musicola. Curr. Res. Environ. Appl. Mycol.* **2015**, *5*, 277–284. [CrossRef]

plants

MDPI

Article

Chemical Composition, *In Vitro* and *In Situ* Antimicrobial and Antibiofilm Activities of *Syzygium aromaticum* (Clove) Essential Oil

Miroslava Kačániová [1,2,*], Lucia Galovičová [1], Petra Borotová [3], Veronika Valková [1], Hana Ďúranová [3], Przemysław Łukasz Kowalczewski [4], Hussein A. H. Said-Al Ahl [5], Wafaa M. Hikal [6,7], Milena Vukic [8], Tatsiana Savitskaya [9], Dzmitrij Grinshpan [9] and Nenad L. Vukovic [8,*]

1 Institute of Horticulture, Faculty of Horticulture and Landscape Engineering, Slovak University of Agriculture, Tr. A. Hlinku 2, 94976 Nitra, Slovakia; l.galovicova95@gmail.com (L.G.); veronika.valkova@uniag.sk (V.V.)
2 Department of Bioenergy, Food Technology and Microbiology, Institute of Food Technology and Nutrition, University of Rzeszow, 4 Zelwerowicza Str., 35-601 Rzeszow, Poland
3 AgroBioTech Research Centre, Slovak University of Agriculture, Tr. A. Hlinku 2, 94976 Nitra, Slovakia; petra.borotova@uniag.sk (P.B.); hana.duranova@uniag.sk (H.Ď.)
4 Department of Food Technology of Plant Origin, Poznań University of Life Sciences, 31 Wojska Polskiego Str., 60-624 Poznań, Poland; przemyslaw.kowalczewski@up.poznan.pl
5 Medicinal and Aromatic Plants Research Department, National Research Centre, 33 El-Bohouth Str., Dokki, Giza 12622, Egypt; shussein272@yahoo.com
6 Department of Biology, Faculty of Science, University of Tabuk, P.O. Box 741. Tabuk 71491, Saudi Arabia; wafaahikal@gmail.com
7 Water Pollution Research Department, Environmental Research Division, National Research Centre, 33 El-Bohouth Str., Dokki, Giza, 12622, Egypt
8 Department of Chemistry, Faculty of Science, University of Kragujevac, 34000 Kragujevac, Serbia; milena.vukic@pmf.kg.ac.rs
9 Research Institute for Physical Chemical Problems, Belarusian State University, Leningradskaya Str. 14, 220030 Minsk, Belarus; savitskayaTA@bsu.by (T.S.); grinshpan@bsu.by (D.G.)
* Correspondence: miroslava.kacaniova@gmail.com (M.K.); nvchem@yahoo.com (N.L.V.)

Citation: Kačániová, M.; Galovičová, L.; Borotová, P.; Valková, V.; Ďúranová, H.; Kowalczewski, P.Ł.; Said-Al Ahl, H.A.H.; Hikal, W.M.; Vukic, M.; Savitskaya, T.; et al. Chemical Composition, *In Vitro* and *In Situ* Antimicrobial and Antibiofilm Activities of *Syzygium aromaticum* (Clove) Essential Oil. *Plants* **2021**, *10*, 2185. https://doi.org/10.3390/plants10102185

Academic Editor: Petko Denev

Received: 19 September 2021
Accepted: 13 October 2021
Published: 15 October 2021

Publisher's Note: MDPI stays neutral with regard to jurisdictional claims in published maps and institutional affiliations.

Abstract: The essential oil of *Syzygium* (*S.*) *aromaticum* (CEO) is known for its good biological activity. The aim of the research was to evaluate *in vitro* and *in situ* antimicrobial and antibiofilm activity of the essential oil produced in Slovakia. The main components of CEO were eugenol 82.4% and (E)-caryophyllene 14.0%. The antimicrobial activity was either weak or very strong with inhibition zones ranging from 4.67 to 15.78 mm in gram-positive and gram-negative bacteria and from 8.22 to 18.56 mm in yeasts and fungi. Among the tested bacteria and fungi, the lowest values of MIC were determined for *Staphylococcus* (*S.*) *aureus* and *Penicillium* (*P.*) *expansum*, respectively. The vapor phase of CEO inhibited the growth of the microscopic filamentous fungi of the genus *Penicillium* when tested *in situ* on bread. The strongest effect of mycelia inhibition in a bread model was observed against *P. expansum* at concentrations of 250 and 500 µL/mL. The best antimicrobial activity of CEO in the carrot model was found against *P. chrysosenum*. Differences between the mass spectra of *Bacillus* (*B.*) *subtilis* biofilms on the tested surfaces (wood, glass) and the control sample were noted from the seventh day of culture. There were some changes in mass spectra of *Stenotrophomonas* (*S.*) *maltophilia*, which were observed in both experimental groups from the fifth day of culture. These findings confirmed the impact of CEO on the protein structure of older biofilms. The findings indicate that, besides being safe and sensorially attractive, *S. aromaticum* has antimicrobial activity, which makes it a potential substitute for chemical food preservatives.

Keywords: *Syzygium aromaticum*; *in vitro*; *in situ*; antimicrobial activity; antibiofilm activity

1. Introduction

Syzygium aromaticum is known for its use as spice in the preparation of food. Besides being valued for its flavoring properties, it can be used as an anti-cancer agent and a traditional remedy for many diseases such as asthma; digestive system, dental, and respiratory disorders; headaches; and sore throats [1]. It is also used in traditional medicine. Clove is widely used for treating dyspepsia, gastritis, and diarrhea. Furthermore, it was found that it has antipyretic, aphrodisiac, appetizing, expectorant, antiemetic, anxiolytic, myorelaxant, analgesic, decongestant, anti-inflammatory, and hypnotic effects [2]. The essential oil of *S. aromaticum* also shows anti-inflammatory, cytotoxic, and anesthetic activities besides the reported antimicrobial, antifungal, antiviral, antioxidant, and insecticidal properties [3].

The phenylpropene eugenol is the most important clove component. It is responsible for the characteristic strong aroma [4]. Inhibition of molds, yeast, and bacteria is one of the possible applications of clove essential oil [5]. The clove essential oil was proven to have repressing effects on some microorganisms such as *Alternaria* spp., *Aspergillus* spp., *Canninghamella* spp., *Lactobacillus* spp., *Fusarium* spp., *Clostridium* spp.; *Mucor* spp., *Salmonella* spp., *Penicillium* spp., and *Bacillus* spp. [6].

While being safe, non-toxic, and biodegradable, clove essential oil (CEO) shows a broad-spectrum antibacterial activity and can thus be used as bacteriostatic and anti-biofilm agent. Budri et al. [7] reported that CEO has significant inhibitory effect on biofilm. Rajkowska et al. [8] confirmed the inhibition of *Candida* biofilm on the surface of different materials by CEO. Adil et al. [9] proved that the main component of CEO may inhibit the adhesion of *S. mutans* to glass, prevent the formation of biofilm, and inhibit developed biofilm. While the formation of biofilm is a complex regulatory process, there are studies available that indicate the effectiveness of CEO as an antibiofilm agent.

The aim of the present study was to analyze the chemical composition and *in vitro* antimicrobial and antibiofilm activity of CEO. Molecular changes in bacterial biofilms were studied. The effectiveness of the gas phase of CEO against *Serratia* (*S.*) *marcescens* and *Penicillium* spp. was evaluated using food models.

2. Results

2.1. Chemical Composition of S. aromaticum EO

The essential oil of *S. aromaticum* was analyzed using gas chromatography with mass spectrometric (GC/MS) and flame ionization (GC-FID) detection. The main components were eugenol 82.4% and (E)-caryophyllene 14.0%. (Table 1). Based on the chemical composition, the analyzed oil was classified as thymol chemotype.

Table 1. Chemical composition of essential oil from *S. aromaticum*.

No	RI [a]	Compound [b]	% [c]
1	830	furfural	Tr
2	983	6-methyl-5-hepten-2-one	Tr
3	1190	methyl salicylate	Tr
4	1193	2-allyl-phenol	Tr
5	1360	eugenol	82.4
6	1379	a-copaene	Tr
7	1422	(E)-caryophyllene	14.0
8	1456	a-humulene	1.8
9	1519	eugenol acetate	0.9
10	1583	caryophyllene oxide	0.7
11	1755	benzyl benzoate	Tr
	total		99.7

[a] Values of retention indices on HP-5MS column; [b] identified compounds; [c] tr—compounds identified in amounts less than 0.1%.

2.2. Antimicrobial Activity of *S. aromaticum* EO

The antimicrobial activity of CEO was evaluated by disk diffusion test and the results are presented in Table 2. A weak to moderate inhibitory activity of CEO was observed in the cases of most of the tested bacteria (including biofilm-forming bacteria), and yeast. *S. aureus* was the most susceptible of all the tested bacteria, with CEO showing a very strong inhibitory activity. MIC 50 and MIC 90 were determined by analysis of the minimum inhibitory concentrations. There were low values of MIC 50 (85.46 μL/mL), and MIC 90 (93.35 μL/mL) values were found for *S. aureus*. The highest MIC 50 and MIC 90 values were determined for *Y. enterocolitica*. Moderate MIC 50 and MIC 90 values were determined for *B. subtilis* and *B. subtilis* biofilm. Strong antimicrobial activity of CEO against all tested penicillia was observed, with *P. expansum* being the most susceptible with inhibition zone 18.56 mm and MIC 50 64.25 and MIC 90 75.12 μL/mL. Details of the results of antimicrobial activity and minimum inhibitory concentrations are given in Table 2.

Table 2. Antimicrobial activity of *S. aromaticum* essential oil.

Microorganism	Zone Inhibition (mm)	Activity of EO	MIC 50 (μL/mL)	MIC 90 (μL/mL)	ATB
S. enteritidis	9.44 ± 1.01	*	216.23	265.41	29.00 ± 0.10
P. aeruginosa	5.22 ± 1.20	*	223.38	284.56	28.00 ± 0.06
Y. enterocolitica	7.67 ± 1.41	*	224.46	278.92	28.00 ± 0.05
S. marcescens	10.44 ± 1.01	**	112.13	145.25	27.00 ± 0.02
S. aureus	15.78 ± 0.67	***	85.46	93.35	29.00 ± 0.03
B. subtilis	11.22 ± 0.83	**	121.13	138.91	26.00 ± 0.05
E. faecalis	4.67 ± 1.32	*	218.36	264.55	29.00 ± 0.03
C. albicans	8.22 ± 1.53	*	221.43	226.32	28.00 ± 0.06
C. krusei	9.89 ± 1.76	*	211.82	265.33	26.00 ± 0.12
C. tropicalis	9.33 ± 0.58	*	214.54	271.36	29.00 ± 0.02
C. glabrata	9.33 ± 1.00	*	217.22	264.36	26.00 ± 0.03
Biofilm *S. maltophilia*	8.89 ± 1.36	*	223.18	284.32	27.00 ± 0.05
Biofilm *B. subtilis*	10.44 ± 0.88	**	103.64	128.64	26.00 ± 0.03
P. commune	16.52 ± 0.15	***	74.32	81.26	25.00 ± 0.05
P. expansum	18.56 ± 0.22	***	64.25	75.12	26.00 ±0.11
P. glabrum	18.32 ± 0.21	***	68.41	78.12	27.00 ± 0.06
P. chrysogenum	17.36 ± 0.08	***	75.11	86.92	26.00 ± 0.09

* Weak antimicrobial activity (zone 5–10 mm). ** Moderate inhibitory activity (zone > 10 mm). *** Very strong inhibitory activity (zone > 15 mm). ATB—antibiotics, positive control (Cefoxitin for G^-, Gentamicin for G^+, Fluconazole for yeast and fungi).

2.3. Antimicrobial Analysis In Situ Using a Food Model

In situ antimicrobial analysis using bread showed that microscopic filamentous fungi of the genus *Penicillium* were inhibited by CEO vapors in all tested concentrations (Table 3). At the lowest concentration of the oil (62.5 μL/L), *P. glabrum* was the most strongly inhibited fungus (42.91%). The weakest inhibition at this concentration was observed for *P. commune*. The inhibition of *P. expansum* was 90.49% at the concentration of 125 μL/L and 100% at 250 μL/L and 500 μL/L. A significant difference in inhibition was observed for this species between 62.5 μL/L and the remaining concentrations. Significant differences in inhibition of *P. glabrum* and *P. chrysogenum* were found between 62.5 μL/L and 250 μL/L or 500 μL/L. These findings confirm the inhibitory effects of CEO against the potentially pathogenic fungi tested.

Table 3. *In situ* analysis of the antifungal activity of the vapor phase of CEO in bread.

Fungi	Mycelial Growth Inhibition [%]			
	Concentration of EO			
	62.5 μL/L	125 μL/L	250 μL/L	500 μL/L
P. commune	16.41 ± 6.84 ^{a}	45.72 ± 5.49 ^{b}	96.22 ± 1.20 ^{c}	98.78 ± 1.72 ^{cd}
P. expansum	16.73 ± 6.38 ^{a}	90.49 ± 12.26 ^{b}	100.00 ± 0.00 ^{bc}	100.00 ± 0.00 ^{bd}
P. glabrum	42.91 ± 5.87 ^{a}	54.50 ± 9.01 ^{a}	93.42 ± 9.30 ^{b}	100.00 ± 0.00 ^{bc}
P. chrysogenum	30.31 ± 9.70 ^{a}	12.21 ± 7.87 ^{a}	97.90 ± 1.49 ^{b}	99.45 ± 0.78 ^{bc}

Means ± standard deviation. Values followed by different superscript within the same row are significantly different ($p < 0.05$).

The effect of the vapor phase of CEO against *P. chrysogenum*, *P. expansum* was recorded in *in situ* analysis using carrots at all tested concentrations, while *S. marcescens* was only inhibited when the highest concentration of the oil was used (Table 4). The highest inhibition of *P. chrysogenum* (93.53%) and *P. expansum* (75.94%) was observed at the concentration of 500 μL/L. The lowest concentration of CEO yielded the lowest inhibition results for both molds. The vapor phase of CEO was shown to have the best inhibitory effect on penicillia in the *in situ* tests with the food model.

Table 4. Results of *in situ* analysis of antimicrobial activity of the vapor phase of CEO on carrots.

Microorganisms	Bacterial Growth Inhibition [%]			
	Concentration of EO			
	62.5 μL/L	125 μL/L	250 μL/L	500 μL/L
P. chrysogenum	16.93 ± 0.37 ^{a}	51.04 ± 4.43 ^{ab}	61.22 ± 15.73 ^{bc}	93.53 ± 5.11 ^{c}
P. expansum	19.43 ± 0.89 ^{a}	59.81 ± 4.33 ^{b}	64.98 ± 2.81 ^{b}	75.94 ± 0.91 ^{c}
S. marcescens	0.00 ± 0.00 ^{a}	0.00 ± 0.00 ^{a}	0.00 ± 0.00 ^{a}	49.11 ± 13.73 ^{b}

Means ± standard deviation. MGI: mycelial growth inhibition, BGI: bacterial growth inhibition; values followed by different superscript within the same row are significantly different ($p < 0.05$).

2.4. Analysis of Biofilm Developmental Phases and Evaluation of Molecular Differences on Different Surfaces Using MALDI-TOF MS Biotyper

The effect of CEO on *B. subtilis* and *S. maltophilia* biofilms was analyzed using MALDI TOF MS Biotyper in order to observe changes in molecular structure that accompany growth inhibition. The spectra of biofilms and planktonic cells in the control group developed identically and, therefore, the spectra of planktonic cells were used instead of the control spectrum for improved clarity. The two experimental spectra from different surfaces (glass, wood) and the planktonic spectrum representing the development of the control group are shown for each day of the experiment. Figure 1 shows the mass spectra of *B. subtilis* biofilm during individual days of the experimental evaluation.

Mass spectra after days 3 and 5 (Figure 1A,B) of the culture showed the same peaks, indicating production of the same proteins by new biofilms and control planktonic cells. There were no changes observed in bacterial cultures at the protein level. Differences between the mass spectra of biofilms on the tested surfaces (wood, glass) and the control sample stared to emerge on day 7 (Figure 1C–F). This indicated changes in the protein profile of the CEO-treated biofilm. It seems that CEO can influence the homeostasis of a bacterial biofilm formed on wooden and glass surfaces.

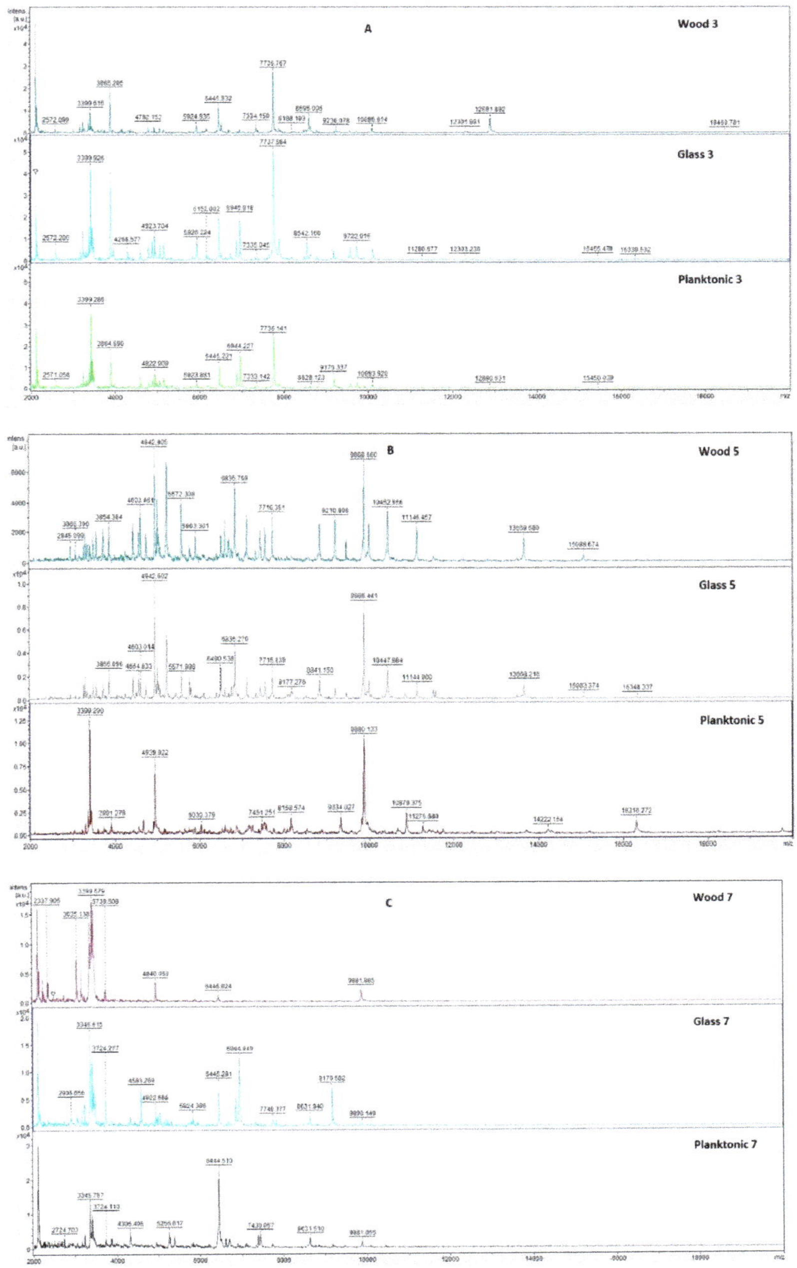

Figure 1. *Cont.*

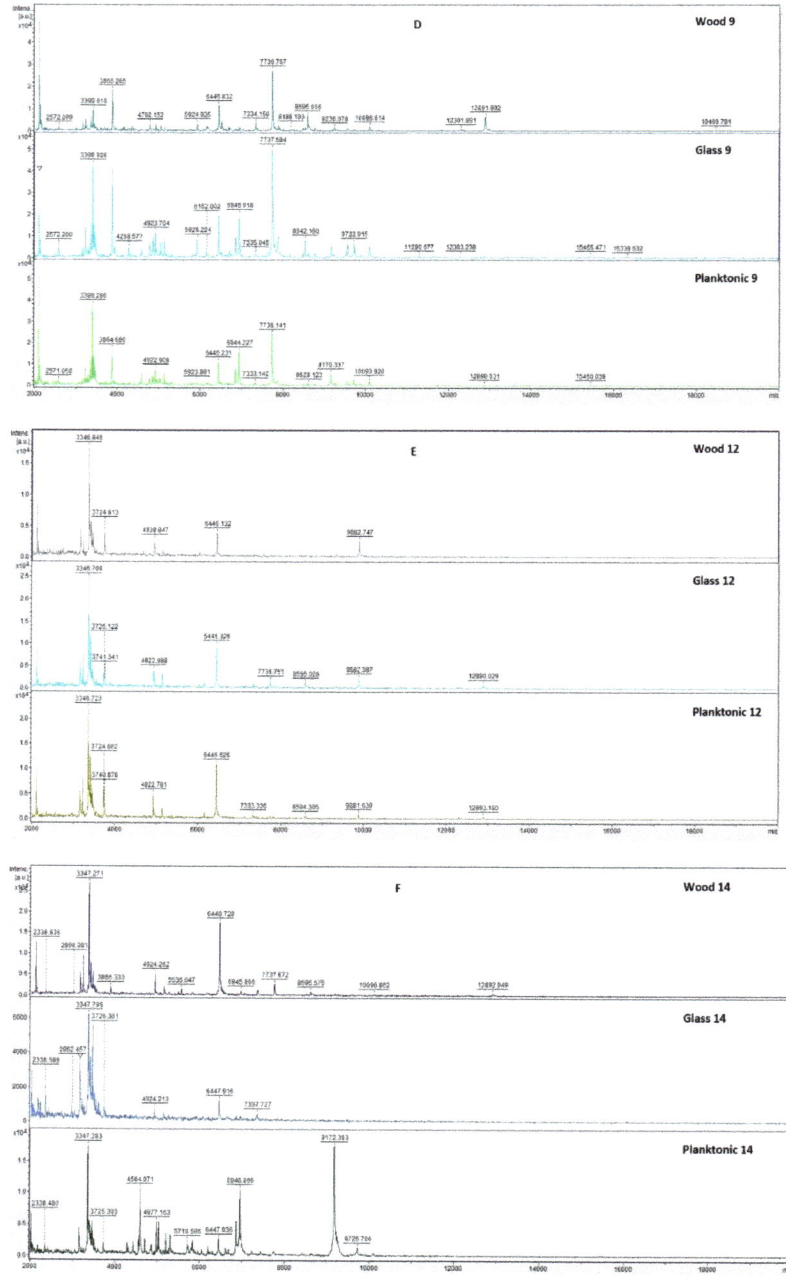

Figure 1. MALDI-TOF mass spectra of *B. subtilis* biofilm during development after the addition of CEO: (**A**) 3rd day, (**B**) 5th day, (**C**) 7th day, (**D**) 9th day, (**E**) 12th day, and (**F**) 14th day.

Using the data obtained with mass spectrometry, a dendrogram was created to visualize similarities in biofilm structure that are based on MSP distance (Figure 2). It can be stated from the dendrogram that the planktonic stage (P) together with the control groups and new biofilms had the shortest distance during the third and fifth days (BSG 3, BSW 3, BSG 5, BSW 5). The similarity in the protein profile of the control groups was confirmed by the short distances of the MSP. New biofilms and control planktonic cells also showed short MSP lengths that corresponded to similar mass spectra. The distance of the experimental groups of SMEs increased gradually over time. The mass spectra analyzed on the 12th and 14th days of the experiment had the longest MSP lengths, indicating changes in the molecular profile of *B. subtilis.*

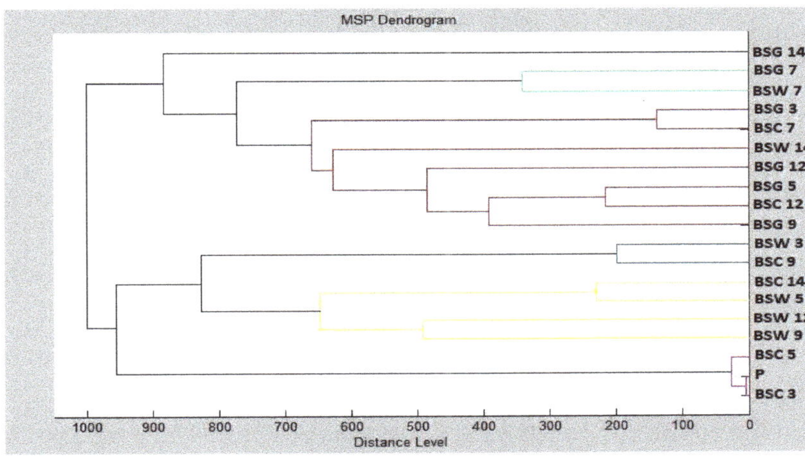

Figure 2. Dendrogram of *B. subtilis* generated using MSPs of the planktonic cells and the control. BS, *B. subtilis*; C, control; G, glass; W, wood; and P, planktonic cells.

Figure 3 shows the spectra of developmental stages of *S. maltophilia* biofilm over the entire duration of the experiment.

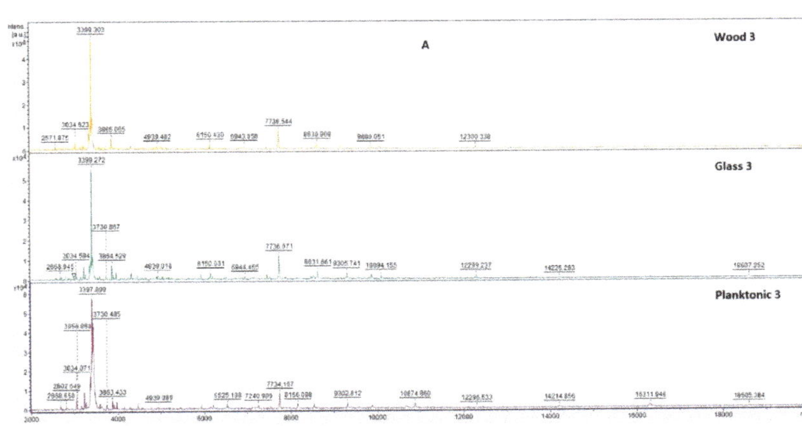

Figure 3. *Cont.*

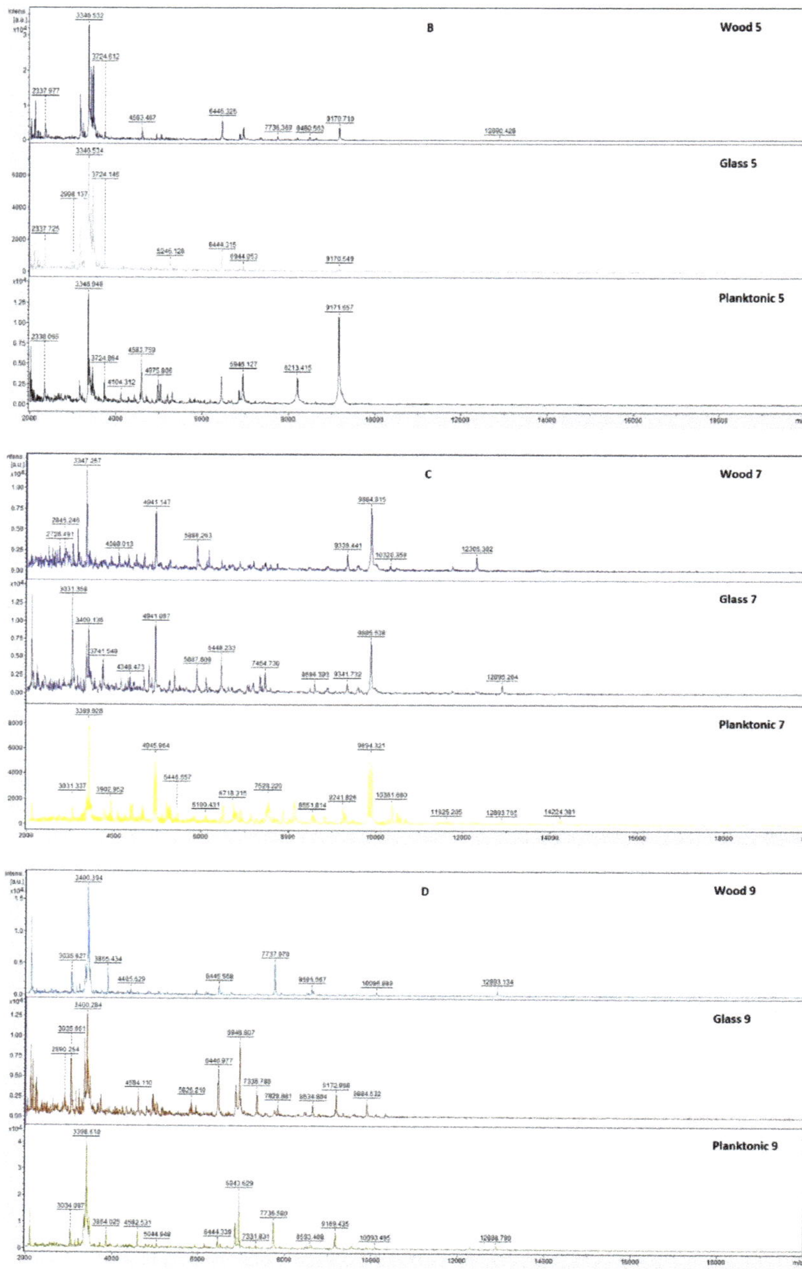

Figure 3. *Cont.*

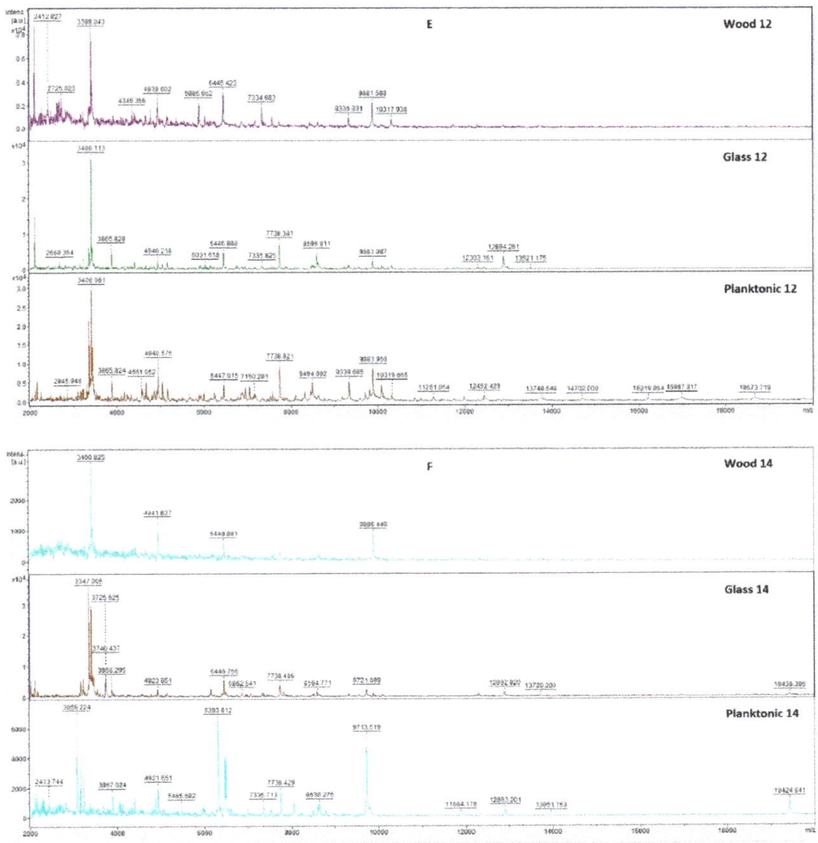

Figure 3. MALDI-TOF mass spectra of *S. maltophilia* during biofilm development: (**A**) 3rd day, (**B**) 5th day, (**C**) 7th day, (**D**) 9th day, (**E**) 12th day, and (**F**) 14th day.

Analysis of *S. maltophilia* protein spectra by MALDI-TOF MS Biotyper showed similarity between the experimental spectra and the control planktonic spectrum on the third day of the experiment (Figure 3A), indicating that the bacterial cells in biofilm developed similarly to the planktonic cells. Changes in mass spectra were observed in both experimental groups from the fifth day (Figure 3B–F). This finding confirms the effect of CEO on the protein structure of older biofilms.

Mass spectra dendrogram also confirmed the similarity of the new biofilms to planktonic cells and control cells (Figure 4). The changing distance of MSP during the progression of the experiment indicates differences in protein production due to the effect of the addition of CEO.

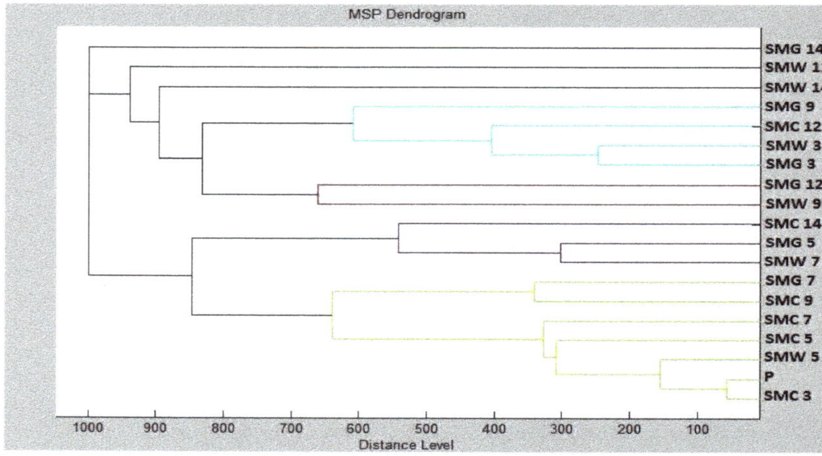

Figure 4. Dendrogram of *S. maltophilia* generated using MSPs of the planktonic cells and the control. SM, *S. maltophilia;* C, control; G, glass; W, wood; and P, planktonic cells.

3. Discussion

Essential oils are potential sources of novel antimicrobial compounds, especially those active against bacterial pathogens and microscopic filamentous fungi. Chemical composition analysis revealed that the tested essential oil was of the eugenol chemotype. Eugenol was the major compound (90.3%), accompanied by β-caryophyllene (4.83%) and eugenol acetate (1.87%), in an essential oil extracted from the south of Brazil [10]. In another study where essential oil of clove from the south of Brazil was analyzed, greater quantities of caryophyllene (39.63%) and less of eugenol (56.06%) were detected [11]. CEO from China [12] and Italy [13] also contained eugenol as their major compound, with 90.84% and 77.9%, respectively. Similar amounts were found in the present study. The amounts of eugenol and accompanying compounds in the essential oil can be directly related to the different geographic areas of origin of the plant material. They can also be influenced by biotic and abiotic factors such as seasonality, stage of development, age of the plant, and climatic conditions [14]. In addition, the extraction method, such as distillation, which is used for obtaining the oil, can also have an effect on its chemical composition. Moreover, it was found that storage conditions may influence the content of its volatile components [15].

CEO, which is used as an antiseptic in oral infections, inhibits Gram-negative and Gram-positive bacteria as well as microscopic filamentous fungi. There are many reports that prove that clove oil and its main active component, eugenol, affect common, food-derived, Gram-negative bacteria such as *Escherichia coli, Salmonella, P. aeruginosa*, etc., and Gram-positive bacteria such as *Staphylococcus, Streptococcus, Listeria*, etc. This conclusion is based on the reported inhibiting effects on migration, adhesion, expression of virulence factors, and biofilm formation in these bacteria. Clove oil and eugenol have good prospects for application in the food antisepsis field [16]. With the disk diffusion method, we found the antimicrobial activity against the tested microorganisms, including biofilm-forming bacteria, to be either weak or very strong. *S. aureus* was found to be the most susceptible to CEO among the bacteria tested. In another study, CEO was active against standard strains of Gram-negative and Gram-positive bacteria, showing inhibition zones between 14 and 25 mm in the disk diffusion test [17]. The authors reported intermediate bacterial activity of the oil against food bacteria with inhibition diameters of 14.6 $\pm$ 1.7 mm for *P. putida*, 15 $\pm$ 0.66 mm for *E. coli*, and 15.6 $\pm$ 0.4 mm for *S. aureus*. In a different research, inhibition zones ranging from 21.3 $\pm$ 2.2 mm for *S. aureus* ATCC and 35.6 $\pm$ 2.7 mm for *P. aeruginosa* ATCC were observed [18]. The results of our study differed from the ones referenced where the zones of inhibition were generally smaller. Cimanga et al. [19] reported that CEO could be a strong inhibitor of *S. aureus* and *E. coli* and a moderate inhibitor of *P. aeruginosa*. Using the microdilution method, Gislene et al. [20], Hili et al. [21], and Nzeako et al. [22] found that CEO manifested antimicrobial activity against *S. aureus*, *E. coli*, and *C. albicans* at various concentrations of the extracts. The sensitivity to the antimicrobial effect of CEO (in descending order) was *E. coli* with MIC 2 μL/mL, *C. albicans* with MIC 16 μL/mL, *E. faecalis* with MIC 32 μL/mL, and *S. aureus* with MIC 32 μL/mL [23]. Different results were found using microdilution methods in our study. Clove oil has a strong antimicrobial activity. Its minimum inhibitory concentration in the range of 0.2–0.625 mg/mL was recorded against the bacteria *S. aureus*, *Staphylococcus epidermidis*, *E. coli*, and *P. aeruginosa* [24–26].

From among the microscopic fungi tested in our study, *P. expansum* was found to be the most susceptible to CEO. Among the tested yeasts, *C. krusei* was the least resistant. Eugenol from the clove is the essential ingredient that is responsible for its antifungal activity [27]. A strong antifungal activity of CEO was reported against opportunistic fungal pathogens such as *C. albicans*, *Cryptococcus neoformans*, and *Aspergillus fumigatus*. A considerable variation in inhibition zone sizes, ranging from 12 to 22 mm, was observed among fungal isolates depending on their sensitivity to CEO [28]. Nunez et al. [29] reported a strong fungicidal effect of eugenol against *C. albicans*, *P. citrinum*, *A. niger*, and *T. mentagrophytes*. Ahmad et al. [30] also reported strong antifungal activity of clove oil against *C. albicans*, *C. neoformans*, and *A. fumigatus*. Pinto et al. [31] determined MIC to evaluate the antifungal activity of the clove oil and its main component, eugenol, against *Candida*, *Aspergillus*, and clinical and ATCC strains of dermatophytes. Similar to our results, the essential oil and eugenol showed inhibitory activity against all the tested strains. Chami et al. [32] confirmed carvacrol and eugenol as strong potential antifungal agents and suggested their use as therapeutic agents for oral candidiasis.

The effect of the vapor phase of CEO against *P. chrysogenum* and *P. expansum* was recorded at all tested concentrations in *in situ* analysis using carrots. Noteworthy, *S. marcescens* was inhibited only when the oil was applied in the concentration of 500 μL/L. The investigation of inhibition of the microscopic filamentous by vapor phase CEO also indicated the susceptibility of penicillia. In the bread model studies, the best inhibition result was observed against *P. expansum* at concentrations of 250 and 500 μL/mL. In the carrot model studies, *P. chrysogenum* was found to be the most susceptible. There exists a different study in which the antifungal activities of CEO and its volatile vapor against dermatophytic fungi, including *C. albicans*, *Epidermophyton floccosum*. *Microsporum audouinii*, *Trichophyton mentagrophytes*, and *Trichophyton rubrum*, were studied. Spore germination and mycelial growth were found to be strongly inhibited by both the oil and its volatile vapor. Moreover, the volatile vapor of CEO showed fungistatic activity, whereas the direct application of the oil resulted in a fungicidal effect [33]. Jain and Agrawal [34] noticed fungistatic activity of volatile vapor of several essential oils. No other authors focused on antifungal activity of volatile vapor of clove oil. There are studies in which the differences in inhibitory activity of essential oils in different food model systems were investigated [35]. It was found that CEO did not have a significant effect on the count of *E. coli* O157: H7 relative to control; however, it showed significant inhibitory effect in blanched spinach at similar concentration [36]. Aguilar-Gonzalez et al. [37] evaluated the antifungal activity of the vapor phase of clove and mustard EOs, individually and in combination, against gray mold (*Botrytis cinerea*) in strawberries, and noted inhibitory activity.

Bacterial biofilms can be characterized as non-homogenous structures with a high surface roughness, diversification of cells' metabolic activity, and presence of extracellular polymeric substances as well as cells' and spores' liberation from the upper parts of the biofilm matrix [38]. The difference between the mass spectra of *B. subtilis* biofilms on the tested surfaces (wood, glass) and the control sample was noticeable from the seventh day in our study. There were some changes in mass spectra of *S. maltophilia* in both experimental groups from the fifth day. These results confirmed the effect of CEO on the protein structure of older biofilms. CEO is one of the most bioactive essential oils due to high antimicrobial activity of eugenol [39]. The main compound of the oil used in the presented research was eugenol, which constituted 86.99% of it. As demonstrated in another study, the activity of CEO against *Alicyclobacillus* biofilm may be related to both its bactericidal effect, resulting in a decrease in the number of planktonic cells, and the changes in the adherence capability of cells [40]. MALDI-TOF MS Biotyper has only been used in a few cases. Li et al. [41] analyzed *B. subtilis* biofilm and determined the spatial distribution of specific peptides and lipopeptides that are produced in biofilms by using this method. Kubesová et al. [42] used MALDI-TOF MS to analyze the biofilm produced by the genus *Candida*. Rams et al. [43] found out that the phenotypic identification of culturable *P. gingivalis* biofilms is 100% accurate using MALDI-TOF MS because of the differences in the protein profile. The changes in the mass spectra profile were demonstrated by MALDI-TOF in Kačániová et al. [44,45], where inhibitory effects on the biofilm of *Coriandrum sativum* and *Citrus aurantium* EOs were detected. The use of MALDI-TOF can be a fast and easy method for the assessment of the biofilm growth and degradation due to structural and molecular changes.

4. Materials and Methods

4.1. Essential Oil

The essential oil of *S. aromaticum* was obtained from Hanus, s.r.o. (Nitra, Slovakia). It was a product of steam distillation of fresh leaves from 2021. It was stored in the dark at 4 °C for the whole time. The analyses were carried out in 2021.

4.2. Chemical Characterization of Essential Oil Samples by Gas Chromatography/Mass Spectrometry (GC/MS) and Gas Chromatography (GC-FID)

GC/MS analysis of CEO was performed using an Agilent 6890N gas chromatograph (Agilent Technologies, Santa Clara, CA, USA) coupled to a quadrupole mass spectrometer 5975B (Agilent Technologies, Santa Clara, CA, USA) equipped with an HP-5MS capillary column (30 m × 0.25 mm × 0.25 μm). The temperature ramp was from 60 °C to 150 °C (increasing rate 3 °C/min) and from 150 °C to 280 °C (increasing rate 5 °C/min). The total run time of the program was 60 min. Helium 5.0 was used as the carrier gas with a flow rate of 1 mL/min. The injection volume was 1 μL (EO sample was diluted in pentane), while the split/splitless injector temperature was set at 280 °C. Split mode injection with split ratio at 40.8:1 was used. The electron-impact mass spectrometric data (EI-MS; 70 eV) were acquired in scan mode over the m/z range 35–550. MS ion source and MS quadrupole temperatures were 230 °C and 150 °C, respectively. Acquisition of data started after 3 min of solvent delay time.

GC-FID analyses were performed on Agilent 6890N gas chromatograph coupled to FID detector. Column (HP-5MS) and chromatographic conditions were the same as for GC-MS. The temperature of the FID detector was set at 300 °C.

The individual volatile constituents of *S. aromaticum* EO sample were identified according to their retention indices [46] and they were compared with the reference spectra (Wiley and NIST databases). The retention indices were determined experimentally by standard method, which included retention times of n-alkanes (C6–C34), injected under the same chromatographic conditions [47]. The percentages of the identified compounds (amounts higher than 0.1%) were derived from their GC peak areas.

4.3. Tested Microorganisms

Gram-positive bacteria (*B. subtilis* CCM 1999, *E. faecalis* CCM 4224, *S. aureus* subsp. *aureus* CCM 8223), gram-negative bacteria (*P. aeruginosa* CCM 3955, *S. enterica* subsp. *enterica* ser. Enteritidis CCM 4420, *S. marcescens* CCM 8588 *Y. enterocolitica* CCM 7204), and yeasts (*C. krusei* CCM 8271, *C. albicans* CCM 8261, *C. tropicalis* CCM 8223, *C. glabrata* CCM 8270) were obtained from the Czech collection of microorganisms (Brno, Czech Republic). The biofilm-forming bacteria *B. subtilis* and *S. maltophilia* were obtained from the dairy industry. Bacteria were identified with 16S rRNA sequencing and MALDI-TOF MS Biotyper. The fungi *P. glabrum*, *P. chrysogenum*, *P. expansum*, and *P. commune* were obtained from grape samples, and were identified by 16S rRNA sequencing and MALDI-TOF MS Biotyper.

4.4. Antimicrobial Activity: Disc Diffusion Method

Antimicrobial activity of *S. aromaticum* EO was determined by the disc diffusion method. The microbial inoculum was cultivated for 24 h on Tryptone soya agar (TSA, Oxoid, Basingstoke, UK) at 37 °C for bacteria and Sabouraud Dextrose Agar (SDA, Oxoid, Basingstoke, UK) at 25 °C for yeasts. The inoculum was adjusted to optical density 0.5 McFarland standard (1.5×10^8 CFU/mL) and 100 μL was added on plates with Mueller Hinton agar (MHA, Oxoid, Basingstoke, UK). Sterile 6-mm discs saturated with 10 μL of *S. aromaticum* EO were placed on the layer of agar containing the suspension of microorganisms. The samples were incubated for 24 h at 37 °C for bacteria and 25 °C for yeasts. Two antibiotics (Cefoxitin, Gentamicin, Oxoid, Basingstoke, UK), and one antifungal (Fluconazole, Oxoid, Basingstoke, UK) were used as positive controls for gram-negative and gram-positive bacteria and yeasts, respectively. Disks impregnated with 0.1% DMSO (dimethylsulfoxid, Centralchem, Bratislava, Slovakia) served as negative control. The antimicrobial activity was categorized as either very strong, moderate, or weak when the zone of growth inhibition was larger than 15, 10, and 5 mm, respectively. The analyses were performed in triplicate.

4.5. Minimum Inhibitory Concentrations (MIC)

Microbial inoculum was cultivated for 24 h in Mueller Hinton Broth (MHB, Oxoid, Basingstoke, UK) at 37 °C for bacteria and Sabouraud Dextrose Broth (SDB, Oxoid, Basingstoke, UK) at 25 °C for yeasts. An aliquot of 50 μL of inoculum with an optical density of 0.5 McFarland was added to a 96-well microtiter plate. Subsequently, the *S. aromaticum* EO was prepared by serial dilution to a concentration range from 400 μL/mL to 0.2 μL/mL in MHB/SDB, and 100 μL of suspension was thoroughly mixed with bacterial inoculum in wells. Bacterial samples were incubated for 24 h at 37 °C. Yeast samples were incubated for 24 h at 25 °C. MHB/SDB with EO was used as a negative control and MHB/SDB with inoculum was used as positive control, representing uninhibited growth.

For non-adherent microorganisms, the absorbance was measured after an incubation period at 570 nm by Glomax spectrophotometer (Promega Inc., Madison, WI, USA). The MIC of biofilm-forming bacteria was measured with the use of crystal violet. The suspension with non-attached cells was discarded, the wells were washed with distilled water three times, and left to dry at room temperature. Following the addition of 200 μL of 0.1% (*w*/*v*) crystal violet to the wells, the plates were incubated for 15 min. Subsequently, the wells were repeatedly washed and dried. Stained biofilms were solubilized with 200 μL of 33% acetic acid [48], and absorbance was measured at 570 nm. Minimum inhibitory concentration was determined as the concentration of CEO at which absorbance was lower than the absorbance of the maximal growth control. The test was performed in triplicate.

4.6. Antimicrobial Analysis In Situ on a Food Model

Antifungal effect of *S. aromaticum* EO vapor phase was evaluated in 0.5-L sterile, glass jars (Bormioli Rocco, Fidenza, Italy) on a bread surface used as a food model. The fungi of *Penicillium* genus were cultivated for 5 days on Sabouraud Dextrose agar (SDA, Oxoid, Basingstoke, UK) at the temperature 25 °C. The cultures were applied to the bread slices ($15 \times 15 \times 1.5$ cm) by three stabs. A 6-cm sterile, filter paper was placed to the jar lid and 100 μL of CEO (62.5, 125, 250, and 500 μL/L diluted in ethyl acetate) were applied. The control group was left untreated. The jars were hermetically sealed and they were incubated in the dark for 14 days at 25 °C ± 1 °C.

Antibacterial analysis of the vapor phase of CEO *in situ* was tested on *S. marcescens*. Warm MHA was poured into 60-mm Petri dishes (PD) and in the lid too. Sliced carrots (0.5 mm) were placed on agar. Then, an inoculum was prepared, as previously described. CEO was diluted in ethyl acetate to 500, 250, 125, and 62.5 μL/L and applied to sterile filter paper. After evaporation of the remaining ethyl acetate (1 min), the dishes were sealed and incubated at 37 °C for 7 days.

Inhibition of the fungal growth was evaluated by stereological methods. Volume density (Vv) of fungi was estimated using ImageJ software. The stereological grid points of the colonies (P) and substrate (p) were counted. The density of fungal growth was calculated as percentage (%) according to the formula Vv = P/p × 100. The antifungal activity of EO was expressed as mycelial growth inhibition in percentage (%) (MGI): MGI = [(C − T)/C] × 100, where C was density of the fungal growth in the control group and T was density of the fungal growth in the treatment group [49,50].

In situ bacterial growth was determined using stereological methods. In this concept, the volume density (Vv) of bacterial colonies was first estimated using ImageJ software counting the points of the stereological grid hitting the colonies (P) and those (p) falling to the reference space (growth substrate used). The volume density of bacterial colonies was consequently calculated as follows: Vv (%) = P/p. The antibacterial activity of EO was defined as the percentage of bacterial growth inhibition (BGI) BGI = [(C − T)/C] × 100, where C and T were bacterial growth (expressed as Vv) in the control group and the treatment group, respectively. Negative results represented growth stimulation.

4.7. Analysis of Differences in Biofilm Development with MALDI-TOF MS Biotyper

The changes of protein spectra during biofilm development after CEO addition were evaluated by MALDI-TOF MS Biotyper. *S. maltophilia* was used as a representative of gram-negative biofilm-forming bacteria and *B. subtilis* was representative of gram-positive biofilm-forming bacteria, and both bacterial strains were isolated from the milk industry. The biofilm-forming bacteria were added to 50-mL polypropylene tubes with 20 mL of MHB; subsequently, a wooden toothpick and a glass slide were added as models of different surfaces. The experimental groups were treated with 0.1% CEO, and control group samples were left untreated. The samples were incubated at 37 °C on a shaker at 170 rpm. The samples were analyzed after 3, 5, 7, 9, 12, and 14 days. The biofilm samples were taken from a glass slide and wooden toothpick with a sterile cotton swab, and they were imprinted onto a MALDI TOF metal target plate. The planktonic cells were obtained from 300 µL of culture medium, cells were centrifuged for 1 min at 12,000 rpm, and the supernatant was discarded. The pellet was resuspended in 30 µL of ultrapure water and the suspension was centrifuged for 1 min at 12,000 rpm. This cycle was repeated three times. Then 1 µL of the so-prepared planktonic cell suspension was applied to a target plate. The target plate was dried and 1 µL of α-Cyano-4-hydroxycinnamic acid matrix (10 mg/mL) was applied. The samples were processed with MALDI-TOF MicroFlex (Bruker Daltonics) linear and positive mode for the range of m/z 200–2000 after crystallization. The spectra were obtained by an automatic analysis and the same sample similarities were used to generate the standard global spectrum (MSP) and the 19 MSP was generated from the spectra by MALDI Biotyper 3.0 and were grouped into dendrograms using Euclidean distance [44].

4.8. Statistical Data Evaluation

SAS® software version 8 was used for data processing. The results of the MIC value (concentration that caused 50% and 90% inhibition in bacterial growth) were determined by logit analysis.

5. Conclusions

The main components of the essential oil of *S. aromaticum* were eugenol 82.4% and (E)-caryophyllene 14.0%. A very strong antifungal effect of the vapor phase of the essential oil against the genus *P. expansum* was observed in the experiments that employed the food model (bread). Strong inhibiting effects of the vapor phase against *P. chrysogenum* were also noted in the test with carrots. The results indicate that CEO has potential for future use in extension of the shelf life of bakery products and that it could find application in the storage of root vegetables. The tested oil had moderate antimicrobial as well as antibiofilm effects that were observed on various surfaces and detected by MALDI-TOF MS Biotyper. In the case of EOs with a dominant proportion of volatile components, a stronger effect was often observed in the vapor phase in comparison to direct contact application. The decrease in microbial population caused by the addition of CEO depended on the concentration, and the inhibitory activity of the oil was lower in food systems when compared to *in vitro* systems.

Author Contributions: Conceptualization, L.G. and M.K.; Data curation, L.G., P.B., V.V., H.Ď., N.L.V., M.V. and M.K.; Methodology, L.G., P.B., N.L.V., M.V., H.Ď., V.V. and M.K.; Supervision, M.K.; Writing—original draft, L.G., P.B., P.Ł.K., H.A.H.S.-A.A., W.M.H., T.S., D.G. and M.K. All authors have read and agreed to the published version of the manuscript.

Funding: This work was supported by the grant APVV SK-BY-RD-19-0014 "The formulation of novel compositions and properties study of the polysaccharides based edible films and coatings with antimicrobial and antioxidant plant additives".

Institutional Review Board Statement: Not applicable.

Informed Consent Statement: Not applicable.

Data Availability Statement: Data are contained within the article.

Acknowledgments: This work was supported by the grants of the VEGA no. 1/0180/20.

Conflicts of Interest: The authors declare no conflict of interest.

References

1. Singh, J.; Baghotia, A.; Goel, S.P. *Eugenia caryophyllata* Thunberg (Family Myrtaceae): A review. *Int. J. Res. Pharm. Biomed. Sci.* **2012**, *3*, 1469–1475.
2. Lee, S.; Najiah, M.; Wendy, W.; Nadirah, M. Chemical composition and antimicrobial activity of the essential oil of *Syzygium aromaticum* flower bud (Clove) against fish systemic bacteria isolated from aquaculture sites. *Front. Agric. China* **2009**, *3*, 332–336. [CrossRef]
3. Chaieb, K.; Zmantar, T.; Ksouri, R.; Hajlaoui, H.; Mahdouani, K.; Abdelly, C.; Bakhrouf, A. Antioxidant properties of the essential oil of *Eugenia caryophyllata* and its antifungal activity against a large number of clinical *Candida* species. *Mycoses* **2007**, *50*, 403–406. [CrossRef]
4. Shobana, S.; Naidu, K.A. Antioxidant activity of selected Indian spices. *Prostaglandins Leukot. Essent. Fat. Acids (PLEFA)* **2000**, *62*, 107–110. [CrossRef] [PubMed]
5. Burt, S. Essential oils: Their antibacterial properties and potential applications in foods—A review. *Int. J. Food Microbiol.* **2004**, *94*, 223–253. [CrossRef] [PubMed]
6. Soliman, K.M.; Badeaa, R.I. Effect of oil extracted from some medicinal plants on different mycotoxigenic fungi. *Food Chem. Toxicol.* **2002**, *40*, 1669–1675. [CrossRef]
7. Budri, P.E.; Silva, N.C.C.; Bonsaglia, E.C.R.; Júnior, A.F.; Júnior, J.P.A.; Doyama, J.T.; Gonçalves, J.L.; Santos, M.V. Effect of essential oils of *Syzygium aromaticum* and *Cinnamomum zeylanicum* and their major components on biofilm production in *Staphylococcus aureus* strains isolated from milk of cows with mastitis. *J. Dairy Sci.* **2015**, *98*, 5899–5904. [CrossRef] [PubMed]
8. Rajkowska, K.; Nowicka-Krawczyk, P.; Kunicka-Styczyńska, A. Effect of clove and thyme essential oils on *Candida* biofilm formation and the oil distribution in yeast cells. *Molecules* **2019**, *24*, 1954. [CrossRef]
9. Adil, M.; Singh, K.; Verma, P.K.; Khan, A.U. Eugenol-induced suppression of biofilm-forming genes in *Streptococcus mutans*: An approach to inhibit biofilms. *J. Glob. Antimicrob. Resist.* **2014**, *2*, 286–292. [CrossRef] [PubMed]
10. Silvestri, J.D.F.; Paroul, N.; Czyewski, E.; Lerin, L.; Rotava, I.; Cansian, R.L.; Mossi, A.; Toniazzo, G.; de Oliveira, D.; Treichel, H. Chemical composition and antioxidant and antibacterial activities of clove essential oil (*Eugenia caryophyllata* Thunb). *Rev. Ceres* **2018**, *57*, 589–594. [CrossRef]
11. Radünz, M.; da Trindade, M.L.M.; Camargo, T.M.; Radünz, A.L.; Dellinghausen Borges, C.; Gandra, E.A.; Helbig, E. Antimicrobial and antioxidant activity of unencapsulated and encapsulated clove (*Syzygium aromaticum*, L.) essential oil. *Food Chem.* **2019**, *276*, 180–186. [CrossRef]
12. Zhang, Y.; Wang, Y.; Zhu, X.; Cao, P.; Wei, S.; Lu, Y. Antibacterial and antibiofilm activities of eugenol from essential oil of *Syzygium aromaticum* (L.) merr. & L. M. Perry (clove) leaf against periodontal pathogen *Porphyromonas gingivalis*. *Microb. Pathog.* **2017**, *113*, 396–402. [CrossRef] [PubMed]
13. Ebani, V.V.; Najar, B.; Bertelloni, F.; Pistelli, L.; Mancianti, F.; Nardoni, S. Chemical composition and in vitro antimicrobial efficacy of sixteen essential oils against *Escherichia coli* and *Aspergillus fumigatus* isolated from poultry. *J. Vet. Sci.* **2018**, *5*, 62. [CrossRef] [PubMed]

14. Andrade, B.F.M.T.; Nunes Barbosa, L.; Probst, I.P.; Fernandes Júnior, A. Antimicrobial activity of essential oils. *J. Essent. Oil Res.* **2017**, *26*, 34–40. [CrossRef]
15. Besten, M.A.; Jasinski, V.C.; Costa, Â.D.G.; Nunes, D.S.; Sens, S.L.; Wisniewski, A., Jr.; Simionatto, E.L.; Riva, D.; Dalmarco, J.B.; Granato, D. Chemical composition similarity between the essential oils isolated from male and female specimens of each five *Baccharis* species. *J. Braz. Chem. Soc.* **2020**, *23*, 1041–1047. [CrossRef]
16. Hu, Q.; Zhou, M.; Wei, S. Progress on the Antimicrobial Activity Research of Clove Oil and Eugenol in the Food Antisepsis Field. *J. Food Sci.* **2018**, *83*, 1476–1483. [CrossRef]
17. Teles, A.M.; Silva-Silva, J.V.; Fernandes, J.M.P.; Abreu-Silva, A.L.; Calabrese, K.D.S.; Mendes Filho, N.E.; Mouchrek, A.N.; Almeida-Souza, F. GC-MS Characterization of Antibacterial, Antioxidant, and Antitrypanosomal Activity of *Syzygium aromaticum* Essential Oil and Eugenol. *Evid. Based Complement Alternat. Med.* **2021**, *2021*, 6663255. [CrossRef]
18. Lalami, A.E.; Moukhafi, K.; Bouslamti, R.; Lairini, S. Evaluation of antibacterial and antioxidant effects of cinnamon and clove essential oils from Madagascar. *Mater. Today Proc.* **2019**, *13*, 762–770. [CrossRef]
19. Cimanga, K.; Kambu, K.; Tona, L.; Apers, S.; De Bruyne, T.; Hermans, N.; Totté, J.; Pieters, L.; Vlietinck, A.J. Correlation between chemical composition and antibacterial activity of essential oils of some aromatic medicinal plants growing in the democratic republic of Congo. *J. Ethnopharmacol.* **2020**, *79*, 213–220. [CrossRef]
20. Gislene, G.F.; Paulo, C.; Giuliana, L. Antibacterial activity of plant extracts and phytochemicals on antibiotic resistant bacteria. *Braz. J. Microbiol.* **2000**, *31*, 314–325. [CrossRef]
21. Hili, P.; Evans, C.S.; Veness, R.G. Antimicrobial action of essential oils: The effect of dimethylsulphoxide on the activity of cinnamon oil. *Lett. Appl. Microbiol.* **1997**, *24*, 269–275. [CrossRef] [PubMed]
22. Nzeako, B.C.; Al-Kharousi, Z.S.; Al-Mahrooqui, Z. Antimicrobial activities of clove and thyme extracts. *Sultan Qaboos Univ. Med. J.* **2006**, *6*, 33–39. [PubMed]
23. Thosar, N.; Basak, S.; Bahadure, R.N.; Rajurkar, M. Antimicrobial efficacy of five essential oils against oral pathogens: An in vitro study. *Eur. J. Dent.* **2013**, *07*, S071–S077. [CrossRef]
24. Fu, Y.; Zu, Y.; Chen, L.; Shi, X.; Wang, Z.; Sun, Z.; Effert, T. Antimicrobial activity of clove and rosemary essential oils alone and in combination. *Phytother. Res.* **2007**, *21*, 989–994. [CrossRef]
25. Nuñez, L.; D'Aquino, M. Microbicide activity of clove essential oil (*Eugenia caryophyllata*). *Braz. J. Microbiol.* **2012**, *43*, 1255–1260. [CrossRef] [PubMed]
26. Xu, J.G.; Liu, T.; Hu, Q.P.; Cao, X.M. Chemical composition, antibacterial properties and mechanism of action of essential oil from clove buds against *Staphylococcus aureus*. *Molecules* **2016**, *21*, 1194. [CrossRef]
27. Panizzi, L.; Falmini, G.; Cioni, P.L.; Morelli, I. Composition and antimicrobial properties of essential oil of four Mediterranean Lamiaceae. *J. Ethnopharm.* **1993**, *39*, 167–170. [CrossRef]
28. Rana, I.S.; Rana, A.S.; Rajak, R.C. Evaluation of antifungal activity in essential oil of the *Syzygium aromaticum* (L.) by extraction, purification and analysis of its main component eugenol. *Braz. J. Microbiol.* **2011**, *42*, 1269–1277. [CrossRef] [PubMed]
29. Nunez, L.; D'aquino, M.; Chirife, J. Antifungal properties of clove oil (*Eugenia caryophylata*) in sugar solution. *Braz. J. Microbiol.* **2001**, *32*, 123–126. [CrossRef]
30. Ahmad, N.; Alam, M.K.; Shehbaz, A.; Khan, A.; Mannan, A.; Rashid, H.S.; Bisht, B.; Owais, M. Antimicrobial activity of clove oil and its potential in the treatment of vaginal candidiasis. *J. Drug Target.* **2005**, *13*, 555–561. [CrossRef] [PubMed]
31. Pinto, E.; Vale-Silva, L.; Cavaleiro, C.; Salgueiro, L. Antifungal activity of the clove essential oil from *Syzygium aromaticum* on *Candida*, *Aspergillus* and dermatophyte species. *J. Med. Microbiol.* **2009**, *58*, 1454–1462. [CrossRef]
32. Chami, N.; Chami, F.; Bennis, S.; Trouillas, J.; Remmal, A. Antifungal treatment with carvacrol and eugenol of oral candidiasis in immunosuppressed rats. *Braz. J. Infect. Dis.* **2004**, *8*, 217–226. [CrossRef] [PubMed]
33. Chee, H.Y.; Lee, M.H. Antifungal activity of clove essential oil and its volatile vapour against dermatophytic fungi. *Mycobiology* **2007**, *35*, 241–243. [CrossRef] [PubMed]
34. Jain, S.K.; Agrawal, S.C. Fungistatic activity of some perfumes against otomycotic pathogens. *Mycoses* **2002**, *45*, 88–90. [CrossRef] [PubMed]
35. Gutierrez, J.; Barry-Ryan, C.; Bourke, P. Antimicrobial activity of plant essential oils using food model media, efficacy, synergistic potential and interactions with food components. *Food Microbiol.* **2009**, *26*, 142–150. [CrossRef]
36. Moreira, M.R.; Ponce, A.G.; Del Valle, C.E.; Roura, S.I. Effects of clove and tea tree oils on *Escherichia coli* O157:H7 in blanched spinach and minced cooked beef. *J. Food Process. Preserv.* **2007**, *31*, 379–391. [CrossRef]
37. Aguilar-Gonzalez, A.E.; Palou, E.; Lopez-Malo, A. Antifungal activity of essential oils of clove (*Syzygium aromaticum*) and/or mustard (*Brassica nigra*) in vapor phase against gray mold (*Botrytis cinerea*) in strawberries. *Innov. Food Sci. Emerg. Technol.* **2015**, *32*, 181–185. [CrossRef]
38. Lindsay, D.; Brözel, V.S.; von Holy, A. Spore formation in *Bacillus subtilis* biofilms. *J. Food Protect.* **2005**, *68*, 860–865. [CrossRef]
39. Kalemba, D.; Kunicka, A. Antibacterial and antifungal properties of essential oils. *Curr. Med. Chem.* **2003**, *10*, 813–829. [CrossRef]
40. Kunicka-Styczyńska, A.; Tyfa, A.; Laskowski, D.; Plucińska, A.; Rajkowska, K.; Kowal, K. Clove Oil (*Syzygium aromaticum* L.) Activity Against *Alicyclobacillus acidoterrestris* Biofilm on Technical Surfaces. *Molecules* **2020**, *25*, 3334. [CrossRef]
41. Li, T.; Wang, D.; Liu, N.; Ma, Y.; Ding, T.; Mei, Y.; Li, J. Inhibition of Quorum Sensing-Controlled Virulence Factors and Biofilm Formation in *Pseudomonas fluorescens* by Cinnamaldehyde. *Int. J. Food Microbiol.* **2018**, *269*, 98–106. [CrossRef]

42. Kubesová, A.; Šalplachta, J.; Horká, M.; Růžička, F.; Šlais, K. *Candida "Psilosis"*—Electromigration Techniques and MALDI-TOF Mass Spectrometry for Phenotypical Discrimination. *Analyst* **2012**, *137*, 1937–1943. [CrossRef] [PubMed]
43. Rams, T.E.; Sautter, J.D.; Getreu, A.; van Winkelhoff, A.J. Phenotypic Identification of Porphyromonas Gingivalis Validated with Matrix-Assisted Laser Desorption/Ionization Time-of-Flight Mass Spectrometry. *Microb. Pathog.* **2016**, *94*, 112–116. [CrossRef] [PubMed]
44. Kačániová, M.; Terentjeva, M.; Galovičová, L.; Ivanišová, E.; Štefániková, J.; Valková, V.; Borotová, P.; Kowalczewski, P.Ł.; Kunová, S.; Felšöciová, S.; et al. Biological Activity and Antibiofilm Molecular Profile of *Citrus aurantium* Essential Oil and Its Application in a Food Model. *Molecules* **2020**, *25*, 3956. [CrossRef]
45. Kačániová, M.; Galovičová, L.; Ivanišová, E.; Vukovic, N.L.; Štefániková, J.; Valková, V.; Borotová, P.; Žiarovská, J.; Terentjeva, M.; Felšöciová, S.; et al. Antioxidant, Antimicrobial and Antibiofilm Activity of Coriander (*Coriandrum Sativum* L.) Essential Oil for Its Application in Foods. *Foods* **2020**, *9*, 282. [CrossRef] [PubMed]
46. Adams, R.P. *Identification of Essential Oil Components by Gas Chromatography/Mass Spectrometry*; Allured Publishing Corporation: Carol Stream, IL, USA, 2007; ISBN 978-1-93233-11-4.
47. Van Den Dool, H.; Kratz, P.D. A Generalization of the Retention Index System Including Linear Temperature Programmed Gas—Liquid Partition Chromatography. *J. Chromatogr. A* **1963**, *11*, 463–471. [CrossRef]
48. Dao, T.; Bensoussan, M.; Gervais, P.; Dantigny, P. Inactivation of Conidia of *Penicillium chrysogenum*, *P. digitatum* and *P. italicum* by Ethanol Solutions and Vapours. *Int. J. Food Microbiol.* **2008**, *122*, 68–73. [CrossRef]
49. Talibi, I.; Askarne, L.; Boubaker, H.; Boudyach, E.H.; Msanda, F.; Saadi, B.; Aoumar, A.A.B. Antifungal activity of some Moroccan plants against *Geotrichum candidum*, the causal agent of postharvest citrus sour rot. *Crop Prot.* **2013**, *35*, 41–46. [CrossRef]
50. Aman, M.; Rai, V.R. Antifungal activity of fungicides and plant extracts against yellow sigatoka disease causing *Mycosphaerella musicola*. *Curr. Res. Environ. Appl. Mycol.* **2015**, *5*, 277–284. [CrossRef]

plants

Article

Artemisia annua Growing Wild in Romania—A Metabolite Profile Approach to Target a Drug Delivery System Based on Magnetite Nanoparticles

Adina-Elena Segneanu [1,*], Catalin Nicolae Marin [1], Ioan Ovidiu-Florin Ghirlea [2], Catalin Vladut Ionut Feier [2], Cornelia Muntean [3] and Ioan Grozescu [3]

[1] Faculty of Physics, West University of Timisoara, Blvd. V. Parvan 4, 300223 Timisoara, Romania; catalin.marin@e-uvt.ro
[2] Faculty of Medicine, Victor Babes University of Medicine and Pharmacy Timisoara, 2 P-ta E. Murgu, 300041 Timisoara, Romania; ovidiu.ghirlea@gmail.com (I.O.-F.G.); catalinfeier10@gmail.com (C.V.I.F.)
[3] Faculty of Industrial Chemistry and Environmental Engineering, Polytechnic University Timisoara, 2 P-ta Victoriei, 300006 Timisoara, Romania; cornelia.muntean@upt.ro (C.M.); ioangrozescu@gmail.com (I.G.)
* Correspondence: adina.segneanu@e-uvt.ro

Abstract: The metabolites profile of a plant is greatly influenced by geographical factors and the ecological environment. Various studies focused on artemisinin and its derivates for their antiparasitic and antitumoral effects. However, after the isolation and purification stage, their pharmaceutical potential is limited due to their low bioavailability, permeability and lifetime. The antibacterial activity of essential oils has been another topic of interest for many studies on this plant. Nevertheless, only a few studies investigate other metabolites in *Artemisia annua*. Considering that secondary metabolites act synergistically in a plant, the existence of other metabolites with antitumor and high immunomodulating activity is even more important. Novel nano-carrier systems obtained by loading herbs into magnetic nanoparticles ensures the increase in the antitumor effect, but also, overcoming the barriers related to permeability, localization. This study reported the first complete metabolic profile from wild grown Romanian *Artemisia annua*. A total of 103 metabolites were identified under mass spectra (MS) positive mode from 13 secondary metabolite categories: amino acids, terpenoids, steroids, coumarins, flavonoids, organic acids, fatty acids, phenolic acids, carbohydrates, glycosides, aldehydes, hydrocarbons, etc. In addition, the biological activity of each class of metabolites was discussed. We further developed a simple and inexpensive nano-carrier system with the intention to capitalize on the beneficial properties of both components. Evaluation of the nano-carrier system's morpho-structural and magnetic properties was performed.

Keywords: secondary metabolites; GC-MS; mass-spectra; bioactive compounds; nano-carrier system; magnetic nanoparticles

Citation: Segneanu, A.-E.; Marin, C.N.; Ghirlea, I.O.-F.; Feier, C.V.I.; Muntean, C.; Grozescu, I. *Artemisia annua* Growing Wild in Romania—A Metabolite Profile Approach to Target a Drug Delivery System Based on Magnetite Nanoparticles. *Plants* **2021**, *10*, 2245. https://doi.org/10.3390/plants10112245

Academic Editors: Ivayla N. Dincheva, Ilian Badjakov and Bistra Galunska

Received: 18 September 2021
Accepted: 13 October 2021
Published: 21 October 2021

Publisher's Note: MDPI stays neutral with regard to jurisdictional claims in published maps and institutional affiliations.

1. Introduction

Romanian phytotherapy has an ancient and very rich tradition based on a very wide diversity of medicinal plants. Thus, in the spontaneous flora of Romania there are about 800 species of medicinal plants. Additionally, plants of the genus Artemisia (*Asteraceae*) form part of this phytopharmacological treasure.

Artemisia annua (common name: wormwood or năfurica in Romanian) is one of the ancient healing plants recognized in traditional medicine from Europe and Asia. Romanian traditional medicine has exploited its therapeutic properties: antihemorrhagic, antiseptic, antioxidant, digestive, antipyretic, immunomodulatory, antibacterial and antitumoral [1–5]. *Artemisia annua* is used also in Romania to prepare a digestive wine [1–5]. In traditional Asian pharmacopoeia, it was recommended, especially for the treatment of fevers and colds.

Chemical screening of bioactive compounds isolated from this plant has highlighted the rich content in volatile compounds, terpenes, sesquiterpenes, alkaloids, flavonoids, coumarins and phenolic compounds.

Studies on biological properties of secondary metabolites isolated from extracts of *Artemisia annua* have confirmed its therapeutic properties (antioxidant, antidiabetic, antiviral, antitumor, immunomodulatory, antiparasitic, antibacterial and antifungal activity) described in traditional medicine [4–10].

Additionally, the latest studies in the field aim to evaluate its potential application in the prevention and treatment of SARS-CoV 19 [6–23].

In recent years, numerous studies have been conducted on the main classes of phytoconstituents (phenolic compounds, alkaloids, terpenes and volatile compounds) in sweet wormwood. Special attention was paid to the isolation and chemical synthesis of artemisinin, considered the main bioactive compound of the plant, with antiparasitic activity. This was an important step in modern research in medicinal plants and in the fight against malaria, an infectious disease that affects the lives of millions of people [6–20]. According to World Health Organization estimates, malaria killed more than 400 million people globally in 2019 alone [24].

Artemisinin is widely used for malaria treatment, and recent studies demonstrated its antitumoral activity on several cancer cell lines. However, at present, the production costs of artemisinin are high. Moreover, its low bioavailability, permeability and life-time in biological media represent the main biomedical limitations [25].

It is known that the biological activity of a plant is due to the synergistic effect of different types of phytoconstituents [26,27]. Therefore, the chemical screening and complete identification of bioactive components from a plant is especially important.

Recent studies have shown that the antitumor activity of the plant is not only due to artemisinin. This might be just one of the plant metabolites with high biological activity [6].

Although the plant has been studied for a long time, its chemical composition is influenced by many understudied factors, among which we only mention geographical position, climate, soil pH and so on [28].

For these reasons, studies on this plant are particularly important given its special therapeutic potential.

Additionally, some small peptides in the composition of this plant could be another key constituent to antitumor activity.

Currently, there are several therapeutic approaches to cancer, including drugs, genetics and anticancer peptides. The results of studies regarding peptide anticancer therapy showed that the sequence of these small anticancer peptides can include several amino acids, such as: arginine, glycine, lysine and leucine, glutamic and aspartic, tyrosine, phenylalanine, proline and protonated histidine. Moreover, recent research has showed that arginine has a key role in the function of the immune system and antitumor activity [29–31].

Nevertheless, there are very few studies that have investigated the amino acid composition of Artemisia genus [32].

Studies regarding wild Romanian Artemisia genus are few and targeted, especially on bioactive components of *Artemisia absinthium* [33,34]. Furthermore, regarding Romanian *Artemisia annua* wild plant, the present research only investigates the content of artemisinin and volatile oils [35–37].

Despite its high therapeutic potential, the chemical screening of the biologically active compounds of this medicinal plant from the spontaneous flora of Romania has not been performed yet.

On the other hand, it should be mentioned that several herbal supplements of *Artemisia annua* are marketed globally, which are recommended, according to the manufacturers, for malaria, arthritis and even cancer.

Recently, several cases of liver disease have been reported, especially following self-medication with herbal supplements to manage cancer or prevent malaria [38,39].

The development of highly efficient, selective, simple and inexpensive nano-carrier systems could be an effective method to avoid these risks while ensuring the controlled intake of phytoconstituents with biological activity [40,41].

Modern drug delivery systems based on magnetic nanoparticles could easily accomplish these requirements. Moreover, the latest developments regarding magnetite nanoparticles have demonstrated their benefits and recommend their use in drug delivery and other different biomedical applications (magnetic resonance imaging, tumor therapeutic hyperthermia, etc.) [25,42].

Magnetic/superparamagnetic nanoparticles could represent a more than interesting alternative due to their advantages: their capability for local delivery and the ability to act selectively. However, the possibility of an immune response is the main drawback of these nanoparticles. Therefore, the design of new drug delivery systems based on magnetic/superparamagnetic nanoparticles for use as early detection methods and in the diagnosis, prognosis and monitoring of the evolution of the cancer treatment is required given the social impact of these diseases [25,43–49].

Recent studies have shown that nano-carriers based on magnetic nanoparticles lead to a high drug tissue permeability and retention effect and thus enhance the beneficial therapeutic effects [45–47].

To our best knowledge, this study investigates the metabolite profile of *Artemisia annua* grown wild in Romania for the first time. Subsequently, a simple and inexpensive nano-carrier system that capitalizes both the therapeutic properties of *Artemisia annua* (whole plant) and magnetic Fe_3O_4 nanoparticles was developed.

2. Results and Discussion

Extensive research in the field of plants and especially on those with high therapeutic potential has shown that they have a very complex composition of compounds with high biological activity that act synergistically in the body [26,27].

Additionally, a full description of a general metabolic profile for a specific herb is all the more difficult, as significant differences in secondary metabolites was reported among the same plants harvested from various geographic regions of the world. Studies in this area confirmed that the content of specific plant secondary metabolites is the result of several environmental stress factors (climate, soil and biological conditions) which directly influence plant growth, development and topography distribution [28,50–52].

The pharmacological properties of different plant secondary metabolites were extensively investigated. However, their therapeutic benefits have not been completely understood [11,21,22]. Plant metabolites with peptide structures are just an example of this [51].

Bioactive metabolite chemical screening of sweet wormwood (năfurica) was tentatively carried out via gas-chromatography coupled with mass spectroscopy (GC-MS) and electrospray ionization–quadrupole time-of-flight mass spectrometry (ESI-QTOF-MS) analysis.

In addition, the amino acid profile was investigated using GC-MS techniques (Figure 1). The mass spectra of components identified were determined via comparison of their retention indices and mass spectra with those of NIST/EPA/NIH, the Mass Spectral Library 2.0 data base, as well as by reviewing the literature [32,53].

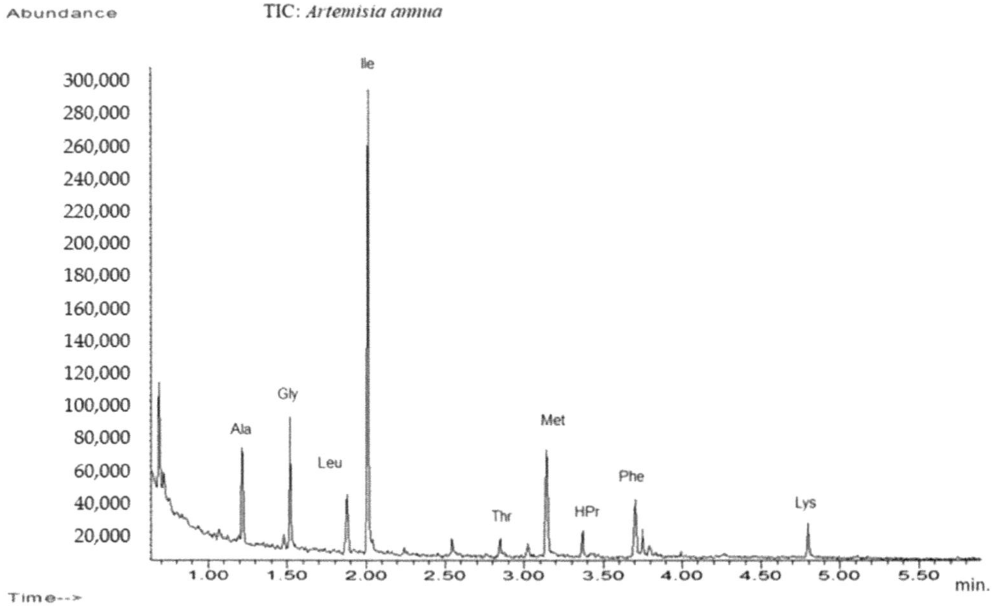

Figure 1. TIC chromatograms of GC-MS for *Artemisia annua*.

The results are listed in Table 1, which presents tentative amino acid identification via GC-MS corresponding to *Artemisia annua* sample [32,54,55].

Table 1. Main compounds identified by GC-MS analysis of plant extract.

Proposed Structure	Abbreviation	SIM (Selected Ion Monitoring)
Alanine	Ala	130, 70
Leucine	Leu	172, 86
Glycine	Gly	116, 74
Isoleucine	Ile	172, 130
Methionine	Met	203, 277
Phenylalanine	Phe	206, 190
Lysine	Lys	170, 128
Threonine	Thr	160, 101
4-Hydroxyproline	HYP	172, 86

2.1. *Mass Spectrometry Analysis of Romanian Artemisia annua*

The plant sample was diluted in methanol and analyzed using ESI-TOF mass spectroscopy (ESI-QTOF-MS). The plant extract sample analysis was carried out in the positive mode. The mass spectra (Figure 2) showed the presence of complex metabolite composition. A total of 103 compounds were detected and identified, which covered various chemical categories, including amino acids, sterols, terpenoids, flavonoids, coumarins, alcohols, aldehydes, glycosides, carbohydrates, fatty acids and so on, which confirmed the data reported in the literature [6–24,28,31–39,46,53–60]. Additionally, the presence of amino acids identified through GC-MS was confirmed via ESI-QTOF-MS analysis. However, only a few studies reported the amino acid profile of Artemisia annua [32].

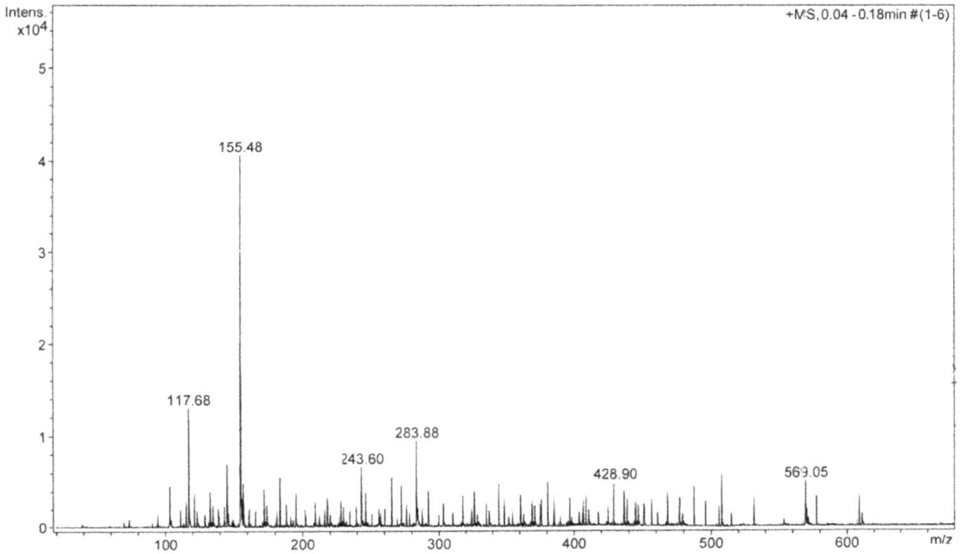

Figure 2. Positive ion mode MS-TOF of *Artemisia annua* sample.

The identified metabolites are listed in Table 2 and classified according to their m/z ratio (mass-to-charge-ratio) (theoretical and measured), chemical name and formula and the related literature.

Table 2. Phytoconstituents identified in *Artemisia annua* sample through MS analysis.

Compound Name	m/z Detected	Theoretic m/z	Formula	Tentative Identification	References
1	76.09	76.07	$C_2H_5NO_2$+	Glycine	[32,61]
2	90.11	90.097	$C_3H_7O_2$+	Alanine	[32,61]
3	103.19	103.17	$C_6H_{14}O$+	Hexanol	[7,8,10,18,23,28,33,36,37,62]
4	106.12	106.09	$C_3H_7NO_3$+	Serine	[32,61]
5	113.10	113.09	$C_4H_4N_2O_2$+	Uracil	[7,8,10,18,24,30,35,38,39,61]
6	118.13	118.14	$C_5H_{11}NO_2$+	Valine	[32,61]
7	120.05	120.03	$C_4H_9NO_3$+	Threonine	[32,61]
8	129.23	129.21	$C_8H_{16}O$+	Caprylaldehyde	[7,8,18,23,28,32,36,37,62]
9	132.14	132.13	$C_5H_9NO_3$+	L-hydroxyproline	[32,61]
10	132.17	132.18	$C_6H_{13}NO_2$+	Leucine	[32,61]
11	134.09	134.10	$C_4H_7NO_4$+	Aspartic acid	[32,61]
12	135.23	135.22	$C_{10}H_{14}$+	p-Cymene	[63]
13	136.21	136.19	C_7H_5NS+	Benzothiazole	[7,8]
14	137.25	137.23	$C_{10}H_{16}$+	Limonene	[9–12,14,19,20,22,23,46,52,59,63–65]
15	139.15	139.12	$C_7H_6O_3$+	Salicylic acid	[7–9,18,23,28,33,36,37,62]
16	147.16	147.14	$C_9H_6O_2$+	Coumarin	[9–12,14,19,20,22,23,46,52,59,60,63,64]
17	147.20	147.19	$C_6H_{14}N_2O_2$+	Lysine	[32,61]
18	148.12	148.13	$C_5H_9NO_4$+	Glutamic acid	[32,61]
19	149.21	149.205	$C_{10}H_{12}O$+	4-Isopropylbenzaldehyde	[7,8,18,23,28,33,36,37]

Table 2. *Cont.*

Compound Name	m/z Detected	Theoretic m/z	Formula	Tentative Identification	References
20	150.22	150.20	$C_6H_{11}NO_2S$+	Methionine	[32,61]
21	151.19	151.22	$C_{10}H_{14}O$+	Cuminol	[7,8,18,23,28,33,36,37,62]
22	153.25	153.233	$C_{10}H_{16}$+	Artemisia ketone	[5–7,9,11–14,16,19–23,35,46,58–60,63,65]
23	155.23	155.21	$C_{12}H_{10}$+	Capillene	[7,8,18,23,28,33,36,37,62]
24	155.27	155.25	$C_{10}H_{18}O$+	Geraniol	[9–12,14,19,20,22,23,46,52,59,61,63,64]
25	156.17	156.16	$C_6H_9N_3O_2$+	Histidine	[32,61]
26	157.29	157.26	$C_{10}H_{20}O$+	Menthol	[9–12,14,19,20,22,23,46,52,59,63,64,66]
27	162.15	162.13	$C_9H_5O_3$+	4-Hydroxycoumarin	[7,8,10,18,23,28,33,36,37,62]
28	165.19	165.16	$C_9H_8O_3$+	4-Hydroxycinnamic acid	[7,8,10,18,23,28,33,36,37,62]
29	165.18	165.20	$C_{10}H_{12}O_2$+	Eugenol	[9–12,14,19,20,22,23,46,52,59,63,64,66]
30	166.22	166.19	$C_9H_{11}NO_2$+	L-phenylalanine	[32,61]
31	171.29	171.33	$C_{12}H_{26}$+	Dodecane	[7,8,18,23,28,33,36,37,62]
32	175.19	175.20	$C_6H_{14}N_4O_2$+	L-arginine	[32,61]
33	181.17	181.16	$C_9H_8O_4$+	Caffeic acid	[7,8,18,23,28,33,36,37,62]
34	183.19	183.17	$C_9H_{10}O_4$+	2,4-Dihydroxy-6 methoxyacetophenone	[7,8,18,23,28,33,35,37]
35	193.15	193.17	$C_{10}H_8O_4$+	Scopoletin	[9–12,14,19,20,22,23,46,52,59,63,64,66]
36	197.21	197.20	$C_{10}H_{12}O_4$+	Xanthoxylin	[7,8,18,23,28,33,36,37,62]
37	197.30	197.29	$C_{12}H_{20}O_2$+	Artemisyl acetate	[5–7,9,11–14,16,19–22,35,46,58–60,63,65]
38	207.21	207.19	$C_{11}H_{10}O_4$+	Scoparone	[9–12,14,19,20,22,23,46,52,58,63,64,66]
39	205.33	205.35	$C_{15}H_{24}$+	Germacrene D	[9–12,14,19,20,22,23,46,52,58,63,64,66]
40	207.39	207.37	$C_{15}H_{26}$+	Cadinene	[9–12,14,19,20,22,23,46,52,58,63,64,66]
41	220.23	220.24	$C_{10}H_{13}N_5O$+	Zeatin	[7,8,18,23,28,33,36,37,62]
42	221.36	221.35	$C_{15}H_{24}O$+	Spathulenol	[9–12,14,19,20,22,23,46,52,58,63,64,66]
43	223.38	223.37	$C_{15}H_{26}O$+	Farnesol	[9–12,14,19,20,22,23,46,52,58,63,64,66]
44	229.35	229.37	$C_{14}H_{28}O_2$+	Myristic acid	[7,8,18,23,28,33,36,37,62]
45	233.30	233.32	$C_{15}H_{20}O_2$+	Alantolactone	[64]
46	235.29	235.33	$C_{15}H_{22}O_2$+	Artemisinic acid	[5–7,9,11–14,16,19–22,35,46,58–60,63,65]
47	237.37	237.35	$C_{15}H_{24}O_2$+	Corymbolone	[9–12,14,19,20,22,23,46,52,59,63,64,66]
48	246.33	246.32	$C_{15}H_{19}NO_2$+	Rupestine	[9–12,14,19,20,22,23,46,52,59,63,64,66]
49	265.31	265.32	$C_{15}H_{20}O_4$+	Abscisic acid	[5–7,9,11–14,16,19–22,35,46,58–60,63,65]
50	241.48	241.5	$C_{17}H_{36}$+	Heptadecane	[7,8,18,23,28,33,36,37,62]
51	247.29	247.30	$C_{15}H_{18}O_3$+	α-Santonin	[64]

Table 2. *Cont.*

Compound Name	m/z Detected	Theoretic m/z	Formula	Tentative Identification	References
52	249.30	249.32	$C_{15}H_{20}O_3$+	Arteannuin B	[5–7,9,11–14,16,19–22,35,46,58–60,63,65]
53	255.35	255.5	$C_{18}H_{38}$+	n-Octadecane	[7,8,18,23,28,33,36,37,62]
54	257.43	257.42	$C_{16}H_{32}O_2$+	Palmitic acid	[7,8,18,23,28,33,36,37,62]
55	263.42	263.40	$C_{17}H_{26}O_2$+	α-Bergamotol acetate	[7,8,18,23,28,33,36,37,62]
56	267.30	267.33	$C_{15}H_{22}O_4$+	Germacranolide	[64]
57	271.39	271.40	$C_{15}H_{10}O_5$+	Apigenin	[56–60,67–71]
58	273.27	273.25	$C_{15}H_{12}O_5$+	Naringenin	[56–60,67–71]
59	281.34	281.32	$C_{15}H_{20}O_5$+	Artemisitene	[5–7,9,11–14,16,19–22,35,46,58–60,63,65]
60	283.34	283.33	$C_{15}H_{22}O_5$+	Artemisinine	[5–7,9,11–14,16,19–22,35,46,58–60,63,65]
61	285.27	285.26	$C_{16}H_{12}O_5$+	Acacetin	[56–60,67–71]
62	287.26	287.24	$C_{15}H_{10}O_6$+	Luteolin	[56–60,67–71]
63	289.27	289.25	$C_{15}H_{12}O_6$+	Eriodyctiol	[56–60,67–71]
64	297.51	297.50	$C_{20}H_{40}O$+	Phytol	[9–12,14,19,20,22,23,46,52,59,63,64,66]
65	301.28	301.26	$C_{16}H_{12}O_6$+	Rhamnocitrin	[56–60,67–71]
66	303.21	303.23	$C_{15}H_{10}O_7$+	Quercetin	[56–60,68,69]
67	305.22	305.25	$C_{15}H_{12}O_7$+	Taxifolin	[56–60,64,67–71]
68	315.31	315.29	$C_{17}H_{14}O_6$+	Cirsimaritin	[56–60,64,67–71]
69	317.24	317.26	$C_{16}H_{12}O_7$+	Rhamnetin	[56–60,67–71]
70	317.37	317.40	$C_{19}H_{24}O_4$+	Capillartemisin B	[56–60,64]
71	319.20	319.23	$C_{15}H_{10}O_8$+	Quercetagetin	[56–60,64,67–71]
72	324.57	324.60	$C_{22}H_{44}O$+	2-Docosanone	[7,8,18,23,28,33,36,37,62]
73	331.26	331.29	$C_{17}H_{14}O_7$+	Rhamnazin	[56–60,64,67–71]
74	333.24	333.26	$C_{16}H_{12}O_8$+	Laricitrin	[56–60,64,67–71]
75	339.67	339.70	$C_{24}H_{50}$+	n-Tetracosane	[7,8,18,23,28,33,36,37,62]
76	345.32	345.30	$C_{18}H_{16}O_7$+	Eupatorine	[56–60,64,67–71]
77	347.27	347.30	$C_{17}H_{14}O_8$+	Syringetin	[56–60,64,67–71]
78	353.68	353.70	$C_{25}H_{52}$+	n-Pentacosane	[7,8,18,23,28,33,36,37,62]
79	354.29	354.31	$C_{16}H_{18}O_9$+	Scopoline	[9–12,14,19,20,22,23,46,52,59,63,64,66]
80	359.31	359.30	$C_{19}H_{18}O_7$+	Retusin	[56–60,64,67–71]
81	361.33	361.30	$C_{18}H_{16}O_8$+	Chrysosplenol D	[56–60,64,67–71]
82	367.71	367.7	$C_{26}H_{54}$+	n-Hexacosane	[7,8,18,23,28,33,36,37,62]
83	375.28	375.30	$C_{19}H_{18}O_8$+	Chrysosplenetin	[56–60,64,67–71]
84	389.39	389.40	$C_{20}H_{20}O_8$+	Artemitin	[9,16,17,21,23,56–60,65]
85	375.28	375.30	$C_{19}H_{18}O_8$+	Casticin	[56–60,64,67–71]
86	377.41	377.40	$C_{20}H_{24}O_7$+	Euparotin	[56–60]
87	411.68	411.70	$C_{30}H_{50}$+	Squalene	[9–12,14,19,20,22,23,46,52,59,63,64,66]
88	413.72	413.70	$C_{29}H_{48}O$+	Stigmasterol	[7,8,18,24,30,35,38,39]
89	415.67	415.70	$C_{29}H_{50}O$+	β-Sitosterol	[7,8,18,23,28,33,36,37,62]
90	425.69	425.70	$C_{30}H_{48}O$+	Taraxasterone	[9–12,14,19,20,22,23,46,52,59,63,64,66]
91	427.71	427.70	$C_{30}H_{50}O$+	Beta-amyrin	[9–12,14,19,20,22,23,46,52,59,63,64,66]

Table 2. *Cont.*

Compound Name	m/z Detected	Theoretic m/z	Formula	Tentative Identification	References
92	433.37	433.40	$C_{21}H_{20}O_{10}+$	Apigenin 7-*O*-glucoside	[64]
94	443.52	443.50	$C_{25}H_{30}O_{7}+$	Tomentin A	[9–12,14,19,20,22,23,46,52,59,63, 64,66]
95	447.38	447.40	$C_{22}H_{22}O_{10}+$	Kaempferide 3-rhamnoside	[57]
96	449.37	449.40	$C_{21}H_{20}O_{11}+$	Cymaroside	[56–60,64,67–71]
97	457.71	457.70	$C_{30}H_{48}O_{3}+$	Oleanic acid	[7,8,18,22,27,31,34,35,60]
98	465.38	465.40	$C_{21}H_{20}O_{12}+$	Isoquercetin	[56–60,64,67–71]
99	495.43	495.40	$C_{22}H_{22}O_{13}+$	Patuletin 3-glucoside	[64]
100	505.37	505.40	$C_{18}H_{32}O_{16}+$	Sophorotriose	[64]
101	517.42	517.40	$C_{25}H_{24}O_{12}+$	Cynarine	[7,8,18,22,27,31,34,35,60]
102	577.77	577.80	$C_{35}H_{60}O_{6}+$	Daucosterol	[7,8,18,22,27,31,34,35,60]
103	611.51	611.50	$C_{27}H_{30}O_{16}+$	Rutin	[56–60,64,67–72]

2.2. Screening and Classification of the Differential Metabolites

The 103 phytochemicals identified through mass spectroscopy were assigned to different chemical classes: terpenoids and sesquiterpenoids (27.2%), flavonoids (24.2%), amino acids (12.6%), hydrocarbons (6.8%), coumarins (4.85%), phenolic acids (2.9%), sterol and steroids (2.9%), fatty acids (2.9%), glycosides (1.9%), hydrocarbons (6.8%), organic acids and esters (3.8%), carbohydrates (0.97%) and miscellaneous (Table 3). Terpenoids and sesquiterpenoids, flavonoids and amino acids constitute the largest group of bioactive compounds from *Artemisia annua*. The distribution of identified metabolites in various chemical categories is listed in Table 3.

Table 3. Classification of metabolites from *Artemisia annua* sample on various chemical categories.

Chemical Class	Metabolite Name
Amino Acids	Glycine
	Alanine
	Serine
	Valine
	Threonine
	L-hydroxyproline
	Leucine
	Aspartic acid
	Lysine
	Glutamic acid
	Methionine
	Hystidine
	L-phenylalanine
	L-arginine

Table 3. *Cont.*

Chemical Class	Metabolite Name
Terpenoids and Sesquiterpenoids	Artemisine
	Artemisinine
	Limonele
	p-Cymene
	Beta-amyrin
	Artesimic acid
	Eugenol
	Menthol
	Artemisia ketone
	Spathulenol
	Artemisyl acetate
	Artemisinic acid
	Phytol
	Rupestine
	α-Santonin
	Arteannuin B
	Farnesol
	Corymbolone
	Abscisic acid
	Alantolactone
	Artemisitene
	Geraniol
	Squalene
	Taraxasterone
	Beta-amyrin
	Germacranolide
	Germacrene D
	Cadinene
Coumarins	Scopoletin
	Tomentin A
	Coumarin
	Scopolin
	Scoparone

Table 3. *Cont.*

Chemical Class	Metabolite Name
Flavonoids	Apigenin
	Chrysosplenetin
	Rhamnazin
	Luteolin
	Naringenin
	Capillartemisin B
	Rutin
	Quercetin
	Quercetagetin
	Acacetin
	Rhamnetin
	Eupatorin
	Syringetin
	Laricitrin
	Eriodictiol
	Casticin
	Chrysosplenol D
	Retusin
	Cynaroside
	Artemitin
	Taxifolin
	Isoquercetin
	Rhamnocitrin
	Kaempferide 3-rhamnoside
	Cirsimaritin
Phenolic Acids	4-Hydroxycoumarin
	4-Hydroxycinnamic acid
	Caffeic acid
Sterol and Steroids	β-Sitosterol
	Stigmasterol
	Daucosterol
Fatty Acid	Oleanic acid
	Palmitic acid
	Myristic acid
Hydrocarbons	n-Octadecane
	Heptadecane
	n-Tetracosane
	n-Hexacosane
	Dodecane
	n-Pentacosane
	Capillene
Glycoside	Patuletin 3-glucoside
	Apigenin 7-*O*-glucoside
Carbohydrates	Sophorotriose
Aldehyde and Ketone	Caprylaldehyde
	4-Isopropylbenzaldehyde
	Hexanol
	2-Docosanone
	2,4-Dihydroxy-6-methoxyacetophenone
Organic Acids and Esters	Salicylic acid
	α-Bergamotol acetate
	Xanthoxylin
	Cynarine

Table 3. *Cont.*

Chemical Class	Metabolite Name
Other	Uracil
	Cuminol
	Benzothiazole
	Zeatin

On the basis of the data analysis reported in Table 3, the metabolites classification chart was obtained, represented in Figure 3.

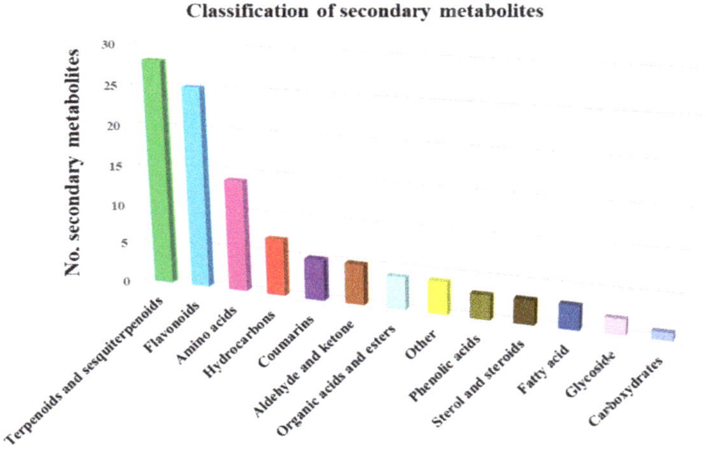

Figure 3. *Artemisia annua* metabolite classification.

Amino acids a total of 14 different amino acids were identified in the plant extract. Additionally, the essential amino acids (valine, leucine, methionine, hystidine and L-phenylalanine) represent 35% of them. The non-essential amino acids are present in a much larger proportion (65%), being representatives of: glycine, alanine, serine, arginine, threonine, acid aspartic, lysine and glutamic acid [31–33,73]. As was reported, amino acids exhibit antitumoral, antiproliferative and immunomodulant activity [31–33,73].

Terpenoids found in the *Artemisia annua* sample are the one of the major constituents of the total identified metabolites. Previous studies on the therapeutic effect of terpenoids have demonstrated their antimicrobial, antibacterial, antifungal, analgesic and anti-insect activity [74].

Sesquiterpenes, another important class of metabolites from *Artemisia annua*, were shown to have antitumoral, antiplasmodial, anti-inflammatory and anti-allergic properties. Sesquiterpenes lactones isolated from *Artemisia annua* are used in antimalaria drugs [75,76].

Coumarins are metabolites highly relevant to human health. Recent studies on coumarins isolated from plants have shown antioxidant, antimicrobial, antiviral, antifungal, and antiparasitic, anti-diabetic, analgesic, anti-neurodegenerative, and anti-inflammatory activity. Moreover, coumarins have been demonstrated to stimulate the immunologic response and are used in the therapy of different tumors: leukemia, renal and prostate tumors, melanoma and breast cancer [77,78].

Flavonoids were another major category of metabolites identified in the plant sample. A total of 25 different flavonoids were found in the *Artemisia annua* sample. These compounds exhibit antioxidant, antitumoral, anti-inflammatory, antimicrobial, anti-cholinesterase, neurodegenerative disease (Alzheimer) and atherosclerosis prevention effects [9–81].

Phenolic acids have shown anti-inflammatory, antioxidant, antimicrobial, neuroprotective, antidiabetic and anticancer effects [82,83].

Sterol and steroids from herbs act as antitumoral, anti-inflammatory, antioxidant antiatherosclerotic agents [84].

Fatty acids are involved in neuroprotection and cardiovascular protection mechanisms. Recent studies reported their beneficial role in autoimmune and neurodegenerative diseases, including Alzheimer disease (AD) [85].

Carbohydrates have shown anti-inflammatory, antioxidant, antiviral, antibacterial, antidiabetic, antitumoral, immunomodulatory and cardioprotective activity [86–88].

Glycosides from herbs showed antitumoral activity, mainly on leukemia and gastric cancer [89].

2.3. Nano-Carrier System Based on Magnetic Nanoparticles of Fe_3O_4

The development of an efficient and selective drug nano-carrier system required an optimal ratio between the herb and magnetic nanoparticles in order to provide the highest biological activity and functionality (selectivity and vectorization). Recent studies regarding types of nano-drug systems have reported the specific bioactive phytochemicals that were loaded into the magnetic nanoparticles [90,91].

2.4. FT-IR Spectroscopy

The incorporation of herb phytochemicals into the pores of Fe_3O_4 nanoparticles was successfully achieved and was confirmed through FT-IR spectroscopy. Figure 4 presents the spectra of the herb, Fe_3O_4 nanoparticles and the nano-carrier system.

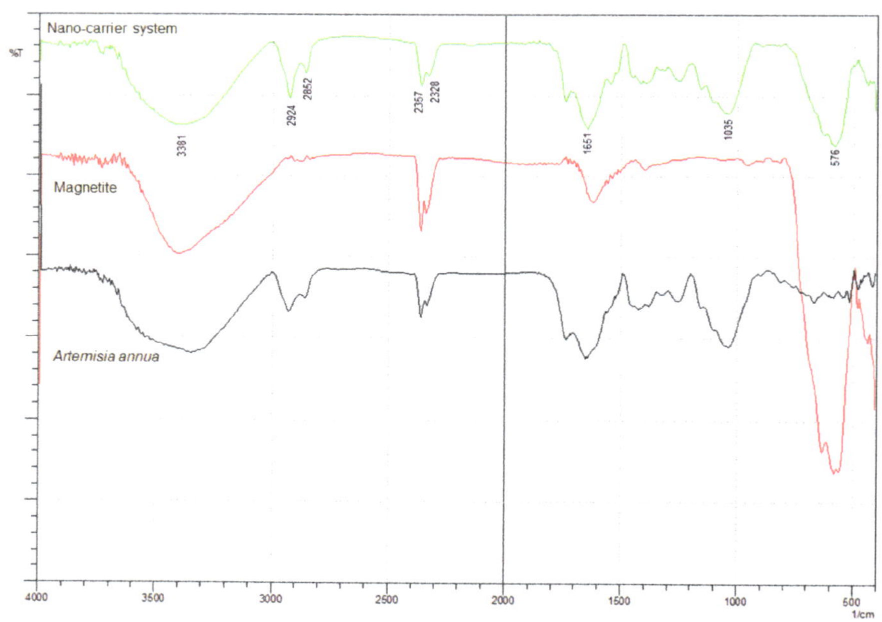

Figure 4. FT-IR spectra of *Artemisia annua*, magnetite nanoparticles and nano-carrier system.

The FT-IR spectra of the herb display the characteristic absorption peaks of *Artemisia annua* (Figure 4). The characteristic group frequencies of different organic molecules detected in *Artemisia annua* can be attributed to flavonoids (1703, 1580, 1460, 630 and 575 cm^{-1}), amino acids (1651, 1580, 1555 and 1545 cm^{-1}), terpenoids (1740, 1651 and

810 cm^{-1}), carbohydrates (3381, 1462, 1126 and 840 cm^{-1}) and fatty acids (2925, 2852, 1250 and 720 cm^{-1}) [47,92–101].

Withal, the band at 1737 cm^{-1} is the characteristic absorption peak of *Artemisia annua* assigned to δ-lactone group [47].

Two main broad metal–oxygen bands are seen in the IR spectra of Fe_3O_4 nanoparticles (Figure 4) in the range 400–600 cm^{-1}. The highest vibration band at 576 cm^{-1} is assigned to the stretching vibrations of $M_{tetra}O$ bond in the tetrahedral voids, and the lowest band at 410 cm^{-1} (partially visible) corresponds to the stretching vibrations of the $M_{octa}O$ bond in the octahedral void peak [102–104].

The spectra of the nano-carrier system (Figure 4) display the characteristic peaks of the herb as well as the metal–oxygen vibration bands at 576 cm^{-1} and at 410 cm^{-1}, which confirm the incorporation of the herb into the pores of Fe_3O_4 nanoparticles [47].

2.5. X-ray Diffraction Spectroscopy

Figures 5–8 present the XRD patterns of Fe_3O_4 nanoparticles, the herb and the nano-carrier system.

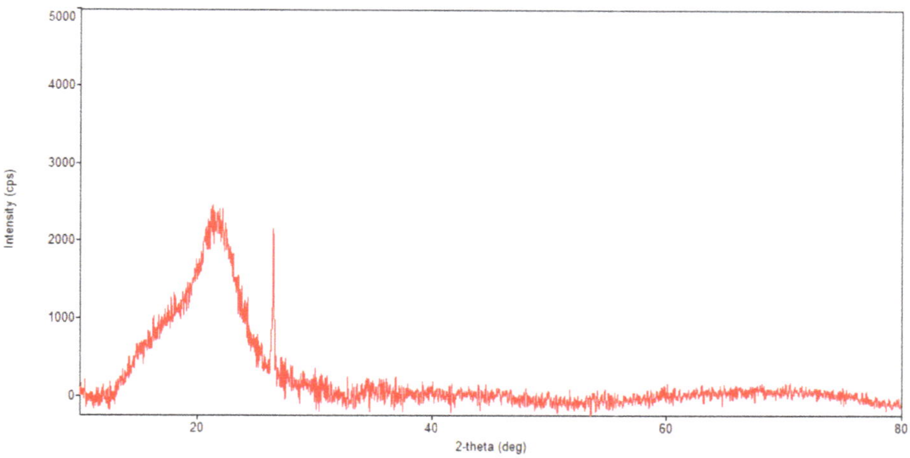

Figure 5. Powder XRD patterns of *Artemisia annua*.

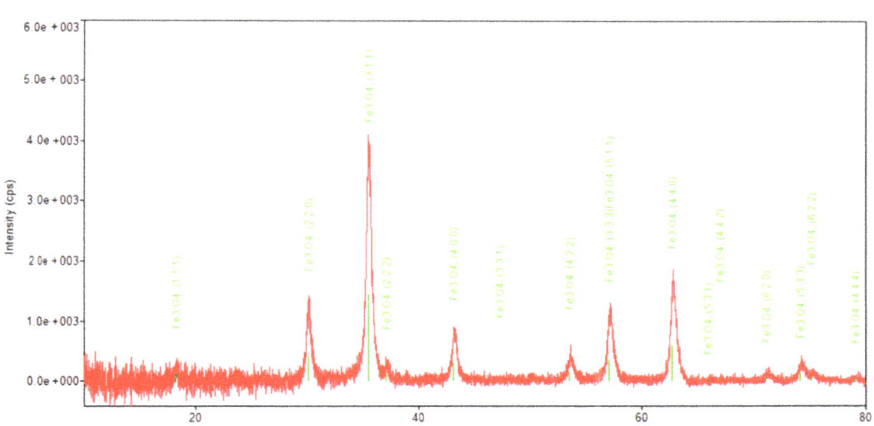

Figure 6. Powder XRD patterns of Fe_3O_4 nanoparticles.

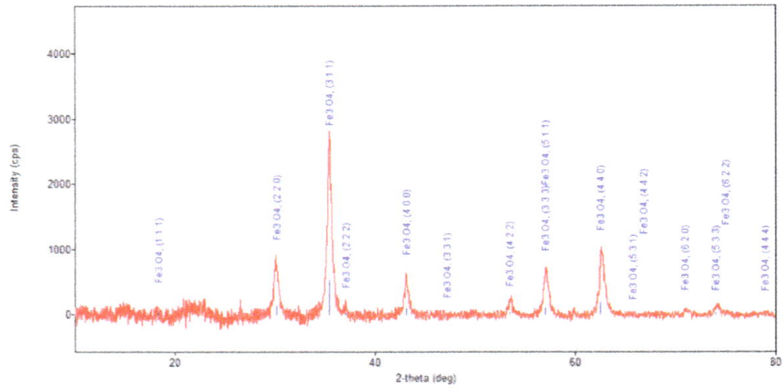

Figure 7. Powder XRD patterns of nano-carrier system.

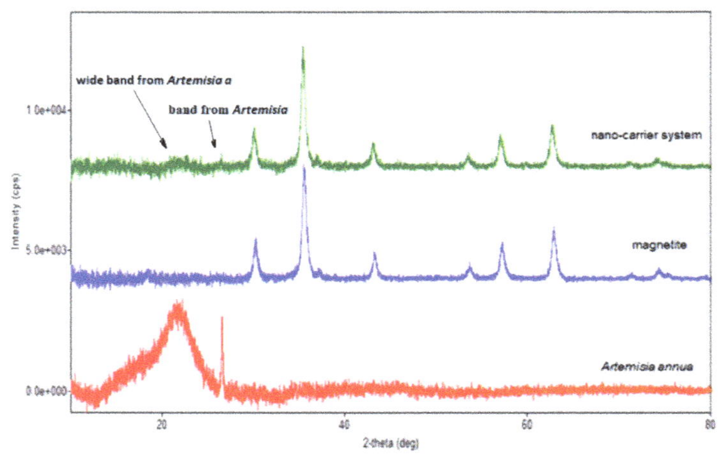

Figure 8. Overlap of the XRD patterns of herb, magnetite and nano-carrier system.

Regarding the diffraction pattern of the herb (Figure 5), in the range of 13–26°, a wide band that is characteristic of some amorphous phases can be observed. This wide band is also found attenuated in the diffraction pattern of the nano-carrier system (Figure 7 and Figure 8). Additionally, in the diffraction pattern of the Fe_3O_4 nanoparticles (Figure 6), only the peaks of the single crystalline spinel phase Fe_3O_4 (average crystallite size 10.9 nm) are present.

In the diffraction pattern of the mixture (Figures 7 and 8), crystalline spinel phase Fe_3O_4 nanoparticles with an average crystallite size of 12.9 nm were identified. A peak at ~26.5° and a band between 13–26° were also present but much attenuated in the spectrum of *Artemisia annua*.

2.6. Scanning Electron Microscopy (SEM)

The SEM micrographs of the herb, magnetic nanoparticles and the nano-carrier system are shown in Figures 9–11.

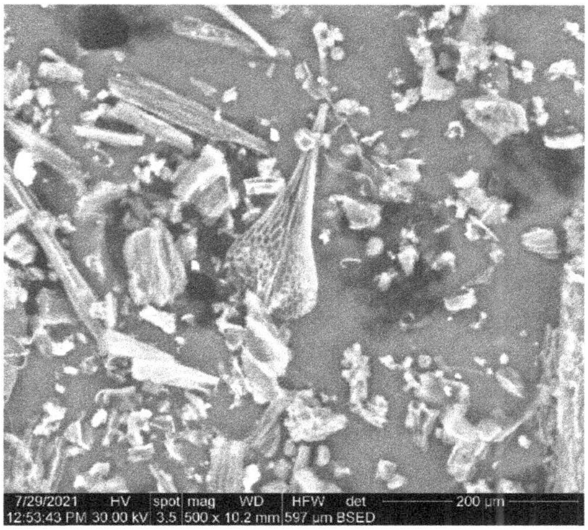

Figure 9. High-resolution SEM images of *Artemisia annua*.

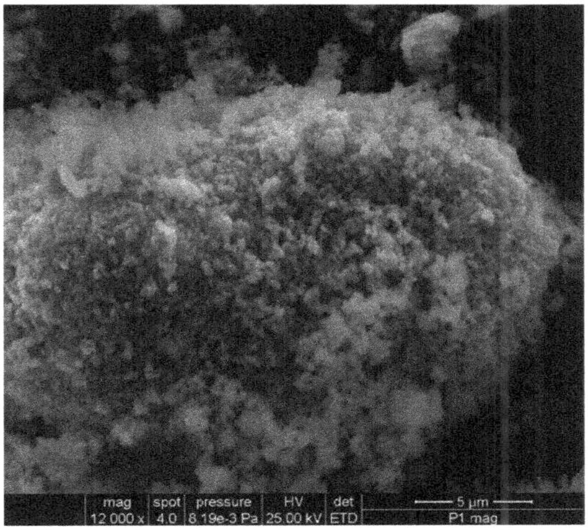

Figure 10. High-resolution SEM images of Fe_3O_4 nanoparticles.

As can be seen in the SEM image (Figure 9), the particles of *Artemisia anuua* shown are in the form of micron-sized fibers and irregular shape particles. The incorporation of herb phytochemicals into the pores of Fe_3O_4 nanoparticles was also confirmed via the scanning electron microscopy (SEM) images of Fe_3O_4 nanoparticles (Figure 10) and Fe_3O_4 loaded with the herb (Figure 11). The SEM image of Fe_3O_4 nanoparticles loaded with the herb (Figure 11b) shows a powder that consists of agglomerations of round nanoparticles with dimensions between 5 and 30 nm, as well as irregular shapes with dimensions greater than 60 nm. In the SEM image at high magnification (Figure 11a), irregular shape particles in Fe_3O_4 nanoparticles' surface modification with herbs can be seen.

(**a**) (**b**)

Figure 11. (**a**) SEM images of Fe_3O_4 nanoparticles loaded with herb at high magnification. (**b**) SEM images of Fe_3O_4 nanoparticles loaded with herb at low magnification.

2.7. Magnetic Properties

The magnetic properties of the Fe_3O_4 nanoparticles and the nano-carrier system were investigated at a low-frequency driving field (50 Hz) by means of an induction hysteresis graph [105]. It was found that both samples exhibit ferromagnetic behavior with narrow hysteresis loops (see Figures 12 and 13).

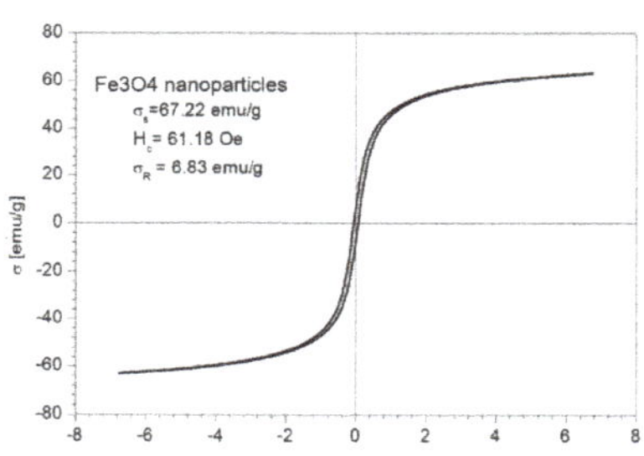

Figure 12. The hysteresis loop of Fe_3O_4 nanoparticles sample.

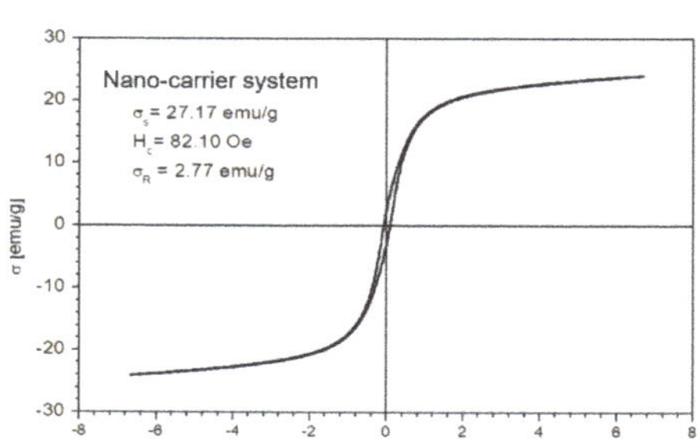

Figure 13. The hysteresis loop of nano-carrier system.

From the measured hysteresis loops, the saturation magnetization (σ_S), the coercive field (H_c) and the remanent magnetization (σ_R) were determined. The results are presented in Table 4.

Table 4. Magnetic properties of Fe_3O_4 nanoparticles and nano-carrier system.

Sample	σ_S (emu/g)	H_C (Oe)	σ_R (emu/g)
Fe_3O_4 nanoparticles	67.22	61.18	6.83
Nano-carrier system	27.17	82.10	2.77

As expected, the saturation magnetization of the sample Fe_3O_4 nanoparticles is larger than that of nano-carrier system. Both samples have small values regarding the remnant ratio σ_R/σ_S (in order of 0.1), which is an indication of the ease with which the magnetization reorients to the nearest easy axis magnetization direction after the removal of the magnetic field. The frequency dependence on the complex magnetic permeability of the samples (Equation (1)) over the frequency range of 1 kHz to 2 MHz was measured at room temperature, and the obtained results are presented in Figure 14 [106].

$$\mu(f) = \mu'(f) - i\mu''(f) \tag{1}$$

where $\mu'(f)$ is the real part;$\mu''(f)$ is the imaginary part.

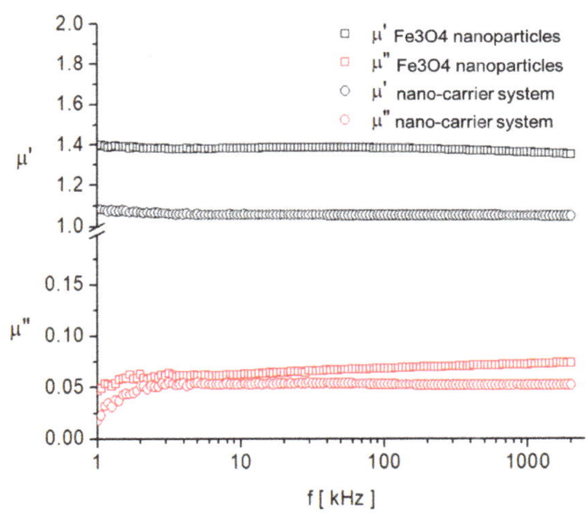

Figure 14. Frequency dependence of the real, μ', and imaginary, μ'', components of the complex magnetic permeability of samples.

In the investigated frequency range, there are no magnetic relaxation peaks that provide clues about the characteristic magnetization processes. However, given the small sizes of the particles and also the low value of the imaginary component of complex magnetic permeability, it can be assumed that the dominant magnetization mechanism is the Neel process, correlated with the rotation of the magnetic moment inside the particles by overcoming the magneto-crystalline anisotropy barrier [107,108].

The obtained results indicate that the nano-carrier system, wherein the selected ratio of plant:magnetic nanoparticles is 3:1, exhibits magnetic properties.

3. Materials and Methods

All used reagents were GC grade. Methanol and chloroform were purchased from VWR (Wien, Austria). The Fe_3O_4 nanoparticles (nanoparticle size: 23 nm) were provided by the National Research and Development Institute for Non-Ferrous and Rare Metals, Pantelimon, Romania. The plant samples (whole plant) were collected in August 2020 from the area of Timis, Romania and were taxonomically authenticated at Victor Babes University of Medicine and Pharmacy, Timisoara, Romania. The plant samples were rapidly frozen in liquid nitrogen (−194 °C), ground and sieved to obtain a particle size lower than 0.5 mm, and they were kept at −80 °C to avoid enzymatic conversion or metabolite degradation. For each analysis, 1.8 g of dried sample was subject to sonication extraction in 25 mL of solvent (methanol/chloroform = 1:1) for 25 min at 40 °C with a frequency of 50 kHz. The solution was concentrated using a rotavapor and the residue was dissolved in MeOH. The extract was centrifuged, and the supernatant was filtered through a 0.2 µm syringe filter and stored at –18 °C until analysis.

3.1. Nano-Carrier System Preparation

For each analysis, 1.5 g of sample was prepared from dried herb (whole plant, ground and sieved to obtain a particle size lower than 0.5 mm), and Fe_3O_4 nanoparticles were added (herb/Fe_3O_4 nanoparticles = 3:1). The obtained mixture was subjected to micronization at room temperature for 5 min.

3.2. GC-MS Analysis

Gas chromatography was carried out using ClarusSQ8 GC/MS (Perkin Elmer) apparatus with a ZB-AAA GC column (10 m × 0.25 mm) (Phenomenex, Torrance, CA, USA); carrier gas, He; flow rate, 1 mL/min, following 3.3. GC-MS Separation Conditions (the standard conditions provided with the EZ: faast GC-MS free amino acids kit).

The oven temperature was 80 °C (held for 9 min) to 220 °C (held for 5 min) at 320 °C/min (held for 10 min); the equilibration time was 1 min. The injection parameters were: split 1:5; 250 °C; 2.5 µL. The carrier gas was helium; 1.1 mL/min; 110 °C. The inlet pressure was 5.9 kPa/min; the detector used was: MS; mode: Scan Transfer Line; temperature: 250 °C; analyzer type: MS; electron energy: 70 eV.

3.3. Mass Spectrometry

MS experiments were conducted using EIS-QTOF-MS from Bruker Daltonics, Bremen, Germany. All mass spectra were acquired in the positive ion mode within a mass range of (100–3000) m/z, with a scan speed of 2.1 scans/s. The source block temperature was kept at 80 °C. The reference provided a spectrum in positive ion mode with fair ionic coverage of the m/z range scanned in full-scan MS. The resulting spectrum was a sum of scans over the total ion current (TIC) acquired at 25–85 eV collision energy to provide the full set of diagnostic fragment ions.

3.4. Identification of Metabolites

The total ion current (TIC) and selected ion monitoring (SIM) values were compared with those from Phenomenex-EZ: faast amino acid analysis user guide and the results are presented in Table 1. The metabolites were identified via comparison of their mass spectra with those of the standard library NIST/NBS-3 (National Institute of Standards and Technology/National Bureau of Standards) spectral database, and the identified phytoconstituents are presented in Table 2.

3.5. FT-IR Spectroscopy

The FT-IR spectrum of the sample was recorded via KBr pellet using a Shimadzu Prestige-21 spectrometer in the range 400–4000 cm^{-1}, with a resolution of 4 cm^{-1}.

3.6. XRD Spectroscopy

The phase composition of the sample was determined via powder X-ray diffractometry (XRD) using monochromatic CuKα radiation (λ = 1.5406Å) on a Rigaku Ultima IV diffractometer equipped with a D/teX Ultra detector and operating at 40 kV and 40 mA.

The analysis was performed in the 2θ range of 10–80°, with a scan speed of 5 °/min and a step size of 0.01° 2θ. The average crystallite size was calculated using the whole pattern profile fitting method (WPPF). The XRD patterns were compared with those from the ICDD Powder Diffraction Database (ICDD file 04-015-9120).

3.7. Scanning Electronic Microscopy (SEM)

The SEM analyses were performed using an SEM-EDS system (QUANTA INSPECT F50) equipped with a field emission gun (FEG), 1.2 nm resolution and an energy dispersive X-ray spectrometer (EDS) with an MnK resolution of 133 eV.

3.8. Magnetization Experiments

The frequency dependence of the Fe_3O_4 nanoparticles and nano-carrier system was measured using an Agilent LCR-meter (E-4980A type) at room temperature over the frequency range (1 kHz to 2 MHz) and various values of polarizing field. The duration of the measurement into a constant magnetic field over the entire frequency range was about 40 s.

Complex magnetic susceptibility measurements were made using the short-circuited coaxial transmission line technique at different values of the polarizing field, H, over the range 0–170 kA/m and at the frequency range (100 MHz–6 GHz). The static magnetization measurements for the Fe_3O_4 nanoparticle sample and the nano-carrier system were performed using a ballistic galvanometer.

4. Conclusions

In the current study, the complete metabolite profiling of *A. annua* growing wild in Romania was accomplished. A total of 14 amino acids were identified for the for the first time in plant samples. The biological activities were discussed for each metabolite category. Furthermore, a simple and economical nano-carrier system was developed. A ratio of herb:magnetic Fe_3O_4 nanoparticles was used, which allowed for the synergic effect of *A. annua* bioactive compounds and its inorganic component properties to be taken advantage of. The morpho-structural characterization of the nano-carrier system was performed. In addition, the magnetic properties of the nano-carrier were evaluated. Further studies are necessary to investigate the biological properties and the bioavailability of the new nano-carrier system.

Author Contributions: Conception and design of study: A.-E.S. and I.G.; methodology: A.-E.S.; acquisition of data: C.M. and C.N.M.; analysis and interpretation of data: C.M. and C.N.M.; writing—original draft preparation: I.O.-F.G.; writing—review and editing: A.-E.S. investigation: I.O.-F.G. and C.V.I.F. All authors have read and agreed to the published version of the manuscript.

Funding: Grant CNFIS-FDI-2021-0414, titled: "Centers of excellence with the support of UVT researchers" (Centre de excelenţă prin sprijinul cercetatorilor din UVT).

Institutional Review Board Statement: Not applicable.

Informed Consent Statement: Not applicable.

Data Availability Statement: All data are contained within the article.

Acknowledgments: The National Center for Micro and Nanomaterials (the center is part of the Department of Science and Engineering of Oxide and Nanomaterials Materials of the Faculty of Applied Chemistry and Materials Science of the Politehnica University of Bucharest). Grant CNFIS-FDI-2021-0414, titled: "Centers of excellence with the support of UVT researchers" (Centre de excelenţă prin sprijinul cercetatorilor din UVT).

Conflicts of Interest: The authors declare no conflict of interest.

References

1. Dihoru, G.; Boruz, V. *The List of Main Spontaneous Medicinal Plants from Romania, Annals of the University of Craiova—Agriculture, Montanology, Cadastre Series*; XLIV 328; University of Craiova: Craiova, Romania, 2014.

2. Pantu Zach, C. *Plantele Cunoscute de Poporul Roman-Vocabular Botanic Cuprinzand Numirile Romane, Franceze, Germane si Stiintifice*; Institutul de Arte Grafice, Ed.; Minerva: Bucuresti, Romania, 1906.
3. Cântar, I.C.; Dincă, M. *Curiosities about the Artemisia Collections from "Alexandru Beldie" Herbarium, Annals of the University of Craiova—Agriculture, Montanology, Cadastre Series*; XLIX; University of Craiova: Craiova, Romania, 2019.
4. Clinciu Radu, R.A.; Zaharia, M.S.; Lungoci, C.; Vodă, A.D.; Nițu, S.; Robu, T. Studies Regarding the Growth Dynamics of Some Species from *Artemisia* Genus, Annals of the University of Oradea. *Fascicle* **2017**, *29*, 29–34.
5. Stan, R.L.; Sevastre, B.; Ionescu, C.; Olah, N.K.; Vicaș, L.G.; Páll, E.; Moisa, C.; Hanganu, D.; Sevastre-Berghian, A.C.; Andrei, S.; et al. *Artemisia Annua L.* extract: A new phytoproduct with sod-like and antitumour activity. *Farmacia* **2020**, *68*, 812–821. [CrossRef]
6. Weathers, P.J.; Arsenault, P.R.; Covello, P.S.; McMickle, A.; Teoh, K.H.; Reed, D.W. Artemisinin production in *Artemisia annua*: Studies in planta and results of a novel delivery method for treating malaria and other neglected diseases. *Phytochem. Rev.* **2011**, *10*, 173–183. [CrossRef] [PubMed]
7. Mamatova, A.S.; Korona-Glowniak, I.; Skalicka-Woźniak, K.; Józefczyk, A.; Wojtanowski, K.K.; Baj, T.; Sakipova, Z.B.; Malm, A. Phytochemical composition of wormwood (*Artemisia gmelinii*) extracts in respect of their antimicrobial activity. *BMC Complement. Altern. Med.* **2019**, *19*, 288. [CrossRef] [PubMed]
8. Deb, M.; Kumar, D. Chemical composition and bioactivity of the essential oils derived from *Artemisia Annua* against the red flour beetle. *Biosci. Biotech. Res. Asia* **2019**, *16*, 463–476. [CrossRef]
9. Mesa, L.E.; Lutgen, P.; Velez, I.D.; Segura, A.M.; Robledo, S.M. *Artemisia annua* L., Potential source of molecules with pharmacological activity in human diseases. *Am. J. Phytomed. Clin. Ther.* **2015**, *3*, 436–450.
10. Iqbal, S.; Younas, U.; Chan, K.W.; Zia-Ul-Haq, M.; Ismail, M. Chemical composition of *Artemisia annua* L. leaves and antioxidant potential of extracts as a function of extraction solvents. *Molecules* **2012**, *17*, 6020–6032.
11. Nigam, M.; Atanassova, M.; Mishra, A.P.; Pezzani, R.; Devkota, H.P.; Plygun, S.; Salehi, B.; Setzer, W.N.; Sharifi-Rad, J. Bioactive compounds and health benefits of *Artemisia* species. *Nat. Prod. Commun.* **2019**, *14*, 1934578X19850354.
12. Sadiq, A.; Hayat, M.Q.; Ashraf, M. Ethnopharmacology of *Artemisia annua* L.: A Review. In *Artemisia Annua-Pharmacology and Biotechnology*; Aftab, T., Ferreira, J.F.S., Khan, M.M.A., Naeem, M., Eds.; Springer: Berlin/Heidelberg, Germany, 2014; pp. 9–25.
13. Badshah, S.L.; Ullah, A.; Ahmad, N.; Almarhoon, Z.M.; Mabkhot, Y. Increasing the strength and production of artemisinin and its derivatives. *Molecules* **2018**, *23*, 100. [CrossRef]
14. Septembre-Malaterre, A.; Lalarizo Rakoto, M.; Marodon, C.; Bedoui, Y.; Nakab, J.; Simon, E.; Hoarau, L.; Savriama, S.; Strasberg, D.; Guiraud, P.; et al. *Artemisia annua*, a traditional plant brought to light. *Int. J. Mol. Sci.* **2020**, *21*, 4986. [CrossRef]
15. Farid-ul-Haq Muhammad Haseeb, M.T.; Hussain, M.A.; Ashraf, M.U.; Naeem-ul-Hassan, M.; Hussain, S.Z.; Hussain, I. A smart drug delivery system based on *Artemisia vulgaris* hydrogel: Design, on-off switching, and real-time swelling, transit detection, and mechanistic studies. *J. Drug Deliv. Sci. Technol.* **2020**, *58*, 101795. [CrossRef]
16. Li, Z.; Li, Q.; Wu, J.; Wang, M.; Yu, J. Artemisinin and its derivatives as a repurposing anticancer agent: What else do we need to do? *Molecules* **2016**, *21*, 1331. [CrossRef] [PubMed]
17. Shahrajabian, M.H.; Sun, W.; Cheng, Q. Exploring *Artemisia annua L.*, artemisinin and its derivatives, from traditional chinese wonder medicinal science. *Not. Bot. Horti Agrobot. Cluj-Napoca* **2020**, *48*, 1719–1741. [CrossRef]
18. Bilia, A.R.; Santomauro, F.; Sacco, C.; Bergonzi, M.C.; Donato, R. Essential oil of *Artemisia annua* L.: An Extraordinary component with numerous antimicrobial properties. *Evid.-Based Complement. Altern. Med.* **2014**, *2014*, 159819. [CrossRef]
19. Bora, K.S.; Sharma, A. The genus *Artemisia*: A comprehensive review. *Pharm. Biol.* **2011**, *49*, 101–109. [CrossRef] [PubMed]
20. Das, S. *Artemisia annua* (Qinghao): A pharmacological review. *Int. J. Pharm. Sci. Res.* **2012**, *3*, 4573–4577.
21. Numonov, S.; Sharopov, F.; Salimov, A.; Sukhrobov, P.; Atolikshoeva, S.; Safarzoda, R.; Habasi, M.; Aisa, H.A. Assessment of artemisinin contents in selected *Artemisia* Species from Tajikistan (Central Asia). *Medicines* **2019**, *6*, 23. [CrossRef]
22. Abad, M.J.; Bedoya, L.M.; Apaza, L.; Bermejo, P. The *Artemisia L.* genus: A review of bioactive essential oils. *Molecules* **2012**, *17*, 2542–2566. [CrossRef]
23. Lang, S.J.; Schmiech, M.; Hafner, S.; Paetz, C.; Steinborn, C.; Huber, R.; El Gaafary, M.; Werner, K.; Schmidt, C.Q.; Syrovets, T.; et al. Antitumor activity of an *Artemisia annua* herbal preparation and identification of active ingredients. *Phytomedicine* **2019**, *62*, 152962. [CrossRef]
24. World Malaria Report. In *20 Years of Global Progress and Challenges*; World Health Organization: Geneva, Switzerland, 2020; ISBN 978-92-4-001579-1.
25. Charlie-Silva, I.; Fernandes Fraceto, L.; Ferreira Silva de Melo, N. Progress in nano-drug delivery of artemisinin and its derivatives: Towards to use in immunomodulatory approaches. *Artif. Cells Nanomed. Biotechnol.* **2018**, *46*, S611–S620. [CrossRef]
26. Zhao, Q.; Luan, X.; Zheng, M.; Tian, X.H.; Zhao, J.; Zhang, W.D.; Ma, B.L. Synergistic mechanisms of constituents in herbal extracts during intestinal absorption: Focus on natural occurring nanoparticles. *Pharmaceutics* **2020**, *12*, 128. [CrossRef]
27. Hussein, R.A.; El-Anssary, A.A. *Plants Secondary Metabolites: The Key Drivers of the Pharmacological Actions of Medicinal Plants, Herbal Medicine*; Builders, P.F., Ed.; IntechOpen Limited: London, UK, 2018.
28. Tzenkova, R.; Kamenarska, Z.; Draganov, A.; Atanassov, A. Composition of *Artemisia annua* essential oil obtained from species gowing wild in Bulgaria. *Biotechnol. Biotechnol. Equip.* **2010**, *24*, 1833–1835. [CrossRef]
29. Albaugh, V.L.; Pinzon-Guzman, C.; Barbul, A. Arginine-Dual roles as an onconutrient and immunonutrient. *J. Surg. Oncol.* **2017**, *115*, 273–280. [CrossRef] [PubMed]

30. Sun-Hee, K.; Roszik, J.; Grimm, E.A.; Ekmekcioglu, S. Impact of L-Arginine metabolism on immune response and anticancer immunotherapy. *Front. Oncol.* **2018**, *8*, 67.
31. Chiangjong, W.; Chutipongtanate, S.; Hongeng, S. Anticancer peptide: Physicochemical property, functional aspect and trend in clinical application (Review). *Int. J. Oncol.* **2020**, *57*, 678–696. [CrossRef] [PubMed]
32. Ochkur, A.V.; Kovaleva, A.M.; Kolesnik, Y.S. Amino-acid composition of subgenus *Artemisia* Herbs. *Chem. Nat. Compd.* **2013**, *49*, 589–591. [CrossRef]
33. Berechet, M.D.; Stelescu, M.D.; Manaila, E.; Craciun, G. Chemical composition of the essential oil of *Artemisia absinthium* from Romania. *Rev. Chim.* **2015**, *66*, 1814–1818.
34. Moacă, E.A.; Pavel, I.Z.; Danciu, C.; Crăiniceanu, Z.; Minda, D.; Ardelean, F.; Antal, D.S.; Ghiulai, R.; Cioca, A.; Derban, M.; et al. Romanian wormwood (*Artemisia absinthium* L.): Physicochemical and nutraceutical screening. *Molecules* **2019**, *24*, 3087. [CrossRef]
35. Ivanescu, B.; Vlase, L.; Corciova, A.; Lazar, M.I. Artemisinin evaluation in Romanian *Artemisia annua* wild plants using a new HPLC/MS method. *Nat. Prod. Res.* **2011**, *25*, 716–722. [CrossRef]
36. Marinas, I.C.; Oprea, E.; Chifiriuc, M.C.; Badea, I.A.; Buleandra, M.; Lazar, V. Chemical composition and antipathogenic activity of *Artemisia annua* essential oil from Romania. *Chem. Biodivers.* **2015**, *12*, 1554–1564. [CrossRef]
37. Toth, E.T.; Dezso, A.C.; Kapas, A.; Pako, J.; Ichim, M.C. Comparison of chemical composition of *Artemisia annua* volatile oil from Romania, Chemical composition and antipathogenic activity of *Artemisia annua* essential oil from Romania. *Planta Med.* **2011**, *77*, PL91. [CrossRef]
38. Ruperti-Repilado, F.J.; Haefliger, S.; Rehm, S.; Zweier, M.; Rentsch, K.M.; Blum, J.; Jetter, A.; Heim MLeuppi-Taegtmeyer, A.; Terracciano, L.; Bernsmeier, C. Danger of herbal tea: A case of acute cholestatic hepatitis due to *Artemisia annua* tea. *Front. Med.* **2019**, *6*, 221. [CrossRef] [PubMed]
39. Alsanad, S.; Howard, R.; Williamson, E. An assessment of the impact of herb-drug combinations used by cancer patients. *BMC Complementary Altern. Med.* **2016**, *16*, 393. [CrossRef]
40. Fouad, D.; Bachra, Y.; Ayoub, G.; Ouaket, A.; Bennamara, A.; Knouzi, N.; Berrada, M. *A Novel Drug Delivery System Based on Nanoparticles of Magnetite Fe3O4 Embedded in an Auto Cross-Linked Chitosan [Online First]*; IntechOpen Limited: London, UK, 2020.
41. Rahman, H.S.; Othman, H.H.; Hammadi, N.I.; Yeap, S.K.; Amin, K.M.; Abdul Samad, N.; Alitheen, N.B. Novel drug delivery systems for loading of natural plant extracts and their biomedical applications. *Int. J. Nanomed.* **2020**, *15*, 2439–2483. [CrossRef]
42. Mohammadi, M.; Pourseyed Aghaei, F. Magnetite Fe_3O_4 surface as an effective drug delivery system for cancer treatment drugs: Density functional theory study. *J. Biomol. Struct. Dyn.* **2021**, *39*, 2798–2805. [CrossRef]
43. Gómez-Sotomayor, R.; Ahualli, S.; Viota, J.L.; Rudzka, K.; Delgado, Á.V. Iron/Magnetite nano-particles as magnetic delivery systems for antitumor drugs. *J. Nanosci. Nanotechnol.* **2015**, *15*, 3507–3514. [CrossRef]
44. Socoliuc, V.; Peddis, D.; Petrenko, V.I.; Avdeev, M.V.; Susan-Resiga, D.; Szabó, T.; Turcu, R.; Tombácz, E.; Vékás, L. Magnetic nanoparticle systems for nanomedicine-A materials science perspective. *Magnetochemistry* **2020**, *6*, 2. [CrossRef]
45. Aderibigbe, B.A. Design of drug delivery systems containing artemisinin and its derivatives. *Molecules* **2017**, *22*, 323. [CrossRef]
46. Elfawal, M.A.; Towler, M.J.; Reich, N.G.; Golenbock, D.; Weathers, P.J.; Rich, S.M. Dried whole plant *Artemisia annua* as an antimalarial therapy. *PLoS ONE* **2012**, *7*, e52746. [CrossRef]
47. Chen, J.; Guo, Z.; Wang, H.-B.; Zhou, J.-J.; Zhang, W.-J.; Chen, Q.-W. Multifunctional mesoporous nanoparticles as pH-responsive Fe^{2+} reservoirs and artemisinin vehicles for synergistic inhibition of tumor growth. *Biomaterials* **2014**, *35*, 6498–6507. [CrossRef]
48. Wang, Y.; Han, Y.; Yang, Y.; Yang, J.; Guo, X.; Zhang, J.; Pan, L.; Xia, G.; Chen, B. Effect of interaction of magnetic nanoparticles of Fe_3O_4 and artesunate on apoptosis of K562 cells. *Int. J. Nanomed.* **2011**, *6*, 1185–1192.
49. Akbari, M.; Mora, R.; Maaza, M. First principle study of silver nanoparticle interactions with antimalarial drugs extracted from *Artemisia annua* plant. *J. Nanopart. Res.* **2020**, *22*, 331. [CrossRef] [PubMed]
50. Chen, G.; Wang, S.; Huang, X.; Hong, J.; Du, L.; Zhang, L.; Ye, L. Environmental factors affecting growth and development of Banlangen (*Radix Isatidis*) in China. *Afr. J. Plant Sci.* **2015**, *9*, 421–426.
51. Pang, Z.; Chen, J.; Wang, T.; Gao, C.; Li, Z.; Guo, L.; Xu, J.; Cheng, Y. Linking plant secondary metabolites and plant microbiomes: A Review. *Front. Plant Sci.* **2021**, *12*, 621276. [CrossRef] [PubMed]
52. Zhang, X.; Zhao, Y.; Guo, L.; Qiu, Z.; Huang, L.; Qu, X. Differences in chemical constituents of *Artemisia annua* L from different geographical regions in China. *PLoS ONE* **2017**, *12*, e0183047. [CrossRef]
53. Adams, R.P. *Identification of Essential Oil Components by Gas Chromatography/Mass Spectroscopy*; Allured Publishing Corporation: Carol Stream, IL, USA, 2007.
54. Segneanu, A.-E.; Grozescu, I.; Cziple, F.; Berki, D.; Damian, D.; Niculite, C.M.; Florea, A.; Leabu, M. *Helleborus purpurascens*—Amino acid and peptide analysis linked to the chemical and antiproliferative properties of the extracted compounds. *Molecules* **2015**, *20*, 22170–22187. [CrossRef] [PubMed]
55. Phenomenex, E.Z. *Faast-Easy Fast Amino Acid Sampling Testing Kit-User Guide*; Torrance, CA, USA; Available online: http://www.phenomenex.com (accessed on 12 October 2021).
56. Czechowski, T.; Rinaldi, M.A.; Toyin, F.M.; Van Veelen, M.; Larson, T.R.; Thilo, W.; Rathbone, D.A.; David, H.; Horrocks, P.; Graham, I.A. Flavonoid versus artemisinin anti-malarial activity in *Artemisia annua* whole-leaf extracts. *Front. Plant Sci.* **2019**, *10*, 984. [CrossRef] [PubMed]

57. Ferreira, J.F.S.; Luthria, D.L.; Sasaki, T.; Heyerick, A. Flavonoids from *Artemisia annua* L. as antioxidants and their potential synergism with artemisinin against malaria and cancer. *Molecules* **2010**, *15*, 3135–3170. [PubMed]
58. Silveira de Almeida, L.M.; Soares Aleixo de Carvalho, L.; Gazolla, M.C.; Silva Pinto, P.L.; Nascimento da Silva, M.P.; de Moraes, J.; Da Silva Filho, A.A. Flavonoids and sesquiterpene lactones from *Artemisia absinthium* and *Tanacetum parthenium* against *Schistosoma mansoni* worms. *Evid.-Based Complement. Altern. Med.* **2016**, *2016*, 9521349.
59. Baraldi, R.; Isacchi, B.; Predieri, S.; Marconi, G.; Vincieri, F.F.; Bilia, A.R. Distribution of artemisinin and bioactive flavonoids from *Artemisia annua* L. during plant growth. *Biochem. Syst. Ecol.* **2008**, *36*, 340–348. [CrossRef]
60. Chu, Y.; Wang, H.; Chen, J.; Hou, Y. New sesquiterpene and polymethoxy-flavonoids from *Artemisia annua* L. *Phcog. Mag.* **2014**, *10*, 213–216. [CrossRef]
61. Brisibe, E.A.; Umoren, E.U.; Brisibe, F.; Magalhäes, P.M.; Ferreira, J.F.S.; Luthria, D.; Wu, X.; Prior, R. Nutritional characterisation and antioxidant capacity of different tissues of *Artemisia annua* L. *Food Chem.* **2009**, *115*, 1240–1246. [CrossRef]
62. Sharopov, F.S.; Salimov, A.; Numonov, S.; Safomuddin, A.; Bakri, M.; Salimov, T.; Setzer, W.N.; Habasi, M. Chemical composition, antioxidant, and antimicrobial activities of the essential oils from *Artemisia annua* L. growing wild in Tajikistan. *Nat. Prod. Commun.* **2020**, *15*, 1934578X20927814.
63. Bhakuni, R.; Jain, D.; Sharma, R.; Kumar, S. Secondary metabolites of *Artemisia annua* and their biological activity. *Curr. Sci.* **2001**, *80*, 35–48.
64. Yang, S. Phytochemical Studies of *Artemisia annua* L. Ph.D. Thesis, Department of Pharmacognosy, The School of Pharmacy, The University of London, ProQuest LLC, London, UK, 2017.
65. Trendafilova, A.; Moujir, L.M.; Sousa, P.M.C.; Seca, A.M.L. Research advances on health effects of edible *Artemisia* species and some sesquiterpene lactones constituents. *Foods* **2021**, *10*, 65. [CrossRef] [PubMed]
66. Dawood, H.M.; Shawky, E.; Hammoda, H.M.; Metwally, A.M.; Ibrahim, R.S. Chemical constituents from *Artemisia annua* and Vitex agnus-castus as new aromatase inhibitors: In-vitro and in-silico studies. *J. Mex. Chem. Soc.* **2020**, *64*, 316–326. [CrossRef]
67. Brown, G.D. The biosynthesis of artemisinin (Qinghaosu) and the phytochemistry of *Artemisia annua* L. (Qinghao). *Molecules* **2010**, *15*, 7603–7698.
68. Nikolova, M.; Gevrenova, R.; Ivancheva, S. High-performance liquid chromatographic separation of surface flavonoid aglycones in *Artemisia annua* L. and *Artemisia vulgaris* L. *J. Serb. Chem. Soc.* **2004**, *69*, 571–574. [CrossRef]
69. Nikolova, M. Infraspecific variability in the flavonoid composition of *Artemisia vulgaris* L. *Acta Bot. Croat.* **2006**, *65*, 13–18.
70. Valant-Vetscheraa, K.M.; Wollenweberb, E. Flavonoid aglycones from the leaf surfaces of some *Artemisia* spp. (Compositae-Anthemideae). *Z. Naturforsch.* **1995**, *50*, 353–357. [CrossRef]
71. Rodríguez, E.; Carman, N.J.; Vander Velde, G.; McReynolds, J.H.; Mabry, T.J.; Irwin, M.A.; Geissman, T.A. Methoxylated flavonoids from *Artemisia*. *Phytochemistry* **1972**, *11*, 3509–3514. [CrossRef]
72. Chougouo, R.D.; Nguekeu, Y.M.; Dzoyem, J.P.; Awouafack, M.D.; Kouamouo, J.; Tane, P.; McGaw, L.J.; Eloff, J.N. Anti-inflammatory and acetylcholinesterase activity of extract, fractions and five compounds isolated from the leaves and twigs of *Artemisia annua* growing in Cameroon. *Springerplus* **2016**, *5*, 1525. [CrossRef] [PubMed]
73. Lieu, E.L.; Nguyen, T.; Rhyne, S.; Kim, J. Amino acids in cancer. *Exp. Mol. Med.* **2020**, *52*, 15–30. [CrossRef] [PubMed]
74. Cox-Georgian, D.; Ramadoss, N.; Dona, C.; Basu, C. Therapeutic and medicinal uses of terpenes. *Med. Plants* **2019**, *12*, 333–359.
75. Fongang Fotsing, Y.S.; Bankeu Kezetas, J.J. Terpenoids as important bioactive constituents of essential oils. In *Essential Oils—Bioactive Compounds, New Perspectives and Application*; Santana de Oliveira, M., Almeida da Costa, W., Gomes Silva, S., Eds.; IntechOpen Limited: London, UK, 2020; ISBN 978-1-83962-698-2.
76. Jahangeer, M.; Fatima, R.; Ashiq, M. Therapeutic and biomedical potentialities of terpenoids—A Review. *J. Pure Appl. Microbiol.* **2021**, *15*, 471–483. [CrossRef]
77. Küpeli Akkol, E.; Genç, Y.; Karpuz, B.; Sobarzo-Sánchez, E.; Capasso, R. Coumarins and coumarin-related compounds in pharmacotherapy of cancer. *Cancers* **2020**, *12*, 1959. [CrossRef]
78. João Matos, M.; Santana, L.; Uriarte, E.; Abreu, O.A.; Molina, E.; Guardado Yordi, E. *Coumarins—An Important Class of Phytochemicals in Phytochemicals-Isolation, Characterisation and Role in Human Health*; Venket Rao, A., Rao, L.G., Eds.; IntechOpen Limited: London, UK, 2015.
79. Kozłowska, A.; Szostak-Wegierek, D. Flavonoids-food sources and health benefits. *Rocz. Panstw. Zakl. Hig.* **2014**, *65*, 79–85.
80. Panche, A.N.; Diwan, A.D.; Chandra, S.R. Flavonoids: An overview. *J. Nutr. Sci.* **2016**, *5*, e47. [CrossRef]
81. Kozłowska, A.; Szostak-Węgierek, D. Flavonoids–food sources, health benefits, and mechanisms involved. In *Bioactive Molecules in Food*; Mérillon, J.M., Ramawat, K., Eds.; Reference Series in Phytochemistry; Springer: Cham, Switzerland, 2018.
82. Kumar, N.; Goel, N. Phenolic acids: Natural versatile molecules with promising therapeutic applications. *Biotechnol. Rep.* **2019**, *24*, e00370. [CrossRef]
83. Minatel, I.O.; Vanz Borges, C.; Ferreira, M.I.; Gomez, H.A.G.; Chen, C.Y.O. Pace Pereira Lima G., Phenolic compounds: Functional properties, impact of processing and bioavailability. In *Phenolic Compounds-Biological Activity*; Soto-Hernandez, M., Palma-Tenango, M., del Rosario Garcia-Mateos, M., Eds.; IntechOpen Limited: London, UK, 2017.
84. Salehi, B.; Quispe, C.; Sharifi-Rad, J.; Cruz-Martins, N.; Nigam, M.; Mishra, A.P.; Konovalov, D.A.; Orobinskaya, V.; Abu-Reidah, I.M.; Zam, W.; et al. Phytosterols: From preclinical evidence to potential clinical applications. *Front. Pharmacol.* **2021**, *11*, 599959. [CrossRef]

85. Nagy, K.; Tiuca, I.-D. *Importance of Fatty Acids in Physiopathology of Human Body in Fatty Acids*; Catala, A., Ed.; IntechOpen Limited: London, UK, 2017.
86. Kilcoyne, M.; Joshi, L. Carbohydrates in therapeutics. *Cardiovasc. Hematol. Agents Med. Chem.* **2007**, *5*, 186–197. [CrossRef]
87. Singh, D.; Rajput, A.; Bhatia, A.; Kumar, A.; Kaur, H.; Sharma, P.; Kaur, P.; Singh, S.; Attri, S.; Buttar, H.; et al. Plant-based polysaccharides and their health functions. *Funct. Foods Health Dis.* **2021**, *11*, 179–200.
88. Dhama, K.; Karthik, K.; Khandia, R.; Munjal, A.; Tiwari, R.; Rana, R.; Khurana, S.K.; Ullah, S.; Khan, R.U.; Alagawany, M.; et al. Medicinal and therapeutic potential of herbs and plant metabolites/Extracts countering viral pathogens-Current knowledge and future prospects. *Curr. Drug Metab.* **2018**, *19*, 236–263. [CrossRef] [PubMed]
89. Khan, H.; Saeedi, M.; Nabavi, S.; Mubarak, M.; Bishayee, A. Glycosides from medicinal plants as potential anticancer agents: Emerging trends towards future drugs. *Curr. Med. Chem.* **2018**, *26*, 2389–2406. [CrossRef]
90. Mohammed, L.; Gomaa, H.G.; Ragab, D.; Zhu, J. Magnetic nanoparticles for environmental and biomedical applications: A review. *Particuology* **2017**, *30*, 1–14. [CrossRef]
91. Wahajuddin Arora, S. Superparamagnetic iron oxide nanoparticles: Magnetic nanoplatforms as drug carriers. *Int. J. Nanomed.* **2012**, *7*, 3445–3471. [CrossRef]
92. Nunes Oliveira, R.; Cordeiro Mancini, M.; Salles de Oliveira, F.C.; Marques Passos, T.; Quilty, B.; da Silva Moreira Thiré, R.M.; McGuinness, G.B. FTIR analysis and quantification of phenols and flavonoids of five commercially available plants extracts used in wound healing. *Matéria* **2016**, *21*, 767–779.
93. Kristoffersen, K.A.; van Amerongen, A.; Böcker, U.; Lindberg, D.; Wubshet, S.G.; de Vogel-van den Bosch, H.; Horn, S.J.; Afseth, N.K. Fourier-transform infrared spectroscopy for monitoring proteolytic reactions using dry-films treated with trifluoroacetic acid. *Sci. Rep.* **2020**, *10*, 7844. [CrossRef]
94. Ercioglu, E.; Velioglu, H.M.; Boyaci, I.H. Determination of terpenoid contents of aromatic plants using NIRS. *Talanta* **2018**, *178*, 716–721. [CrossRef]
95. Mawa, S.; Jantan, I.; Husain, K. Isolation of terpenoids from the stem of Ficus aurantiaca Griff and their effects on reactive oxygen species production and chemotactic activity of neutrophils. *Molecules* **2016**, *21*, 9. [CrossRef]
96. Sai Prakash Chaturvedula, V.; Mubarak, C.; Prakash, I. IR spectral analysis of diterpene glycosides isolated from Stevia rebaudiana. *Food Nutr. Sci.* **2012**, *3*, 1467–1471.
97. Forfang, K.; Zimmermann, B.; Kosa, G.; Kohler, A.; Shapaval, V. FTIR Spectroscopy for evaluation and monitoring of lipid extraction efficiency for oleaginous fungi. *PLoS ONE* **2017**, *12*, e0170611. [CrossRef] [PubMed]
98. Mehrotra, R. Infrared spectroscopy, gas chromatography/Infrared in food analysis. In *Encyclopedia of Analytical Chemistry*; Meyers, R.A., Ed.; Wiley: Hoboken, NJ, USA, 2006; ISBN 978047197670.
99. Iranshahi, M.; Emami, S.A.; Mahmoud-Soltani, M. Detection of sesquiterpene lactones in ten *Artemisia* species population of Khorasan Provinces. *Iran. J. Basic Med Sci.* **2007**, *10*, 183–188.
100. Limem, S.; Maazaoui, R.; Mani Kongnine, D.; Mokhtar, F.; Karmous, T. Preliminary identification of Citrullus colocynthis from Togo by FT-IR and Raman Spectroscopy. *Int. J. Adv. Res.* **2015**, *3*, 354–360.
101. Segneanu, A.E.; Gozescu, I.; Dabici, A.; Sfirloaga, P.; Szabadai, Z. *Organic Compounds FT-IR Spectroscopy in Macro to Nano Spectroscopy*; Uddin, J., Ed.; InTechOpen: Rijeka, Croatia, 2012; ISBN 978-953-51-0664-7.
102. Monallem-Bahont, M.; Bertrand, S.; Pena, O. Synthesis and characterization of Zn1-xNixFe2O4 spinels prepared by a citrate precursor. *J. Solid State Chem.* **2005**, *178*, 1080–1086. [CrossRef]
103. Ahn, Y.; Choi, E.J.; Kim, E.H. Superparamagnetic relaxation in cobalt ferrite nano-particles synthesized from hydroxide carbonate precursors. *Rev. Adv. Mater. Sci.* **2003**, *5*, 477–480.
104. Naseri, M.G.; Saion, E.B.; Hashim, M.; Shaari, A.H.; Ahangar, H.A. Synthesis and characterization of zinc ferrite nanoparticles by a thermal treatment method. *Solid State Commun.* **2011**, *151*, 1031–1035. [CrossRef]
105. Mihalca, I.; Ercuta, A. Structural relaxation in Fe70Cr10.5P11.5Mn1.5C6.5 amorphous alloys. *J. Optoelectron. Adv. Mater.* **2003**, *5*, 245–249.
106. Marin, C.N.; Fannin, P.C.; Malaescu, I. Time solved susceptibility spectra of magnetic fluids. *J. Magn. Magn. Mater.* **2015**, *388*, 45–48. [CrossRef]
107. Fannin, P.C.; Marin, C.N. Determination of the Landau-Lifshitz damping parameter by means of complex susceptibility measurements. *J. Magn. Magn. Mater.* **2006**, *299*, 425–429. [CrossRef]
108. Mălăescu, I.; Marin, C.N. Study of magnetic fluids by means of magnetic spectroscopy. *Phys. B Condens. Matter* **2005**, *365*, 134–140.

plants MDPI

Article

Chemical Composition, Antioxidant, and Antimicrobial Activity of *Dracocephalum moldavica* L. Essential Oil and Hydrolate

Milica Aćimović [1,*], Olja Šovljanski [2], Vanja Šeregelj [2], Lato Pezo [3], Valtcho D. Zheljazkov [4], Jovana Ljujić [5], Ana Tomić [2], Gordana Ćetković [2], Jasna Čanadanović-Brunet [2], Ana Miljković [6] and Ljubodrag Vujisić [5]

1 Institute of Field and Vegetable Crops Novi Sad, Maksima Gorkog 30, 21000 Novi Sad, Serbia
2 Faculty of Technology, University of Novi Sad, Bulevar cara Lazara 1, 21000 Novi Sad, Serbia; oljasovljanski@uns.ac.rs (O.Š.); vanjaseregelj@tf.uns.ac.rs (V.Š.); anav@uns.ac.rs (A.T.); cetkovic@tf.uns.ac.rs (G.Ć.); jasnab@uns.ac.rs (J.Č.-B.)
3 Institute of General and Physical Chemistry, Studentski trg 12-16, 11000 Belgrade, Serbia; latopezo@yahoo.co.uk
4 College of Agricultural Sciences, Oregon State University, Corvallis, OR 97331, USA; valtcho.jeliazkov@oregonstate.edu
5 Faculty of Chemistry, University of Belgrade, Studentski trg 12-16, 11000 Belgrade, Serbia; jovanalj@chem.bg.ac.rs (J.L.); ljubaw@chem.bg.ac.rs (L.V.)
6 Faculty of Medicine, University of Novi Sad, Hajduk Veljkova 3, 21000 Novi Sad, Serbia; ana.miljkovic@mf.uns.ac.rs
* Correspondence: milica.acimovic@ifvcns.ns.ac.rs

Abstract: Steam distillation was used for the isolation of *Dracocephalum moldavica* L. (Moldavian dragonhead) essential oil (DMEO). This aromatic herbaceous plant is widespread across the Northern Hemisphere regions and has been utilized in health-improving studies and applications. In addition to the DMEO, the hydrolate (DMH), a byproduct of the distillation process, was also collected. The DMEO and DMH were analyzed and compared in terms of their chemical composition, as well as their in vitro biological activities. The main component in DMEO was geranyl acetate, while geranial was dominant in DMH. The DMEO demonstrated better antioxidant and antimicrobial activities compared with the DMH against *Staphylococcus aureus*, *Escherichia coli*, *Salmonella* Typhimurium, and *Listeria monocytogenes*, which represent sources of food-borne illness at the global level. The DMEO and DMH show promise as antioxidant and antimicrobial additives to various products.

Keywords: Moldavian dragonhead; phytochemicals; large-scale distillation; in vitro biological activity; time-kill kinetics modeling

Citation: Aćimović, M.; Šovljanski, O.; Šeregelj, V.; Pezo, L.; Zheljazkov, V.D.; Ljujić, J.; Tomić, A.; Ćetković, G.; Čanadanović-Brunet, J.; Miljković, A.; et al. Chemical Composition, Antioxidant, and Antimicrobial Activity of *Dracocephalum moldavica* L. Essential Oil and Hydrolate. *Plants* **2022**, *11*, 941. https://doi.org/10.3390/plants11070941

Academic Editor: Fabrizio Araniti

Received: 25 February 2022
Accepted: 14 March 2022
Published: 31 March 2022

Publisher's Note: MDPI stays neutral with regard to jurisdictional claims in published maps and institutional affiliations.

1. Introduction

Medicinal and aromatic plants have been utilized in traditional medicine around the world for millennia and continue to play an important role in treating diseases and providing nutritive support [1]. *Dracocephalum moldavica* L., known as Moldavian dragonhead, or Moldavian balm, is an aromatic herbaceous plant native to the temperate climate of Asia, however, nowadays it is found across the Northern Hemisphere [2]. *D. moldavica* essential oil (DMEO) has a citrus-like flavor due to the contents of geranial, neral, and geranyl acetate, and reassembles other lemon-scented plants such as lemon balm and lemon catnip [3,4]. Previous research showed that this plant had antioxidant [5–9] and antimicrobial properties [10–13]. In addition, the plant also displayed sedative [14,15], antidepressant [16], antinociceptive [17], anti-inflammatory [18,19], as well as neuroprotective [20,21] and cardioprotective effects [22–25].

Recently, there has been an increased interest in its therapeutic benefits due to reports on its biological activities. Indeed, medicinal plants are being used in healthcare and everyday nutrition, mainly as functional food and nutraceuticals [26,27], but also as natural preservatives [28,29]. Medicinal plants can be used fresh or dry, in the form of herbal tea, extract, essential oil, or pharmaceutical formulations (tablets, capsules, etc.). Plant essential oils possess a wide spectrum of biological activities, and are used in the food industry, but also in human and veterinary medicine [30]. Hydrolates (hydrosols and distillate waters) are byproducts of steam distillation. Generally, hydrolates contain a small amount of dissolved essential oil and other constituents and possess some biological activities [31]. Therefore, hydrolates may have potential application in the food industry for flavoring, preservation, and in soft drinks, but can also be applied in aromatherapy, cosmetics, agriculture, and veterinary medicine [32].

This investigation aimed to determine the composition of essential oil and hydrolate of *D. moldavica* (DMEO and DMH, respectively), grown as an essential-oil-bearing crop in the Republic of Serbia and distilled under semi-industrial conditions, and the in vitro antioxidant and antimicrobial activities.

2. Results

2.1. Volatile Compounds of Essential Oil and Hydrolate

The main compounds detected in the DMEO and DMH are presented in Table 1 (GC-FID chromatograms of essential oil and hydrolate are given in the Supplementary Materials, Figures S1 and S2, respectively). The most abundant compounds in the DMEO (20 compounds, comprising 98.1%) were geranyl acetate (53.2%), followed by geranial (16.8%), and neral (10.7%), while in the hydrolate, there were 23 identified compounds (comprising 96.0%), and the most abundant were geranial (23.4%), neral (22.4%), and geraniol (21.3%) (Table 1).

Table 1. Chemical composition of *D. moldavica* essential oil (DMEO) and hydrolate (DMH).

No	Compound	RI	DMEO	DMH
1	1-octen-3-ol	974	-	0.5
2	6-methyl-5-hepten-2-one	986	0.1	3.3
3	dehydro-1,8-cineole	988	-	1.3
4	3-octanol	995	-	0.2
5	1,8-cineole	1028	-	0.6
6	Benzene acetaldehyde	1041	-	0.6
7	*cis*-linalool oxide (furanoid)	1069	-	1.7
8	*trans*-linalool oxide (furanoid)	1086	-	0.7
9	Linalool	1096	1.6	8.7
10	Camphor	1141	-	0.5
11	*trans*-chrysanthemal	1147	-	0.3
12	Nerol oxide	1149	0.1	0.4
13	Borneol	1159	0.6	2.9
14	Terpinen-4-ol	1171	0.1	1.7
15	Cryptone	1185	-	0.1
16	α-Terpineol	1188	-	0.9
17	*trans*-Isocitral	1777	0.4	-
18	Nerol	1227	-	2.2
19	Neral	1234	10.7	22.4
20	Linalool acetate	1247	7.9	-
21	Geraniol	1254	-	21.3
22	Geranial	1266	16.8	23.4
23	Lavandulyl acetate	1285	0.1	-
24	Methyl geranate	1318	0.1	-
25	Eugenol	1356	-	0.5
26	Neryl acetate	1358	4.6	0.1
27	α-Copaene	1369	0.2	-
28	Geranyl acetate	1379	53.2	1.7
29	*trans*-caryophyllene	1412	0.6	-
30	α-humulene	1447	0.1	-
31	*trans*-β-farnesene	1451	0.1	-
32	γ-muurolene	1474	0.4	-
33	E,E-α-farnesene	1502	0.1	-
34	Caryophyllene oxide	1575	0.3	-
	Total		98.1	96.0

RI—Retention Indices on non-polar capillary column HP-5MS.

2.2. Antioxidant Activity

The results of in vitro antioxidant activity of DMEO and DMH are shown in Table 2. As a consequence of the high content of geranyl acetate, the DMEO exhibited a significantly stronger antioxidant potential than the DMH. The DMEO at concentration of 250 mg/mL showed the highest scavenging activity against lipid radicals (397.20 µmolTE/100 mL), followed by ABTS$^{•+}$ (312.54 µmolTE/100 mL), superoxide anion (294.77 µmolTE/100 mL), and DPPH$^{•}$ (246.39 µmolTE/100 mL). The reducing power of DMEO was lower than presented scavenging abilities but still with a high value of 171.46 µmolTE/100 mL.

Table 2. In vitro antioxidant activity of *D. moldavica* essential oil (DMEO) and hydrolate (DMH).

Antioxidant Activity (µmolTE/100 mL)	DMEO	DMH
DPPH$^{•}$	246.39 ± 1.17^{b}	8.82 ± 0.36^{a}
ABTS$^{•+}$	312.54 ± 11.63^{b}	25.44 ± 1.98^{a}
SOA	294.77 ± 13.29^{b}	19.58 ± 0.11^{a}
BCB	397.20 ± 36.12^{b}	41.63 ± 2.17^{a}
RP	171.46 ± 2.56^{b}	9.50 ± 0.64^{a}

Results are expressed as mean ± standard deviation ($n = 3$). Values in the row with different superscripts are significantly different at $p < 0.05$ according to Fisher's least significant differences (LSD) test. DPPH$^{•}$—2,2-diphenyl-1-picrylhydrazyl; ABTS$^{•+}$—2,2′-azino-bis-3-ethylbenzothiazoline-6-sulphonic acid; SOA—superoxide anion; BCB—β-carotene bleaching; and RP—reducing power.

2.3. Antimicrobial Activity

According to the obtained results of in vitro antimicrobial activity, a significant antimicrobial effect against almost all of the tested bacteria was gained, but not against the tested strains of yeasts and fungi (Table 3). The tested DMEO did not show an antimicrobial effect against *B. cereus* and *P. aeruginosa* strains, while in the case of the tested DMH, the antimicrobial effect was also absent against *S.* Typhimurium. *Pseudomonas aeruginosa* represents a very resistant strain, well-known as bacteria resistant to numerous antibiotics, and indeed, showed resistance to the antibiotics in this study.

Table 3. Assessment of the antimicrobial effect.

Test Organisms	The Inhibition Zone * (mm)			
	D. moldavica		Controls	
	DMEO	DMH	Antibiotic	Actidione
B. cereus ATCC 11778	16.0 ± 0.0	11.0 ± 0.0	27.0 ± 0.0	-
S. aureus ATCC 25923	40.0 ± 0.0	28.33 ± 0.0	28.0 ± 0.0	-
L. monocytogenes ATCC 35152	40.0 ± 0.0	27.0 ± 0.0	26.3 ± 0.6	-
E. coli ATCC 25922	40.0 ± 0.0	34.33 ± 0.58	27.0 ± 0.0	-
P. aeruginosa ATCC 27853	21.0 ± 0.0	nd	21.0 ± 0.0	-
S. Typhimurium ATCC 13311	40.0 ± 0.0	10.33 ± 0.58	29.33 ± 0.6	-
S. cerevisiae ATCC 9763	11.33 ± 0.58	nd	-	34.0 ± 0.0
C. albicans ATCC 10231	nd	nd	-	37.0 ± 0.0
A. brasiliensis ATCC 16404	nd	nd	-	27.3 ± 0.6

* Mean value diameter of zone including disc (6 mm) ± standard deviation; nd—not detected.

The antimicrobial potential of DMEO and DMH toward sensitive microorganisms was conducted by the microdilution method after the satisfactory results of the preliminary screening. As shown in Table 4, the obtained MICs of DMEO showed relatively low activity against all bacteria (MIC ≤ 3.125%). Conversely, higher MICs (between 3.125–12.5%) of DMH were noted for sensitive bacteria. Consequently, the tested essential oil and hydrolate can be utilized as antimicrobial agents for the worldwide struggle for prolonged shelf-life of food, absence of food-borne pathogens, and epidemic crisis, but also can be used as an eco-based substance in antimicrobial formulations in the pharmaceutical and cosmetics industries.

Table 4. Minimal inhibitory concentration (%).

Test Organisms	DMEO	DMH
B. cereus ATCC 11778	>100 *	>100
S. aureus ATCC 25923	0.78	12.5
L. monocytogenes ATCC 35152	1.56	6.25
E. coli ATCC 25922	3.125	3.215
P. aeruginosa ATCC 27853	>100	>100
S. Typhimurium ATCC 13311	0.78	>100
S. cerevisiae ATCC 9763	>100	>100
C. albicans ATCC 10231	>100	>100
A. brasiliensis ATCC 16404	>100	>100

* According resistance on the initial concentration (see Table 3).

The in vitro antimicrobial potential of tested substances can be further clarified by the time-kill or pharmacodynamics kinetics monitoring. In this way, in vitro examination of antimicrobial substances can be quantified in view of the path of antimicrobial activity as a function of contact time between sensitive microorganisms and targeted concentration of tested essential oil and hydrolate [33]. Therefore, a pharmacodynamics pathway of the antimicrobial effect of tested samples was conducted for all sensitive bacteria. The first step was to determine bacterial profile growth curves without the addition of either DMEO nor DMH. Non-treated bacterial suspensions were verified at the same time as the essential oil- or hydrolate-treated samples. Bacterial growth profile curves (Figure 1) indicated the number of live bacterial cells over an incubation period. There are noticeably three growth phases for all four tested bacteria: lag phase, exponential (log), and stationary phase. The initial phase is especially emphasized for *L. monocytogenes* and *S. aureus*, while this period of cell adaption is minimal for the other two sensitive bacteria. The differences are detected for maximum yield, which was obtained after approximately 12 h of incubation. Briefly, the highest concentration of bacterial cells was 7.6 CFU/mL, 8 CFU/mL, 8.4 CFU/mL, and 8.8 log CFU/mL for *S. aureus*, *L. monocytogenes*, *S.* Typhimurium, and *E. coli*, respectively. Regression coefficients for the obtained growth profile curves are shown in Table 5. Additionally, the fit between experimental and model calculated results are given in the Supplementary Materials, Table S1, indicating a very good predictive capacity of the obtained models, with a coefficient of determination of 0.99 for all tested bacteria.

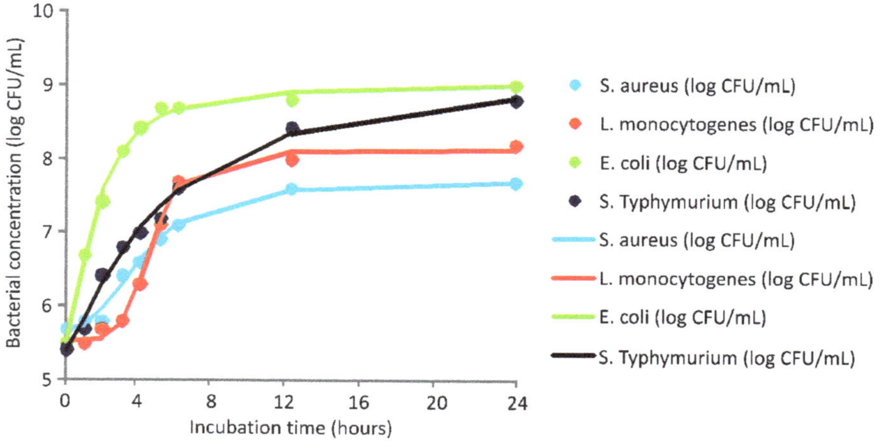

Figure 1. Bacterial growth kinetics—control samples (incubation of sensitive bacteria without the tested antimicrobial substances)—markers signify the experimental data; lines indicate predictive results.

Table 5. Regression coefficients for bacterial growth kinetics (control samples).

Coefficient	Bacterial Concentration (Log CFU/mL)			
	S. aureus	*L. monocytogenes*	*E. coli*	*S.* Typhimurium
d	7.72	8.12	9.04	9.27
a	5.7	5.53	5.52	5.38
c	4.29	4.60	1.64	4.95
b	2.58	5.59	1.65	1.30

The regression coefficients could be depicted as follows: a–minimum of the experimentally gained values (t = 0); d–the maximally gained values (t = ∞); c–the infection point (the point between a and d); b–the steepness of the infection point c.

The pharmacodynamics potential of DMEO at different concentrations is graphically represented in Figure 2. Kinetics profiles for MIC value indicate the biocide effect for *S. aureus* and *S.* Typhimurium after a contact time of 3 h, while the same effect was observed for *L. monocytogenes* and *E. coli* for 4 h and 12 h, respectively. A bactericidal effect was achieved in twice as short a time for *S. aureus* when 2- and 4-time MIC were applied. Interestingly, the effect of double MIC concentration did not decrease contact time which is necessary for complete inhibition of *L. monocytogenes* but quadruple MIC enabled achieving bactericidal effect after only 1 h of contact. Similar behavior was observed for *S.* Typhimurium, but with a biocide effect after 3 h and 2 h for 2- and 4-time MIC, respectively. The killing rate of 2- and 4-time MIC of DMEO for *E. coli* was achieved for 2 h or 4 h shorter contact time compared with the MIC effect.

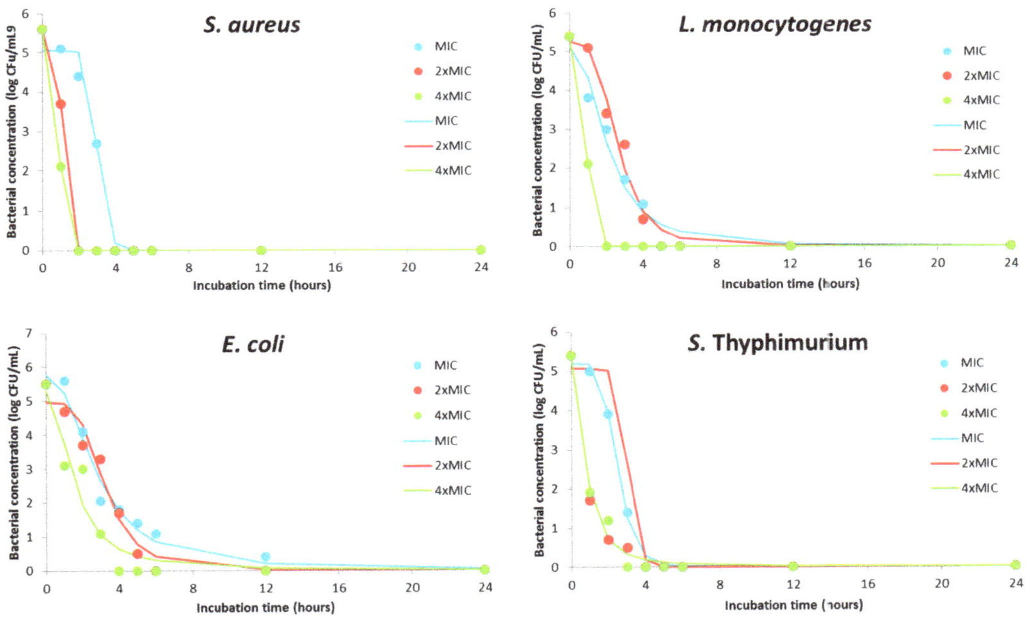

Figure 2. Kinetics modeling for pharmacodynamics potential of antimicrobial activity using MIC, 2×MIC, and 4×MIC of DMEO (markers signify the experimental data; lines indicate predictive results).

Additionally, Table 6 summarizes regression coefficients of the kinetics models. This parameter simplifies the speed and intensity of MIC and multiple MIC values. Furthermore, Table S1 of the Supplementary Materials involves the goodness of fit between experimentally and model obtained results. It can be concluded that the kinetics models (Figure 2) were precise, with high coefficients of determination (0.93–1), and can be used to understand the antimicrobial effect of the DMEO against sensitive bacteria.

Table 6. Regression coefficients for time-kill kinetics models for DMEO-treated samples.

Coefficients	DMEO Concentration											
	S. aureus			*L. monocytogenes*			*E. coli*			*S.* Typhimurium		
	MIC	2×MIC	4×MIC	MIC	2×MIC	4×MIC	MIC	2×MIC	4×MIC	MIC	2×MIC	4×MIC
d	0.00	0.00	0.00	0.00	0.00	0.00	0.00	0.00	0.00	0.00	0.00	0.00
a	5.05	5.6	5.6	5.13	5.27	5.4	5.77	4.97	5.31	5.20	5.05	5.39
c	3.03	1.05	0.96	2.06	2.59	0.96	2.78	3.25	1.53	2.45	3.03	0.77
b	11.64	14.9	11.7	2.34	3.65	11.89	2.26	3.9	2.06	5.72	11.64	1.99

The regression coefficients could be depicted as follows: a–minimum of the experimentally gained values (t = 0); d–the maximally gained values (t = ∞); c–the infection point (the point between a and d); b–the steepness of the infection point c.

As previously noticed, MIC values for DMH are significantly higher compared with DMEO in the case of *L. monocytogenes* and *S. aureus*. The same concentration of DMH and DMEO is necessary for the inhibition of *E. coli* activity. Regardless of the mentioned difference, the same pharmacodynamics study was done for hydrolate-treated samples (Figure 3). The biocide effect of MIC values for all three sensitive bacteria indicated a rapid bacteriostatic effect, showing a decrease in cell viability after the first three hours of contact time. The final biocide effect was observed after 6 h. The biocide effect of MIC and double MIC values was observed in the same contact time for in the case of *L. monocytogenes* and *E. coli*. On the other hand, using a 2-time MIC value reduced the required contact time for *S. aureus*. The complete reduction in bacterial viability was detected for 4 h, 5 h, or 6 h contact time in the case of using 4-time MIC values for *L. monocytogenes*, *S. aureus*, and *E. coli*, respectively. In this study, the antimicrobial effect of DMH as a byproduct in DMEO production was reasonable suggesting new possibilities for the utilization of DMH.

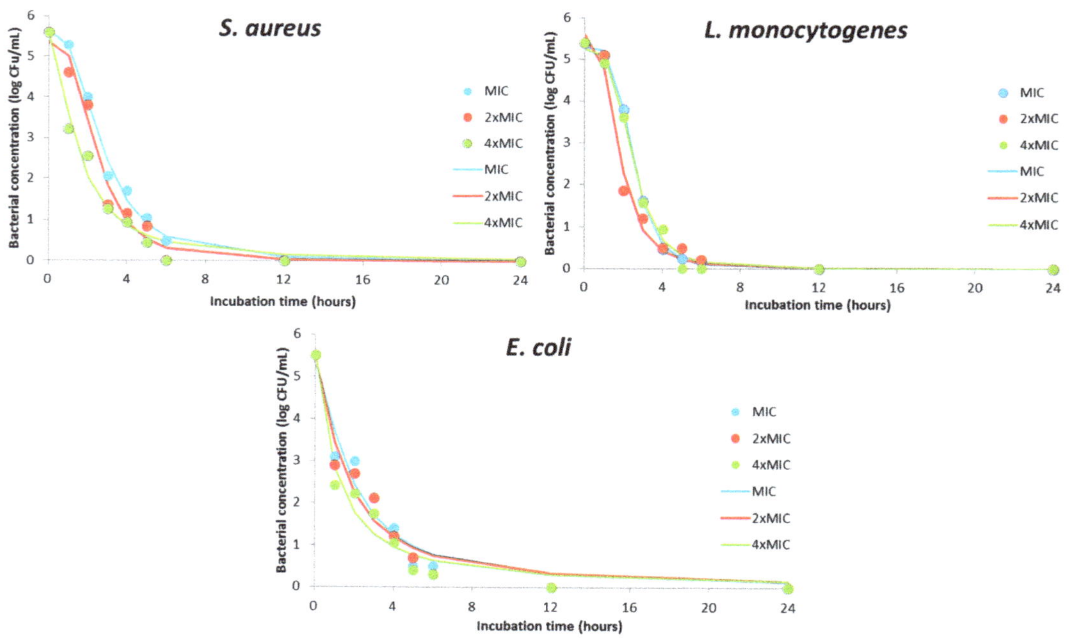

Figure 3. Kinetics modeling for pharmacodynamics potential of antimicrobial activity using MIC, 2×MIC, and 4×MIC of DMH (markers signify the experimental data; lines indicate predictive results).

According to the regression coefficients for the gained kinetics models (Table 7) and the fitness between experimentally and model obtained results (Supplementary Materials, Table S1), it can be concluded that the models were accurate, with high coefficients of determination (0.95–1) and the proposed models fit well with the experimental data. In summary, the obtained models for DMH-treated samples can be used for the prediction of antimicrobial effect based on contact time between bacterial cells and DMH.

Table 7. Regression coefficients for time-kill kinetics models for DMH-treated samples.

Coefficients	DMH Concentration											
	S. aureus			*L. monocytogenes*			*E. coli*			*S. aureus*		
	MIC	2×MIC	4×MIC	MIC	2×MIC	4×MIC	MIC	2×MIC	4×MIC	MIC	2×MIC	4×MIC
d	0.00	0.00	0.00	0.00	0.00	0.00	0.00	0.00	0.00	0.00	0.00	0.00
a	5.65	5.36	5.53	5.31	5.59	5.25	5.36	5.4	5.46	5.65	5.36	5.53
c	2.68	2.40	1.43	2.47	1.76	2.43	1.76	1.52	1.05	2.68	2.40	1.43
b	2.65	3.01	1.65	4.35	3.05	3.85	1.44	1.32	1.16	2.65	3.01	1.65

The regression coefficients could be depicted as follows: a–minimum of the experimentally gained values (t = 0); d–the maximally gained values (t = ∞); c–the infection point (the point between a and d); b–the steepness of the infection point c.

3. Discussion

3.1. Volatile Compounds of Essential Oil and Hydrolate

According to referenced data from the literature about DMEO chemical composition (Table 8), as well as cluster analysis performed using this data for the construction an unrooted phylogenetic tree (Figure 4), it could be assumed that the largest number of accessions belonged to the geranial + neral + geranyl acetate chemotype [2,9,10,12,17,34–46]. The *D. moldavica* plants grown and utilized in this study also belonged to the above chemotype. Other previously reported chemotypes of *D. moldavica* include: the geranyl acetate + geranial + geraniol one [42,47–51], the 1,8 cineole + 4-terpineol [52], and the linalool + geranial + fenchone [53].

Table 8. Chemical composition of DMEO according to the literature data.

No	Reference	1,8-Cineole	4-Terpineol	Fenchone	Geranial	Geraniol	Geranyl Acetate	Linalool	Methyl Chavicol	Neral	Nerol	Neryl Acetate
1	[2]	0.0	0.0	0.0	29.6	5.4	27.2	0.4	0.0	19.4	0.4	3.0
2	[34] *	0.3	0.0	0.1	16.3	22.3	35.6	0.3	0.0	11.9	1.0	2.6
3	[35] *	0.0	0.0	0.0	26.2	4.6	35.0	0.2	0.0	20.7	0.0	4.1
4	[36]	0.0	0.0	0.0	25.5	0.5	15.2	1.3	16.0	9.7	0.3	1.2
5	[37]	0.0	0.0	0.0	27.3	20.7	23.2	0.8	0.0	18.6	0.0	2.1
6	[47]	0.0	0.0	0.0	21.6	39.5	12.4	0.8	0.0	17.1	1.5	1.6
7	[38] *	0.0	0.0	0.0	9.3	16.0	52.7	0.6	0.0	5.1	0.3	2.9
8	[9] *	0.0	0.0	0.0	36.6	2.9	26.7	0.3	0.0	25.7	0.1	1.2
9	[39]	0.0	0.0	0.0	26.3	16.9	22.5	1.5	0.0	21.3	1.0	0.4
10	[12]	0.0	0.0	0.0	28.5	19.6	16.7	0.8	0.0	21.2	1.9	1.8
11	[40] *	0.0	0.0	0.0	23.6	16.8	29.2	2.0	0.0	20.2	1.9	0.0
12	[48]	0.0	0.0	0.0	11.2	24.3	36.6	0.8	0.0	16.3	0.4	0.9
13	[52]	31.3	22.8	0.0	0.0	0.0	0.0	0.0	0.0	0.0	0.0	0.0
14	[41]	0.0	0.0	0.0	19.8	15.1	27.9	2.4	0.0	18.0	2.2	4.2
15	[42]	0.0	0.0	0.0	30.9	34.2	25.4	0.6	0.0	0.3	0.0	1.5
16	[42]	0.0	0.0	0.0	27.8	36.0	27.5	1.0	0.0	0.4	0.0	1.8
17	[43]	0.0	0.0	0.0	8.4	15.9	46.7	0.5	0.0	5.8	0.3	2.6
18	[17]	0.0	0.0	0.0	31.1	0.0	0.0	1.5	0.0	31.1	17.1	4.8
19	[44]	0.0	0.0	0.0	19.1	9.3	30.4	2.7	0.0	17.8	2.9	2.5
20	[10]	0.0	0.0	0.0	21.6	17.6	19.9	1.1	0.0	32.1	0.0	1.6
21	[53] *	0.4	0.0	13.8	15.9	6.9	1.3	28.1	1.2	0.0	1.4	0.9
22	[45] *	0.01	0.0	0.0	50.7	3.4	10.0	0.1	0.0	26.8	0.0	0.0
23	[49]	0.0	0.0	0.0	44.0	28.0	14.0	0.4	0.0	6.3	0.2	2.6
24	[46]	0.3	0.0	0.1	41.9	5.3	19.0	0.5	0.0	25.3	0.3	0.6
25	[50]	0.0	0.0	0.0	18.3	19.5	34.9	0.4	0.0	14.8	0.0	2.9
26	[51]	0.0	0.0	0.0	24.5	8.8	32.6	0.8	0.0	22.7	2.4	3.4
27	TS	0.0	0.1	0.0	16.8	0.0	53.2	1.6	0.0	10.7	0.0	4.6

* Average value from different agrotechnical measures (cropping patterns, fertilization, and irrigation); TS—this study.

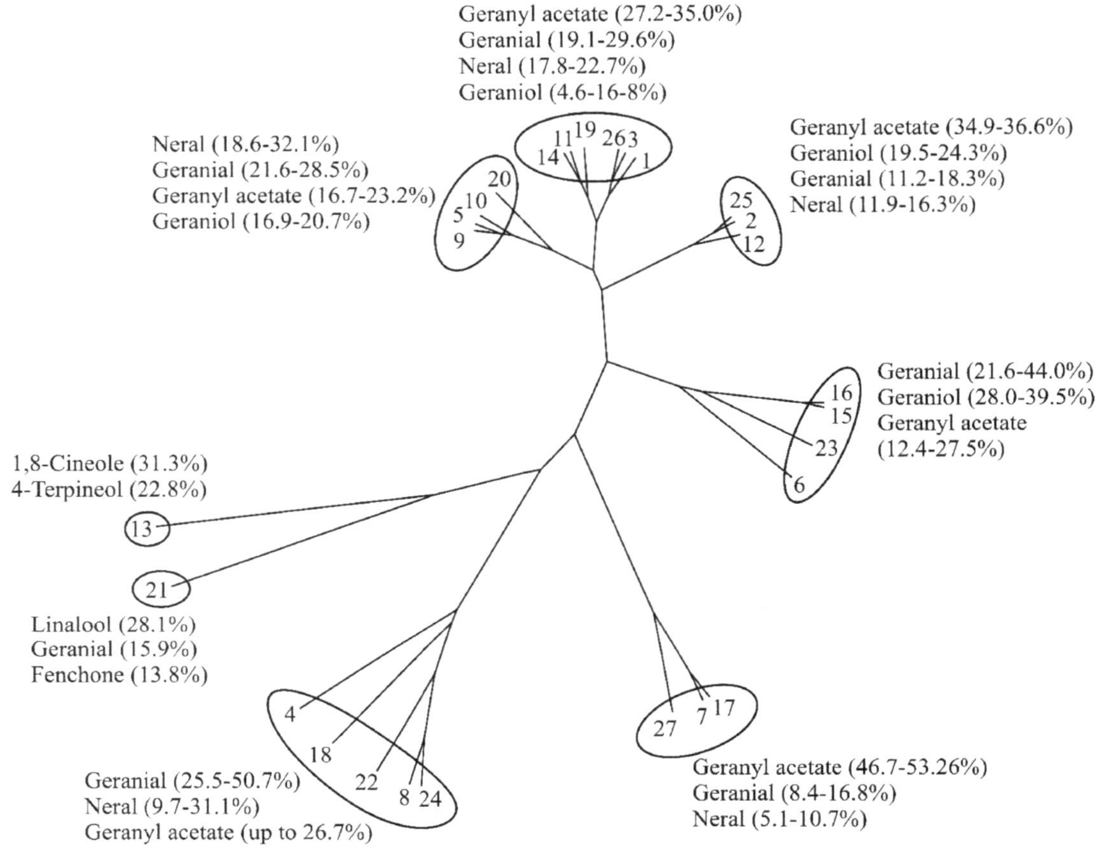

Figure 4. The unrooted phylogenetic tree of *D. moldavica* essential oil (DMEO) according to Table 8.

The phylogenetic cluster tree for DMEO was estimated and drawn using R software 4.0.3, as it is previously described for essential oil of white horehound [54], immortelle [55], hyssop [56], naked catmint [57], and sweet wormwood [58].

3.2. Antioxidant Activity

Plant essential oils represent natural sources of bioactive compounds with several models of action, among them, scavenging of free radicals, prevention of chain reactions initiation, reducing agents, and termination of peroxides, quenching of singlet oxygen, and binding of metal ion catalysts [59]. Different in vitro assays have been used to estimate the antioxidant potential of DMEO, while the data of DMH antioxidant activity have not been previously published in the scientific literature. In general, for complex systems such as essential oils, using at least three methods with different mechanisms is preferable and in some cases required to evaluate antioxidant activities. The DPPH$^{\bullet}$ and ABTS$^{\bullet+}$ assays are efficient tools for the determination of antioxidant activity; although these methods have an identical mechanism, the mediums for the assay are different which means there is a dissimilar solubility of the isolated bioactive compounds [60]. The SOA, like ABTS$^{\bullet+}$ assay, share the water medium, but at a different pH level. Superoxide anion is the most frequent free radical in vivo and deals as a precursor for other reactive oxygen species that possess the capability to induce damage of important biological molecules. In the BCB assay, due to the absence of antioxidants, β-carotene undergoes rapid discoloration; this could be explained by the integrated oxidation of β-carotene and linoleic acid-generating free lipid species. Another defense mechanism in preventing the body from the dangerous effect of free radicals is reducing these molecules by the antioxidants. In the present study, reducing power was monitored through measurement of the ferrous ions transformation in presence of antioxidants.

Geranyl acetate is known for its strong antioxidant properties due to its capacity to reduce free radical stability via electron or hydrogen donating mechanisms [61]. As well, geranyl acetate is insoluble in water but soluble in organic solvents and oil, which is the reason for its low concentration in DMH.

Investigation of the DMEO antioxidant capacity obtained from *D. moldavica* plants grown as a single-crop vs. intercropping systems with soybean in response to the application of chemical fertilizer (urea and triple superphosphate) and organic manure had the IC50 values in the range from 1.45 to 5.28 µg/mL [9]. Furthermore, the evaluation of the antioxidant activity of DMEO using DPPH$^\bullet$, ABTS$^{\bullet+}$, and BCB assays showed that essential oil possesses weaker scavenger activity for DPPH$^\bullet$ and ABTS$^{\bullet+}$ radicals than ascorbic acid and BHT, while higher activity was reported for peroxyl radicals [12]. The results from the latter study were in agreement with the results of the present study. Generally, both DMEO and DMH contain very efficient bioactive compounds such as geranyl acetate, geranial, geraniol, and neral, which are responsible for the antioxidant activity. There is not enough information about antioxidant mechanisms and other biological activities of DMEO and DMH.

3.3. *Antimicrobial Activity*

The gained antimicrobial effect (see Section 2.3) can be the result of the chemical composition of the DMEO and DMH, due to the fact that geranyl acetate, as an ester derived from geraniol, as well as geranial and neral (together known as citral) have good antibacterial properties and good thermal stability [62,63]. The mentioned group of bioactives as well as this group of bioactives are especially dominant in DMEO, but also in DMH (Table 2).

Resistance of *B. cereus* on the tested samples was not in line with the results of El-Baky and El-Baroty [11] that reported the inhibitory effect of DMEO from Egypt in a concentration of 0.07 mg/mL. This may suggest the difference in antimicrobial effect based on the origin of the plant and differences in its chemical profile, which has been reported previously [64–66]. Both DMEO and DMH samples in this study showed an antimicrobial effect on *S. aureus*, *L. monocytogenes*, and *E. coli* strains that is strongly correlated to the results of Eshani et al. [12], which demonstrated a significant antimicrobial effect of DMEO against the mentioned bacteria. Notably, the obtained max of 40 mm was registered for all the tested bacteria in the case of DMEO, while this value for the tested DMH was lower. Moreover, the inhibition zones of both DMEO and DMH were even higher than that in the positive control, i.e., cefotaxime and clavulanic acid combination, indicating the significant potential of using the DMEO and DMH as natural ingredients in various products to improve microbial antibiotic resistance. To the best of our knowledge, the high antimicrobial performance of DMH has not been previously reported. In addition to the antibacterial effect, several scientific groups reported antifungal activity of DMEO [10,67,68] which did not correlate with the results of this research.

Additional observation can be directed to the fact that all four sensitive bacteria represent common pathogenic bacteria that cause foodborne diseases [69], while alimentary infection and intoxications caused by these pathogens represent a growing public health problem [70]. Due to the mentioned facts, the obtained results are promising in view of finding alternative agents with rapid biocide effect for the food or packaging production.

4. Materials and Methods

4.1. Plant Material

D. moldavica was grown during the 2021 cropping season in the Institute of Field and Vegetable Crops Novi Sad (IFVCNS), experimental fields in Bački Petrovac (Vojvodina Province, North Serbia) on *gleyed calcareous chernozem* soil type. The species were confirmed by Milica Rat, research associate at the botanical collection, and deposited under the voucher number 2-1513 at the Herbarium BUNS (Faculty of Sciences, University of Novi Sad). The previous crop was barley. Granular mineral fertilizer (70 kg NPK in formulation 15:15:15) was applied in the previous fall prior to the fall plowing and disking.

The seeds were sown in pots in a greenhouse, in March, and the seedlings were transplanted at the end of April in an experimental plot of 70 m × 10 m, with a 70 cm spacing between rows and 50 cm between plants. During the vegetation period, only hand weeding and hoeing were performed. Plants were harvested in August, at full bloom, by cutting the plants at 5 cm above the ground, dried in a flat-bed solar dryer at temperatures less than 40 °C for two days, and the essential oil was isolated via steam distillation.

4.2. Essential Oil Isolation

The steam distillation of the dried aerial plant parts of *D. moldavica* was performed in a semi-industrial distillation unit at IFVCNS [71]. Briefly, 50 kg of dried biomass was put in the distillation vessel (0.8 m^3), which was supplied with hot dry steam from a separate steam generator. After 20 min, a condensate (essential oil and condensed water) started to accumulate in the glass Florentine flask. After 4 h of distillation, the essential oil and the hydrolate were separated: the DMEO was decanted from the aqueous layer, dried over anhydrous sodium sulfate, while the DMH was purified by filtration using MN 651/120 filter paper. The essential oil yield was 0.65% in dried biomass. In order to prepare the DMH sample for analysis of volatile compounds, 400 mL of hydrolate were extracted by dichloromethane via a Likens–Nickerson apparatus for 2 h.

4.3. Analysis of Volatile Compounds

Gas chromatograph (Agilent 7890A) with two detectors, flame ionization (FID) and mass selective (Agilent 5975C), and non-polar capillary column HP-5MS (30 m × 0.25 mm × 0.25 μm) were used for the analysis of DMEO and DMH. The operating conditions were the same as in our previous works [58,71]. Identification of the components was conducted according to their linear retention indices (RI), and comparison with mass spectral libraries (Adams ver. 4, Wiley ver. 5, and NIST ver. 17). The relative abundance of each detected compound was calculated from GC/FID chromatograms as a percentage area of each peak (only identified compounds are shown).

4.4. In vitro Assessment of Antioxidant Activity

The potential antioxidant activity of DMEO and DMH were assessed using five common in vitro antioxidant assays. The tests were performed with DMEO dissolved in methanol at the concentration of 250 mg/mL. For all assays, the Trolox equivalents were used for expression of antioxidant activities as μmol per 100 mL (μmolTE/100 mL). The DMH was tested in its original eluted form after filtration.

4.4.1. DPPH$^•$ (2,2-diphenyl-1-picrylhydrazyl)

The DPPH assay was performed according to Aborus et al. [72]. Briefly, 250 μL DPPH$^•$ solution in methanol (0.89 mM) was mixed with 10 μL of the sample in a microplate well and left in the dark at ambient temperature for 50 min. Absorbance was measured at 515 nm, and methanol was used as a blank.

4.4.2. ABTS$^{•+}$ (2,2′-azino-bis-3-ethylbenzothiazoline-6-sulphonic Acid)

The ABTS$^{•+}$ radical scavenging assay was evaluated employing the method according to Aborus et al. [72]. The absorbances of 250 μL activated ABTS$^{•+}$ (with MnO_2), before and

35 min (incubated at 25 °C) after the addition of 2 μL of juice were measured at 414 nm. Water was used as blank.

4.4.3. SOA (Superoxide Anion)

Superoxide anion radical scavenging activity was determined by the nitroblue tetrazolium reduction method, adapted for a 96-well microplate [73]. An amount of 50 μL of each sample solution dissolved in phosphate buffer, or 50 μL of buffer (blank test), was mixed with 50 μL of 166 mM nicotinamide adenine dinucleotide (NADH), 150 μL of 43 mM nitrotetrazolium blue (NBT), and 50 μL of phenazine methosulfate (PMS) in triplicate. The tests were conducted at room temperature with two readings of 560 nm, being the initial when PMS is added, and the final after 2 min.

4.4.4. RP (Reducing Power)

Reducing power (RP) was determined by the method of Oyaizu [74] adapted for a 96-well microplate. In brief, a 25 μL sample or 25 μL water (blank test), 25 μL sodium phosphate buffer (pH = 6.6), and 25 μL of 1% potassium iron(III) cyanide were mixed and incubated in a water bath for 20 min at 50 °C. After cooling, 25 μL of 10% trichloroacetic acid was added and solutions were centrifuged at 2470× *g* for 10 min. After centrifugation, 50 μL of supernatant was mixed with 50 μL of distilled water and 10 μL of 0.1% iron(III) chloride in the microplate. Absorbances were measured immediately at 700 nm.

4.4.5. BCB (*β*-Carotene Bleaching)

The *β*-carotene bleaching capacity of samples was evaluated by the *β*-carotene linoleate model system of Al-Saikhan et al. [75]. The absorbances of all the samples were taken at 470 nm at zero time and after 180 min, while during this time the microplate was incubated at 50 °C.

4.5. *In Vitro Assessment of the Antimicrobial Activity*

Observation of the antimicrobial activity of the DMEO and DMH was performed using references strains of bacteria *Bacillus cereus*, *Escherichia coli*, *Listeria monocytogenes*, *Pseudomonas aeruginosa*, *Salmonella* Typhimurium, and *Staphylococcus aureus*, as well as referent representatives of yeasts and fungi (*Aspergillus brasiliensis*, *Candida albicans*, and *Saccharomyces cerevisiae*). All tested strains were obtained from the American Type Culture Collection (ATCC).

4.5.1. Screening of Antimicrobial Effect of DMEO and DMH (Disk Diffusion Method)

Evaluation of antimicrobial activity of the DMEO and DMH was completed by the disk-diffusion method. As Micić et al. [76] reported in detail, the nutrient medium (Müller–Hinton agar or Sobouraud maltose agar) was inoculated with microbial suspensions (approx. 6 log CFU/mL) and the samples (15 μL) were applied onto three sterile cellulose discs. Bacteria were grown on Müller–Hinton agar (HiMedia, Mumbai, India) at 37 °C for 24 h and at 30 °C (*B. cereus*) for 18 h. Yeast and fungi were grown on Sabouraud maltose agar (HiMedia, Mumbai, India) at 25 °C for 48 h. Cells were suspended in a sterile 0.9% NaCl solution. As a negative control, sterile distilled water was used, while positive controls were commercially available antibiotics chloramphenicol and tetracycline (Sigma-Aldrich, St. Louis, MO, USA) as well as actidione (Biochemica, Billingham, U.K.). The obtained results were interpreted as follows: sensitive (diameter of inhibition zone above 26 mm), intermediary (inhibition zone 22–26 mm), and resistance (inhibition zone below 22 mm).

4.5.2. Minimal Inhibitory Concentration (MIC)

The MIC was evaluated for all bacteria, yeast, and fungi that are sensitive to the DMEO and DMH using the microdilution method labeled by Pavlić et al. [77]. The initial concentration was defined as 100%, while other concentrations were prepared using successive dilutions (100–0.39%) using dimethyl sulfoxide (50 mg/mL) for essential oil or sterile distilled water for hydrolate. The used dissolvents were inert for all tested microorganisms and did not have a biocide effect on bacterial and yeast growth [66]. MIC represents the lowest concentration of antimicrobial agents that, under defined in vitro conditions, prevents the appearance of visible growth of a microorganism within a defined period of time. MIC is calculated based on the numbers of cells of positive control and treated samples with DMEO or DMH. The test microtiter plate had a positive control (inoculated media without DMEO or DMH) and a negative control (100 µL of medium mixed with 100 µL of DMEO or DMH).

4.5.3. Assessment of Antimicrobial Activity Using a Time-Kill Procedure

The pharmacodynamics potential of antimicrobial activity was followed using monitoring of the time-kill kinetics as reported by Ferro et al. [78]. All sensitive bacteria (approx. 6 log CFU/mL) were tested during contact time with 1, 2, and 4-time MIC concentrations in several samples times (0 h, 2 h, 3 h, 4 h, 5 h, 6 h, 12 h, and 24 h of incubation for bacteria at 37 °C). An inoculated medium without the sample was positive control, while a non-inoculated medium was a blank. The four-parameter sigmoidal model established by Romano et al. [79] performed kinetic modeling.

4.6. Statistical Analyses

The collected data were processed statistically using the software package STATISTICA 10.0 (StatSoft Inc., Tulsa, OK, USA). All analyses were performed in three replicates. The obtained results were expressed as the mean value with standard deviation (SD). Analysis of variance (ANOVA) with Tukey's HSD post hoc test for comparison of the sample means were used to explore the variations of parameters. All observed samples were checked for variance equality (using Levene's test) and normal distribution (using Shapiro–Wilk's test).

5. Conclusions

Expressed needs for natural materials and phytochemicals, as well as needs to find antimicrobial substances that are substitutes for antibiotics, inspire researchers to look for new sources of these compounds. Herbs, such as *D. moldavica*, could be important raw materials in the pharmaceutical and food industry. The main compounds in the essential oil were geranyl acetate, geranial, and neral, while in the hydrolate these were geranial, neral, and geraniol. Geranyl acetate, a monoterpene, is commonly used at an industrial level in a wide range of products such as powders, soaps, perfumes, as well as flavoring agents, due to its intensely fruity and floral aroma. Due to its low solubility in water, it was not detected in the hydrolate. Citral (a mixture of two monoterpene aldehydes—geranial and neral) is widely used as a flavoring agent in food, beverage, and cosmetic products. *Geraniol* is a commercially important terpene alcohol used in the food, fragrance, and cosmetic industry, as well as insecticidal and repellent compounds in pesticides and household products. *D. moldavica* essential oil (DMEO) and hydrolate (DMH) as significant sources of these compounds are prospective raw materials for many purposes.

Supplementary Materials: The following supporting information can be downloaded at: https://www.mdpi.com/article/10.3390/plants11070941/s1, Table S1: The "goodness of fit" kinetics models; Figure S1: GC-FID Chromatogram of *Dracocephalum moldavica* essential oil (DMEO); and Figure S2: GC-FID Chromatogram of *Dracocephalum moldavica* hydrolate (DMH).

Author Contributions: Conceptualization, M.A. and O.Š.; methodology, V.Š. and J.L.; software, L.P.; validation, A.T., A.M. and L.V.; formal analysis, O.Š., V.Š. and A.T.; investigation, A.M. and J.L.; resources, M.A.; data curation, L.V.; writing—original draft preparation, M.A. and G.Ć.; writing—review and editing, J.Č.-B. and V.D.Z.; visualization, L.P.; supervision, V.D.Z.; project administration, M.A.; funding acquisition, V.D.Z. All authors have read and agreed to the published version of the manuscript.

Funding: This research received no external funding.

Data Availability Statement: The data is contained in the article and supplementary file.

Acknowledgments: This research was supported by the Ministry of Education, Science, and Technological Development of the Republic of Serbia (grant nos. 451-03-68/2022-14/200032 and 451-03-68/2022-14/200134). Also, we acknowledge funding support from the Department of Crop and Soil Science at Oregon State University, United States.

Conflicts of Interest: The authors declare no conflict of interest.

References

1. Xu, Y.; Liang, D.; Wang, G.T.; Wen, J.; Wang, R.J. Nutritional and functional properties of wild food medicine plants from the coastal region of South China. *J. Evid.-Based Integr. Med.* **2020**, *25*, 1–13. [CrossRef] [PubMed]
2. Aćimović, M.; Stanković, J.; Cvetković, M.; Todosijević, M.; Rat, M. The chemical composition of the essential oil of *Dracocephalum moldavica* L. from Vojvodina Province (Serbia). *Biol. Nyssana.* **2019**, *10*, 23–28. [CrossRef]
3. Suschke, U.; Sporer, F.; Schneele, J.; Geiss, H.K.; Reichling, J. Antibacterial and cytotoxic activity of *Nepeta cataria* L., *N. cataria* var. *citriodora* (Beck.) Balb. and *Melissa officinalis* L. essential oils. *Nat. Prod. Commun.* **2007**, *2*, 1277–1286. [CrossRef]
4. Aćimović, M.; Sikora, V.; Brdar-Jokanović, M.; Kiprovski, B.; Popović, V.; Koren, A.; Puvača, N. *Dracocephalum moldavica*: Cultivation, chemical composition and biological activity. *J. Agron. Technol. Eng. Manag.* **2019**, *2*, 153–167.
5. Poviliate, V.; Cuvelier, M.E.; Berset, C. Antioxidant properties of Moldavian dragonhead (*Dracocephalum moldavica* L.). *J. Food Lipids* **2001**, *8*, 45–64. [CrossRef]
6. Mohtashami, S.; Babalar, M.; Mirjalili, M.H. Phenological variation in medicinal traits of *Dracocephalum moldavica* L. (Lamiaceae) under different growing conditions. *J. Herbs. Spices Med. Plants.* **2013**, *19*, 377–390. [CrossRef]
7. Aprotosoaie, A.C.; Mihai, C.T.; Vochita, G.; Rotinberg, P.; Trifan, A.; Luca, S.V.; Petreus, T.; Gille, E ; Miron, A. Antigentoxic and antioxidant activities of a polyphenolic extract from European *Dracocephalum moldavica* L. *Ind. Crops. Prod.* **2016**, *79*, 248–257. [CrossRef]
8. Aslanipour, B.; Heidari, R.; Farnad, N. Phenolic combination and comparison ofantioxidant activity in three different alcoholic extracts of *Dracocephalum moldavica* L. *Turk. J. Agric. Food Sci. Technol.* **2017**, *5*, 199–206.
9. Fallah, S.; Rostaei, M.; Lorigooini, Z.; Sukri, A.A. Chemical compositions of essential oil and antioxidant activity of dragonhead (*Dracocephalum moldavica*) in sole crop and dragonhead-soybean (*Glycine max*) intercropping system under organic manure and chemical fertilizers. *Ind. Crops Prod.* **2018**, *115*, 158–165. [CrossRef]
10. Sonboli, A.; Mojarrad, M.; Gholipour, A.; Ebrahimi, S.N.; Arman, M. Biological activity and composition of the essential oil of *Dracocephalum moldavica* L. grown in Iran. *Nat. Prod. Commun.* **2008**, *3*, 1547–1550. [CrossRef]
11. El-Baky, H.H.; El-Baroty, G.S. Chemical and biological evaluation of the essential oil of Egyptian Moldavian dragonhead (*Dracocephalum moldavica* L.). *Int. J. Integr. Biol.* **2008**, *3*, 202–208.
12. Ehsani, A.; Alizadeh, O.; Hashemi, M.; Afshari, A.; Aminzare, M. Phytochemical, antioxidant and antibacterial properties of *Melissa officinalis* and *Dracocephalum moldavica* essential oils. *Vet. Res. Forum* **2017**, *8*, 223–229. [PubMed]
13. Keikhaie, R.K.; Jahantigh, H.R.; Bagheri, R.; Kehkhaie, R.A. The effect of the ethanol extract of *Dracocephalum moldavica* (Badrashbu) against strains of antibiotic resistant *Escherichia coli* and *Klebsiella pneumoniae*. *Int. J. Infection.* **2018**, *5*, e65295. [CrossRef]
14. Martinez-Vazquez, M.; Estrada-Reyes, R.; Martinez-Laurrabaquio, A.; Lopez-Rubalcava, C.; Heinze, G. Neuropharmacological study of *Dracocephalum moldavica* L. (Lamiaceae) in mice: Sedative effect and chemical analysis of an aqueous extract. *J. Ethnopharmacol.* **2012**, *441*, 908–917. [CrossRef] [PubMed]
15. Olha, B.; Antonina, P.; Mariia, S. Chemical compositions and sedative activities of the *Dracocephalum moldavica* L. and *Ocimum americanum* L. essential oils. *Pharmacol. Online* **2021**, *2*, 179–187.
16. Zuniga, M.I.J.; Mariles, A.J.H.; Flores, J.L.C.; Herrera, J.A.M.; Sotelo, M.G.R.; Montes, G.I.C.; Gomez, Y.I.M.G. Antidepressant-like effects of *Dracocephalum moldavica* L. in mouse models of immobility tests. *Pharmacogn. J.* **2019**, *11*, 976–983. [CrossRef]
17. Maham, M.; Akbari, H.; Delazar, A. Chemical composition and antinococeptive effect of the essential oil of *Dracocephalum moldavica* L. *Pharm. Sci.* **2013**, *18*, 187–192.
18. Jiang, H.; Zeng, L.; Guo, S.; Xing, J.; Li, Z.; Liu, R. Tilianin extracted from Dracocephalum moldavica L. induces intrinsic apoptosis and drives inflammatory microenvironment response on pharyngeal squamous carcinoma cells via regulating TLR4 signaling pathways. *Front. Pharmacol.* **2020**, *11*, 205. [CrossRef]
19. Nie, L.; Li, R.; Huang, J.; Wang, L.; Ma, M.; Huang, C.; Wu, T.; Yan, R.; Hu, X. Abietane diterpenoids from *Dracocephalum moldavica* L. and their anti-inflammatory activities in vitro. *Phytochemistry* **2021**, *184*, 112680. [CrossRef]
20. Sun, Y.; Liu, T.; Jiang, Z.; Gao, Z.; Zhang, M.; Wang, D.; Zheng, Q. Neuroprotective effect of *Dracocephalum moldavica* L. total flavonoids in transient cerebral ischemia in rats. *Annu. Res. Rev. Biol.* **2014**, *4*, 1915–1926. [CrossRef]
21. Li, J.; Xu, S. Tilianin attenuates MPP+-induced oxidative stress and apoptosis of dopaminergic neurons in a cellular model of Parkinson's disease. *Exp. Ther. Med.* **2022**, *23*, 293. [CrossRef]
22. Najafi, M.; Ghasemian, E.; Fatizad, F.; Garjani, A. Effects of total extract of *Dracocephalum moldavica* on ischemia/reperfusion induced arrhythmias and infarct size in the isolated rat heart. *Iran. J. Basic Med. Sci.* **2009**, *11*, 229–235. [CrossRef]
23. Zeng, C.; Jiang, W.; Yang, X.; He, C.; Wang, W.; Xing, J. Pretreatment with total flavonoid extract from *Dracocephalum moldavica* L. attenuates ischemia reperfusion-induced apoptosis. *Sci. Rep.* **2018**, *8*, 17491. [CrossRef]

24. Cao, W.; Yuan, Y.; Wang, Y.; Tian, L.; Wang, X.; Xin, J.; Wang, Y.; Guo, X.; Qin, D. The mechanism study of *Dracocephalum moldavica* L. total flavonoids on apoptosis induced by myocardial ischemia/reperfusion injury in vivo and in vitro. *Biomed. J. Sci. Tech. Research.* **2019**, *20*, 14985–14996. [CrossRef]
25. Jin, M.; Yu, H.; Jin, X.; Yan, L.; Wang, J.; Wang, Z. Dracocephalum moldavica L. extract protect H9c2 cardiomyocytes against H2O2-induced apoptosis and oxidative stress. *Hindawi BioMed Res. Int.* **2020**, *2020*, 8379358. [CrossRef] [PubMed]
26. Aćimović, M.G. Nutraceutical potential of Apiaceae. In *Bioactive Molecules in Food, Reference Series in Phytochemistry*; Mérillon, J.M., Ramawat, K., Eds.; Springer: Cham, Germany, 2017; pp. 1311–1341. [CrossRef]
27. Oniszczuk, T.; Kasprzak-Drozd, K.; Olech, M.; Wojtowicz, A.; Nowak, R.; Rusinek, R.; Szponar, J.; Combrzynski, M.; Oniszczuk, A. The impact of formulation on the content of phenolic compounds in snakes enriched with Dracocephalum moldavica L. seed: Introduction to receiving a new functional food product. *Molecules* **2021**, *26*, 1245. [CrossRef]
28. Nieto, G. How are medicinal plants useful when added to foods? *Medicines* **2020**, *7*, 58. [CrossRef]
29. Amjadi, S.; Nouri, S.; Yorghanlou, R.A.; Roufegarinejad, L. Development of hydroxypropyl methylcellulose/sodium alginate blend active film incorporated with Dracocephalum moldavica L. essential oil for food preservation. *J. Thermoplast. Compos. Mater.* **2020**, 1–7. [CrossRef]
30. Napoli, E.; Di Vito, M. Toward a new future for essential oils. *Antibiotics.* **2021**, *10*, 207. [CrossRef] [PubMed]
31. Rao, R. Hydrosols and water-soluble essential oils: Medicinal and biological properties. In *Recent Progress in Medicinal Plants, Essential oils I*; Govil, J.N., Bhattacharya, S., Eds.; Studium Press LLC: Houston, TX, USA, 2013; Volume 36, pp. 119–140.
32. Aćimović, M.; Tešević, V.; Smiljanić, K.; Cvetković, M.; Stanković, J.; Kiprovski, B.; Sikora, V. Hydrolates–by-products of essential oil distillation: Chemical composition, biological activity and potential uses. *Adv. Technol.* **2020**, *9*, 54–70. [CrossRef]
33. Sim, J.H.; Jamaludin, N.S.; Khoo, C.H.; Cheah, Y.K.; Halim, S.N.B.A.; Seng, H.L.; Tiekink, E.R.T. In vitro antibacterial and time-kill evaluation of phosphanegold(I) dithiocarbamates, R3PAu[S2CN(iPr)CH2CH2OH] for R = Ph, Cy and Et, against a broad range of Gram-positive and Gram negative bacteria. *Gold Bull.* **2014**, *47*, 225–236. [CrossRef]
34. Nasiri, Y. Crop productivity and chemical compositions of dragonhead (*Dracocephalum moldavica* L.) essential oil under different cropping patterns and fertilization. *Ind. Crops Prod.* **2021**, *171*, 113920. [CrossRef]
35. Faridvand, S.; Rezaei-Chiyaneh, E.; Battaglia, M.L.; Gitari, H.I.; Raza, M.A.; Siddique, K.H.M. Application of bio and chemical fertilizers improves yield, and essential oil quantity and quality of Moldavian balm (*Dracocephalum moldavica* L.) intercropped with mung bean (*Vigna radiate* L.). *Food Energy Secur.* **2021**, e319. [CrossRef]
36. Fattahi, A.; Shakeri, A.; Tayarani-Najaran, Z.; Kharbach, M.; Segers, K.; Heyden, V.Y.; Taghizadeh, S.F.; Rahmani, H.; Asili, J. UPLC- PDA- ESI- QTOF- MS/MS and GC- MS analysis of Iranian *Dracocephalum moldavica* L. *Food Sci. Nutr.* **2021**, *9*, 4278–4286. [CrossRef] [PubMed]
37. Rudy, S.; Dziki, D.; Biernacka, B.; Krykowski, A.; Rudy, M.; Gawlik-Dziki, U.; Kachel, M. Drying characteristics of *Dracocephalum moldavica* leaves: Drying kinetics and physiochemical properties. *Processes* **2020**, *8*, 509. [CrossRef]
38. Alaei, S. Essential oil content and composition of *Dracocephalum moldavica* under different irrigation regimes. *Int. J. Hortic. Sci. Technol.* **2019**, *6*, 167–175. [CrossRef]
39. Janmohammadi, M.; Nouraein, M.; Sabaghnia, N. Influence of different weed management techniques on the growth and essential oils of dragonhead (*Dracocephalum moldavica* L.). *Rom. Biotechnol. Lett.* **2017**, *22*, 12950–12960.
40. Hegazy, M.H.; Alzuibr, F.A.; Mahmoud, A.A.; Mohamed, H.F.Y.; Said-Al Ahl, H.A.H. The effects of zinc application and cutting on growth, herb, essential oil and flavonoids in three medicinal Lamiaceae plants. *Eur. J. Med. Plants.* **2016**, *12*, 1–12. [CrossRef]
41. Said-Al Ahl, H.A.H.; Sabra, A.S.; Gendy, A.N.; Aziz, E.E.; Tkachenko, K.G. Changes in content and chemical composition of *Dracocephalum moldavica* L. essential oil at different harvest dates. *J. Med. Plants Stud.* **2015**, *3*, 61–64.
42. Janmohammadi, M.; Sufi-Mahmoudi, Z.; Ahadnezhad, A.; Yousefzadeh, S.; Sabaghnia, N. Influence of chemical and organic fertilizer on growth, yield and essential oil of dragonhead (*Dracocephalum moldavica* L.) plant. *Acta Agric. Slov.* **2014**, *103*, 73–81. [CrossRef]
43. Alaei, S.; Mahna, N. Comparison of essential oil composition in *Dracocephalum moldavica* in greenhouse and field. *J. Essent. Oil-Bear. Plants* **2013**, *16*, 346–351. [CrossRef]
44. Aziz, E.E.; Hussein, M.S.; Wahba, H.E.; Razin, A.M. Essential oil constituents of *Dracocephalum moldavica* L. grown under salt stress and different sources of soil amendment. *Middle-East J. Sci. Res.* **2013**, *16*, 706–713. [CrossRef]
45. Ganbarzadeh, Z.; Mohsenzadeh, S.; Rowshan, V.; Moradshahi, A. Evaluation of the growth, essential oil composition and antioxidant activity of *Dracocephalum moldavica* under water deficit stress and symbiosis with *Claroideoglomus etunicatum* and *Micrococcus yunnanensis*. *Sci. Hortic.* **2019**, *256*, 108652. [CrossRef]
46. Gohari, G.; Mohammadi, A.; Akbari, A.; Panahirad, S.; Dadpour, M.R.; Fotopoulos, V.; Kimura, S. Titanium dioxide nanoparticles (TIO_2 NPs) promote growth and ameliorate salinity stress effects on essential oil profile and biochemical attributes of *Dracocephalum moldavica*. *Sci. Rep.* **2020**, *10*, 912. [CrossRef] [PubMed]
47. Amini, R.; Ebrahimi, A.; Nasab, A.D.M. Moldavian balm (*Dracocephalum moldavica* L.) essential oil content and composition as affected by sustainable weed management treatments. *Ind. Crops Prod.* **2020**, *150*, 112416. [CrossRef]
48. Golparvar, A.R.; Haidipanah, A.; Gheisari, M.M.; Khaliliazar, R. Chemical constituents of essential oil of *Dracocephalum moldavica* L. and *Dracocephalum kotschyi* Boiss. from Iran. *Acta Agric. Slov.* **2016**, *107*, 25–31. [CrossRef]

49. Amini, R.; Zafarani-Moattar, P.; Shakiba, M.R.; Sarikhani, M.R. Essential oil yield and composition of Moldavian balm (*Dracocephalum moldavica* L.) as affected by inoculation treatments under drought stress conditions. *J. Essent. Oil-Bear. Plants* **2020**, *23*, 728–742. [CrossRef]
50. Davazdahemami, S.; Allahdadi, M. Essential oil yield and composition of four annual plants (ajowan, dill, Moldavian balm and black cumin) under saline irrigation. *Food Ther. Health Care* **2022**, *4*, 5. [CrossRef]
51. Pouresmaeil, M.; Sabzi-Nojadeh, M.; Movafeghi, A.; Aghbash, B.N.; Kosari-Nasab, M.; Zengin, G.; Maggi, F. Phytotoxic activity of Moldavian dragonhead (*Dracocephalum moldavica* L.) essential oil and its possible use as bio-herbicide. *Process. Biochem.* **2022**, *114*, 86–92. [CrossRef]
52. Ding, S.E.; Cixi, T.; Charles, P. Synthesis and toxicity of Chinese *Dracocephalum moldavica* (Labiatae) fundamental oil against two grain storage insect. *Afr. J. Insects.* **2015**, *2*, 077–081.
53. Hussein, M.S.; El-Sherbeny, S.E.; Khalil, M.Y.; Naguib, N.Y.; Aly, S.M. Growth characters and chemical constituents of *Dracocephalum moldavica* L. plants in relation to compost fertilizer and planting distance. *Sci. Hortic.* **2006**, *108*, 322–331. [CrossRef]
54. Aćimović, M.; Ivanović, S.; Simić, K.; Pezo, L.; Zeremski, T.; Ovuka, J.; Sikora, V. Chemical characterization of *Marrubium vulgare* volatiles from Serbia. *Plants* **2021**, *10*, 600. [CrossRef] [PubMed]
55. Aćimović, M.; Ljujić, J.; Vulić, J.; Zheljazkov, V.D.; Pezo, L.; Varga, A.; Tumbas Šaponjac, V. *Helichrysum italicum* (Roth) G. Don essential oil from Serbia: Chemical composition, classification and biological activity—May it be a suitable new crop for Serbia? *Agronomy* **2021**, *11*, 1282. [CrossRef]
56. Aćimović, M.; Pezo, L.; Zeremski, T.; Lončar, B.; Marjanović Jeromela, A.; Stanković Jeremić, J.; Cvetković, M.; Sikora, V.; Ignjatov, M. Weather conditions influence on hyssop essential oil quality. *Processes* **2021**, *9*, 1152. [CrossRef]
57. Aćimović, M.; Lončar, B.; Pezo, M.; Stanković Jeremić, J.; Cvetković, M.; Rat, M.; Pezo, L. Volatile compounds of *Nepeta nuda* L. from Rtanj Mountain (Serbia). *Horticulturae* **2022**, *8*, 85. [CrossRef]
58. Aćimović, M.; Stanković Jeremić, J.; Todosijević, M.; Kiprovski, B.; Vidović, S.; Vladić, J.; Pezo, L. Comparative study of the essential oil and hydrosol composition of sweet wormwood (*Artemisia annua* L.) from Serbia. *Chem. Biodivers.* **2022**, *19*, e202100954. [CrossRef] [PubMed]
59. Azmir, J.; Zaidul, I.S.M.; Rahman, M.M.; Sharif, K.M.; Mohamed, A.; Sahena, F.; Jahurul, M.H.A.; Ghafoor, K.; Norulaini, N.A.N.; Omar, A.K.M. Techniques for extraction of bioactive compounds from plant materials: A review. *J. Food Eng.* **2013**, *117*, 426–436. [CrossRef]
60. Aćimović, M.; Šeregelj, V.; Šovljanski, O.; Tumbas Šaponjac, V.; Švarc Gajić, J.; Brezo-Borjan, T.; Pezo, L. In vitro antioxidant, antihyperglycemic, anti-inflammatory and antimicrobial activity of *Satureja kitaibelii* Wierzb. ex Heuff. subcritical water extract. *Ind. Crops Prod.* **2021**, *169*, 113672. [CrossRef]
61. Seema Farhath, M.S.; Vijaya, P.P.; Vimal, M. Antioxidant activity of geraniol, geranial acetate, gingerol and eugenol. *Res. Pharm.* **2013**, *3*, 1–6.
62. de Lira, M.H.P.; de Andrade, F.P., Jr.; Moraes, G.F.Q.; Macena, G.S.; Pereira, F.O.; Lima, I.O. Antimicrobial activity of geraniol: An integrative review. *J. Essent. Oil Res.* **2020**, *32*, 187–197. [CrossRef]
63. Celuppi, L.C.M.; Capelezzo, A.P.; Cima, L.B.; Zeferino, R.C.F.; Zanetti, M.; Riella, H.G.; Fiori, M.A. Antimicrobial cellulose acetate films by incorporation of geranyl acetate for active food packaging application. *Res. Soc. Dev.* **2022**, *11*, e40111125141. [CrossRef]
64. Seidel, V.; Peyfoon, E.; Watson, D.G.; Fearnley, J. Comparative study of the antibacterial activity of propolis from different geographical and climatic zones. *Phytother. Res.* **2008**, *22*, 1256–1263. [CrossRef] [PubMed]
65. Al-Ani, I.; Zimmermann, S.; Reichling, J.; Wink, M. Antimicrobial activities of European propolis collected from various geographic origins alone and in combination with antibiotics. *Medicines* **2018**, *5*, 2. [CrossRef] [PubMed]
66. Riabov, P.A.; Micić, D.; Božović, R.B.; Jovanović, D.V.; Tomić, A.; Šovljanski, O.; Đurović, S. The chemical, biological and thermal characteristics and gastronomical perspectives of *Laurus nobilis* essential oil from different geographical origin. *Ind. Crops Prod.* **2020**, *151*, 112498. [CrossRef]
67. Zhou, S.; Wei, C.; Zhang, C.; Han, C.; Kuchkarova, N.; Shao, H. Chemical Composition, Phytotoxic, Antimicrobial and Insecticidal Activity of the Essential Oils of *Dracocephalum integrifolium*. *Toxins* **2019**, *11*, 598. [CrossRef] [PubMed]
68. Abbasi, N.; Fattahi, M.; Youbert, I.; Icon, G.; Sefidkon, F. Volatile Compounds and Antifungal Activity of *Dracocephalum moldavica* L. at different phenological stages. *J. Essent. Oil Res.* **2020**, *34*, 87–95. [CrossRef]
69. Latha, C.; Anu, C.J.; Ajaykumar, V.J.; Sunil, B. Prevalence of *Listeria monocytogenes, Yersinia enterocolitica, Staphylococcus aureus,* and *Salmonella enterica* Typhimurium in meat and meat products using multiplex polymerase chain reaction. *Vet. World* **2017**, *10*, 927–931. [CrossRef]
70. Abebe, E.; Gugsa, G.; Ahmed, M. Review on major food-borne zoonotic bacterial pathogens. *J. Trop. Med.* **2020**, *2020*, 4674235. [CrossRef]
71. Aćimović, M.; Cvetković, M.; Stanković Jeremić, J.; Pezo, L.; Varga, A.; Čabarkapa, I.; Kiprovski, B. Biological activity and profiling of *Salvia sclarea* essential oil obtained by steam and hydrodistillation extraction methods via chemometrics tools. *Flavour Fragr. J.* **2022**, *37*, 20–32. [CrossRef]
72. Aborus, N.E.; Tumbas Šaponjac, V.; Čanadanović-Brunet, J.; Ćetković, G.; Hidalgo, A.; Vulić, J.; Šeregelj, V. Sprouted and freeze-dried wheat and oat seeds—Phytochemical profile and in vitro biological activities. *Chem. Biodivers.* **2018**, *15*, e1800119. [CrossRef]

73. Girones-Vilaplana, A.; Valentão, P.; Moreno, D.A.; Ferreres, F.; García-Viguera, C.; Andrade, P.B. New beverages of lemon juice enriched with the exotic berries maqui, açaí, and blackthorn: Bioactive components and in vitro biological properties. *J. Agric. Food Chem.* **2012**, *60*, 6571–6580. [CrossRef] [PubMed]
74. Oyaizu, M. Studies on products of browning reaction–antioxidant activities of products of browning reaction prepared from glucosamine. *Jpn. J. Nutr.* **1986**, *44*, 307–315. [CrossRef]
75. Al-Saikhan, M.S.; Howard, L.R.; Miller, J.C., Jr. Antioxidant activity and total phenolics in different genotypes of potato (*Solanum tuberosum*, L.). *J. Food Sci.* **1995**, *60*, 341–343. [CrossRef]
76. Micić, M.; Đurović, S.; Riabov, P.; Tomić, A.; Šovljanski, O.; Filip, S.; Tosti, T.; Dojčinović, B.; Božović, R.; Jovanović, D.; et al. Rosemary essential oils as a promising source of bioactive compounds: Chemical composition, thermal properties, biological activity, and gastronomical perspectives. *Foods* **2021**, *10*, 2734. [CrossRef] [PubMed]
77. Pavlić, B.; Teslić, N.; Vidaković, A.; Vidović, S.; Veličanski, A.; Versari, A.; Radosavljević, R.; Zeković, Z. Sage processing from by-product to high-quality powder: I. Bioactive potential. *Ind. Crops. Prod.* **2017**, *107*, 81–89. [CrossRef]
78. Ferro, B.E.; van Ingen, J.; Wattenberg, M.; van Soolingen, D.; Mouton, J.W. Time–kill kinetics of antibiotics active against rapidly growing mycobacteria. *J. Antimicrob. Chemother.* **2015**, *70*, 811–817. [CrossRef] [PubMed]
79. Romano, A.; Toraldo, G.; Cavella, S.; Masi, P. Description of leavening of bread dough with mathematical modelling. *J. Food Eng.* **2007**, *83*, 142–148. [CrossRef]

plants MDPI

Article

Chemical Profile and Antimicrobial Activity of the Essential Oils of *Helichrysum arenarium* (L.) Moench. and *Helichrysum italicum* (Roth.) G. Don

Valtcho D. Zheljazkov [1,*], Ivanka Semerdjieva [2,3], Elina Yankova-Tsvetkova [3], Tess Astatkie [4], Stanko Stanev [5], Ivayla Dincheva [6] and Miroslava Kačániová [7,8]

1 Crop and Soil Science Department, Oregon State University, 3050 SW Campus Way, 109 Crop Science Building, Corvallis, OR 97331, USA
2 Department of Botany and agrometeorology, Agricultural University, Mendeleev 12, 4000 Plovdiv, Bulgaria; v_semerdjieva@abv.bg
3 Department of Plant and Fungal Diversity, Division of Flora and Vegetation, Institute of Biodiversity and Ecosystem Research, BAS, 2, Gagarin Str., 1113 Sofia, Bulgaria; e_jankova@abv.bg
4 Faculty of Agriculture, Dalhousie University, P.O. Box 550, Truro, NS B2N 5E3, Canada; astatkie@dal.ca
5 Institute of Roses, Essential and Medical Plants, Agricultural Academy, bul "Osvobozhdenie" 49, 6100 Kazanlak, Bulgaria; sdstanev@abv.bg
6 Department of Agrobiotechnologies, AgroBioInstitute, Agricultural Academy, 8 Dragan Tsankov blvd., 1164 Sofia, Bulgaria; ivadincheva@yahoo.com
7 Institute of Horticulture, Faculty of Horticulture and Landscape Engineering, Slovak University of Agriculture, Tr. A. Hlinku 2, 94976 Nitra, Slovakia; kacaniova.miroslava@gmail.com
8 Department of Bioenergetics and Food Analysis, Institute of Food Technology and Nutrition, University of Rzeszow, 35-601 Rzeszow, Poland
* Correspondence: valtcho.jeliazkov@oregonstate.edu

Citation: Zheljazkov, V.D.; Semerdjieva, I.; Yankova-Tsvetkova, E.; Astatkie, T.; Stanev, S.; Dincheva, I.; Kačániová, M. Chemical Profile and Antimicrobial Activity of the Essential Oils of *Helichrysum arenarium* (L.) Moench. and *Helichrysum italicum* (Roth.) G. Don. *Plants* **2022**, *11*, 951. https://doi.org/10.3390/plants11070951

Academic Editor: Suresh Awale

Received: 2 February 2022
Accepted: 29 March 2022
Published: 31 March 2022

Abstract: This study compared the essential oils (EO) composition of *Helichrysum arenarium* (Bulgarian populations) with that of the cultivated species *H. italicum*. The EO composition of *H. arenarium* and *H. italicum* were analyzed via gas chromatography. In general, 75 components were identified in *H. arenarium* EO and 79 in *H. italicum* EO. The predominant constituents in *H. arenarium* EO were α-pinene (34.64–44.35%) and sabinene (10.63–11.1%), which affirmed the examined population as a new chemical type. Overall, the main EO constituents of *H. italicum* originating in France, Bosnia and Corsica were neryl acetate (4.04–14.87%) and β-himachalene (9.9–10.99%). However, the EOs profile of *H. italicum* introduced from the above three countries differed to some extent. D-limonene (5.23%), italicene, α-guaiene and neryl acetate (14.87%) predominated in the *H. italicum* introduced from France, while α-pinene (13.74%), δ-cadinene (5.51%), α-cadinene (3.3%), β-caryophyllene (3.65%) and α-calacorene (1.63%) predominated in plants introduced from Bosnia. The EOs of the plants introduced from France and Corsica had similar chemical composition and antimicrobiological activity.

Keywords: Bulgaria; Bosnia; Corsica; France; essential oil; Helichrysum; protected species; α-pinene; sabinene; neryl acetate

1. Introduction

Plants species belonging to *Helichrysum* Mill. genus (Asteraceae) have long been known for their healing properties, and preparations based on *Helichrysum* species have been and continue to be used around the world [1]. The pharmaceutical, cosmetic and perfume industries have taken a strong interest in *Helichrysum* species because of the specific essential oil (EO) aroma and composition [2]. Extracts from *Helichrysum* species possess a wide range of pharmacological activities such as antioxidant, antimicrobial, antiatherosclerotic, antiproliferative, antidiabetic, neuroprotective and antiinflammatory activities [3–6]. There are 16 species in the *Helichrysum* genus spread across Europe [7]. Two species, *H. arenarium* (L.) Moench. and *H. plicatum* DC. [8], are naturally occurring in

Bulgaria, while *H. italicum* (Roth) G. Don. is an introduced cultivated species in this country. Products derived from *H. italicum* are widely used in the traditional medicine, cosmetics and the food industry and are particularly popular in the Mediterranean countries [9,10]. In recent years, there has been increasing interest in products from *H. italicum*. As a result, the species has been commercially cultivated in France [11], Portugal [12], Bosnia and Herzegovina [2,13], Italy [14,15], Serbia [10] and recently in Bulgaria. There has been significant interest in the phytochemical composition and pharmacological activity of *H. italicum* during the last decade and a half [13,15–18].

Helichrysum arenarium has a long tradition as a medicinal plant in the European ethnomedicine [19]. Medicines based on *Helichrysi* flos were enlisted in the State Pharmacopoeia of the USSR [20], Pharmacopoeia Helvetica [21] the Polish Pharmacopoeia [22], as well as in a herbal monograph on *H. arenarium* [23]. Because of its healing properties, *H. arenarium* has been collected from its natural populations; hence, wild collection has the potential to disturb stable populations of this species. In some European and Asian countries, such as Sweden, Poland, Kazakhstan and Serbia, the species is protected and cultivated [24–26]. In Bulgaria, *H. arenarium* is protected according to the Biodiversity Act, included in Annex 4 of this act and in the List of Species of Medicinal Plants under special regimen of conservation [27,28].

Research studies related to its phytochemical composition have focused mainly on the content of phenols and flavonoids [29–52] (Table 1). This is not accidental because phenolic compounds, including flavonoids (like in *Helichrysi* flos), used in traditional medicine (biological source *H. arenarium*) have been demonstrated to have cholagogue, choleretic, hepatoprotective and inhibitory effects on tumor necrosis; in addition, these compounds are used to make a detoxifying herbal drug [19,30,40]. Studies on EO composition of *H. arenarium* are limited and the existing data for EO composition diverge widely [4,30–33,35] (Table 1). For example, in a study of the Hungarian population, the predominant EO constituents were linalool, carvacrol, anethole, anisaldehyde and thymol [31,32]; in Serbia, major EO constituents included diepi-α-cedrene, α-ylangene, cyclosativene and limonene [4]; in Iran, spathulenol, β-pinene, limonene, alpha-cadinol and borneol were observed [43,45]. The latter authors concluded that the observed differences in the EO composition were due to the different geographical habitats of the species [4,31–33,43,45]. So far, phytochemical studies of the Bulgarian wild populations of *H. arenarium* have not been conducted. To preserve the natural population of the species, it may need to be cultivated ex situ through its development as a cultivated crop. Phytochemical studies are necessary for the selection of accessions possessing high content and desirable composition of the EO. Overall, information on the EO composition of *H. italicum* cultivated in Bulgaria is limited.

Table 1. Literature data of phytochemical research on *Helichrysum arenarium*.

Reference	Main Compounds	Country
Rančić et al. [4]	diepi-α-cedrene (17.9%), α-ylangene (13.9%), cyclosativene (11.9%), limonene (11.4%)	Serbia
Mao et al. [5]	narirutin, naringin, eriodictyol, luteolin, galuteolin, astragalin, kaempferol	China
Smirnova and Pervykh [29]	flavonoids-astragalin, luteolin, kaempferol	Russian Federation
Czinner et al. [30]	phenolic compound	Hungary
Czinner et al. [31]	linalool (1.7%), anethole (3.2%), carvacrol (3.6%), α-muurolol (1.3%), 1.5% of β-asarone	Hungary
Lemberkovics et al. [32]	linalool, alpha-terpineol, carvone monoterpenes; anethole, anisaldehyde, thymol, carvacrol, eugenol, beta-asarone, butylhydroxyanisole aromatic components; alpha-humulene, beta-caryophyllene, gamma-muurolene, delta-cadinene, copaene, alpha-gurjunene, caryophyllenol, delta-cadinol and globulol sesquiterpenes, caprylic acid, pelargonic, caprinic, lauric acids, methyl palmitate	Hungary

Table 1. *Cont.*

Reference	Main Compounds	Country
Judzentiene and Butkiene [33]	β-caryophyllene; δ-cadinene; octadecane; heneicosane	Lithuania
Bryksa-Godzisz et al. [34]	phenolic compounds	Poland
Radušienė and Judžentienė [35]	1.8-cineole (2.3–7%); α-copaene (2.2–3.6%); trans-caryophyllene (4.4–8.8%); epi-a-cadinol (2–4%); m/z-149 (phthalide)(0.6–5.6%); heneicosane (1.5–5.1%)	Lithuania
Yang et al. [36]	Flavonoids (naringenin-7-O-β-d-glycoside, isoquercitrin, astragalin)	China
Lv et al. [37]	prenylated phthalide glycosides	China
Zhang et al. [38]	6,7-dimethoxy-4-hydroxy-1-naphthoic acid (1),(Z)-5-hydroxy-7-methoxy-4-[3-methyl-4-(O-β-d-xylopyranosyl)but-2-enyl]isobenzofuran-1(3H)-one (2).	China
Eshbakova and Aisa [39]	naringenin, helichrysum phthalide, diosmin, oleanolic acid	Republic of Uzbekistan
Morikawa et al. [40]	naringenin 7-O-β-D-glucopyranoside, apigenin 7-O-β-D-glucopyranoside, apigenin 7-O-gentiobioside, apigenin 7,4′-di-O-β-D-glucopyranoside	cultivated in Poland purchased Tochimoto Tenkaido Co., Ltd., Osaka, Japan
Albayrak et al. [41]	phenolic compounds	Turkey
Yong et al. [42]	β-sitosterol, stigmasterol, β-sitosterol, β-D-glucopyranoside, stigmasterol, caffeic acid ethyl ester.	China
Oji et al. [43]	limonene (21.2%), alpha-cadinol (18.2%), borneol (11.9%), delta-cadinene (9%), bornyl acetate (8%), alpha-humulene (7.3%).	Iran
Gradinaru et al. [44]	caffeic acid; flavonoids (apigenin, naringenin, apigenin-7-O-glucoside, naringenin-O-hexosides)	Romania
Moghadam et al. [45]	spathulenol (36.6%), β-pinene (12.5%)	Iran
Bandeira Reidel et al. [46]	β-caryophyllene (27–46%); (E)-2-hexenal; β-pinene (7.4%);	Italy
Bandeira Reidel et al. [47]	β-pinene (7.4%); β-caryophyllene (27.5%); δ-cadinene (3.2%); pentadecanoic acid, methyl ester (31%)	Italy
Babotă et al. [48]	phenolic compound; methoxylated flavone; sterolic compound;	Romania
Judzentiene et al. [49]	1,8-cineole (8.9%, one sample), β-caryophyllene (5.8–36.2%, 14 oils), γ- and δ-cadinene (5.8% and 9%); octadecane (7.1–22.3%).	Lithuania
Liu et al. [50]	linalool (2.81%); 4-acetyl-1-methylcyclohexene (1.88%); β-spathulenol (24.03%); caryophyllene oxide (3.05%); ledol (6.22%); hinesol (3.86%); β-eudesmol (2.56%); α-eudesmol (4.37%); α-cadinol (7.76%); α-bisabolol (5.71%)	Inner Mongolia, China
Stankov et al. [51]	oleic acid (30.28%), ethyl hexadecanoate (20.19%), linoleic acid (18.89%), sclareol (4.22%)	Turkey
Ivanović et al. [52]	phenolic compounds	Slovenia

Previous research showed that *H. italicum* EO exhibited antioxidant, antimicrobial, antiviral, anti-inflammatory, and antiproliferative activity [13]. *H. italicum* showed low or no activity against tested bacteria. However, for all Gram-negative bacteria (*E. coli, P. aeruginosa, Salmonella typhimurium, S. enteritidis, K. aerogenes* and *P. hauseri*) minimum inhibitory concentration (MIC) and minimum bactericidal concentration (MBC) values were higher than 454.5 μL/mL EO. For the Gram-positive bacteria (*B. cereus, L. monocytogenes, R. equi,* and *S. epidermidis*) MIC and MBC values was 454.5 μL/mL, while for other (*B. spizizenii, E. faecalis, L. innocua, L. ivanovii,* and *S. aureus*) MIC and MBC values were higher than 454.5 μL/mL of EO [10].

The present study analyzed the EO composition of the Bulgarian population of *H. arenarium* and compared it with the EO of *H. italicum*, which already has an established international market. The working hypothesis was that EO composition of *H. italicum* and *H. arenarium* would be similar.

2. Results

2.1. Qualitative Composition of the Essential Oil (EO)

2.1.1. *Helichrysum arenarium*

The analysis of variance (ANOVA) results that identified significant (bold) and non-significant differences among the mean constituents of *H. arenarium* are shown in Table 2. Data from EO analysis of *H. arenarium* are presented in a Supplementary Table S1. Overall, 75 EO constituents were identified and grouped into the following classes: monoterpenes, sesquiterpenes, diterpenoids and long-chain alkane, in total 90.82–94.4% of the total oil. The monoterpenes predominated in the three tested samples (65.72–73.99%) (Table 3): α-Pinene (34.64–44.35%) and sabinene (10.63–11.1%) were predominant in the three samples and β-pinene, trans-verbenol and D-limonene were observed in similar quantity (Table 3) (Figure 1). With respect to the sesquiterpenes in *H. arenarium* EO, (16.08–19.41%), germacrene D (3.56–4.86%) and β-gurjunene (3.61%) were predominant (Table 3). The concentrations of sabinene, D-limonene, trans-verbenol, n-tetradecane and β-gurjunene in *H. arenarium* EO were not significantly different between the three locations; the overall means of these compounds are shown in Table 4.

Table 2. ANOVA *p*-values showing the significance of the differences among the three collections in terms of *H. arenarium* concentrations.

Constituent	*p*-Value	Constituent	*p*-Value	Constituent	*p*-Value
α-pinene	**0.033** *	β-gurjunene	0.322	1-terpinen-4-ol	**0.002**
sabinene	0.716	germacrene D	**0.027**	long-chain alkane	**0.017**
β-pinene	**0.073**	germacra-4(15),5,10(14)-trien-1	**0.089**	n-tetradecane	0.138
D-limonene	0.403	monoterpenes	**0.004**	diterpenoids	0.355
trans-verbenol	0.400	sesquiterpenes	**0.084**		

* Significant effects that require multiple means comparison are shown in bold.

Table 3. Mean *H. arenarium* concentration (%) of α-pinene, β-pinene, 1-terpinen-4-ol, germacrene D, germacra-4(15),5,10(14)-trien, monoterpenes, sesquiterpenes, diterpenoids and long-chain alkane obtained from three collections (Location 1, Location 2 and Location 3) within the natural populations.

Collection	α-Pinene	β-Pinene	1-Terpinen-4-ol	Germacrene D	Germacra-4(15),5,10(14)-trien
689 Location 1	44.35 a *	2.85 a	0.92 c	3.56 b	1.14 ab
691 Location 2	36.60 b	2.30 b	2.11 a	5.33 a	1.09 b
699 Location 3	34.64 b	2.67 ab	1.26 b	4.83 a	1.37 a
Collection		Monoterpenes	Sesquiterpenes	Long-chain alkane	Diterpenoids
689 Location 1		73.99 a	16.08 b	4.33 b	3.25 b
691 Location 2		68.96 b	18.01 ab	6.23 a	3.45 ab
699 Location 3		65.72 c	19.41 a	5.69 a	4.27 a

* Within each constituent, means sharing the same letter are not significantly different.

Table 4. Overall mean *H. arenarium* concentration (%) of sabinene, D-limonene, trans-verbenol, n-tetradecane and β-gurjunene where there was no significant difference among the three collections.

Constituent	Overall Mean Concentration	$\sqrt{MSE}$ *
Sabinene	10.80	0.607
D-limonene	2.11	0.151
trans-verbenol	3.18	0.217
n-tetradecane	2.36	0.135
β-gurjunene	3.61	0.259

* $\sqrt{MSE}$ = square root of the mean square error (*MSE*) that estimates the common standard deviation (σ).

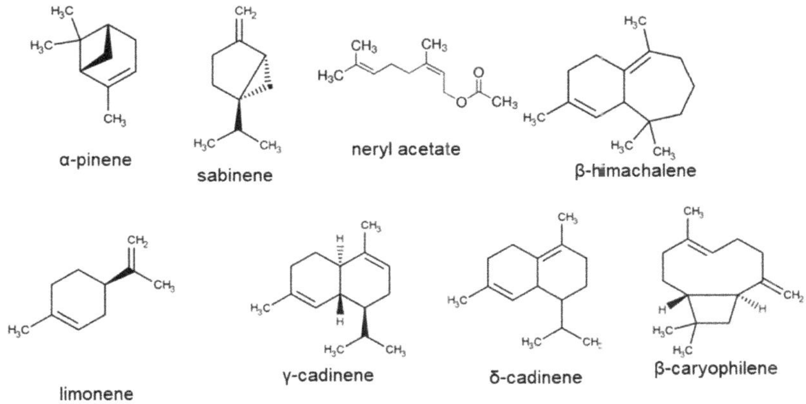

Figure 1. Structural formulas of the main compounds of *Helichrysum arenarium* and *H. italicum*.

2.1.2. Helichrysum italicum

The ANOVA results that identified significant (bold) and nonsignificant differences among the mean constituents of *H. italicum* are shown in Table 5. The chemical composition of EO of *H. italicum* is presented in Supplementary Table S2, Tables 6 and 7. The Gas chromatography-mass spectrometry (HS/GC-MS) analyses showed 79 EO constituents (Supplementary Table S2). Generally, sesquiterpenes were predominant in the EO from all three locations: 45.23% (France), 47.9% (Corsica) and 54.8% (Bosnia). *β*-Himachalene was found to be present in the three analyzed samples in the range of 9.9% to 10.9% (Table 6), although EO of each of those samples possessed a specific profile. For example, italicene and *α*-guaiene were present in higher quantity in the plants introduced from Corsica and France (Table 6). The EO of plants introduced from Bosnia was characterized by higher content of *δ*-cadinene (5.51%), *α*-cadinene (3.3%), *β*-caryophyllene (3.65%) and *α*-calacorene (1.63%) compared with the other samples (Supplementary Table S2). The structural formulas of some of the main compounds are presented on Figure 1.

Table 5. ANOVA *p*-values showing the significance of the differences among the three countries in terms of *H. italicum* concentrations.

Constituent	*p*-Value	Constituent	*p*-Value
α-Pinene	**<0.001 ***	*β*-Caryophyllene	**0.001**
D-Limonene	**0.019**	*p*-Cymen-7-ol acetate	**0.004**
2-Methyl butyl-2-methyl butyrate	**<0.001**	*α*-Guaiene	**<0.001**
Isoamyl tiglate	**0.004**	*γ*-Curcumene	**0.002**
1-Terpinen-4-ol	**0.001**	*β*-Himachalene	0.467
Nerol	**0.001**	*β*-Curcumene	0.232
Neryl acetate	**0.001**	Germacrene D-4-ol	**0.025**
α-Copaene	**0.049**	*γ*-Eudesmol	**0.027**
Italicene	**0.016**	tau.-Muurolol	0.751
α-cis-Bergamotene	**0.002**	*β*-Eudesmol	0.120
Sesquiterpenes	**0.048**	Monoterpenes	**<0.001**
Ester	0.669	Long-chain alkane	**0.069**

* Significant effects that require multiple means comparison are shown in bold

Table 6. Mean *H. italicum* concentration (%) of α-pinene, D-limonene, 2-methyl butyl-2-methyl butyrate, isoamyl tiglate, 1-terpinen-4-ol, nerol, neryl acetate, α-copaene, italicene, α-cis-bergamotene, β-caryophyllene, p-cymen-7-ol acetate, α-guaiene, γ-curcumene, ger-macrene D-4-ol, γ-eudesmol, β-himachalene, monoterpenes, sesquiterpenes, ester and long-chain alkane obtained from three countries.

Country	α-Pinene	D-Limonene	2-Methyl butyl-2-methyl butyr	Isoamyl tiglate	1-Terpinen-4-ol	Nerol	Neryl acetate
Bosnia	13.74 a *	3.37 b	0.087 c	0.83 b	0.29 c	0.19 b	4.04 c
France	4.84 b	5.23 a	4.31 a	1.74 a	1.37 b	2.26 a	14.87 a
Corsica	2.83 c	4.94 a	3.44 b	1.93 a	1.67 a	2.50 a	12.37 b
Country	**α-Copaene**	**Italicene**	**α-cis-Bergamotene**	**β-Caryophyllene**	**p-Cymen-7-ol acetate**	**α-Guaiene**	**γ-Curcumene**
Bosnia	2.38 a	2.93 b	0.38 b	3.65 a	5.27 a	1.71 b	2.46 a
France	1.99 ab	4.67 a	1.19 a	0.38 b	2.47 b	3.98 a	0.93 b
Corsica	1.71 b	4.23 a	1.19 a	0.48 b	2.50 b	4.08 a	0.65 b
Country	**Germacrene D-4-ol**	**γ-Eudesmol**	**β-Himachalene**	**Monoterpenes**	**Sesquiterpenes**	**Long-chain alkane**	**Ester**
Bosnia	2.49 a	1.71 b	10.80 ab	29.83 c	54.80 a	0.74 b	7.84 ab
France	0.80 b	3.35 a	9.90 b	37.30 a	45.23 b	0.84 ab	7.96 a
Corsica	1.81 a	3.47 a	10.99 a	35.29 b	47.90 ab	0.99 a	7.09 c

* Within each constituent, means sharing the same letter are not significantly different.

Table 7. Overall mean *H. italicum* concentration (%) of β-Himachalene, β-Curcumene. tau.-Muurolol, β-Eudesmol, and Ester where there was no significant difference among the three collections.

Constituent	Overall Mean Concentration	$\sqrt{MSE}$
β-Himachalene	10.57	0.832
β-Curcumene	2.05	0.411
tau.-Muurolol	1.13	0.083
β-Eudesmol	1.44	0.208
Ester	7.63	0.982

$\sqrt{MSE}$ = square root of the mean square error (MSE) that estimates the common standard deviation (σ).

Monoterpenes represented the second major class of the *H. italicum* EO. In the EO of plants introduced from France, monoterpenes were observed in the greatest quantity (37.3%), with D-limonene (5.23%) and neryl acetate (14.87%) being the predominant monoterpenes (Table 6). The main EO constituents of plants introduced from Bosnia were α-pinene (13.74%) and *p*-cymen-7-ol acetate (5.27%) (Table 6).

2.1.3. Antimicrobial Activity of the *H. italicum* EO

The *H. italicum* EO of plants introduced from Bosnia, Corsica, and France were tested for antimicrobial activity against nine microorganisms using the disc diffusion method. Antimicrobial activity of different microorganisms and location ranged from 2.33 to 14.67 mm. Overall, the EO of *H. italicum* from all locations was more effective against *S. aureus* and ranged between 9.33 and 14.67 mm (Table 8). Moderate antimicrobial effect was found against *C. krusei* and *C. tropicalis*. The lowest antimicrobial activity was found against *Y. eneterocolitca*. In general, the tested EOs were more effective against Gram-positive bacteria.

Table 8. Mean antimicrobial activities obtained from the three locations where *H. italicum* were collected.

Location	SA	EF	SP	PA	YE	SE	CA	CK	CT
Bosnia	9.33 b	4.00 b	8.33 b	2.33 b	2.33 b	5.67 a	5.33 a	4.67 b	5.67 a
Corsica	14.67 a	1.67 c	6.67 c	2.67 b	5.33 a	3.33 b	5.33 a	5.67 b	4.33 a
France	14.67 a	12.33 a	10.67 a	5.33 a	5.00 a	4.00 b	5.67 a	10.67 a	6.33 a

Within each constituent, means sharing the same letter are not significantly different. SA-*Staphylococcus aureus* subs. *aureus* CCM 4223, EF-*Enterococcus faecalis* CCM 4224, SP-*Streptococcus pneumonia* CCM 4501, PA-*Pseudomonas aeroginosa* CCM 1959, YE-*Yersinia enterocolitica* CCM 5671, SE-*Salmonella enterica* subsp. *enterica* CCM 3807, CA-*Candida albicans* CCM 8186, CK-*C. krusei* CCM 8271 and CT-*C. tropicalis* CCM 8223 (CT).

3. Discussion

3.1. Helichrysum arenarium

Results of the present study on *H. arenarium* show that α-pinene (34.64–44.35%), sabinene (10.63–11.1%), germacrene D (3.56–4.86%), β-gurjunene (3.61%), β-pinene, trans-verbenol and D-limonene were the predominant constituents of *H. arenarium* EO. Unlike in previously published data on the species, the results from this study affirmed a new chemical type (chemotype) of *H. arenarium* in Bulgaria.

In Bulgaria, *H. arenarium* grows on sandy and coastal habitats at up to 500 m above sea level: the Black Sea coast, the Danube Plain (central part), north-eastern Bulgaria and south-eastern Bulgaria [8]. Previously, Czinner et al. [31] analyzed steam-distilled EO of *H. arenarium* plants collected in the Caucasus region and established that the largest group of compounds was the aliphatic acids (34.6%), among which were dodecanoic acid (11.9%) and decanoic acid (9.8%), followed by ester methyl palmitate (28 5%) and further aromatic compounds (10.2%) such as carvacrol and anethole (3.6 and 3.2%, respectively). On the other hand, Lemberkovics et al. [32], using the same analytical approach, reported that the predominant compound in the EOs of Polish and Hungarian commercial samples was methyl palmitate (21.7–28.5%), while caprinic acid (19.8%) was the main EO constituent in a cultivated plant sample from Hungary. These discrepancies in chemical profiles could be a consequence of different environmental factors, such as isolation, soil type, precipitation, etc. Furthermore, Judzentiene and Butkiene [33] reported chemical profiles of *H. arenarium* EOs from inflorescences and leaves of yellow and orange flowering plants. Apparently, the EO from inflorescences of both types of plants, yellow and orange, had two dominant constituents, β-caryophyllene and heneicosane, followed by α-copaene (9–25.6%, 3–32.1% and 1.5–7.2%, respectively). One of the main constituents in the EOs extracted from leaf in both plant types with yellow and orange inflorescences, excluding β-caryophyllene, was δ-cadinene (9.8–22.3% and 6.6–11.8%, respectively). Other EO constituents included 1,8-cineole, α-copaene, (E)-β-ionone, γ-cadinene, selina-3,7(11)-diene, epi-α-cadinol, α-cadinol, octadecane, isophytol and tricosane [33].

Analyses of the composition of the EO from Central European samples were conducted by several authors [31,33,35]. These analyses also reported differences between EO obtained from different geographic locations. Samples from the Caucasus region analyzed by Czinner et al. [31] showed the presence of 1.5% of β-asarone, which was not found in the samples from Central Europe.

3.2. Helichrysum italicum

Helichrysum italicum is a thermophilic plant species, which is among the most frequently studied species [13]. The plants introduced from Bosnia, France and Corsica had differing and specific composition of EO. For example, the basic constituents in EO of the plants introduced from Bosnia were α-pinene, β-caryophyllene, p-cymen-7-ol acetate and β-himachalene (Table 5), while the predominant constituents in EO of the plants introduced from France and Corsica were D-limonene, neryl acetate, nerol, italicene, α-guaiene, γ-eudesmol, 2-methyl butyl-2-methyl butyr and β-himachalene. The established differences could be due to the fact that *H. italicum* is characterized by high polymorphism, spontaneous hybridization and variations in EO composition [3,13]. Review of the literature provided evidence that variation in the *H. italicum* EO composition could be due to numerous factors, such as geographical origin, ecological factors, geographical features of the habitat, the sampled part of the plant, the relevant stage of growth and the extraction methods [10,47,53,54]. Depending on the geographical origin, some researchers reported several chemotypes of *H. italicum*. For example, based on literature data, Ninčević et al. [13] named the following chemotypes for *H. italicum*: (1) EO from Corsica (neryl acetate, neryl propionate, aliphatic ketones and β-diketones); (2) EO from Serbia (α-pinene, then y-curcumene, β-selinene, neryl acetate and β-caryophyllene); (3) EO from the Adriatic Coast (α-curcumene or γ-curcumene or α-pinene, neryl acetate); (4) EO from Greece (geraniol, geranil acetat and nerolidol); and (5) EO from Tuscana (α-pinen, or neryl acetate, β-selinene, β-cariophylene and α-selinene).

Another classification of the chemotypes of *H. italicum* was generated by Aćimović et al. [10]. Depending on the main constituents, the authors identified 10 chemotypes of *H. italicum*, namely (1) high neryl acetate chemotype (50.5–83.4%), (2) moderate neryl acetate chemotype (19.5–48%), (3) neryl acetate + ar-curcumene (3.9–20.3% and 0.8–14.5%, respectively), (4) ar-curcumene + γ-curcumene (17.9–28.6% and 12–22%, respectively), (5) γ-curcumene (13.6–27.7%), (6) high α-pinene chemotype (25.2–53.5%), (7) moderate α-pinene (5.6–20%), (8) juniper camphor (25.3–45.1%), (9) β-selinene (11.6–38%) and (10) italidiones chemotype [10].

The samples examined in this study showed different EO profiles and could not be assigned to any of the above-mentioned chemotypes of *H. italicum*. The predominant constituents in EO of plants introduced from Bosnia were α-pinene (13.74%), δ-cadinene (5,51%), α-cadinene (3.3%), β-himachalene (9.9%) and β-caryophyllene (3.65%). The EOs of the plant samples introduced from France and Corsica had similar profiles; these contained neryl acetate (12.37–14.87%), β-himachalene (9.9–10.99%) and D-limonene (5.23–4.94%).

3.3. Comparing the EOs between H. arenarium and H. italicum

The species from the genus *Helichrysum* are widely used in traditional medicine worldwide and they are known as everlasting flowers [1]. As noted in the introduction, some *Helichrysum* species are widespread and others are cultivated in the Mediterranean, Iberian Peninsula and Eastern Europe [13,19,55]. The EO composition of *H. italicum* has been studied widely, while *H. arenarium* has been studied mainly for the content of phenols and flavonoids (Table 1). Data comparing the composition of the EO between the two species have not been published. The samples of *H. arenarium* collected and analyzed in this study contain a specific EO profile that differs from the EO composition of *H. italicum*. Monoterpenes (α-pinene, sabinene) were the predominant class of compounds in the EO of *H. arenarium*, while the EO of *H. italicum* was dominated by the class of sesquiterpenes (neryl acetate and β-himachalene). However, the EO samples of *H. italicum* from three regions differed from each other to some extent. D-limonene (5.23%), italicene, α-guaiene and neryl acetate (14.87%) predominated in the *H. italicum* plants introduced from France, while the *H. italicum* plants introduced from Bosnia had predominantly α-pinene (13.74%) and δ-cadinene (5, 51%). This difference in the EO composition between *H. arenarium* and *H. italicum* refutes our working hypothesis.

3.4. Antimicrobial Activity of the H. italicum EO

Djihane et al. [56] studied antimicrobial activity of *H. italicum* EO against Gram-positive bacteria (G^+), Gram-negative (G^-) bacteria and fungi. Gram positive bacteria were more sensitive to the presence of EO than G^- bacteria or fungi. The results from this study were similar with our results.

In the current study, fungi were the most resistant class of microorganisms to the presence of the EO. Oliva et al. [57] tested *H. italicum* EO against methicillin-sensitive *Staphylococcus aureus* (ATCC 29213), *Escherichia coli* (ATCC 25922), *Candida albicans* (ATCC 14053) and the clinical strains of methicillin-resistant *S. aureus*, carbapenem-resistant *Klebsiella pneumoniae*, carbapenem-resistant *Acinetobacter baumannii* and carbapenem-resistant *Pseudomonas aeruginosa*. Interestingly, fungicidal/bactericidal potency against *C. albicans* and carbapenem-resistant *A. baumannii* was revealed at a concentration of 5% v/v. Staver et al. [58] conducted antimicrobial assays and showed that EO had weak to moderate antimicrobial potential with *S. aureus* and *S. epidermidis* as the most sensitive bacterial strains. In the study of Dzamic et al. [59], the most sensitive bacteria to *H. italicum* EO were *Bacillus cereus* and *Salmonella typhimurium*, while the most sensitive fungus was yeast, *Candida albicans*. Similar to our study, Mollova et al. [60] showed *H. italicum* EO from France had more pronounced antimicrobial activity against the G^+ bacteria *Staphylococcus aureus*, *Bacillus subtilis*, and the fungus *Aspergillus brasiliensis*, as well as a stronger antioxidant potential compared with the other EOs. The obtained results from this study are in good agreement with the findings of Cantore et al. [61] who reported that G^+ bacteria are more sensitive to plant EOs than G^- bacteria. Mesic et al. [62] reported that immortelle essential oil inhibited only Gram-positive bacteria and possessed antifungal effects.

With respect to the antibacterial properties of *H. italicum* EO and its related constituents, Rossi et al. [63] demonstrated that the EO obtained from endemic plants of Corsica was more effective against the G^+ bacterium *S. aureus* than against the G^- strains *E. coli*, *Enterobacter aerogenes* and *P. aeruginosa*. In our study, the most resistant microorganisms tested were G^- bacteria. It is commonly known that G^- bacteria are less susceptible to EO than G^+ bacteria, and this is directly connected to the bacterial cell wall structure. In G^- bacteria, the cell wall is a complex envelope constituted by the cytoplasmic membrane, the periplasm and the outer membrane. Results reported in Cantore et al. [61] and Rossi et al. [63] are consistent with those in our study. Antimicrobial activity of *H italicum* EO from Algeria with α-cedrene, α-curcumene and geranyl acetate as dominant compounds assayed by disk diffusion method inhibited growth of *S. aureus*, *M. luteus*, *E. cereus*, *B. cereus*, *S. epidermidis*, *B. subtilis*, *P. aeruginosa*, *E. faecalis* and *P. mirabilis*, but did not affect *E. coli*, *K. pneumonia* and *L. monocytogenes*. In addition, yeasts (*C. albicans* and *S. cervisae*), as well as fungi (*F. solani*, *A. niger*, *A. alternata* and *A. rabiei*), were also inhibited by *H. italicum* EO [64].

4. Materials and Methods

4.1. Plant Material

The materials utilized in this study were aerial parts in full flowering. The plant materials of *H. arenarium* were collected from three locations (numbered 689; 691; 699) of population Pobitite kamani, near Varna town (Figure 2A) (43.228196 N; 27.705116 E; 114 masl) with an official permit (#790/19.04.2019 of MOCB). The collected samples were air-dried at room temperature until a constant weight. Voucher specimens of *H. arenarium* were deposited at the Herbarium of the Agricultural University, Plovdiv, Bulgaria (SOA) [65].

Figure 2. Wild population of *Helichrysum arenarium* of population Pobitite kamani, near Varna town (**A**), cultivated *Helichrysum italicum* (**B**).

Samples of *H. italicum* introduced from Bosnia, France and Corsica and grown side by side in Bulgaria were obtained from experimental fields at the Institute of Roses, Essential and Medical Plants in Kazanlak, Bulgaria (Figure 2B). These plantations were established via vegetative propagation/rooting of fresh green cuttings prepared from the original imported plants. The plants originating from Bosnia and Herzegovina and Corsica were imported to Bulgaria as seedlings in trays, while the plants from France were imported and grown from collected seeds.

4.2. Essential Oil (EO) Extraction

The EO of all samples was extracted via hydrodistillation in a 2 L Clevenger-type apparatus (Laborbio Ltd. Sofia, Bulgaria, laborbio.com, accessed on 12 June 2021). Since the samples from *H. arenarium* were much smaller, the EO of *H. arenarium* was isolated from 45 g of flowering aerial parts of each accession by hydrodistillation in a Clevenger-type distillation unit for 2 h plus in 0.8 L of water. The duration of the hydrodistillation was 2 h and all samples were extracted in two replicates.

Samples from the introduced and grown in Bulgaria materials of *H. italicum* were 1000 g of fresh aboveground plant material in the full flowering stage. In addition, to obtain larger EO samples for biological activity testing, steam distillation was performed in 5 L metal cylindrical containers using 1.5 L of water under the grate on which the raw material was placed. The steam distillation time was 1.5 h. The EO extraction was done at the Institute of Roses, Essential and Medical Plants in Kazanlak, Bulgaria, and each extraction was performed in two replicates. After isolation of each subsample, EO volume and weight were measured, and the EO samples were stored in a freezer at 4 °C for further analyses.

4.3. Gas Chromatography (GC) Flame Ionization Detection (FID) and Gas Chromatography–Mass Spectroscopy (MS) Analyses of the Essential Oils (EO)

The chemical profiles of the *H. italicum* and *H. arenarium* EO, in two replications, were determined by GC-FID and GC/MS techniques using a 7890A gas chromatograph (Agilent Technologies Inc., Santa Clara, CA, USA), according to the methods described in our previous study [66]. The GC-MS analysis was performed on a 7890A gas chromatograph (Agilent Technologies Inc., Santa Clara, CA, USA) coupled directly to an Agilent mass selective detector (MSD-5975C). The system was equipped with a HP-5ms fused silica capillary column (5% phenyl 95% dimethylpolysiloxane, 30 m × 0.32 mm i.d., film thickness 0.25 μm, Agilent Technologies, USA). The oven temperature was programmed from 40 °C to 300 °C at a rate of 5 °C/min, and held for 10 min. The temperatures of the injector, the MS quadrupole and the ion source were 250 °C, 150 °C and 230 °C, respectively. The MSD transfer line was maintained at 270 °C.

All mass spectra were acquired in the EI mode (scan range of m/z 50–500 at 1 s/decade; ionization energy of 70 eV). Split ratio was 1:10. The constituents present in the EO samples were identified by comparing their linear retention indices, estimated using a mixture of a homologous series of aliphatic hydrocarbons from C8 to C40 and MS fragmentation patterns with those from an Adams mass spectra library and NIST′08 (National Institute of Standards and Technology).

The GC analysis was performed on an Agilent GC-7890A gas chromatograph (Agilent Technologies, USA) equipped with a flame ionization detector (FID) and HP-5 silica fused capillary column (30 m length × 0.32 mm i.d. × 0.25 μm film thickness) under the same conditions as described above. The FID temperature was maintained at 280 °C for the oil analyses. The relative composition of the investigated samples was calculated on the basis of the GC-FID peak areas (measured using the HP-5 ms column) without using a correction factor.

The GC-FID analysis of the EO was performed with a gas chromatograph 7890A gas chromatograph (Agilent Technologies Inc., Santa Clara, CA, USA) coupled to a flame ionization detector (FID) and HP-5 silica fused capillary column (30 m length × 0.32 mm i.d. × 0.25 μm film thickness). The oven temperature was programmed as mentioned above. The detector and injector temperatures were 280 °C and 220 °C, respectively. The carrier gas was helium at a flow rate of 1 mL/ min. Essential oil samples (1 μL) were injected using the split mode. The percentage composition of EO samples was calculated using the peak normalization method.

4.4. Method for Testing Antimicrobial Activity

The EO of *H. italicum* (plant material originated in Bosnia, France and Corsica) was tested against nine microorganisms with an agar disc diffusion method according to in our previous study [66]. In this study, 0.1 mL of microbial suspension was spread on the Mueller Hinton Agar (MHA, Oxoid, UK) for bacteria and Sabouraud Dextrose agar (SDA, Oxoid, UK) for yeasts. Six mm diameter filter paper discs were used for testing. The filter paper was impregnated with 15 µL of EO and placed on MHA, SDA, respectively, with a microbial inoculum. The MHA was maintained at 4 °C for 2 h and then at 37 °C for 24 h and SDA was maintained at 4 °C for 2 h and then at 25 °C for 24 h. After a 24 h incubation period, the diameter of the inhibition zones was measured (in mm). Chloramphenicol (30 µg, Oxoid, UK) and fluconazole (25 µg, Oxoid, UK) served as positive antimicrobial controls. Antimicrobial activity was measured in triplicate.

Microorganisms

Nine strains of microorganisms were used to determine antimicrobial activity of the EOs, including three Gram-positive bacteria (SA-*Staphylococcus aureus* subs. *aureus* CCM 4223, EF-*Enterococcus faecalis* CCM 4224, SP-*Streptococcus pneumonia* CCM 4501), Gram-negative bacteria (PA-*Pseudomonas aeroginosa* CCM 1959, YE-*Yersinia enterocolitica* CCM 5671, SE-*Salmonella enterica* subsp. *enterica* CCM 3807), and yeasts (CA-*Candida albicans* CCM 8186, CK-*C. krusei* CCM 8271, CT-*C. tropicalis* CCM 8223 (CT)). The microorganisms were obtained from the Czech Collection of Microorganisms (Brno, Czech Republic).

4.5. Statistical Analyses of the Data

One-way analysis of variance was conducted to determine the effect of (1) collection location of *H. arenarium* on the concentration (%) of α-pinene, sabinene, β-pinene, D-limonene, trans-verbenol, 1-terpinen-4-ol, n-tetradecane, β-gurjunene, germacrene D, germacra-4(15),5,10(14)-trien-1, monoterpenes, sesquiterpenes, long-chain alkane and diterpenoids, and (2) country of origin of *H. italicum* (Bosnia, France and Corsica) on the concentration (%) of α-pinene, D-limonene, 2-methyl butyl-2-methyl butyrate, isoamyl tiglate, 1-terpinen-4-ol, nerol, neryl acetate, α-copaene, italicene, α-cis-bergamotene, β-caryophyllene, *p*-cymen-7-ol acetate, α-guaiene, γ-curcumene, β-himachalene, β-curcumene, germacrene D-4-ol, γ-eudesmol, tau.-muurolol, β-eudesmol, monoterpenes, sesquiterpenes, ester and long-chain alkane.

One-way analysis of variance was also conducted to determine if there were significant differences among the three locations where *H. italicum* was collected in terms of nine antimicrobial activities (SA, EF, SP, PA, YE, SE, CA, CK and CT).

For each response variable, the validity of model assumptions was verified by examining the residuals as described in Montgomery [67]. When the effect was either marginally significant ($0.05 < p$-value < 0.1) or significant (p-value < 0.05), multiple means comparison was completed using Fisher's LSD at the 5% level of significance, and letter groupings were generated. The analysis was completed using the GLM Procedure of SAS [68].

5. Conclusions

This study assessed the chemical profile of *Helichrysum arenarium* (Bulgarian populations) with that of the cultivated species *H. italicum* introduced from three different countries and grown side by side in Bulgaria. The main components in *H. arenarium* EO were α-pinene (34.64–44.35%) and sabinene (10.63–11.1%), indicating a possible new chemotype not previously reported in the literature. The chemical profile of *H. italicum* EO originating in France, Bosnia and Corsica were neryl acetate (4.04–14.87%) and β-himachalene (9.9–10.99%); however, there were differences between the EO from plants introduced from the above countries. The *H. italicum* EO plants originating in France, Bosnia and Corsica were evaluated for antimicrobial activity and it was revealed that the EO of plants from France and Corsica had similar composition and antimicrobial activity.

Supplementary Materials: The following are available online at https://www.mdpi.com/article/10.3390/plants11070951/s1, Table S1: Constituents and concentrations of *Helichrysum arenarium* from Bulgaria. The min-max range represents the *H. arenarium* essential oil constituents from the three locations in Bulgaria. Table S2: Constituents and concentrations of *Helichrysum italicum* introduced from France, Corsica and Bosnia. The min-max range includes variation in concentrations of the essential oil constituents from all three locations; France, Corsica, and Bosnia.

Author Contributions: Conceptualization, V.D.Z. and I.S.; methodology, I.D., T.A., S.S., M.K.; software, T.A.; formal analysis, T.A.; investigation, V.D.Z., I.S., E.Y.-T., I.D., S.S.; resources, V.D.Z., I.S., E.Y.-T., I.D., S.S., T.A.; writing—original draft preparation, V.D.Z. and I.S.; writing—review and editing, V.D.Z., I.S., E.Y.-T., I.D., S.S., T.A.; supervision, V.D.Z.; project administration, E.Y.-T. All authors have read and agreed to the published version of the manuscript.

Funding: This research was funded by the National Science Fund Bulgaria (Grant KП-06-H26/6/13.12.2018).

Data Availability Statement: Data are contained within the article.

Acknowledgments: The authors are grateful for the financial support provided by the National Science Fund (Grant KП-06-H26/6/13.12.2018), led by Elina Yankova-Tsvetkova.

Conflicts of Interest: The authors declare no conflict of interest.

References

1. Akaberi, M.; Sahebkar, A.; Azizi, M.; Emami, S.M. Everlasting flowers: Phytochemistry and pharmacology of the genus *Helichrysum*. *Ind. Crops Prod.* **2019**, *138*, 111471. [CrossRef]
2. Talić, S.; Odak, I.; Lukic, T.; Brkljaca, M.; Bevanda, A.M.; Lasić, A. Chemodiversity of *Helichrysum italicum* (Roth) G. Don subsp. *italicum* essential oils from Bosnia and Herzegovina. *Fresenius Environ. Bull.* **2021**, *30*, 2492–2502.
3. Kramberger, K.; Jenko Pražnikar, Z.; Baruca Arbeiter, A.; Petelin, A.; Bandelj, D.; Kenig, S. A Comparative Study of the Antioxidative Effects of *Helichrysum italicum* and *Helichrysum arenarium* Infusions. *Antioxidants* **2021**, *10*, 380. [CrossRef] [PubMed]
4. Rančić, A.; Soković, M.; Vukojević, J.; Simić, A.; Marin, P.; Duletić-Laušević, S.; Djoković, D. Chemical composition and antimicrobial activities of essential oils of *Myrrhis odorata* (L.) Scop, *Hypericum perforatum* L and *Helichrysum arenarium* (L.) Moench. *J. Essent. Oil Res.* **2005**, *17*, 341–345. [CrossRef]
5. Mao, Z.; Gan, C.; Zhu, J.; Ma, N.; Wu, L.; Wang, L.; Wang, X. Antiatherosclerotic activities of flavonoids from the flowers of *Helichrysum arenarium* L. MOENCH through the pathway of anti-inflammation. *Bioorg. Med. Chem. Lett.* **2017**, *27*, 2812–2817. [CrossRef]
6. Les, F.; Venditti, A.; Cásedas, G.; Frezza, C.; Guiso, M.; Sciubba, F.; Serafini, M.; Bianco, A.; Valero, M.S.; López, V. Everlasting flower (*Helichrysum stoechas* Moench) as a potential source of bioactive molecules with antiproliferative, antioxidant, antidiabetic and neuroprotective properties. *Ind. Crops Prod.* **2017**, *108*, 295–302. [CrossRef]
7. Clapham, A.R. Genus *Helichrysum*. In *Flora Europaea*; Tutin, T.G., Heywood, V.H., Burges, N.A., Moore, D.M., Valentine, D.H., Walters, S.M., Webb, D.A., Eds.; Cambridge University Press: London, UK; New York, NY, USA; Melbourne, Australia, 1976; Volume 4, pp. 128–131.
8. Kuzmanov, B.; Gussev, C. Genus *Helichrysum*. In *Flora Flora in Bulgaria*; Anchev, M., Kozhuharov, S., Eds.; Acad. Publ. House Prof. Marin Drinov: Sofia, Bulgaria, 2013; Volume 11, p. 525. ISBN 978-954-322-522-4.
9. Viegas, D.A.; Palmeira-de-Oliveira, A.; Salgueiro, L.; Martinez-de-Oliveira, J.; Palmeira-de-Oliveira, R. *Helichrysum italicum*: From traditional use to scientific data. *J. Ethnopharmacol.* **2014**, *151*, 54–65. [CrossRef]
10. Aćimović, M.; Ljujić, J.; Vulić, J.; Zheljazkov, V.D.; Pezo, L.; Varga, A.; Tumbas Šaponjac, V. *Helichrysum italicum* (Roth) G. Don essential oil from Serbia: Chemical composition, classification and biological activity—May it be a suitable new crop for Serbia? *Agronomy* **2021**, *11*, 1282. [CrossRef]
11. Morone-Fortunato, I.; Montemurro, C.; Ruta, C.; Perrini, R.; Sabetta, W.; Blanco, A.; Lorusso, E.; Avato, P. Essentialoils, genetic relationship sand in vitro establishment of *Helichrysum italicum* (Roth) G. Don ssp. *italicum* from wild Mediterrane an germplasm. *Ind. Crops Prod.* **2010**, *32*, 639–649. [CrossRef]
12. Costa, P.; Miguel Loureiro, J.; Teixeira, M.A.; Rodrigues, A.E. Extraction of aromatic volatiles by hydrodistillation and super critical fluid extraction with CO_2 from *Helichrysum italicum* subsp. *picardii* growingin Portugal. *Ind. Crops Prod.* **2015**, *77*, 680–683. [CrossRef]
13. Ninčević, T.; Grdiša, M.; Šatović, Z.; Jug-Dujaković, M. *Helichrysum italicum* (Roth) G. Don: Taxonomy, biological activity, biochemical and genetic diversity. *Ind. Crops Prod.* **2019**, *138*, 111487. [CrossRef]
14. Paolini, J.; Desjobert, J.M.; Costa, J.; Bernardini, A.F.; Castellini, C.B.; Cioni, P.L.; Guido Flamini, G.; Morelli, I. Composition of essential oils of *Helichrysum italicum* (Roth) G. Don fil subsp. *italicum* from Tuscan archipelago islands. *Flavour Frag. J.* **2006**, *21*, 805–808. [CrossRef]
15. Leonardi, M.; Ambryszewska, K.E.; Melai, B.; Flamini, G.; Cioni, P.L.; Parri, F.; Pistelli, L. Essential oil composition of *Helichrysum italicum* (Roth) G.Don ssp. *italicum* from Elba Island (Tuscany, Italy). *Chem. Biodivers.* **2013**, *10*, 343–355. [CrossRef] [PubMed]

16. Mastelić, J.; Politeo, O.; Jerkovic, I.; Radosevic, N. Composition and antimicrobial activity of *Helichrysum italicum* essential oil and its terpene and terpenoid fractions. *Chem. Nat. Compd.* **2005**, *41*, 35–40. [CrossRef]
17. Mastelić, J.; Politeo, O.; Jerković, I. Contribution to the analysis of the essential oil of *Helichrysum italicum* (Roth) G. Don.–Determination of ester bonded acids and phenols. *Molecules* **2008**, *13*, 795–803. [CrossRef]
18. Gismondi, A.; Di Marco, G.; Antonella Canini, A. *Helichrysum italicum* (Roth) G. Don essential oil: Composition and potential antineoplastic effect. *S. Afr. J. Bot.* **2020**, *133*, 222–226. [CrossRef]
19. Pljevljakušić, D.; Bigović, D.; Janković, T.; Jelačić, S.; Šavikin, K. Sandy everlasting (*Helichrysum arenarium* (L.) Moench): Botanical, chemical and biological properties. *Front. Plant Sci.* **2018**, *9*, 1123. [CrossRef]
20. *The State Pharmacopoeia of the USSR*, 11th ed.; Part 2; USSR: Moscow, Russia, 1999.
21. *Pharmacopoeia Helvetica*, 7th ed.; Département Fédéral de L'intérieur: Berne, Switzerland, 1987; Volume 1.
22. *Farmakopea Polska–Wydanie VI/Polish Pharmacopoeia*, 6th ed.; Rzeczpospolita Polska Minister Zdrowa, Urzad Rejestracji Produktów Leczniczych: Warszawa, Poland, 2002; Wyrobów Medycznych i Produktów Biobójczych.
23. *WHO Monographs on Medicinal Plants Commonly Used in the Newly Independent States (NIS)*; World Health Organization: Geneva, Switzerland, 2015.
24. Butorac, B. *Helichrysum arenarium* (L.) Moench subsp. *arenarium*. In *Red Data Book of Flora of Serbia*; 1. Extinct and critically endangered taxa; Stevanović, V., Ed.; Ministry of Environment of the Republic of Serbia: Belgrade, Serbia, 1999; pp. 294–296.
25. Olsson, K.; Pihlik, U.; Radušiene, J.; Wedelsbäck, B.K. *Helichrysum arenarium (L.) Moench (Everlasting) in Spice and Medicinal Plants in the Nordic and Baltic Countries Conservation of Genetic Resources*; The SPIMED-Project Group at the Nordic Gene Bank: Alnarp, Sweden, 2005; pp. 55–65.
26. Sawilska, A.K.; Jendrzejczak, E. Efficiency of Sandy Everlasting [*Helichrysum arenarium* (L.) Moench] cultivation from in vitro seedlings and achenes. *Ind. Crops Prod.* **2013**, *43*, 50–55. [CrossRef]
27. Biological Diversity Act (Act on Amending and Supplementing). Decree No 354 accepted by the 40th National Assembly on 1st November 2007–State Gazette, No 94/16.11.2007, 2–44 2007. Available online: https://www.lex.bg/bg/laws/ldoc/2135456926 (accessed on 12 June 2021). (In Bulgarian).
28. Medicinal Plant Act. Decree No 29 accepted by the National Assembly on 23 March 2000, State Gazette, No 29/07.04.2000, 9–21. Available online: https://lex.bg/laws/ldoc/2134916096 (accessed on 12 June 2021). (In Bulgarian).
29. Smirnova, L.P.; Pervykh, L.N. Quantitative determination of the total content of flavonoids in the flowers of immortelle *Helichrysum arenarium*. *Pharm. Chem. J.* **1998**, *32*, 321–324. [CrossRef]
30. Czinner, E.; Hagymási, K.; Blázovics, A.; Kéry, A.; Szoke, E.; Lemberkovics, A. In vitro antioxidant properties of *Helichrysum arenarium* (L.) Moench. *J. Ethnopharmacol.* **2000**, *73*, 437–443. [CrossRef]
31. Czinner, E.; Lemberkovics, E.; Bihátsi-Karsai, E.; Vitányi, G.; Lelik, L. Composition of the essential oil from the Inflorescence of *Helichrysum arenarium* (L.) Moench. *J. Essent. Oil Res.* **2000**, *12*, 728–730. [CrossRef]
32. Lemberkovics, E.; Czinner, E.; Balázs, A.; Bihátsi-Karsai, E.; Vitányi, G.; Lelik, L.; Bernáth, J.; Szóke, E. New data on composition of esssential oil from inflorescence of everlasting (*Helichrysum arenarium* (L.) Moench.). *Acta Pharm. Hung.* **2001**, *71*, 187–191. [PubMed]
33. Judzentiene, A.; Butkiene, R. Chemical composition of the essential oils of wild *Helichrysum arenarium* (L.) with differently colored inflorescences from eastern Lithuania. *J. Essent. Oil Res.* **2006**, *18*, 80–83. [CrossRef]
34. Bryksa-Godzisz, M.; Weglarz, Z.; Przybyl, J. Phenolic compounds in yellow everlasting (*Helichrysum arenarium* (L.) Moench) growing wild in the middle part of the Bug river valley. *Herba Pol.* **2006**, *52*, 26–31.
35. Radušienė, J.; Judžentienė, A. Volatile composition of *Helichrysum arenarium* field accessions with differently coloured inflorescences. *Biologija* **2008**, *54*, 116–120. [CrossRef]
36. Yang, Y.; Huang, Y.; Gu, D.; Yili, A.; Sabir, G.; Aisa, H.A. Separation and purification of three flavonoids from *Helichrysum arenarium* (L.) Moench by HSCCC. *Chromatographia* **2009**, *69*, 963–967. [CrossRef]
37. Lv, H.; Sabir, G.; Kungurhan, B.; Liu, Y.; Aisa, H.A. New phthalide glycosides from *Helichrysum arenarium* (L.) Moench. *J. Asian Nat. Prod. Res.* **2009**, *11*, 352–356. [CrossRef]
38. Zhang, Y.W.; Sun, W.X.; Li, X.; Zhao, C.C.; Meng, D.L.; Li, N. Two new compounds from *Helichrysum arenarium* (L.). *J. Asian Nat. Prod. Res.* **2009**, *11*, 289–293. [CrossRef]
39. Eshbakova, K.A.; Aisa, N.A. Component of *Helichrysum arenarium*. *Chem. Nat. Compd.* **2009**, *45*, 929–930. [CrossRef]
40. Morikawa, T.; Wang, L.-B.; Nakamura, S.; Ninomiya, K.; Yokoyama, E.; Matsuda, H.; Muraoka, O.; Wu, L.-J.; Yoshikawa, M. Medicinal flowers. XXVII. new flavanone and chalcone glycosides, arenariumosides I, II, III, and IV, and tumor necrosis factor-α inhibitors from everlasting, flowers of *Helichrysum arenarium*. *Chem. Pharm. Bull.* **2009**, *57*, 361–367. [CrossRef]
41. Albayrak, S.; Aksoy, A.; Sağdiç, O.; Budak, U. Phenolic compounds and antioxidant and antimicrobial properties of helichrysum species collected from eastern Anatolia, Turkey. *Turk. J. Biol.* **2010**, *34*, 463–473. [CrossRef]
42. Yong, F.; Aisa, H.A.; Mukhamatkhanova, R.F.; Shamyanov, I.D.; Levkovich, M.G. New flavanone and other constituens of *Helichrysum arenarium* to China. *Chem. Nat. Compd.* **2011**, *46*, 872–875. [CrossRef]
43. Oji, K.A.; Shafaghat, A. Constituents and antimicrobial activity of the essential oils from flower, leaf and stem of *Helichrysum armenium* in NPC. *Nat. Prod. Commun.* **2012**, *7*, 671–674. [CrossRef] [PubMed]
44. Gradinaru, A.C.; Silion, M.; Trifan, A.; Miron, A.; Aprotosoaie, A.C. *Helichrysum arenarium* subsp. *arenarium* phenolic composition and antibacterial activity against lower respiratory tract pathogens. *Nat. Prod. Res.* **2014**, *28*, 2076–2080. [CrossRef]

45. Moghadam, H.D.; Sani, A.; Sangatash, M.M. Inhibitory effect of *Helichrysum arenarium* essential oil on the growth of food contaminated microorganisms. *J. Essent. Oil Bear. Plants* **2014**, *17*, 911–921. [CrossRef]
46. Bandeira Reidel, R.V.; Cioni, P.; Cervelli, C.; Ruffoni, B.; Pistelli, L. Essential oil composition from three *Helichrysum* species. *Planta Med.* **2016**, *82* (Suppl. 1), S1–S381. [CrossRef]
47. Bandeira Reidel, R.V.; Cioni, P.L.; Ruffoni, B.; Cervelli, C.; Pistelli, L. Aroma profile and essential oil composition of *Helichrysum* species. *Nat. Prod. Commun.* **2017**, *12*, 1507–1512. [CrossRef]
48. Babotă, M.; Mocan, A.; Vlase, L.; Crișan, O.; Ielciu, I.; Gheldiu, A.-M.; Vodnar, D.C.; Crișan, G.; Păltinean, R. 2018. Phytochemical analysis, antioxidant and antimicrobial activities of *Helichrysum arenarium* (L.) Moench. and *Antennaria dioica* (L.) Gaertn. Flowers. *Molecules* **2018**, *23*, 409. [CrossRef]
49. Judzentiene, A.; Charkova, T.; Misiūnas, A. Chemical composition of the essential oils from *Helichrysum arenarium* (L.) plants growing in Lithuanian forests. *J. Essent. Oil Res.* **2019**, *31*, 305–311. [CrossRef]
50. Liu, X.; Xuemin Jing, X.; Li, G. A process to acquire essential oil by distillation concatenated liquid-liquid extraction and flavonoids by solid-liquid extraction simultaneously from *Helichrysum arenarium* (L.) Moench inflorescences under ionic liquid-microwave mediated. *Sep. Purif. Technol.* **2019**, *209*, 164–174. [CrossRef]
51. Stankov, S.; Fidan, H.; Petkova, N.; Stoyanova, A.; Dincheva, I.; Dogan, H.; Cosge, B.; Senkal, C.B.; Uskutoglu, T.; Bas, H.; et al. Phytochemical composition of *Helichrysum arenarium* (L.) Moench essential oil (aerial parts) from Turkey. *Food Technol. Ukr. Food J.* **2020**, *9*, 503–512. [CrossRef]
52. Ivanović, M.; Albreht, A.; Krajnc, P.; Vovk, I.; Razboršek, M.I. Sustainable ultrasound-assisted extraction of valuable phenolics from inflorescences of *Helichrysum arenarium* L. using natural deep eutectic solvents. *Ind. Crops Prod.* **2021**, *160*, 113102. [CrossRef]
53. Melito, S.; Petretto, G.L.; Podani, J.; Foddai, M.; Maldini, M.; Chessa, M.; Pintore, G. Altitude and climate influence *Helichrysum italicum* subsp. *microphyllum* essential oils com-position. *Ind. Crops Prod.* **2016**, *80*, 242–250. [CrossRef]
54. Usai, M.; Foddai, M.; Bernardini, A.F.; Muselli, A.; Costa, J.; Marchetti, M. Chemical composition and variation of the essential oil of wild Sardinian *Helichrysum italicum* G. Don subsp. *microphyllum* (Willd.) Nym from vegetative period to post-blooming. *J. Essent. Oil Res.* **2010**, *22*, 373–380. [CrossRef]
55. Benítez, G.; González-Tejero, M.R.; Molero-Mesa, J. Pharmaceutical ethnobotany in the western part of granada province (southern Spain): Ethnopharmacological synthesis. *J. Ethnopharmacol.* **2010**, *129*, 87–105. [CrossRef] [PubMed]
56. Djihane, B.; Wafa, N.; Elkhamssa, S.; Pedro, H.J.; Maria, A.E.; Mohamed Mihoub, Z. Chemical constituents of *Helichrysum italicum* (Roth) G. Don essential oil and their antimicrobial activity against Gram-positive and Gram-negative bacteria, filamentous fungi and *Candida albicans*. *Saudi Pharm. J.* **2017**, *25*, 780–787. [CrossRef]
57. Oliva, A.; Garzoli, S.; Sabatino, M.; Tadić, V.; Costantini, S.; Ragno, R.; Božović, M. Chemical composition and antimicrobial activity of essential oil of *Helichrysum italicum* (Roth) G. Don fil. (Asteraceae) from Montenegro. *Nat. Prod. Res.* **2020**, *34*, 445–448. [CrossRef]
58. Staver, M.M.; Gobin, I.; Ratkaj, I.; Petrovic, M.; Vulinovic, A.; Sablic, D.M.; Broznic, D. In vitro antiproliferative and antimicrobial activity of the essential oil from the flowers and leaves of *Helichrysum italicum* (Roth) G. Don growing in Central Dalmatia (Croatia). *J. Essent. Oil Bear. Plants* **2018**, *21*, 77–91. [CrossRef]
59. Dzamic, A.M.; Mileski, K.S.; Ciric, A.D.; Ristic, M.S.; Sokovic, M.D.; Marin, P.D. Essential oil composition, antioxidant and antimicrobial properties of essential oil and deodorized extracts of *Helichrysum italicum* (Roth) G. Don. *J. Essent. Oil Bear. Plants* **2019**, *22*, 493–503. [CrossRef]
60. Mollova, S.; Fidan, H.; Antonova, D.; Bozhilov, D.; Stanev, S.; Kostova, I.; Stoyanova, A. Chemical composition and antimicrobial and antioxidant activity of *Helichrysum italicum* (Roth) G.Don subspecies essential oils. *Turk. J. Agric. For.* **2020**, *44*, 371–378. [CrossRef]
61. Cantore, L.P.; Iacobellis, S.N.; Marco, D.A.; Capasso, F.; Senatore, F. Antibacterial activity of *Coriandrum sativum* L. and *Foeniculum vulgare* Miller essential oil. *J. Agric. Food Chem.* **2004**, *52*, 7862–7866. [CrossRef] [PubMed]
62. Mesic, A.; Mahmutović-Dizdarević, I.; Tahirović, E.; Durmišević, I.; Eminovic, I.; Jerković-Mujkić, A.; Bešta-Gajević, R. Evaluation of toxicological and antimicrobial activity of lavender and immortelle essential oils. *Drug Chem. Toxicol.* **2021**, *44*, 190–197. [CrossRef] [PubMed]
63. Rossi, P.G.; Berti, L.; Panighi, J.; Luciani, A.; Maury, J.; Muselli, A.; De Rocca Serra, D.; Gonny, M.; Bolla, J.M. Antibacterial action of essential oils from Corsica. *J. Essen. Oils Res.* **2007**, *19*, 176–182. [CrossRef]
64. Combes, C.; Legrix, M.; Rouquet, V.; Rivoire, S.; Grasset, S.; Cenizo, V.; Moga, A.; Portes, P. 166 *Helichrysum italicum* essential oil prevents skin lipids peroxidation caused by pollution and UV. *J. Investig. Dermatol.* **2017**, *137*, S221. [CrossRef]
65. Thiers, B. *Index Herbariorum: A Global Directory of Public Herbaria and Associated Staff*; New York Botanical Garden's Virtual Herbarium: New York, NY, USA, 2019; Available online: http://sweetgum.nybg.org/ih (accessed on 25 June 2021).
66. Zheljazkov, V.D.; Semerdjieva, I.B.; Dincheva, I.; Kacaniova, M.; Astatkie, T.; Radoukova, T.; Schlegel, V. Antimicrobial and antioxidant activity of *Juniper* galbuli essential oil constituents eluted at different times. *Ind. Crops Prod.* **2017**, *109*, 529–537. [CrossRef]
67. Montgomery, D.C. *Design and Analysis of Experiments*, 10th ed.; Wiley & Sons: New York, NY, USA, 2020.
68. SAS Institute Inc. *SAS/STAT®9.4 User's Guide*; SAS Institute Inc.: Cary, NC, USA, 2014.

plants

MDPI

Article

Phytochemical Composition, Anti-Inflammatory and ER Stress-Reducing Potential of *Sambucus ebulus* L. Fruit Extract

Oskan Tasinov [1,*], Ivayla Dincheva [2], Ilian Badjakov [2], Yoana Kiselova-Kaneva [1], Bistra Galunska [1], Ruben Nogueiras [3,4] and Diana Ivanova [1]

1 Department of Biochemistry, Molecular Medicine and Nutrigenomics, Medical University of Varna, 84B Tzar Osvoboditel Blvd., 9002 Varna, Bulgaria; yoana.kiselova@mu-varna.bg (Y.K.-K.); bistra.galunska@gmail.com (B.G.); divanova@mu-varna.bg (D.I.)
2 AgroBioInstitute, Agricultural Academy, 8 Dr. Tsankov Blvd., 1164 Sofia, Bulgaria; ivadincheva@yahoo.com (I.D.); ibadjakov@gmail.com (I.B.)
3 Center for Research in Molecular Medicine and Chronic Diseases (CiMUS), Department of Physiology, University of Santiago de Compostela-Instituto de Investigación Sanitaria, 15782 Santiago de Compostela, Spain; ruben.nogueiras@usc.es
4 CIBER Fisiopatología de la Obesidad y Nutrición (CIBERobn), 15706 Santiago de Compostela, Spain
* Correspondence: oskan.tasinov@gmail.com; Tel.: +359-896-036961

Abstract: *Sambucus ebulus* L. (SE) fruits are used for their immunostimulation, hematopoietic and antiviral potential. Recently, we focused on analyzing the mechanism underlying SE fruit aqueous extract's (FAE) immunomodulation and anti-inflammatory activities, with attention to its endoplasmic reticulum (ER) stress-reducing potential. J774A.1 macrophages were treated with SE FAE alone or in conditions of lipopolysaccharides (LPS) stimulation. Using GC–MS and LC–MS/MS, its phytochemical composition was analyzed. To measure transcription and protein levels, we used qPCR and Western blot, respectively. The prevailing phytochemicals in SE FAE were hydroxycinnamic acids, proanthocyanidins and anthocyanins. The content of some amino acids, organic acids, alcohols, fatty acids and esters were newly reported. Extracts exerted an immunostimulation potential by stimulating *IL-6, TNFα, Ccl2, COX2* and *iNOS* transcription, without inducing ER stress. SE FAE suppressed the LPS-induced transcription of inflammation related genes (*IL-1β, IL-6, TNFα, Ccl2, Icam-1, Fabp4, COX2, iNOS, Noxo1, IL-1ra, Sirt-1*) and reduced the protein levels of iNOS, peIF2α, ATF6α and CHOP. The effects were comparable to that of salicylic acid. SE suppresses LPS-stimulated inflammatory markers on the transcription and translation levels. Targeting ER stress is possibly another mechanism underlying its anti-inflammatory potential. These findings reveal the potential of SE fruits as a beneficial therapeutic of inflammation and ER stress-related pathological conditions.

Keywords: *Sambucus ebulus* L.; phytochemical composition; anti-inflammatory; ER stress; lipopolysaccharides; macrophages

Citation: Tasinov, O.; Dincheva, I.; Badjakov, I.; Kiselova-Kaneva, Y.; Galunska, B.; Nogueiras, R.; Ivanova, D. Phytochemical Composition, Anti-Inflammatory and ER Stress-Reducing Potential of *Sambucus ebulus* L. Fruit Extract. *Plants* **2021**, *10*, 2446. https://doi.org/10.3390/plants10112446

Academic Editor: Othmane Merah

Received: 7 October 2021
Accepted: 9 November 2021
Published: 12 November 2021

1. Introduction

Traditional medicine is a good source of knowledge about therapeutics, which are consequently researched and successfully implicated in modern pharmaceutical preparations. *Sambucus ebulus* L. (SE), also known as dwarf elder or dwarf elderberry, is a widely used as wound-healing, anti-nociceptive, anti-rheumatoid, anti-influenza, antibacterial and diuretic medicinal plant in Bulgaria, Turkey, Iran, Lebanon, Romania and Bosnia–Herzegovina [1–5]. Fresh fruits, jam, tea or decoction of SE fruits are used as immunostimulating and hematopoietic herbal preparations, as well as for the treatment of rheumatoid arthritis and gastrointestinal disorders [1,2,6]. The number of modern studies focusing on SE biological activities are growing, but there is still insufficient knowledge regarding molecular mechanisms of action of fresh or dry fruits and various fruit extracts.

Only ripe fruits are used in traditional medicine recipes and the chemical content varies depending on the types of the extract [3,7]. Data from phytochemical analyses in the literature reveal that SE fruits are high in polyphenolics, especially anthocyanins and proantchocyanidins, phenolic acids, hydroxycinnamic acids, flavonol glycosides, as well as organic acids, tannins, pectins, resins, vitamin C, volatile substances (eugenol, valeric acid, citronellal etc.), amino acids (including some essential ones), and plant sterols [3,7–16]. Many chromatographic analyses of SE fruit extracts have been carried out up to date, and, still, the information about the presence of certain specific organic compounds remains unclear, especially with regard to soil characteristics, variety of extragents used for sample preparation. Therefore, a detailed phytochemical analysis could be useful, especially in examining the molecular mechanisms of SE fruits on human health.

Numerous studies have established the strong in-vitro antioxidant activity of SE fruit extracts, analyzing its iron chelating, NO radical scavenging, and ABTS cation radical decolorization activity, and their interrelations with polyphenolic and anthocyanin content [3,7,8]. The presence of different functional groups in polyphenolics and organic acids found in the tested SE fruit extracts is considered to determine, to a great extent, their antioxidant and anti-inflammatory activities. In oxidatively challenged 3T3-L1 preadipocytes, SE fruit aqueous extract (FAE) acts as modulator of antioxidant genes' transcription [17]. In macrophages treated with ethanol- or lipopolysaccharides (LPS), SE FAE suppresses the ethanol- and LPS-stimulated transcription of glutamate–cysteine ligase, glutathione peroxidase and nuclear factor kappa B (NFκB) [9,18]. Acetone extracts, hydrophilic and anthocyanin-rich fractions of SE fruits possessing high in-vitro antioxidant activity protect macrophages from the oxidative stress-mediated cytotoxicity caused by *tert*-Butyl hydroperoxide [19]. Ethyl acetate fraction of SE fruits possesses cytoprotective and anti-inflammatory activity reducing ethanol-induced cell death, proinflammatory gene transcription in macrophages [9]. Methanolic extracts of SE fruits reduce carrageenan-induced paw edema in rats [20]. Others describe the antiemetic, neuroprotective and anti-herpes-simplex-virus activities of SE fruit extracts [12,21].

In an intervention study on healthy adult volunteers, SE fruit tea enhances serum antioxidant potential, improves lipid profile [22], decreases serum CRP, IL-1β, leptin and adiponectin levels [23], thus indicating an immune- and fat metabolism-modulating activity. A clinical trial reported the effectiveness of SE fruit ethanol extract for the treatment of paederus dermatitis, proving its anti-inflammatory and wound healing potential [24].

LPS-stimulated macrophages are widely used in-vitro models for testing anti-inflammatory activity of medicinal plant extracts. The macrophages are source of a variety of pro-inflammatory cytokines, chemokines, and may act in a paracrine and endocrine mode. In low grade inflammation, such as in adiposity, where the activation of chemokine release is associated with macrophage recruitment and unlocking a self-feeding inflammatory process that leads to such complications as insulin resistance and related atherosclerosis [25]. The released cytokines and chemokines, such as TNFα, IL-6, IL-1β, NO, as a product of iNOS, activate signaling pathways mediated by Jun N-terminal kinase (JNK), the inhibitor of κB-kinase (IKK)β and other serine kinases [25–28], and resulting in NFκB activation. The latter stimulates the transcription of pro-inflammatory genes [29].

Along with the protein synthesis, endoplasmic reticulum (ER) plays an important role in sensing nutrients and responds to different stress conditions by activating the unfolded protein response and subsequently implicating it into insulin resistance and cardiovascular diseases [30,31]. ER stress can promote inflammation, and vice versa [32,33]. ER stress-related inflammation could be mediated by iNOS [34]. Therefore, the enzyme iNOS as a cross point of inflammation and ER stress could be a possible therapeutic target.

There are data that ER stress and inflammation in different pathological conditions could be reduced by compounds such as resveratrol [35,36], epigallocatechin gallate [37] and proanthocyanidins found in herbal extracts [38]. SE fruits, being rich polyphenolics, anthocyanins and stilbenes, could be effective in combating ER stress and inflammation.

We aimed to analyze the phytochemical composition of SE FAE and to test its immune- and ER stress-modulating potential in a model of unstimulated and LPS-challenged J774A.1 mouse macrophages. The phytochemical analysis of SE FAE revealed the presence of numerous compounds with anti-inflammatory and ER stress-reducing activity. For first time it was established that the transcription-modulating effect of SE FAE on inflammatory cytokines, chemokines, and enzymes in non-stimulated macrophages. In LPS-challenged macrophages, SE FAE suppresses the translation of iNOS and ER stress-related proteins.

2. Results

2.1. Phytochemical Content and Composition

Among the phytochemical compounds identified in the tested SE FAE 15 amino acids (AAs), 10 organic acids (OAs), 36 sugar acids and alcohols, 25 mono-, di- and trisaccharides, 13 fatty acids (saturated and unsaturated) and their esters (Table 1), and 38 phenolic compounds were detected and quantified (Table 2).

Table 1. List of polar phytochemicals identified in the analyzed polar fraction (A) of SE FAE using GC-MS technique. The concentration was given in μg/mL extract. Results are presented as mean ± standard deviation.

Compound	Content, μg/mL
Amino Acids	
L-Valine	3.02 ± 0.21
L-Leucine	8.06 ± 0.56
L-Isoleucine	8.48 ± 0.59
L-Proline	20.01 ± 1.40
L-Threonine	3.89 ± 0.27
L-Phenylalanine	10.25 ± 0.72
L-Lysine	4.37 ± 0.31
Glycine	3.78 ± 0.26
Serine	2.59 ± 0.18
L-Aspartic acid	16.32 ± 1.14
L-Asparagine	6.19 ± 0.43
L-Glutamic acid	1.34 ± 0.09
L-Glutamine	22.99 ± 1.61
DL-Ornithine	12.36 ± 0.86
L-Tyrosine	2.66 ± 0.19
Organic Acids	
Succinic acid	12.64 ± 0.88
Fumaric acid	6.61 ± 0.46
Malic acid	9.22 ± 0.65
Pyroglutamic acid (5-oxoproline)	33.63 ± 2.35
4-Aminobutyric acid	5.69 ± 0.40
2-Hydroxyglutaric acid	4.07 ± 0.29
2-Ketoglutaric acid	8.02 ± 0.56
Phenylpyruvic acid	2.18 ± 0.15
2,3-Dihydroxybutanedioic acid	10.49 ± 0.73
Isocitric acid	18.12 ± 1.27
Sugar Acids and Alcohols	
Glycerol	36.12 ± 2.53
Digalactosylglycerol	6.99 ± 0.63
Glyceric acid	17.05 ± 1.19
Threitol	7.66 ± 0.54
Erythreol	2.09 ± 0.15
Erithreonic acid	2.65 ± 0.19
Threonic acid	8.40 ± 0.59

Table 1. *Cont.*

Compound	Content, μg/mL
Sugar Acids and Alcohols	
Xylitol	4.20 ± 0.29
Arabinitol	34.65 ± 2.43
Pentonic acid	7.69 ± 0.54
L-Glycerol-3-phosphate	17.72 ± 1.24
Ribonic acid	4.76 ± 0.33
Manitol	2.98 ± 0.21
Sorbitol	49.26 ± 3.45
Glucuronic acid isomer	8.49 ± 0.59
Galactitol	1.91 ± 0.13
Galacturonic acid isomer	15.91 ± 1.11
Glucuronic acid isomer	13.03 ± 0.91
Gluconic acid isomer	1.78 ± 0.12
Galacturonic acid isomer	2.89 ± 0.20
Glucuronic acid isomer	3.87 ± 0.27
Galactonic acid	6.33 ± 0.44
Gluconic acid isomer	3.71 ± 0.26
Glucaric acid	14.00 ± 0.98
Galactaric acid	3.38 ± 0.24
Myo-inositol	6.71 ± 0.47
Galactosylglycerol	22.50 ± 1.58
Sorbitol-6-phosphate	43.32 ± 3.03
myo-Inositol-1-phosphate isomer	5.64 ± 0.39
myo-Inositol-2-phosphate isomer	7.43 ± 0.52
Gluconic acid-6-phosphate	1.54 ± 0.11
myo-Inositol-1-phosphate isomer	3.30 ± 0.23
myo-Inositol-2-phosphate isomer	6.87 ± 0.48
Maltitol; alpha-D-Glc-(1,4)-D-sorbitol	4.90 ± 0.34
Galactinol isomer; alpha-D-Gal-(1,3)-myo-Inositol	0.69 ± 0.05
Galactinol isomer; alpha-D-Gal-(1,3)-myo-Inositol	3.67 ± 0.26
Saccharides (mono-, di-, and tri-)	
Xylose methoxyamine	5.94 ± 0.42
Arabinose methoxyamine	12.65 ± 0.89
Fructose isomer	14.31 ± 1.00
Fructose isomer	18.89 ± 1.32
Sorbose isomer	28.11 ± 1.97
Sorbose isomer	21.35 ± 1.49
Galactose isomer	35.19 ± 2.46
Galactose isomer	13.86 ± 0.97
Glucose isomer	17.34 ± 1.21
Glucose isomer	13.59 ± 0.95
Fructose-6-phosphate isomer	16.20 ± 1.13
Mannose-6-phosphate isomer	3.47 ± 0.24
Galactose-6-phosphate isomer	18.79 ± 1.32
Glucose-6-phosphate isomer	30.27 ± 2.12
Fructose-6-phosphate isomer	5.81 ± 0.41
Galactose-6-phosphate isomer	3.32 ± 0.23
Glucose-6-phosphate isomer	4.52 ± 0.32
Sucrose; alpha-D-Glc-(1,2)-beta-D-Fru isomer	24.81 ± 1.74
Trehalose; alpha-D-Glc-(1,1)-alpha-D-Glc isomer	10.10 ± 0.71
Melibiose isomer; alpha-D-Gal-(1,6)-D-Glc isomer	18.59 ± 1.30
Melibiose isomer; alpha-D-Gal-(1,6)-D-Glc isomer	18.80 ± 1.32
Sucrose; alpha-D-Glc-(1,2)-beta-D-Fru isomer	20.55 ± 1.44
Trehalose; alpha-D-Glc-(1,1)-alpha-D-Glc isomer	16.13 ± 1.13
Raffinose; alpha-D-Gal-(1,6)-alpha-D-Glc-(1,2)-beta-D-Fru isomer	12.91 ± 0.90
Raffinose; alpha-D-Gal-(1,6)-alpha-D-Glc-(1,2)-beta-D-Fru isomer	25.61 ± 1.79

Table 1. *Cont.*

Compound	Content, µg/mL
Saturated, unsaturated acids and esters	
9-(E)-Hexadecenoic acid	8.52 ± 0.77
9-(Z)-Hexadecenoic acid	6.57 ± 0.59
Heptadecanoic acid	7.56 ± 0.68
Hexadecatrienoic acid	4.85 ± 0.44
Hexadecanoic acid (Palmitic acid)	6.56 ± 0.59
Heptadecanoic acid	6.06 ± 0.55
9,12-(Z,Z)-Octadecadienoic acid (Linoleic acid)	9.69 ± 0.87
9,12,15-(Z,Z,Z)-Octadecatrienoic acid (Linolenic acid)	8.42 ± 0.76
Octadecanoic acid (Stearic acid)	11.12 ± 1.00
(2E,4E)-2,4-Octadecadienoic acid	15.65 ± 1.41
1-Monopalmitin	13.80 ± 1.24
Monooctadecanoylglycerol	8.62 ± 0.78
beta-Sitosterol	15.22 ± 1.37

All metabolites are trimethylsilyl derivatives; SE FAE–*Sambucus ebulus* L. fruit aqueous extract; essential AAs are given in *italic*. Additional data regarding chromatographic parameters and total ion chromatogram of tested polar compounds are given in Table S1 and Figure S1, respectively.

Table 2. Polyphenolic compounds identified in non-anthocyanin fraction (B) and anthocyanin fraction (C) of the SE FAE using LC-PDA-ESI-MS/MS technique. The concentrations are given in µg/mL extract. Results are presented as mean ± standard deviation.

Compound	Content, µg/mL
Anthocyanins	
Cyanidin-3-O-galactoside (idaein)	382.15 ± 13.19
Cyanidin-3-O-glucoside (chrysanthemin)	31.07 ± 1.10
Cyanidin-3-O-arabinoside	85.87 ± 2.80
Cyanidin-3-O-xyloside	14.35 ± 0.53
Proanthocyanidin monomers	
Catechin	40.19 ± 1.33
Epicatechin	322.37 ± 11.75
Proanthocyanidin dimers	
EC→EC (1)	171.40 ± 6.23
EC→EC (2)	169.24 ± 6.15
EC→EC (3)	189.86 ± 6.90
EC→EC (4)	157.91 ± 5.74
Proanthocyanidin trimers	
EC→EC→EC (1)	225.23 ± 8.16
EC→EC→EC (2)	242.27 ± 8.78
EC→EC→EC (4)	198.92 ± 7.21
EC→EC→EC (4)	249.36 ± 9.04
Stilbenes	
trans-Resveratrol-3-O-glucoside	51.92 ± 1.94
Cyclohexanecarboxylic acid	
Quinic acid	108.00 ± 4.02

Table 2. *Cont.*

Compound	Content, μg/mL
Hydroxycinnamic acids	
3-O-Caffeoylquinic acid (chlorogenic acid)	567.06 ± 20.55
Caffeic acid-O-galactoside	98.72 ± 3.58
Caffeic acid-O-glucoside	74.66 ± 2.71
5-O-Caffeoylquinic acid (neochlorogenic acid)	906.08 ± 32.84
p-Coumaric acid-O-glucoside	236.37 ± 8.57
3-O-p-Coumaroylquinic acid	399.47 ± 14.48
Feruloylquinic acid	248.93 ± 9.02
4-O-p-Coumaroylquinic acid	219.83 ± 7.97
Ferulic acid-O-galactoside	131.66 ± 4.77
Ferulic acid-O-glucoside	122.26 ± 4.43
Flavonol glycosides	
Quercetin-3-O-rhamnosyl-galactoside	25.57 ± 0.93
Quercetin-3-O-galactoside (hyperoside)	29.17 ± 1.06
Kaempferol-3-O-galactoside	11.15 ± 0.40
Quercetin-3-O-rhamnosyl-glucoside	20.35 ± 0.74
Quercetin-3-O-glucoside (isoquercetin)	22.80 ± 0.83
Kaempferol-3-O-glucoside (astragalin)	9.94 ± 0.36
Quercetin-3-O-arabinoside (guaiaverin)	16.77 ± 0.61
Quercetin-3-O-xyloside	13.97 ± 0.51
Kaempferol-3-O-rhamnosyl-galactoside	12.52 ± 0.45
Kaempferol-3-O-rhamnosyl-glucoside	9.15 ± 0.33
Kaempferol-3-O-arabinoside	11.15 ± 0.40
Kaempferol-3-O-xyloside	12.80 ± 0.46
Total analyzed polyphenols	5840.50

EC–epicatechin; SE FAE–*Sambucus ebulus* L. fruit aqueous extract. Additional data regarding precursor ion and fragment ion mass-to-charge ratios (m/z) of the analyzed polyphenols are given in Table S2. Representative LC-PDA-ESI-MS/MS chromatograms of detected polyphenols are given in Figure S2 (anthocyanins), Figure S3 (proanthocyanidin monomers), Figure S4 (proanthocyanidin proanthocyanidin di- and trimers), Figure S5 (stilbenes), Figure S6 (hydroxycinnamic acids), Figure S7 (hydroxycinnamic acids), Figures S8–S11 (flavonols).

2.1.1. Polar Compounds

The most abundant AAs were L-glutamine (18.20% of AAs content) followed by L-proline (15.84% of AAs), L-aspartic acid (12.92% of AAs), DL-ornithine (9.78% of AAs). Six of all fifteen identified AAs were essential (Val, Leu, Ile, Thr, Phe and Lys). Among them phenylalanine was found to be in highest concentration (10.25 ± 0.75 μg/mL), followed by isoleucine (8.48 ± 0.49 μg/mL) and leucine (8.06 ± 0.56 μg/mL) (Table 1). All essential AA comprise 30% of all detected AA content in SE extract.

Among the identified polar OAs, pyroglutamic acid (30.39% of OAs content) and isocitric acid (16.37% of OAs) are found to be in highest concentrations (Table 1). Sorbitol and its 6-phosphate are pre-dominant (92.58 ± 3.24 μg/mL, 34.47% of alcohols) alcohols in the SE FAE, followed by glycerol and its 3-phosphate (60.84 ± 3.35 μg/mL, 22.65% of alcohols) and arabinitol (34.65 ± 2.43 μg/mL, 12.90% of alcohols). Glucuronic and galacturonic acid isomers and glyceric acid were among the prevailing sugar acids (Table 1).

Dominating saccharides were galactose and its 6-phosphate form (57.30 μg/mL), followed by glucose and its 6-phosphate form (52.13 μg/mL) and sorbose (49.46 μg/mL). Sucrose (45.36 μg/mL) was the prevailing disaccharide in the extract. In total, the amount of saccharides was 411.11 μg/mL (Table 1).

Among the tested fatty acids and fatty esters, we found the highest concentration for octadecadienoic acid (15.65 ± 1.41 μg/mL, 18.41% of fatty acids) and for beta-sitosterol (15.22 ± 1.37 μg/mL, 40.43% of esters and sterols) (Table 1).

2.1.2. Polyphenolic Content

S. ebulus L. fruits are a rich source of polyphenolics especially anthocyanins, proanthocyanidins and phenolic acids (Table 2). Cyanidin-3-O-galactoside (382.15 μg/mL, 74.43% of anthocyanins) was at highest concentration among anthocyanins and epicatechin was the major proanthocyanidin (322.37 μg/mL). Among proanthocyanidin polymers dominate epicatechin dimers (688.42 μg/mL) and trimers (915.79 μg/mL) comprising 11.79% and 15.68%, respectively, of all detected polyphenolics in our samples. All together anthocyanins and proanthocyanidins were 1966.76 μg/mL representing 33.67% of all analyzed polyphenols. Trans-Resveratrol-3-O-glucoside (51.93 μg/mL) was the only detected stilbene. Dominating hydroxycinnamic acids found in *S. ebulus* fruits are neochlorogenic and chlorogenic acid, followed by 3-O-p-coumaroylquinic acid and feruloylquinic acid. In total, the amount of hydroxycinnamic acids in the tested SE extract was 3005.02 μg/mL (300.5 mg/g DW) and represented 51.45% of all detected polyphenols. Hyperoside was one the major flavonol detected in our samples.

2.2. Investigation of Inflammation Related Biomarkers in a Model of LPS-Stimulated J774A.1 Macrophages

Aiming to study the anti-inflammatory action of the aqueous extract of dwarf elderberry under conditions of LPS-stimulated inflammatory response in J774A.1 mouse macrophages, the transcriptional levels of genes coding for proteins mediating and involved in the inflammatory process as well as the translation levels of iNOS were analyzed. Macrophage cells were pre-treated with increasing concentrations of 2.5%, 5% and 10% *v/v* (0.25 mg DW/mL, 0.5 mg DW/mL, 1 mg DW/mL respectively) SE FAE or salicylic acid (SA) for 24 h followed by LPS stimulation for an additional 24 h, and included respective control treatments. It was previously reported that the concentrations of extract used in recent experiment are non-toxic for the J774A.1 cell line [18]. The analyzed genes included interleukin 1 beta (*IL-1β*), interleukin 6 (*IL-6*), tumor necrosis factor alpha (*TNFα*), monocyte chemoattractant protein-1 (*MCP-1*, chemokine nomenclature: C–C motif chemokine ligand 2 (*Ccl2*)), intercellular adhesion molecule-1 (*Icam1*), fatty acid binding protein 4 (*Fabp4*, adipocyte protein 2 (*aP2*), prostaglandin-endoperoxide synthase 2 (*Ptgs2*, cyclooxygenase-2 (*COX2*)), inducible NO synthase (*iNOS*), NADPH oxidase organizer 1 (*Noxo1*), interleukin 1 beta receptor antagonist (*IL-1ra*) and sirtuin 1 (*Sirt-1*). The intracellular iNOS protein levels were analyzed as well.

2.2.1. The Effect of LPS-Stimulation on Inflammation Related Biomarkers in J774A.1 Macrophages

As an inflammatory agent, LPS increased transcription levels of *IL-1β, IL-6, TNFα, Ccl2, Icam1* and *Fabp4* by fold-changes of 182 ($p < 0.001$), 27 ($p < 0.001$), 6 ($p < 0.001$), 14.9 ($p < 0.001$), 7 ($p < 0.01$) and 1.9 ($p < 0.001$), respectively (Figures 1a–c and 2a–c). Similarly, LPS-stimulated transcription of *COX2, iNOS*, and of *Noxo1* by 18 ($p < 0.001$), 18 ($p < 0.001$), 3.4 ($p < 0.05$) folds, respectively, and of iNOS protein levels as well by 11.7 ($p < 0.01$) folds (Figure 3a–d). Concerning anti-inflammatory genes' expression, we observed a 12- ($p < 0.001$) and 5-fold ($p < 0.01$) increase for *IL-1ra* and *Sirt-1*, respectively (Figure 4a,b).

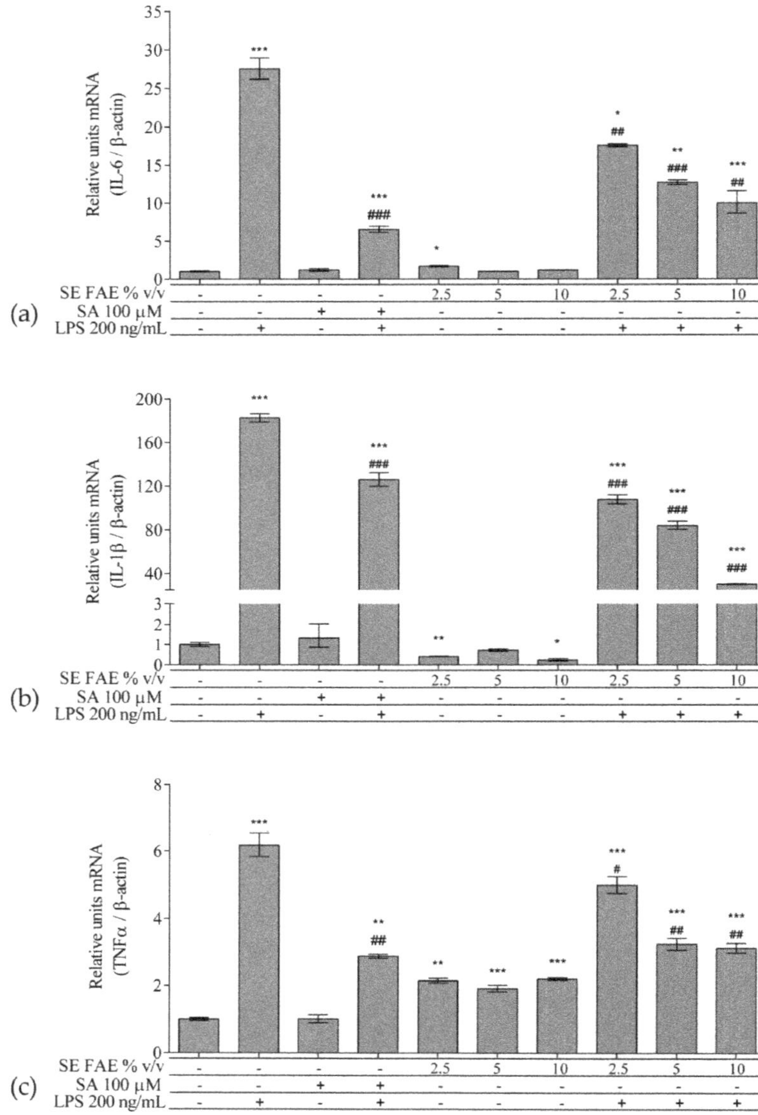

Figure 1. Changes in mRNA levels of *IL-1β* (**a**), *IL-6* (**b**), and *TNFα* (**c**) in J774A.1 mouse macrophages pre-treated with increasing concentrations (2.5%, 5%, 10% *v*/*v*) of SE FAE or with SA for 24 h and subsequently stimulated or not with LPS. Results were obtained using qPCR technique. Data are presented as mean ± SEM. Legend: SE FAE–*Sambucus ebulus* L. fruit aqueous extract; SA–100 μM salicylic acid; LPS–200 ng/mL lipopolysaccharides. * $p < 0.05$, ** $p < 0.01$, *** $p < 0.001$ vs. untreated cells; # $p < 0.05$, ## $p < 0.01$, ### $p < 0.001$ vs. LPS.

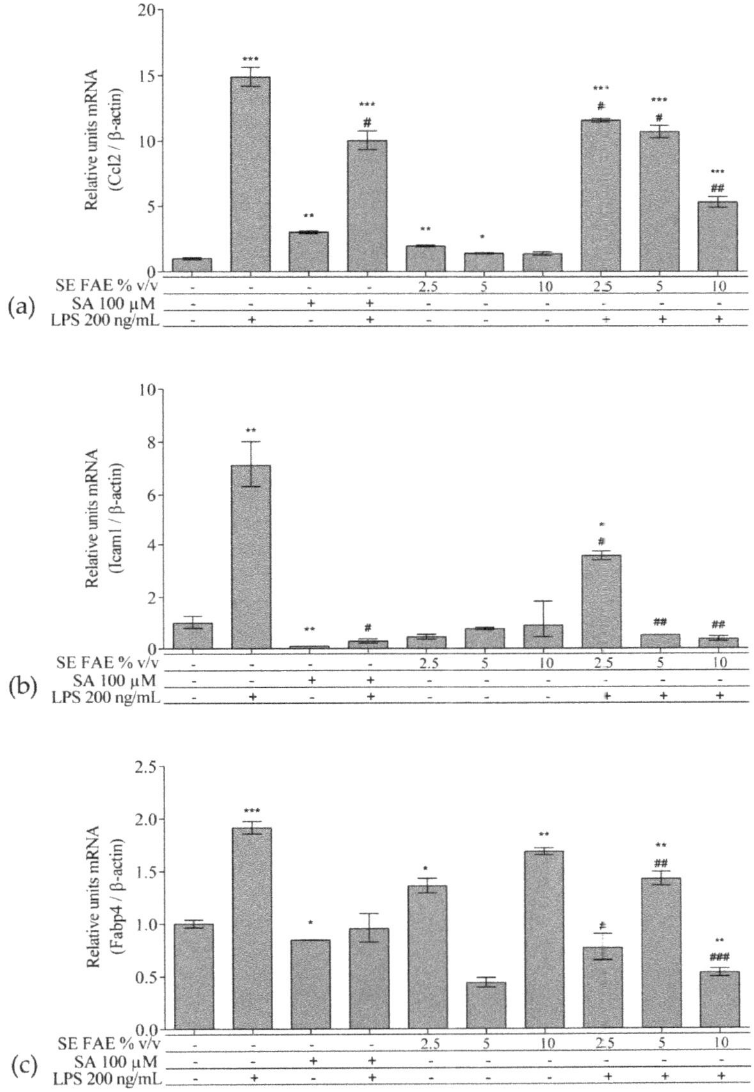

Figure 2. Changes in mRNA levels of *Ccl2* (**a**), *Icam1* (**b**), and *Fabp4* (**c**) in J774A.1 mouse macrophages pre-treated with increasing concentrations (2.5%, 5%, 10% v/v) of SE FAE or with SA for 24 h and subsequently stimulated or not with LPS. Results were obtained using qPCR technique. Data are presented as mean ± SEM. Legend: SE FAE–*Sambucus ebulus* L. fruit aqueous extract; SA–100 µM salicylic acid; LPS–200 ng/mL lipopolysaccharides. * $p < 0.05$, ** $p < 0.01$, *** $p < 0.001$ vs. untreated cells; # $p < 0.05$, ## $p < 0.01$, ### $p < 0.001$ vs. LPS.

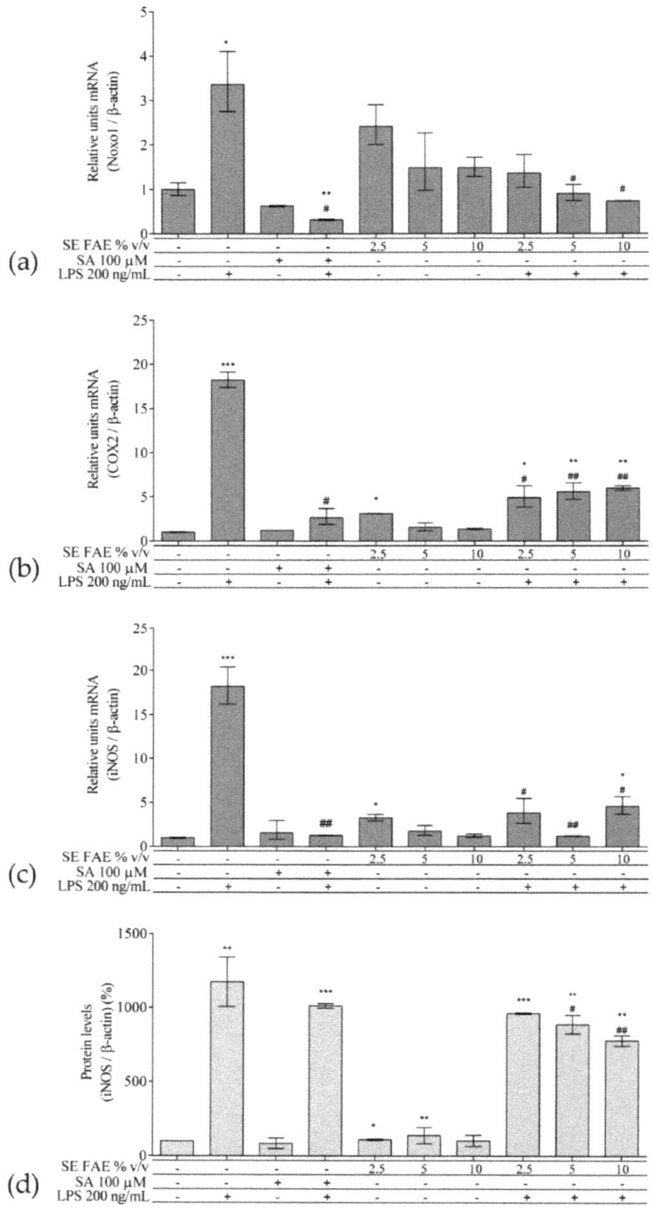

Figure 3. Changes in mRNA levels of *COX2* (**a**), *iNOS* (**b**), *Noxo1* (**c**), and of protein levels of iNOS (**d**) in J774A.1 mouse macrophages pre-treated with increasing concentrations (2.5%, 5%, 10% v/v) of SE FAE or with SA for 24 h and subsequently stimulated or not with LPS. Results were obtained using qPCR ((**a**), (**b**) and (**c**)) or western blot technique (**d**). Data are presented as mean ± SEM. Legend: SE FAE–*Sambucus ebulus* L. fruit aqueous extract; SA–100 µM salicylic acid; LPS–200 ng/mL lipopolysaccharides. * $p < 0.05$, ** $p < 0.01$, *** $p < 0.001$ vs. untreated cells; # $p < 0.05$, ## $p < 0.01$ vs. LPS treatment.

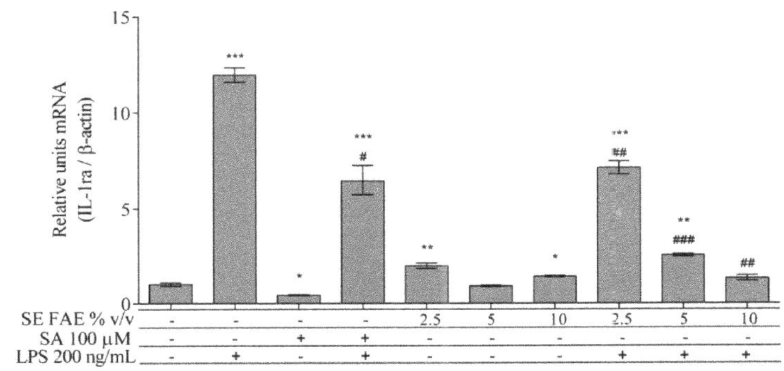

(a)

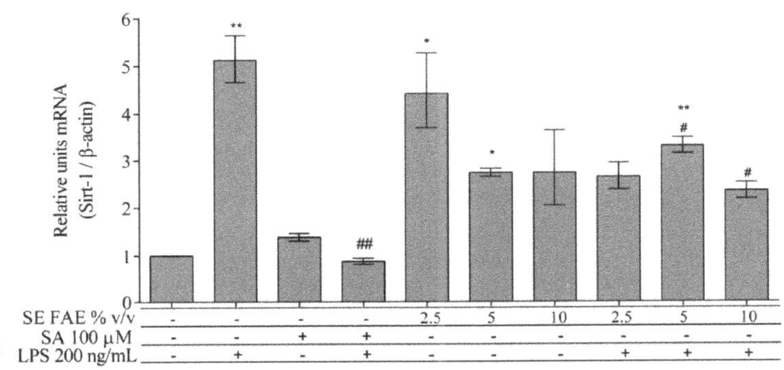

(b)

Figure 4. Changes in mRNA levels of *IL-1ra* (**a**) and of *Sirt-1* (**b**) in J774A.1 mouse macrophages pre-treated with increasing concentrations (2.5%, 5%, 10% *v*/*v*) of SE FAE or with SA for 24 h and subsequently stimulated or not with LPS. Results were obtained using qPCR technique. Data are presented as mean ± SEM. Legend: SE FAE–*Sambucus ebulus* L. fruit aqueous extract; SA–100 μM salicylic acid; LPS–200 ng/mL lipopolysaccharides. * $p < 0.05$, ** $p < 0.01$, *** $p < 0.001$ vs. untreated cells; # $p < 0.05$, ## $p < 0.01$, ### $p < 0.001$ vs. LPS treatment.

2.2.2. The Effect of SE FAE on Inflammation-Related Biomarkers in Non-Stimulated J774A.1 Macrophages

When applied alone, 2.5% *v*/*v* and 10% *v*/*v* SE FAE slightly reduced the gene expression of *IL-1β* by 60% ($p < 0.01$) and 77% ($p < 0.05$), respectively, as compared to untreated cells (Figure 1a). While 2.5% *v*/*v* of herbal extract induced the gene expression of *IL-6* (by 67%, $p < 0.05$), *TNFα* (by 115%, $p < 0.01$), *Ccl2* (by 95%, $p < 0.01$), and *Fabp4* (by 36%, $p < 0.05$) (Figures 1b,c and 2a,c). The higher concentration of SE FAE (5% extract) in culture media stimulated transcription of *TNFα* (by 92%, $p < 0.001$) and of *Ccl2* (by 39%, $p < 0.05$) (Figures 1c and 2a), while the highest concentration (10% extract) induced transcription of *TNFα* (by 121%, $p < 0.001$) and of *Fabp4* (by 68%, $p < 0.01$) (Figures 1c and 2c). SA, applied alone, similarly to SE FAE, it enhanced transcription levels of *Ccl2* (by 200%, $p < 0.01$), but in contrast with SE FAE, it slightly reduced those of *Icam1* (by 91%, $p < 0.01$) and of *Fabp4* (by 16%, $p < 0.05$) (Figure 2a–c), while no significant effects on *IL-1β*, *IL-6* and *TNFα* transcription levels were observed (Figure 1a–c).

The treatment with 2.5% *v*/*v* of SE FAE alone significantly induced the transcription levels of *COX2* (by 210%, $p < 0.05$) and of *iNOS* (by 230%, $p < 0.05$) and both 2.5% *v*/*v* and 5% *v*/*v* of the extract induced iNOS protein levels by 9% ($p < 0.05$) and by 38% ($p < 0.01$), respectively (Figure 3). No effect of SA alone was observed on the gene expression levels of all analyzed inflammation and phagocytosis-related enzymes (Figure 3).

SE FAE in concentrations of 2.5% v/v and 10% v/v induced the transcription levels of *IL-1ra* by 98% ($p < 0.01$) and 41% ($p < 0.05$), respectively (Figure 4a). In contrast, SA treatment reduced *IL-1ra* transcription by 57% ($p < 0.05$) (Figure 4a). Transcription of the so-called longevity gene *Sirt-1* was stimulated upon 2.5% v/v and 5% v/v SE FAE treatment by 343% ($p < 0.05$) and by 274% ($p < 0.05$), respectively (Figure 4b). There was no significant effect of SA applied alone on *Sirt-1* transcription levels (Figure 4b).

2.2.3. The effect of SE FAE on Inflammation-Related Biomarkers in LPS-Stimulated J774A.1 Macrophages

In LPS-stimulated macrophages, the pre-treatment with all three increasing concentrations of SE FAE (2.5% v/v, 5% v/v and 10% v/v), as compared with LPS treatment, significantly reduced the transcription levels of *IL-1β*, *IL-6*, *TNFα*, *Ccl2*, and of *Icam1* with up to 83% ($p < 0.001$), 67.7% ($p < 0.01$), 49% ($p < 0.01$), 64% ($p < 0.01$), and 94.9% ($p < 0.01$), respectively (Figures 1a–c and 2a,b). The effect followed a dose-dependent manner. Similarly, all concentrations reduced LPS-stimulated *Fabp4* mRNA levels, with stronger effect exerted by 2.5% v/v (by 60.2%, $p < 0.05$) and by 10% v/v (by 72.4%, $p < 0.001$) SE FAE (Figure 2c). Considering the effect of SE FAE alone on *Fabp4*, we may recognize that the same concentrations stimulating its gene expression (2.5% v/v and 10% v/v) are the ones exerting the stronger reducing effect in the case of LPS-stimulated cells.

Pre-treatment with the SA as a known anti-inflammatory compound, significantly reversed the LPS stimulation of all genes except *Fabp4*, as follows: *IL-1β* (31%, $p < 0.001$); *IL-6* (76%, $p < 0.001$); *TNFα* (53%, $p < 0.01$); *Ccl2* (32%, $p < 0.05$); *Icam1* (96%, $p < 0.05$) (Figures 1a–c and 2a,b). The inhibitory effect of SE FAE on LPS-stimulated transcription of pro-inflammatory genes was similar to the effect of the positive control SA. In the case of *Icam1*, both the extract and the SA reduced the LPS-induced mRNA levels back to normal. When applied in highest concentration (10% v/v) the herbal extract had a reducing effect on the LPS-stimulated gene expression of *IL-1β*, *Ccl2*, and of *Fabp4*, which was even stronger than that of the SA.

SE FAE significantly inhibited the LPS-stimulated transcription levels of *COX2*, *iNOS* and of *Noxo1*, by up to 73% ($p < 0.05$), 93% ($p < 0.01$) and 78% ($p < 0.05$), respectively, and the protein levels of iNOS by up to 33% ($p < 0.01$) (Figure 3a–d). When SA was applied prior to LPS stimulation mRNA levels of *COX2*, *iNOS* and *Noxo1* were reduced by 85% ($p < 0.05$), 92.9% ($p < 0.01$), and by 90.7% ($p < 0.05$), respectively (Figure 3a–c). The effect shown by SE FAE was similar to that of SA and they both independently reduced LPS-stimulated transcription of *iNOS* and of *Noxo1* back to the normal levels. Pre-treatment with herbal extract showed a stronger iNOS mRNA- and protein levels-reducing effect than the SA did in LPS-challenged cells.

Application of SE FAE suppressed the LPS-induced transcription of *IL-1ra* by up to 88.95% ($p < 0.01$) in a dose-dependent manner and that of *Sirt-1* by up to 54% ($p < 0.05$) (Figure 4). Similar effect was observed in the SA pre-treated cells, where LPS-induced *IL-1ra* and *Sirt-1* mRNA transcription levels were decreased by 46% ($p < 0.05$) and by 82% ($p < 0.01$), respectively (Figure 4). SE FAE exerted stronger reducing activity than that of SA on LPS-stimulated *IL-1ra* transcription, decreasing it to the normal levels.

2.3. Investigation of ER Stress-Related Biomarkers in a Model of LPS-Stimulated J744A.1 Macrophages

Regarding the well-known relationship between inflammation and ER stress, we have analyzed intracellular protein levels of ER stress-related proteins: activating transcription factor 6 alpha (ATF6α), phosphorylated eukaryotic translation initiation factor 2 alpha (peIF2α), and their downstream target gene's product C/EBP homologous protein (CHOP, growth arrest and DNA damage-inducible gene 153 (GADD153)) in a model of LPS-stimulated J744A.1 macrophages (Figure 5). Cells were pre-treated with increasing concentrations of 2.5%, 5% and 10% v/v (0.25 mg DW/mL, 0.5 mg DW/mL, 1 mg DW/mL respectively) SE FAE or SA for 24 h followed by LPS-stimulation for additional 24 h, and respective control treatments were performed as well.

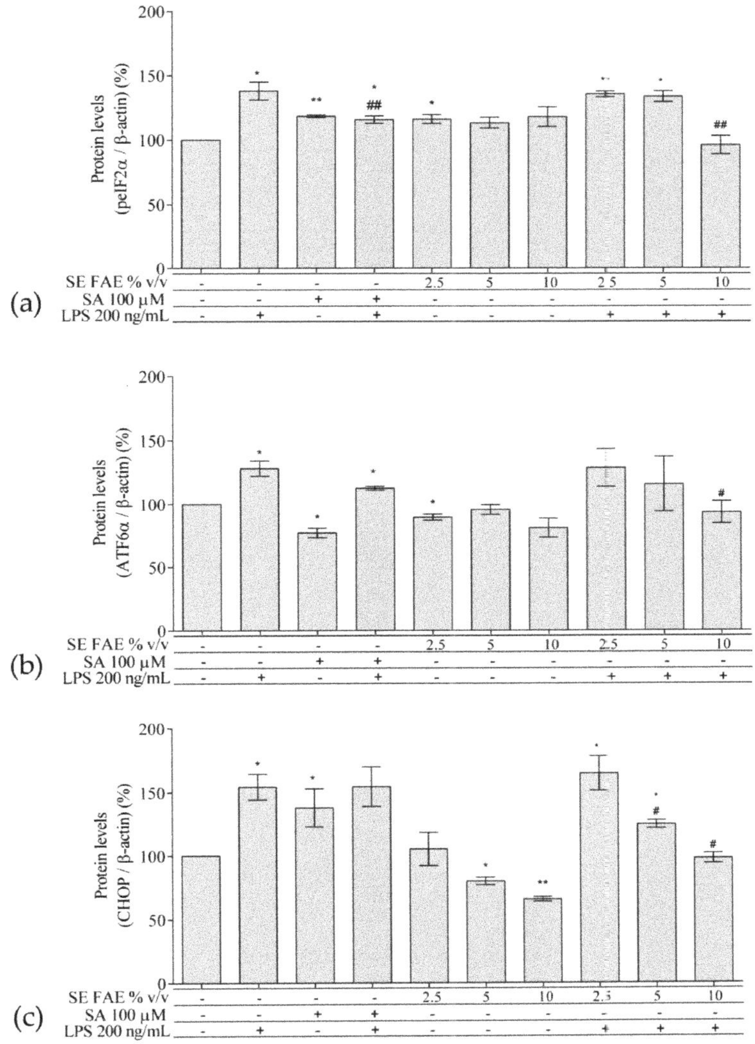

Figure 5. Changes in the protein levels of peIF2α (**a**), ATF6α (**b**), and CHOP (**c**) in J774A.1 mouse macrophages pre-treated with increasing concentrations (2.5%, 5%, 10% *v*/*v*) of SE FAE or with SA for 24 h and subsequently stimulated or not with LPS. Results were obtained using the Western blot technique. Data are presented as mean ± SEM. Legend: SE FAE–*Sambucus ebulus* L. fruit aqueous extract; SA–100 μM salicylic acid; LPS–200 ng/mL lipopolysaccharides. * $p < 0.05$, ** $p < 0.01$ vs. untreated cells; # $p < 0.05$, ## $p < 0.01$ vs. LPS treatment.

2.3.1. The Effect of LPS-Stimulation on ER Stress-Related Biomarkers in J774A.1 Macrophages

LPS treatment significantly induced the levels of peIF2α (Figure 5a), ATF6α (Figure 5b), and CHOP (Figure 5c) proteins by 38% ($p < 0.05$), 28% ($p < 0.05$), and 54% ($p < 0.05$), respectively.

2.3.2. The Effect of SE FAE on ER Stress-Related Biomarkers in Non-Stimulated J744A.1 Macrophages

Applied alone, SE FAE in concentration of 2.5% v/v slightly increased peIF2α (by 16%, $p < 0.05$) (Figure 5a), but also slightly decreased the expression of ATF6α (by 11%, $p < 0.05$), similarly to the effect of SA (Figure 5b). The higher concentrations of the extract, 5% v/v and 10% v/v, reduced, in a dose-dependent manner CHOP levels by 20% ($p < 0.05$) and by 34% ($p < 0.01$), respectively (Figure 5c). The effect of SE FAE on CHOP levels was opposite to the effect of SA in non-stimulated with LPS macrophages. 2.3.3. The effect of SE FAE on ER stress-related biomarkers in LPS-stimulated J744A.1 macrophages.

2.3.3. The Effect of SE FAE on ER Stress-Related Biomarkers in LPS-Stimulated J744A.1 Macrophages

Pre-treatment of macrophages with 10% v/v SE FAE significantly altered the LPS-stimulated levels of peIF2α (by 30%, $p < 0.01$), similarly to the effect of the SA (by 16%, $p < 0.05$). The same concentration of the extract similarly affected the expression of ATF6α (reduced by 27%, $p < 0.05$) (Figure 5b). Both, the 5% v/v and 10% v/v of the extract in the culture media significantly reduced in a dose-dependent manner the LPS-stimulated levels of CHOP by 19% ($p < 0.05$) and by 36%, respectively (Figure 5c). LPS-stimulated ATF6α and CHOP levels were not reduced by the pre-treatment with SA.

The original western blot gels presenting the changes in protein levels of iNOS, peIF2α, ATF6α and CHOP in J774A.1 mouse macrophages pre-treated with increasing concentrations (2.5%, 5%, 10% v/v) of SE FAE or with SA for 24 h and subsequently stimulated or not with LPS, are given in Figure S12.

2.3.4. Correlation Analyzes of ER Stress-Related Biomarkers

Based on the established clear and consistent dose-dependent effect of SE FAE in conditions of ± LPS stimulation, a subsequent correlation analyzes of transcription factors peIF2α and ATF6α and their downstream target CHOP were performed. High positive correlations between ATF6α and CHOP ($r = 0.83$, $p < 0.05$) and between peIF2α and CHOP ($r = 0.67$, $p = 0.08$) were established (Figure 6).

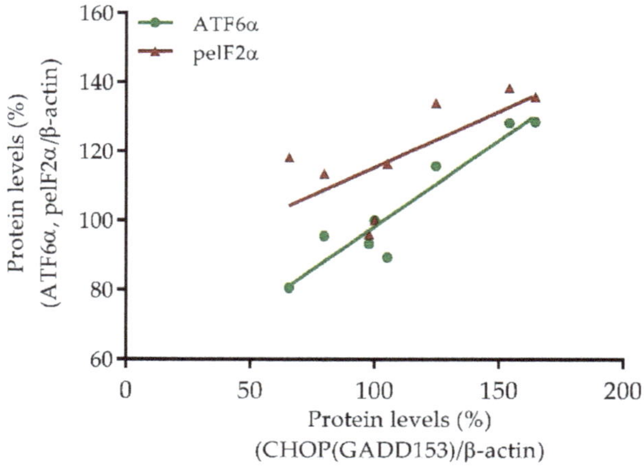

Figure 6. Correlations between protein levels of peIF2α and CHOP (red triangle) and of ATF6α and CHOP (green dot) in J774A.1 mouse macrophages pre-treated with increasing concentrations (2.5%, 5%, 10% v/v) of SE FAE and subsequently stimulated or not with LPS.

3. Discussion

3.1. Newly Detected Phytochemicals in SE Fruit Aqueous Extract

In the recent study we recognized as newly reported 10 AAs including 3 essential AAs, 8 OAs, 12 sugar acids and their phosphates, 16 alcohols, alcohol phosphates and their glycosides, 11 saccharides (mono-, di-, and tri-), 6 saturated and unsaturated fatty acids and 3 of their esters and 3 anthocyanins.

3.1.1. Amino Acids

AAs are important for building cellular proteins, nucleotides, for maintaining acid-base balance and as neurotransmitters. Especially the essential AAs are valuable food components for living organisms. Other authors have reported three essential (Val, Leu and Thr) and three non-essential AAs (Ala, Gln and Tyr) in SE extracts [13]. In our study we have identified all of them except Ala. In addition, we have found three newly detected essential AAs (Ile, Phe and Lys) and six non-essential AAs (Pro, Gly, Ser, Asp, Asn, Gln) and Orn. In total, the content of AAs in the tested SE extract was 126.30 μg/mL, where the essential AAs comprise 30% of all AA content. Considering these findings, it could be concluded that SE fruits are a good source of both essential and non-essential AAs.

3.1.2. Organic Acids

The total amount of all detected OAs was 110.66 μg/mL. We identified, for first time, seven OAs, including isocitric acid (16.37%) and succinic acid (11.42%), important substrates for the normal functioning of citric acid cycle and cell energy production. Fumaric and malic acid were previously reported in SE fruit extract [13]. Additionally, other authors reported the presence of citric acid in SE fruit extract, also important for energy metabolism [13]. Pyroglutamic acid (5-oxoproline) was found in the highest content (30.39% of all OA) in our samples. Pyroglutamic acid is very important as an intermediate in γ-glutamyl cycle involved in transmembrane amino acid transportation and for synthesis of the antioxidant glutathione. We may speculate that the high content of pyroglutamic acid in our samples may be due to a high amount of its keto derivative L-proline (15.84% of AAs) found in SE fruits [39].

3.1.3. Sugar Acids and Alcohols

Regarding the content of sugar acids and sugar alcohols, the presence of pectic acid and sorbitol have been reported in the fruit of *S. nigra* [40,41]. Data in the literature concerning SE fruits content of sugar acids and sugar alcohols are not available.

In our study, the total amount of sugar alcohols was 268.62 μg/mL. For first time, we identified 17 sugar alcohols and their derivatives: sorbitol and its phosphate form (34.47% of sugar alcohols), glycerol, glycerol-3-phosphate and digalactosylglycerol (22.65%) and arabinitol (12.90%). Summarizing, the newly identified sugar alcohols and their derivatives comprise more than 68% of all quantified sugar alcohols in our samples. Therefore, SE fruits should be considered as a good natural source of sorbitol, glycerol and their derivatives.

The total amount of sugar acids in our samples was 115.49 μg/mL. For the first time, 12 sugar acids were identified; of them, the highest content was found for glucuronic acid (21.98%), galacturonic acid (16.28%), and glyceric acid (14.76%). Glucuronic and galacturonic acids are the most abundant sugar acids in the tested SE fruit extract, as they are the major components of plant polysaccharides, like cellulose and pectin.

3.1.4. Saccharides

Literature data provide information mostly about glucose and sucrose content in SE fruit extracts [13]. We report new data regarding the saccharide content of SE fruit extract: the monosaccharides comprise 64.12% (263.61 μg/mL), followed by disaccharides 26.51% (108.98 μg/mL), and trisaccharides (mainly raffinose) 9.37% (38.52 μg/mL). Other identified monosaccharides were fructose, fructose-6-phosphate, arabinose, xylose, and mannose-6-phosphate. Disaccharides were presented by melibiose and trehalose as well.

3.1.5. Fatty Acids and Fatty Esters

Data regarding the lipid composition of SE fruits are very limited. Most studies identified four sterols in SE fruit extract: brassicasterol, campesterol, stigmasterol, β-sitosterol [11,12]. In our samples, β-sitosterol (15.22 μg/mL) was the only sterol we detected. The newly identified fatty esters included 1-monopalmitin and monooctadecanoylglycerol. Other authors provide data regarding the presence of octadecanoic and octadecadienoic acids as well as palmitic, octadecenoic, dehydroabietic, oleic, oleanolic, ursolic, and maslinic acid [12]. We found the highest amount of any fatty acid for octadecadienoic acid (15.65 ±1.41 μg/mL) comprising 18.41% of all fatty acids (84.98 μg/mL), and five newly identified fatty acid (hexadecenoic, heptadecanoic, hexadecatrienoic, hexadecanoic and octadecatrienoic).

Oleanolic, ursolic, and maslinic acid are pentacyclic triterpenes known to possess anticancer properties [42,43]. Ursolic acid, particularly, reduces LPS-stimulated NFκB signaling [44], inflammatory cytokine production by inhibition JNK signaling [45], ER stress induced by high fat diet, and NFκB related inflammation [46]. It is assumed that ursolic acid may also improve insulin sensitivity [47,48]. In our previous study on healthy volunteers we reported that SE fruit tea intake improves lipid profiles, reducing total and LDL cholesterol serum levels and improving volunteers' HDL/LDL ratios [22]. The intake of phytosterols including β-sitosterol may reduce total serum cholesterol [49] and low-density cholesterol [50]. As brassicasterol, campesterol [11] and β-sitosterol are among the phytosterols with known cholesterol-lowering activity [51–53], it is not surprisingly that SE fruit tea exerts a cholesterol-lowering effect.

3.1.6. Phenolic Compounds

Compared with other species of the *Sambucus* genus, such as *S. nigra*, *S. cerulea*, and *S. racemose*, *S. ebulus* is the richest in total hydroxycinnamic acids, catechin, epicatechin and flavonols [15].

Anthocyanins are the predominant colored polyphenols in elderberries. In our study we identified three new anthocyanins: cyanidin-3-O-galactoside (the major anthocyanin in most plants), cyanidin-3-O-arabinoside and cyanidin-3-O-xyloside. In accordance with other studies, we also found in our samples cyanidin-3-O-glucoside, but not cyaniding-3-O-sambubioside [14]. Anthocyanins exerts numerous beneficial health effects including antioxidant, anti-inflammatory, anticancer, antidiabetic, anti-toxic, cardiovascular and nerve-protective capacities [54].

Flavanols, including catechin and epicatechin, were previously reported in SE fruits [15,16], as were proanthocyanidin dimers and trimers [55]. SE is the richest among the *Sambucus* sp. in catechins and epicatechins [15]. Epicatechin is the major proanthocyanidin monomer and a component of proanthocyanidin dimers and trimers. It is considered that one of the richest sources of proanthocyanidins are grape seeds [56]; proanthocyanidin dimer and trimer content in SE fruits is comparable to that in the grape seeds [55].

Resveratrol is the most abundant stilbene in plants. Grape peels are known as one of the best sources of resveratrol, containing on average 0.169 mg/g DW [57]. We found that *trans*-resveratrol-3-O-glucoside represents 5.19 mg/g DW. Thus, SE fruits and its FAE seems to be considerable sources of resveratrol. Resveratrol exerts a wide range of biological activities. It acts as calorie-restriction mimetic, increasing the levels of so-called survival protein SIRT1 and improves energy metabolism, decreases plasma glucose, triglycerides and inflammatory cytokines [58]. Its positive impacts on human heath are complemented by improved plasma antioxidant activity and reduced oxidative stress [59,60]. In obese individuals, resveratrol improves insulin sensitivity [61] and mitochondrial oxidative capacity when used in combination with epigallocatechin gallate [62].

Quinic acid is a compound conjugating with hydroxycinnamic acids to form their esters. Its presence in SE fruit tea (hot extraction) was reported previously by our group [55]. Recently, we also confirmed its availability in SE FAE (cold extraction).

Hydroxycinnamic acids are the most abundant phenolic acids in fruits, vegetables, and coffee beans [63]. They present as esters of hydroxycarboxylic acids, such as quinic acid or as glycosylated derivatives. Among them are caffeic acid, ferulic acid, chlorogenic acid, isoferulic acid and coumaric acid. There are data showing that SE fruits contain highest concentration of neochlorogenic acid and chlorogenic acid among all *Sambucus* sp., followed by *Sambucus cerulea* [15]. The same authors report also presence of p-coumaric acid-O-glucoside, 3-O-p-coumaroylquinic acid, and 4-O-p-coumaroylquinic acid in SE fruits. A study on SE fruit tea confirmed the presence of these hydroxycinnamic acids and those that we have also found in SE FAE [55]. There was no significant difference in hydroxycinnamic acid content between SE fruit tea prepared by hot and cold FAE extraction. In accordance with others, we have also found that the neochlorogenic acid followed by chlorogenic and 3-O-p-coumaroylquinic acid were the main hydroxycinnamic acids in SE FAE. The beneficial effects of hydroxycinnamic acids as potential chemo-preventives are associated to their antioxidant activity [64]. Coumaric and ferulic acid and their amides significantly reduce LPS-stimulated NO synthesis, iNOS protein content and mRNA levels in RAW 264.7 macrophages, thus presuming a mechanisms of their anti-inflammatory activity [65]. Plant extracts rich in neochlorogenic acid possess various biological activities, including antioxidant and anti-inflammatory [66–68]. As SE FAE is rich in hydroxycinnamic acids and their derivatives, it could be suggested that hydroxycinnamic acids are the main bioactive components determining its antioxidant, and anti-inflammatory.

The most abundant flavonol glycoside in *Sambucus* sp. is quercetin-3-O-rutinoside (rutin) [15]. Other flavonol glycosides detected in SE include quercetin glycosides, followed by kaempferol glycosides, and isorhamnetin glycosides [12,15,55]. In our samples the total amount of flavonols was 195.35 μg/mL, comprising only 3.35% of all analyzed polyphenols. In addition, the quercetin glycosides (128.63 μg/mL) dominate over kaempferol glycosides (66.72 μg/mL), representing 65.85% and 34.15% respectively of all identified flavonols in SE FAE. The presence of flavonols, quercetin and kaempferol in SE fruit extracts was widely reported in the literature [16]. Other studies provide data regarding the content of rutin [9,16,55], isoquercetin and hyperoside [9,15,55], kaempferol 3-O-rutinoside [15], isorhamnetin-3-O-laminaribioside [12], isorhamnetin 3-O-rutinoside (narcissin) [12,15], isorhamnetin 3-O-glucoside [9,12], and myricetin [16] in SE fruit extracts. In accordance with the data of others, we have also identified quercetin-3-O-rhamnosyl-galactoside, quercetin-3-O-rhamnosyl-glucoside, guaiaverin, quercetin-3-O-xyloside, kaempferol-3-O-galactoside, astragalin, kaempferol-3-O-rhamnosyl-galactoside, kaempferol-3-O-rhamnosyl-glucoside, kaempferol-3-O-arabinoside, kaempferol-3-O-xyloside in our samples [55]. Flavonoid-rich herb extracts possess strong antioxidant and anti-inflammatory activities [69,70]. Both isoquercetin and hyperoside exert antioxidant and anti-inflammatory [71,72] effect. Similarly, quercetin and rutin exhibit anti-inflammatory, anti-cancer, anti-bacterial and anticonvulsant activities [73–75].

3.2. SE FAE Modulates mRNA and Protein Levels of Inflammation-Related Biomarkers in LPS-Challenged J774A.1 Macrophages

The anti-inflammatory effect of polyphenols is due to the decreased activation of macrophages and T-lymphocytes and the suppressed production of cytokines and chemokines or their receptors. Polyphenols such as resveratrol, catechin and quercetin, found in SE fruits, inhibit NFκB-dependent production of ICAM and VCAM in endothelial cells, as well as the expression of MCP-1 receptors CCR1 and CCR2 [76,77]. Inhibition of the latter reduces the chemotaxis of leukocytes to the site of inflammation and the subsequent increased production of IL-6. Anthocyanin metabolites reduce TNFα-induced expression of MCP-1 and ICAM, and thus combat oxidative stress. In models of LPS-induced inflammatory response of macrophages, anthocyanidin- and anthocyanin-rich extracts inhibit iNOS transcription and iNOS and COX-2 translation by targeting the NFκB and MAPK kinase signaling pathways [78,79]. Karlsen et al. [80] reported that blackcurrant and blackberry polyphenols significantly inhibited NFκB in LPS challanged monocytes isolated from healthy adults.

In our previous study we found that SE FAE reduces LPS activated mRNA expression of NFκB, which correlated with decreased transcription levels of glutamate–cysteine ligase and glutathione peroxidase enzymes [18]. Neochlorogenic and chlorogenic acid, also found in SE FAE, suppress LPS-stimulated activation of NFκB patway resulting in reduced iNOS synthesis and activation of COX-2, thus decreasing NO, prostaglandin E2, TNFα, IL-1β, and IL-6 levels in RAW 267 macrophages [81]. Neochlorogenic and chlorogenic acid-rich plant extracts significantly reduce the carrageenan-induced paw edema in rats, in vivo [81]. Coumaric and ferulic acids were found to reduce LPS-stimulated iNOS protein and mRNA levels [65]. Few studies have reported the strong antioxidant, anti-inflammatory and antidiabetic properties of cyanidin-3-O-galactoside, one of the main anthocyanin in SE FAE [54]. Ursolic acid, found in SE leaves, reveals anti-inflammatory activity by reducing TNFα-induced expression of ICAM-1 and VCAM-1 in human umbilical vein endothelial cells [82]. Ursolic acid reduces LPS-stimulated NFκB [44] and JNK signaling, thus inhibiting inflammatory cytokine production [45]. It was reported, also, that it may combat ER stress- and NFκB-related inflammation in animals on a high-fat diet [46]. We also found an anti-inflammatory effect of SE FAE, which may be due to the presence of ursolic acid in SE fruits reported by others [12].

LPS stimulates the gene expression of cytokines IL-1β, TNFα and IL-6, chemokine ICAM-1 and the enzymes COX-2 and iNOS by activating the NFκB-dependent signaling pathway [83–88]. The activation of iNOS results in increased production of $ONOO^-$ and further stimulation of COX2 gene expression and prostaglandin E2 production [89]. Recently we observed that the pre-treatment with SE FAE significantly reduces LPS-stimulated transcription of pro-inflammatory cytokines IL-1β, TNFα, IL-6, the chemokines MCP-1, ICAM-1, enzymes COX-2, iNOS, as well as the protein levels of iNOS.

The effect was comparable to that of salicylic acid, a known anti-inflammatory agent, used in our study as a positive control. The possible mechanism behind the observed anti-inflammatory effect of SE extract might due to the presence of neochlorogenic acid, chlorogenicacid, ursolic acid, resveratrol, catechin and quercetin, by suppressing the NFκB signaling pathway. Additional mechanism for reducing COX2 activity and prostaglandin production might be the direct $NO^{\cdot}$ radical-scavenging activity of SE FAE [8].

Moreover, when applied alone, the lower doses of SE FAE induce the transcription of IL-6, TNFα and MCP-1 by two-fold; COX2 and iNOS transcription by three-fold and iNOS protein expression ($p < 0.05$). These results support the traditional application of dwarf elderberries in folk medicine as an effective immunostimulant. Our previous study reported increased NFκB, glutamate–cysteine ligase and glutathione peroxidase transcription and thus confirms the immunostimulatory effect of SE FAE [18]. Immunostimulatory effect was also proven for *S. nigra*, another member of the genus *Sambucus* [90].

NADPH oxidase (NOX), is one of the major enzymes in vascular endothelial cells, catalyzing the formation of a superoxide radical anion [91]. Endothelial eNOS, as well as iNOS, produce NO, which reacts with a superoxide radical anion forming highly reactive $ONOO^-$ [92,93] and contributing to the development of oxidative stress. NOX is highly active in activated macrophages, taking part in a respiratory burst for destroying bacterial cell walls [94]. NOX is among the newly established target molecules in the treatment of hypertension and atherosclerosis, and concomitant pathologies such as diabetes and cardiovascular diseases [91,95].

By suppressing the LPS-induced gene expression of NOX subunit Noxo1, SE FAE exhibits strong antioxidant and anti-inflammatory activity. The effect of the extract on LPS-induced Noxo1 transcription was similar to that of SA. Both SE FAE and SA are very effective in completely neutralizing LPS-induced Noxo1 overexpression. Compounds such as epigallocatechin gallate, quercetin and isorhamnetin, the derivatives of which are found in SE fruits, were shown to target Noxo1 [96–101], while resveratrol decreases NOX activity [102]. NOX, and in particular its Noxo1 subunit, has been suggested as playing an important role in the IL-1β-dependent activation of NF-κB [103]. Therefore, the inhibition of Noxo1 gene expression is one potential mechanism by which SE fruits suppress the NFκB-dependent expression of pro-inflammatory genes and proteins such as iNOS.

FABP4 one of the fatty acid-binding proteins is expressed in both adipocytes and macrophages [104,105]. Macrophages are specific target cells for Fabp4 and its deficiency prevents atherosclerosis [100]. High FFA contribute to the development of atherosclerosis, are proinflammatory and activate TLR4 signaling cascades. The same signaling pathway is also activated by LPS [25]. In animal models of obesity and insulin resistance it was shown that the inhibition of FABP4 protein in macrophages reduces inflammatory cytokines (MCP-1, IL-1β, IL-6 and TNFα) and the formation of atherosclerotic lesions and foam cells and improves insulin sensitivity [106,107]. Recently, we studied the effect of SE FAE on the transcription levels of the Fabp4 gene. SE FAE, applied alone, slightly stimulated Fabp4, as was observed for other studied inflammatory genes. On the other hand, in LPS-stimulated cells, pre-treatment with the same concentrations of SE FAE completely prevents LPS-stimulated Fabp4 transcription. These findings suggest another possible anti-inflammatory mechanism of SE FAE action.

SIRT-1 is among the most studied sirtuins from class III histone deacetylases, which activation improves obesity related insulin resistance [108] and possesses anti-inflammatory potential [25]. Sirt-1 activators, such as resveratrol, may inhibit ICAM1 and TNFα induction [109]. In our study, SE FAE induces the expression of Sirt-1. The same result is observed in macrophages pre-treated with the extract and stimulated by LPS. SIRT-1 decreases serine phosphorylation in IRS-1, improves insulin signaling and, as a consequence, increases glucose transport [110]. The same mechanism is involved in improving insulin sensitivity decreased by TNF-alpha. These findings suggest another possible anti-inflammatory and insulin sensing mechanism of SE FAE, although these mechanisms are unclear and require additional studies in future.

3.3. SE FAE Modulates Levels of ER Stress-Related Proteins in LPS-Challenged J774A.1 Macrophages

Activation of ER stress may lead to the phosphorylation of JNK and IKK, which is well-known to promote NFκB signaling and the consequent inflammation [32] accompanied by JNK-mediated phosphorylation of IRS 1/2 [111] to impair insulin signaling. On the other hand, triggered by viral or bacterial infections, the production of TNFα, IL-6, IL-1β, and INFγ may amplify the ER stress in many cell types including macrophages, pancreatic β cells and hepatocytes [112,113]. Since both the processes of inflammation and ER stress may result from each other, we analyzed the expression of three important ER stress-related proteins as potential mechanisms to explain the observed anti-inflammatory potential of SE FAE.

We have observed a significant increase in protein levels of transcription factors ATF6α and peIF2α and their downstream target CHOP in LPS-stimulated macrophages. Applied alone SE FAE downregulated the synthesis of CHOP at a dose-dependent manner and slightly that of ATF6α. SE FAE significantly reduced LPS-stimulated CHOP levels. The decrease in ATF6α levels and the phosphorylated eIF2α in LPS-stimulated macrophages provides evidence for a possible mechanism by which the extract inhibits CHOP synthesis. This cytoprotective mechanism in the case of stimulated ER stress is confirmed by the established high significant correlation between protein levels of CHOP and transcription factors (Figure 5). This may explain the previously reported cytoprotective effects of SE fruits as well [9,19]. SA reduced only the LPS-stimulated peIF2α protein levels. It should be noted that the SE FAE effect was in a similar direction, however it was stronger than that of SA. With regard to the well-known anti-inflammatory activities of SA [114] and the links between inflammation and the activation of ER stress, we expected that SA might have reducing effect on ER stress-related biomarkers. However, SA did not exhibit any protective effect against LPS-stimulated ATF6α and CHOP levels, as SE FAE, in contrast, did. According to this observation, we may suggest that the SE FAE uses mechanisms different from those of SA, resulting not only in reduced transcription of inflammatory markers but also in the translation of ER stress-related ones.

ER stress could be activated by high levels of FFAs, similarly to inflammation, excess nutrients, improperly folded proteins and local hypoxia, which is characteristic of obesity. This results in increased oxidative stress in the liver and in adipose tissue of obese animals [115]. An interesting fact is the established activated expression of Fabp4 in macrophages and its association with the development of ER stress and inflammation [106]. In accordance with previous analyses, the suppression of Fabp4 in macrophages protects cells from the FFA-induced inflammatory process, which may result in increased insulin sensitivity and glucose tolerance [106]. The ability of the SE FAE to inhibit LPS-induced transcription of Fabp4 suggests that it would also have a protective effect in combating ER stress. FFA and glucose activate PERK-mediated phosphorylation and activation of eIF2α and RNA splicing of Xbp-1 in obese rat and human adipocytes [116,117]. CHOP is induced predominantly by the PERK/eIF2α/ATF4 signaling cascade associated with ER stress, as well as by the IRE1α/Xbp-1 signaling pathway and the ATF6α transcription factor in different pathological conditions, including diabetes [118–121].

The induction of CHOP is associated with the activation of apoptosis and DNA damage. Its induction in humans and animal macrophages is associated with the detachment of atherosclerotic plaques in atherosclerosis [122]. The production of superoxide by NOX in atherosclerotic plaque-associated macrophages activates CHOP and subsequent ER stress-mediated cell death [123]. ER stress may stimulate NFκB, by Ca^{2+}- and reactive oxygen species-dependent mechanisms [124] and the activation of PERK/eIF2α-mediated phosphorilation of IKK [125]. Another important mediator of the ER stress-related activation of NFκB signaling and the consequent TNFα, IL-6 and IL-1β cytokines production is iNOS [34]. This transforms iNOS enzyme into a cross point of inflammation and ER stress, and, consequently intoa possible therapeutic targets.

By preventing the LPS-induced transcription of iNOS and Noxo1 and the subsequent translation of iNOS protein, SE FAE may reduce superoxide radical and $ONOO^-$ production, thus reducing the activation of ER stress-related inflammation; whereas, suppressing CHOP synthesis by suppression of peIF2 and ATF6α possess another important mechanism for combating ER stress-related activation of inflammation and cytokine production. These are the first results confirming that SE, and in particular SE fruits, could suppress the induction of inflammation by suppressing the activation of ER stress.

The complex phytochemical composition of SE FAE presumes that it acts at different levels on inflammation and ER stress regulatory cascades. SE FAE acts at different cross-points starting from the direct scavenging of reactive oxygen species, through the regulation of gene transcription, to protein synthesis.

Our results contribute to the relationship between inflammation, ER stress and insulin resistance. We may speculate that SE FAE could have a beneficial effect in preventing atherosclerosis, insulin resistance and diabetes type 2, but this requires additional specific studies to confirm the possible insulin sensing effect of the SE fruits.

4. Materials and Methods

4.1. Plant Material

Well-ripened fruits of *Sambucus ebulus* L. were harvested from North-Eastern Bulgaria in the period August–September, 2014 and were dried in the dark at room temperature. SE FAE was prepared using 150 mg finely grounded dry fruits, extracted three times with 3 mL distilled water for 3 min in a vortex mixer (2000 rpm), at room temperature. After centrifugation (5 min, 3500 rpm) the supernatants were collected and diluted up to 15 mL with PBS buffer (pH = 7.4) for cell-culture experiments or with distilled water for phytochemical analyses. A specimen from *S. ebulus* L. fruits was deposited under No. 108144 in the Herbarium SO (by Index Herbariorum) of Sofia University St. Kliment Ohridski, Faculty of Biology.

4.2. Phytochemical Analysis

4.2.1. Extraction

The samples were filtered through a 0.45-µm PTFE filter (Waters, USA) and the filtrates were loaded onto a reverse phase solid phase extraction column (Discovery® DSC-18, 5 g, 20 mL) (Sigma-Aldrich Co. LLC, St. Louis, MO, USA). The SFE columns were activated rinsed with 40.0 mL distilled water and 1 mL of the filtered sample was loaded onto the SPE column. Three fractions were isolated: anthocyanin fraction (C)—eluted with 2 × 12 mL 0.1% (*v*/*v*) formic acid in acetonitrile; non-anthocyanin fraction (B) containing phenolic acids, flavonols, flavonols, eluted with a 2 × 12-mL ethyl acetate; polar fraction (A), containing organic acids, aminoacids and carbohydrates-eluted with 2 × 12-mL 0.2% (*v*/*v*) formic acid in water. The eluates were evaporated to dryness under reduced pressure at a temperature below 40 °C.

4.2.2. Analysis of Polar Fraction (A)

An aliquot (0.2 mL) of fraction A was submitted to lyophilisation for 6 h at −20 °C. The dry residue was subjected to the following derivatization protocol: 300.0 µL solution of methoxyamine hydrochloride (20.0 mg/mL in pyridine) was added to a residue and the mixture was heated on Thermo-Shaker TS-100 (1 h/70 °C/300 rpm). After cooling, 100.0 µL N,O-Bis (trimethylsilyl)trifluoroacetamide (BSTFA) were added to the mixture, then heated on Thermoshaker, Analytik Jena AG, Jena, Germany (40 min/70 °C/300 rpm). Then, 1.0 µL of the solution was injected in the GC–MS system (Agilent GC 7890, Agilent MD 5975). The separations were done on a chromatographic column HP-5ms (length 30 m, diameter 0.32 mm, film thickness 0.25 µm) at a gradient temperature mode: initial 100 °C for 2 min; ramp up to 180 °C with 15 °C/min for 1 min; ramp up to 300 °C with 5 °C/min for 10 min. Injector and detector temperatures were set at 250 °C; the velocity of the carrier gas helium was set at 1.0 mL/min. The MS scanning was in the range 50–550 m/z.

4.2.3. Analysis of Fractions B and C

Fractions B and C were analyzed using LC-PDA-ESI-MS/MS chromatographic system; in negative ESI mode for fraction B, and in positive ESI mode for fraction C as previously described [55].

For the analysis of polyphenolics, the dry residues of fraction B and C were dissolved in 200 µL metahnol:formic acid, (99:1 *v*/*v*), the solution was filtered through a 0.22 µm PTFE filter and 2 µL of the filtrate were injected into LC-PDA-ESI-MS/MS system.

An LTQ Orbitrap mass spectrometer (Thermo Scientific, Hemel Hempstead, UK) equipped with an ESI source (in negative mode) was used for accurate mass measurements. Operation parameters were as follows: source voltage—4 kV; sheath, auxiliary and sweep gas −20, 10 and 2 arbitrary units, respectively; capillary temperature was275 °C. The samples were analyzed in full-scan mode at a resolution of 30,000 at m/z 400 and data-dependent MS/MS events were acquired at a resolving power of 15,000. The most intense ions were detected during full-scan MS-activated data-dependent scanning. Ions that were insufficiently intense were analyzed in MS2 mode with a resolution power of 15,000 at m/z 400. An isolation width of 100 amu was used. Precursors were fragmented by a collision-induced dissociation with energy of 30 V and an activation time of 10 ms. The mass range in FTMS mode was from m/z 100 to 1000. The data analyses were performed using XCalibur software v2.0.7 (Thermo Fisher Scientific, Hemel Hempstead, UK).

Chromatographic separations were performed on an Accela chromatograph (Thermo Scientific, Waltham, MA, USA) equipped with a quaternary pump, a photodiode array detector (PDA) and a thermostated autosampler. A Kinetex C18 column (100 Å, 2.6 µm, 150 × 2.1 mm) was used to perform chromatographic separations (Phenomenex Inc., Torrance, CA, USA). The elution was done in a gradient mode with water/0.1% formic acid (solvent A) and acetonitrile (solvent B) at a constant flow rate of 0.3 ml/min. The gradient composition of the mobile phase was as follows: 0 min, 10% B; 1 min, 10% B; 15 min, 30%

B; 22 min, 50% B; 28 min, 100% B; 34 min, 100% B, 36 min, 10% B. prior each analysis the column was equilibrated for 6 min. The total run time was 36 min.

4.2.4. Identification and Quantitative Analysis

The identification of compounds in fraction A was done either by comparison the retention times and Kovats indexes (RI) of the tested compounds with the same parameters of the corresponding pure standards or with mass spectra from the Golm Metabolome Database (http://csbdb.mpimp-golm.mpg.de/csbdb/gmd/gmd.html, 30 August 2021) and NIST'08 (National Institute of Standards and Technology, Gaithersburg, MD, USA) libraries.

The quantification of phenolics in fractions B and C was performed by the external standard method as previously described [55]. Fifteen phenolic compounds were confirmed by comparing their retention times, exact masses and fragmentation patterns with corresponding standards. The identification of the remaining compounds without available standards was based on accurate mass measurements of the $[M-H]^-$ ions and the fragmentation patterns, which was compared with the literature data.

4.3. Cell Culture

J774A.1 mouse macrophages were purchased from American Type Culture Collection (ATCC, Manassas, VA, USA). Cells were cultured in 75 cm^3 flasks at 37 °C in a humidified chamber (CO2CELL48, MMM Medcenter Einrichtungen GmbH, Planegg, Germany) with 95% air and 5% CO2 in Dulbecco's Modified Eagle Medium (DMEM, with 4.5 g/L of glucose and L-glutamine) (LONZA, Verviers, Belgium) supplemented with 10% heat-inactivated fetal bovine serum (FBS, Sigma-Aldrich, Taufkirchen, Germany) and 1% antibiotics (100 U/mL penicillin, 100 μg/mL streptomycin) (LONZA, Verviers, Belgium). Cells were sub-cultivated until they reach 80% confluence. Cell counts were prepared in quadruplicate by 0.4% trypan blue exclusion dye (Chemapol, Prague, Czech Republic) using a counting Burker chamber.

4.4. Study Design

The experimental model involved macrophage cells seeded in 6-well plates (2×10^5 cells/well) and allowed to adhere overnight. The study design included the following experimental groups: (1) cells treated only with LPS; (2) cells treated with SE FAE; (3) cells pre-treated with SE FAE and consequently challenged with LPS. For control groups, we used untreated cells (blank); salicylic acid-treated cells (positive, anti-inflammatory control) and cells pre-treated with salicylic acid and consequently challenged with LPS.

Cells were pre-treated with SE FAE with increasing concentrations of 2.5%, 5% and 10% *v*/*v* (0.25 mg DW/mL, 0.5 mg DW/mL, 1 mg DW/mL, respectively) or salicylic acid (100 μM) (Merck, Germany) dissolved in DMEM (with 4.5 g/L glucose, w/o phenol red and L-glutamine) supplemented with 10% heat-inactivated FBS, 100 U/mL penicillin/100 μg/mL streptomycin mixture and 2 mM L-glutamine. After 24 h cells were treated with 200 ng/mL LPS (*Escherichia coli* 026:B6, Sigma-Aldrich, Taufkirchen, Germany) or not, by the simple refreshing of culture media and incubated for additional 24 h. Following the last incubation period, the cells were lysed and total RNA or total protein were extracted and subjected to subsequent analyses. All treatments were performed in triplicate.

4.5. Gene Expression Analysis

4.5.1. RNA Extraction and cDNA Synthesis

Total RNA was extracted using TRI reagent (Ambion, Waltham, MA, USA) according to the manufacturers' requirement. RevertAid First Strand cDNA Synthesis kit (ThermoFisher Scientific, Waltham, MA, USA) was used to reversely transcribe 20 ng of total RNA using oligo $(dT)_{18}$ priming strategy. Following the manufacturers' protocol reaction conditions in final volumes of 10 μL were provided. cDNA synthesis was performed on GeneAmp PCR 7500 thermal cycler (Applied Biosystems, Waltham, MA, USA). After synthesis cDNA was diluted by adding of 30 μL nuclease-free distilled water to each sample and stored at −80 °C.

4.5.2. qPCR Analysis

Gene transcription levels were analyzed using the qPCR method and performed on an ABI PRISM 7500 (Applied Biosystems, Waltham, MA, USA). KAPA SYBR®® FAST qPCR Master Mix (2X) with low ROX (KAPA Biosystems, Cape Town, South Africa) was used. The amplification reaction's final volume was 5 μL in 96-well plates, with 0.39 μL of cDNA template. Final concentration of primers' was 300 nM. Reaction conditions were as follows: 95 °C/5 min; 40 cycles at 95 °C/15 sec and 60 °C/1 min. A dissociation step was added to the instrument's protocol to check for nonspecific amplification. As an internal control, the β-actin gene was used. Relative gene expression levels were calculated using the $2^{-\Delta\Delta Ct}$ method [126]. The used primer sequences (Sigma-Aldrich, Taufkirchen, Germany) for each gene analyzed are presented in Table 3. Expression levels of mRNA are presented as relative units (RU) compared to the untreated control group of cells, where the levels of mRNA expression were considered to be equal to 1. Analyses were performed in triplicate and included all cell groups in one plate while testing the expression of each gene.

Table 3. Oligonucleotide sequences of all used qPCR primer sets.

Gene Name	Forward Primer (5′–3′)	Reverse Primer (5′–3′)
Actb (β-actin)	ACGGCCAGGTCATCACTATTG	CAAGAAGGAAGGCTGGAAAAG
Ptgs2 (COX2)	TGAGCAACTATTCCAAACCAGC	GCACGTAGTCTTCGATCACTATC
iNOS	GGCAGCCTGTGAGACCTTTG	GCATTGGAAGTGAAGCGTTTC
TNFα	CCCTCACACTCAGATCATCTTCT	GCTACGACGTGGGCTACAG
IL-6	GAGTTGTGCAATGGCAATTCTG	GCAAGTGCATCATCGTTGTTCAT
IL-1β	TTCAGGCAGGCAGTATCACTC	CCACGGGAAAGACACAGGTAG
Ccl2 (MCP-1)	AGGTGTCCCAAAGAAGCTGTA	ATGTCTGGACCCATTCCTTCT
Sirt-1	TGATTGGCACCGATCCTCG	CCACAGCGTCATATCATCCAG
IL-1ra	GCTCATTGCTGGGTACTTACAA	CCAGACTTGGCACAAGACAGG
Icam1	GACCCCAAGGAGATCACATTC	GAAGATCGAAAGTCCGGA
Noxo1	AGAGGAGCCCTTATCCCAACC	TGTCCAGAATTTCTTGAGCCTTG
Fabp4 (aP2)	AGTGAAAACTTCGATGATTACATGAA	GCCTGCCACTTTCCTTGTG

4.6. Protein Expression Analysis

4.6.1. Protein Extraction and Quantification

Total protein was extracted using Pierce™ IP Lysis Buffer (25 mM Tris-HCl pH 7.4, 150 mM NaCl, 1 mM EDTA, 1% NP-40 and 5% glycerol) (Thermo Fisher Scientific, Waltham, MA, USA) with freshly added Halt™ Protease and Phosphatase Inhibitor Cocktail (Thermo Fisher Scientific, Waltham, MA, USA), according to the manufacturers' requirement. Prior to our analyses, total protein concentration was measured using a Bradford reagent (Protein assay dye concentrate, Bio-Rad Laboratories; Hercules, CA, USA) and calculated against a standard curve of standard bovine serum albumin (BSA) (Thermo Fisher Scientific, Waltham, MA, USA) dilutions.

4.6.2. Western Blotting

Protein lysates were subjected to SDS-PAGE, electrotransferred to a polyvinylidene difluoride membranes (PVDF; Merck Millipore; Billerica, MA, USA) and subsequently incubated with the following antibodies: ATF6α (90 kDa) (1:1000, sc-166659), CHOP (GADD153) (26 kDa) (1:1000, sc-7351), peIF2α (Ser52) (36 kDa) (1:1000, sc-12412), iNOS (1300 kDa) (1:1000, sc-7271) (Santa Cruz Biotechnology; Santa Cruz, CA, USA) and β-actin (42 kDa) (1:5000; A5316) (Sigma; St. Louis, MO, USA) after incubating the membranes with 3% BSA (β-actin, ATF6α), 5% BSA (peIF2α) or 5% skim milk (CHOP, iNOS) blocking buffer. Specific antigen–antibody bindings were detected using horseradish-peroxidase conjugated secondary antibodies (Dako Denmark; Glostrup, Denmark) and an enhanced chemiluminescence detection method, according to the manufacturer's instructions (Pierce ECL Western Blotting Substrate; Thermo Scientific, Waltham, MA, USA) as described previously [127,128]. Autoradiographic films (Fujifilm; Tokyo, Japan) were scanned and the band's signal was quantified by densitometry using ImageJ-1.53 software (National Institutes of Health, Bethesda, MD, USA). Values were expressed relative to β-actin.

4.7. Statistical Analysis

GraphPad Prism v7.0 software (GraphPad Software, Inc.; La Jolla, CA, USA) was used to perform the statistical analyses (Student's *t*-tests, Spearman correlation, 95% CI). The values of $p < 0.05$ were considered as significant. Data were presented as mean ± SD (concentration of phytochemical) or ±SEM (mRNA and protein expression levels). All analyses and treatments were performed in triplicates.

5. Conclusions

The SE FAE is confirmed to be rich in phytochemicals, predominantly hydroxycinnamic acids, anthocyanins, proanthocyanidins and resveratrol, with strong antioxidant-, anti-inflammatory- and ER stress-reducing potential, as well as in AAs including essential ones, organic acids, alcohols and saturated and unsaturated fatty acids and esters, some of them reported for first time in SE fruits.

Considering the results, we may conclude that SE FAE, applied alone, possesses immunostimulating potential, without promoting any additional stress, such as ER stress. It may reduce the ER stress-related expression of the CHOP protein, independently from its immunostimulating potential. The SE fruit extract exerted significant antioxidant and anti-inflammatory action, decreasing the LPS-stimulated transcription of oxidative stress, inflammation, atherosclerosis and insulin resistance-related cytokines, chemokines and enzymes, as well as the translation of iNOS. The herb extract possesses significant ER stress-reducing potential, by suppressing the LPS-stimulated synthesis of peIF2α, ATF6α and CHOP proteins. Taken together, these results reveal a new possible mechanism explaining the anti-inflammatory potential of SE fruits, by targeting ER stress and related biomarkers.

These findings are in concordance with the traditional usage of SE fruits and its FAE as potential natural immunomodulation preparation, beneficial in the prevention or treatment of oxidative stress- and inflammation-related conditions.

Supplementary Materials: The following are available online at https://www.mdpi.com/article/10.3390/plants10112446/s1, Figure S1: Representative chromatogram of analyzed polar compounds (fraction A) by GC-MS technique, Figure S2: Representative LC-PDA-ESI-MS/MS chromatogram of Sambucus ebulus L. fruit anthocyanins (1-Cyanidin-3-O-Galactoside, 2-Cyanidin-3-O-Glucoside, 3-Cyanidin-3-O-Arabinoside, 4-Cyanidin-3-O-Xyloside), Figure S3: Representative LC-PDA-ESI-MS/MS chromatogram of Sambucus ebulus L. fruit proanthocyanidin monomers, Figure S4: Representative LC-PDA-ESI-MS/MS chromatogram of Sambucus ebulus L. fruit proanthocyanidin di-and trimers, Figure S5: Representative LC-PDA-ESI-MS/MS chromatogram of Sambucus ebulus L. fruit stilbenes, Figure S6: Representative LC-PDA-ESI-MS/MS chromatogram of Sambucus ebulus L. fruit hydroxycinnamic acids (1–3-O-Caffeoylquinic acid, 2–Caffeic acid-O-galactoside, 3–Caffeic acid-O-glucoside, 4–5-O-Caffeoylquinic acid, 5–p-Coumaric acid-O-glucoside, 6–3-O-p-Coumaroylquinic acid, 7–Feruloylquinic acid; 8–4 -O-p-Coumaroylquinic acid; 9–Ferulic acid-O-galactoside; 10–Ferulic acid-O-glucoside), Figure S7: Representative LC-PDA-ESI-MS/MS chromatogram of Sambucus ebulus L. fruit flavonols (1-Kaempferol-3-O-arabinoside, 2-Kaempferol-3-O-xyloside), Figure S8: Representative LC-PDA-ESI-MS/MS chromatogram of Sambucus ebulus L. fruit flavonols (1-Quercetin-3-O-galactoside, 2-Quercetin-3-O-glucoside), Figure S9: Representative LC-PDA-ESI-MS/MS chromatogram of Sambucus ebulus L. fruit flavonols (1-Kaempferol-3-O-rhamnosyl-galactoside, 2-Kaempferol-3-O-rhamnosyl-glucoside), Figure S10: Representative LC-PDA-ESI-MS/MS chromatogram of Sambucus ebulus L. fruit flavonols (1-Quercetin-3-O-rhamnosyl-galactoside, 2-Quercetin-3-O-rhamnosyl-glucoside), Figure S11: Representative LC-PDA-ESI-MS/MS chromatogram of Sambucus ebulus L. fruit flavonols (1-Kaempferol-3-O-galactoside, 2-Kaempferol-3-O-glucoside), Figure S12: Original Western blot gels presenting the changes in protein levels of iNOS, peIF2α, ATF6α and CHOP in J774A.1 mouse macrophages pre-treated with increasing concentrations (2.5%, 5%, 10% v/v) of SE FAE or with SA for 24 h and subsequently stimulated or not with LPS, Table S1: Relative Kovat's retention index (RI) of analyzed polar compounds (fraction A) presented in Table 1, using GC-MS technique, Table S2: Precursor ion and fragment ion mass-to-charge ratios (m/z) of the analyzed polyphenols using the LC-PDA-ESI-MS/MS technique.

Author Contributions: Writing—original draft, methodology, formal analysis, investigation, conceptualization, project administration and funding acquisition O.T.; writing—methodology, resources and formal analysis, I.D. and I.B.; writing—review and editing Y.K.-K.; writing—review and editing, resources, supervision R.N.; writing—review and editing, resources, supervision, conceptualization B.G. and D.I. All authors have read and agreed to the published version of the manuscript.

Funding: This research was funded by Ministry of Education and Science of Bulgaria, grant number 94-14863/01.07.2013, project "Science and business", BG051PO001-3.3-05/0001. The APC was funded by Medical University-Varna, Bulgaria.

Institutional Review Board Statement: Not applicable.

Informed Consent Statement: Not applicable.

Data Availability Statement: The data presented in this study are available in the article.

Acknowledgments: This study was supported by Medical University—Varna, Bulgaria. We would like to acknowledge the support of Carlos Dieguez, and Ruben Nogueiras laboratory team members from CIMUS, for the hospitality and support while maintaining the western blot analyzes.

Conflicts of Interest: The authors declare no conflict of interest.

References

1. Kültür, Ş. Medicinal plants used in Kırklareli Province (Turkey). *J. Ethnopharmacol.* **2007**, *111*, 341–364. [CrossRef] [PubMed]
2. Tasinov, O.; Kiselova-Kaneva, Y.; Ivanova, D. *Sambucus ebulus* from traditional medicine to recent studies. *Scr. Sci. Med.* **2013**, *45*, 36–43. [CrossRef]
3. Ebrahimzadeh, M.A.; Pourmorad, F.; Bekhradnia, A.R. Iron chelating activity, phenol and flavonoid content of some medicinal plants from Iran. *Afr. J. Biotechnol.* **2008**, *7*, 3188–3192. [CrossRef]
4. Šarić-Kundalić, B.; Dobeš, C.; Klatte-Asselmeyer, V.; Saukel, J. Ethnobotanical study on medicinal use of wild and cultivated plants in middle, south and west Bosnia and Herzegovina. *J. Ethnopharmacol.* **2010**, *131*. [CrossRef] [PubMed]
5. Chirigiu, L.; Chirigiu, R.G.; Tircomnicu, V.; Bubulica, M.V. GC-MS analysis of chemical composition of *Sambucus ebulus* leaves. *Chem. Nat. Compd.* **2011**, *47*, 126–127. [CrossRef]
6. Marc, E.B.; Nelly, A.; Annick, D.D.; Frederic, D. Plants used as remedies antirheumatic and antineuralgic in the traditional medicine of Lebanon. *J. Ethnopharmacol.* **2008**, *120*, 315–334. [CrossRef]
7. Tasinov, O.; Kiselova-Kaneva, Y.; Ivanova, D. Antioxidant activity, total polyphenol content and anthocyanins content of *Sambucus ebulus* L. aqueous and aqueous–ethanolic extracts depend on the type and concentration of extragent. *Sci. Technol.* **2012**, *II*, 37–41.
8. Ebrahimzadeh, M.A.; Nabavi, S.F.; Nabavi, S.M.; Pourmorad, F. Nitric oxide radical scavenging potential of some Elburz medicinal plants. *Afr. J. Biotechnol.* **2010**, *9*, 5212–5217. [CrossRef]
9. Tasinov, O.B.; Kiselova-Kaneva, Y.D.; Nazifova-Tasinova, N.F.; Todorova, M.N.; Trendafilova, A.B.; Ivanova, D.G. Chemical composition and cytoprotective and anti-inflammatory potential of *Sambucus ebulus* fruit ethyl acetate fraction. *Bulg. Chem. Commun.* **2020**, *52*, 100–106.
10. Pribela, A.; Durcanska, J.; Piry, J.; Karovicova, J. Volatile substances of dwarf elder (*Sambucus ebulus*) fruits. *Biol. Ser. C* **1992**, *47*, 225–230.
11. Bubulica, M.V.; Chirigiu, L.; Popescu, M.; Simionescu, A.; Anoaica, G.; Popescu, A. Analysis of sterol compounds from *Sambucus ebulus*. *Chem. Nat. Compd.* **2012**, *48*, 520–521. [CrossRef]
12. Zahmanov, G.; Alipieva, K.; Denev, P.; Todorov, D.; Hinkov, A.; Shishkov, S.; Simova, S.; Georgiev, M.I. Flavonoid glycosides profiling in dwarf elder fruits (*Sambucus ebulus* L.) and evaluation of their antioxidant and anti-herpes simplex activities. *Ind. Crops Prod.* **2015**, *63*, 58–64. [CrossRef]
13. Zahmanov, G.; Alipieva, K.; Simova, S.; Georgiev, M.I. Metabolic differentiations of dwarf elder by NMR-based metabolomics. *Phytochem. Lett.* **2015**, *11*, 404–409. [CrossRef]
14. Mikulic-Petkovsek, M.; Schmitzer, V.; Slatnar, A.; Todorovic, B.; Veberic, R.; Stampar, F.; Ivancic, A. Investigation of anthocyanin profile of four elderberry species and interspecific hybrids. *J. Agric. Food Chem.* **2014**, *62*, 5573–5580. [CrossRef]
15. Mikulic-Petkovsek, M.; Ivancic, A.; Todorovic, B.; Veberic, R.; Stampar, F. Fruit Phenolic Composition of Different Elderberry Species and Hybrids. *J. Food Sci.* **2015**, *80*, C2180–C2190. [CrossRef]

16. Vankova, D.V.; Todorova, M.N.; Kisselova-Kaneva, Y.D.; Galunska, B.T. Development of new and robust LC-MS method for simultaneous quantification of polyphenols from *Sambucus ebulus* fruits. *J. Liq. Chromatogr. Relat. Technol.* **2019**, *42*, 408–416. [CrossRef]
17. Tasinov, O.; Kiselova-Kaneva, Y.; Ivanova, D. *Sambucus ebulus* L. fruit aqueous infusion modulates GCL and GPX4 gene expression. *Bulg. J. Agric. Sci.* **2013**, *19*, 143–146.
18. Tasinov, O.B.; Kiselova-Kaneva, Y.D.; Ivanova, D.G. Effects of dwarf elder fruit infusion on nuclear factor kappa B and glutathione metabolism-related genes transcription in a model of lipopolysaccharides challenged macrophages. *Bulg. Chem. Commun.* **2020**, *52*, 68–74.
19. Todorova, M.; Tasinov, O.; Pasheva, M.; Vankova, D.; Ivanova, D.; Galunska, B.; Kiselova-Kaneva, Y. Cytoprotective activity of *Sambucus ebulus* fruit extracts in conditions of oxidative tert-buthyl-hydroperoxyde induced cell toxicity. *Bulg. Chem. Commun.* **2019**, *51*, 125–130.
20. Ebrahimzadeh, M.A.; Mahmoudi, M.; Karami, M.; Saeedi, S.; Ahmadi, A.H.; Salimi, E. Separation of active and toxic portions in *Sambucus ebulus*. *Pak. J. Biol. Sci.* **2007**, *10*, 4171–4173. [CrossRef]
21. Fathi, H.; Ebrahimzadeh, M.A.; Ziar, A.; Mohammadi, H. Oxidative damage induced by retching; antiemetic and neuroprotective role of *Sambucus ebulus* L. *Cell Biol. Toxicol.* **2015**, *31*, 231–239. [CrossRef] [PubMed]
22. Ivanova, D.; Tasinov, O.; Kiselova-Kaneva, Y. Improved lipid profile and increased serum antioxidant capacity in healthy volunteers after *Sambucus ebulus* L. fruit infusion consumption. *Int. J. Food Sci. Nutr.* **2014**, *65*, 740–744. [CrossRef]
23. Ivanova, D.; Nashar, M.; Kiselova-Kaneva, Y.; Tasinov, O.; Vankova, D.; Nazifova-Tasinova, N. The impact of human intervention studies on the evaluation of medicinal plant antioxidant and anti-inflammatory activities. In *Human Health and Nutrition: New Research*; Nova Science Publishers, Inc.: Hauppauge, NY, USA, 2015; pp. 145–166, ISBN 9781634828536.
24. Ebrahimzadeh, M.A.; Rafati, M.R.; Damchi, M.; Golpur, M.; Fathiazad, F. Treatment of paederus dermatitis with *Sambucus ebulus* lotion. *Iran. J. Pharm. Res.* **2014**, *13*, 1065–1072. [CrossRef] [PubMed]
25. Olefsky, J.M.; Glass, C.K. Macrophages, Inflammation, and Insulin Resistance. *Annu. Rev. Physiol.* **2010**, *72*, 219–246. [CrossRef] [PubMed]
26. Hotamisligil, G.S.; Arner, P.; Caro, J.F.; Atkinson, R.L.; Spiegelman, B.M. Increased adipose tissue expression of tumor necrosis factor-alpha in human obesity and insulin resistance. *J. Clin. Investig.* **1995**, *95*, 2409–2415. [CrossRef]
27. Hirosumi, J.; Tuncman, G.; Chang, L.; Görgün, C.Z.; Uysal, K.T.; Maeda, K.; Karin, M.; Hotamisligil, G.S. A central, role for JNK in obesity and insulin resistance. *Nature* **2002**, *420*, 333–336. [CrossRef]
28. Bandyopadhyay, G.K.; Yu, J.G.; Ofrecio, J.; Olefsky, J.M. Increased p85/55/50 expression and decreased phosphotidylinositol 3-kinase activity in insulin-resistant human skeletal muscle. *Diabetes* **2005**, *54*, 2351–2359. [CrossRef]
29. Nguyen, M.T.A.; Favelyukis, S.; Nguyen, A.K.; Reichart, D.; Scott, P.A.; Jenn, A.; Liu-Bryan, R.; Glass, C.K.; Neels, J.G.; Olefsky, J.M. A subpopulation of macrophages infiltrates hypertrophic adipose tissue and is activated by free fatty acids via toll-like receptors 2 and 4 and JNK-dependent pathways. *J. Biol. Chem.* **2007**, *282*, 35279–35292. [CrossRef]
30. Boden, G. Insulin Resistance and Inflammation: Links between Obesity and Cardiovascular Disease. In *Glucose Intake and Utilization in Pre-Diabetes and Diabetes: Implications for Cardiovascular Disease*; Academic Press: Amsterdam, The Netherlands, 2015; pp. 95–101, ISBN 9780128005798.
31. Özcan, U.; Cao, Q.; Yilmaz, E.; Lee, A.H.; Iwakoshi, N.N.; Özdelen, E.; Tuncman, G.; Görgün, C.; Glimcher, L.H.; Hotamisligil, G.S. Endoplasmic reticulum stress links obesity, insulin action, and type 2 diabetes. *Science* **2004**, *306*, 457–461. [CrossRef]
32. Zhang, K.; Kaufman, R.J. From endoplasmic-reticulum stress to the inflammatory response. *Nature* **2008**, *454*, 455–462. [CrossRef]
33. Anthony, T.G.; Wek, R.C. TXNIP switches tracks toward a terminal UPR. *Cell Metab.* **2012**, *16*, 135–137. [CrossRef] [PubMed]
34. Jiang, M.; Wang, H.; Liu, Z.; Lin, L.; Wang, L.; Xie, M.; Li, D.; Zhang, J.; Zhang, R. Endoplasmic reticulum stress-dependent activation of iNOS/NO-NF-κB signaling and NLRP3 inflammasome contributes to endothelial inflammation and apoptosis associated with microgravity. *FASEB J.* **2020**, *34*, 10835–10849. [CrossRef]
35. Sun, Z.M.; Guan, P.; Luo, L.F.; Qin, L.Y.; Wang, N.; Zhao, Y.S.; Ji, E.S. Resveratrol protects against CIH-induced myocardial injury by targeting Nrf2 and blocking NLRP3 inflammasome activation. *Life Sci.* **2020**, *245*, 117362. [CrossRef]
36. Li, Y.; Xu, S.; Giles, A.; Nakamura, K.; Lee, J.W.; Hou, X.; Donmez, G.; Li, J.; Luo, Z.; Walsh, K.; et al. Hepatic overexpression of SIRT1 in mice attenuates endoplasmic reticulum stress and insulin resistance in the liver. *FASEB J.* **2011**, *25*, 1664–1679. [CrossRef]
37. Karthikeyan, B.; Harini, L.; Krishnakumar, V.; Kannan, V.R.; Sundar, K.; Kathiresan, T. Insights on the involvement of (−)-epigallocatechin gallate in ER stress-mediated apoptosis in age-related macular degeneration. *Apoptosis* **2017**, *22*, 72–85. [CrossRef]
38. Fu, K.; Chen, L.; Miao, L.; Guo, Y.; Zhang, W.; Bai, Y. Grape Seed Proanthocyanidins Protect N2a Cells against Ischemic Injury via Endoplasmic Reticulum Stress and Mitochondrial-associated Pathways. *CNS Neurol. Disord.—Drug Targets* **2019**, *18*, 334–341. [CrossRef] [PubMed]
39. NCI Thesaurus. Available online: https://ncit.nci.nih.gov/ncitbrowser/ConceptReport.jsp?dictionary=NCI_Thesaurus&ns=NCI_Thesaurus&code=C82933 (accessed on 14 August 2021).
40. Veberic, R.; Jakopic, J.; Stampar, F.; Schmitzer, V. European elderberry (*Sambucus nigra* L.) rich in sugars, organic acids, anthocyanins and selected polyphenols. *Food Chem.* **2009**, *114*, 511–515. [CrossRef]
41. Washüttl, J.; Riederer, P.; Bancher, E. A qualitative and quantitative study of sugar-alcohols in several foods. *J. Food Sci.* **1973**, *38*, 1262–1263. [CrossRef]

42. Chakravarti, B.; Maurya, R.; Siddiqui, J.A.; Kumar Bid, H.; Rajendran, S.M.; Yadav, P.P.; Konwar, R. In vitro anti-breast cancer activity of ethanolic extract of *Wrightia tomentosa*: Role of pro-apoptotic effects of oleanolic acid and urosolic acid. *J. Ethnopharmacol.* **2012**, *142*, 72–79. [CrossRef]
43. Juan, M.E.; Planas, J.M.; Ruiz-Gutierrez, V.; Daniel, H.; Wenzel, U. Antiproliferative and apoptosis-inducing effects of maslinic and oleanolic acids, two pentacyclic triterpenes from olives, on HT-29 colon cancer cells. *Br. J. Nutr.* **2008**, *100*, 36–43. [CrossRef]
44. Wang, Y.-J.; Lu, J.; Wu, D.-M.; Zheng, Z.H.; Zheng, Y.-L.; Wang, X.-H.; Ruan, J.; Sun, X.; Shan, Q.; Zhang, Z.-F. Ursolic acid attenuates lipopolysaccharide-induced cognitive deficits in mouse brain through suppressing p38/NF-κB mediated inflammatory pathways. *Neurobiol. Learn. Mem.* **2011**, *96*, 156–165. [CrossRef]
45. Kaewthawee, N.; Brimson, S. The effects of ursolic acid on cytokine production via the MAPK pathways in leukemic T-cells. *EXCLI J.* **2013**, *12*, 102–114. [CrossRef] [PubMed]
46. Lu, J.; Wu, D.-M.; Zheng, Y.-L.; Hu, B.; Cheng, W.; Zhang, Z.-F.; Shan, Q. Ursolic acid improves high fat diet-induced cognitive impairments by blocking endoplasmic reticulum stress and IκB kinase β/nuclear factor-κB-mediated inflammatory pathways in mice. *Brain. Behav. Immun.* **2011**, *25*, 1658–1667. [CrossRef]
47. Zhang, W.; Hong, D.; Zhou, Y.; Zhang, Y.; Shen, Q.; Li, J.Y.; Hu, L.H.; Li, J. Ursolic acid and its derivative inhibit protein tyrosine phosphatase 1B, enhancing insulin receptor phosphorylation and stimulating glucose uptake. *Biochim. Biophys. Acta Gen. Subj.* **2006**, *1760*, 1505–1512. [CrossRef]
48. Jayaprakasam, B.; Olson, L.K.; Schutzki, R.E.; Tai, M.H.; Nair, M.G. Amelioration of obesity and glucose intolerance in high-fat-fed C57BL/6 mice by anthocyanins and ursolic acid in cornelian cherry (*Cornus mas*). *J. Agric. Food Chem.* **2006**, *54*, 243–248. [CrossRef]
49. Best, M.M.; Duncan, C.H.; van Loon, E.J.; Wathen, J.D. Lowering of serum cholesterol by the administration of a plant sterol. *Circulation* **1954**, *10*, 201–206. [CrossRef] [PubMed]
50. Kassis, A.N.; Vanstone, C.A.; AbuMweis, S.S.; Jones, P.J.H. Efficacy of plant sterols is not influenced by dietary cholesterol intake in hypercholesterolemic individuals. *Metabolism* **2008**, *57*, 339–346. [CrossRef] [PubMed]
51. Chen, Z.Y.; Ma, K.Y.; Liang, Y.; Peng, C.; Zuo, Y. Role and classification of cholesterol-lowering functional foods. *J. Funct. Foods* **2011**, *3*, 61–69. [CrossRef]
52. Jesch, E.D.; Seo, J.M.; Carr, T.P.; Lee, J.Y. Sitosterol reduces messenger RNA and protein expression levels of Niemann-Pick C1-like 1 in FHs 74 Int cells. *Nutr. Res.* **2009**, *29*, 859–866. [CrossRef]
53. Jones, P.J.H.; MacDougall, D.E.; Ntanios, F.; Vanstone, C.A. Dietary phytosterols as cholesterol-lowering agents in humans. *Can. J. Physiol. Pharmacol.* **1997**, *75*, 217–227. [CrossRef]
54. Liang, Z.; Liang, H.; Guo, Y.; Yang, D. Cyanidin 3-o-galactoside: A natural compound with multiple health benefits. *Int. J. Mol. Sci.* **2021**, *22*, 2261. [CrossRef]
55. Kiselova-Kaneva, Y.; Galunska, B.; Nikolova, M.; Dincheva, I.; Badjakov, I. High resolution LC-MS/MS characterization of polyphenolic composition and evaluation of antioxidant activity of *Sambucus ebulus* fruit tea traditionally used in Bulgaria as a functional food. *Food Chem.* **2021**, *367*, 130759. [CrossRef]
56. Gu, L.; Kelm, M.A.; Hammerstone, J.F.; Beecher, G.; Holden, J.; Haytowitz, D.; Gebhardt, S.; Prior, R.L. Concentrations of Proanthocyanidins in Common Foods and Estimations of Normal Consumption. *J. Nutr.* **2004**, *134*, 613–617. [CrossRef]
57. Rivière, C.; Pawlus, A.D.; Mérillon, J.M. Natural stilbenoids: Distribution in the plant kingdom and chemotaxonomic interest in Vitaceae. *Nat. Prod. Rep.* **2012**, *29*, 1317–1333. [CrossRef] [PubMed]
58. Timmers, S.; Konings, E.; Bilet, L.; Houtkooper, R.H.; Van De Weijer, T.; Goossens, G.H.; Hoeks, J.; Van Der Krieken, S.; Ryu, D.; Kersten, S.; et al. Calorie restriction-like effects of 30 days of resveratrol supplementation on energy metabolism and metabolic profile in obese humans. *Cell Metab.* **2011**, *14*, 612–622. [CrossRef] [PubMed]
59. Espinoza, J.L.; Trung, L.Q.; Inaoka, P.T.; Yamada, K.; An, D.T.; Mizuno, S.; Nakao, S.; Takami, A. The Repeated Administration of Resveratrol Has Measurable Effects on Circulating T-Cell Subsets in Humans. *Oxid. Med. Cell. Longev.* **2017**, *2017*, 6781872. [CrossRef]
60. Yiu, E.M.; Tai, G.; Peverill, R.E.; Lee, K.J.; Croft, K.D.; Mori, T.A.; Scheiber-Mojdehkar, B.; Sturm, B.; Praschberger, M.; Vogel, A.P.; et al. An open-label trial in Friedreich ataxia suggests clinical benefit with high-dose resveratrol, without effect on frataxin levels. *J. Neurol.* **2015**, *262*, 1344–1353. [CrossRef] [PubMed]
61. Crandall, J.P.; Oram, V.; Trandafirescu, G.; Reid, M.; Kishore, P.; Hawkins, M.; Cohen, H.W.; Barzilai, N. Pilot study of resveratrol in older adults with impaired glucose tolerance. *Journals Gerontol. Ser. A Biol. Sci. Med. Sci.* **2012**, *67*, 1307–1312. [CrossRef]
62. Most, J.; Timmers, S.; Warnke, I.; Jocken, J.W.; Van Boekschoten, M.; De Groot, P.; Bendik, I.; Schrauwen, P.; Goossens, G.H.; Blaak, E.E. Combined epigallocatechin-3-gallate and resveratrol supplementation for 12 wk increases mitochondrial capacity and fat oxidation, but not insulin sensitivity, in obese humans: A randomized controlled trial. *Am. J. Clin. Nutr.* **2016**, *104*, 215–227. [CrossRef]
63. Herrmann, K. Occurrence and content of hydroxycinnamic and hydroxybenzoic acid compounds in foods. *Crit. Rev. Food Sci. Nutr.* **1989**, *28*, 315–347. [CrossRef]
64. Weng, C.J.; Yen, G.C. Chemopreventive effects of dietary phytochemicals against cancer invasion and metastasis: Phenolic acids, monophenol, polyphenol, and their derivatives. *Cancer Treat. Rev.* **2012**, *38*, 76–87. [CrossRef] [PubMed]
65. Kim, E.O.; Min, K.J.; Kwon, T.K.; Um, B.H.; Moreau, R.A.; Choi, S.W. Anti-inflammatory activity of hydroxycinnamic acid derivatives isolated from corn bran in lipopolysaccharide-stimulated Raw 264.7 macrophages. *Food Chem. Toxicol.* **2012**, *50*, 1309–1316. [CrossRef]

66. Hamauzu, Y.; Yasui, H.; Inno, T.; Kume, C.; Omanyuda, M. Phenolic profile, antioxidant property, and anti-influenza viral activity of Chinese quince (*Pseudocydonia sinensis Schneid.*), quince (*Cydonia oblonga Mill.*), and apple (*Malus domestica Mill.*) fruits. *J. Agric. Food Chem.* **2005**, *53*, 928–934. [CrossRef] [PubMed]
67. Iwai, K.; Kishimoto, N.; Kakino, Y.; Mochida, K.; Fujita, T. In vitro antioxidative effects and tyrosinase inhibitory activities of seven hydroxycinnamoyl derivatives in green coffee beans. *J. Agric. Food Chem.* **2004**, *52*, 4893–4898. [CrossRef]
68. Liu, S.L.; Peng, B.J.; Zhong, Y.L.; Liu, Y.L.; Song, Z.; Wang, Z. Effect of 5-caffeoylquinic acid on the NF-κB signaling pathway, peroxisome proliferator-activated receptor gamma 2, and macrophage infiltration in high-fat diet-fed Sprague-Dawley rat adipose tissue. *Food Funct.* **2015**, *6*, 2779–2786. [CrossRef] [PubMed]
69. Zou, Y.; Lu, Y.; Wei, D. Antioxidant activity of a flavonoid-rich extract of *Hypericum perforatum* L. in vitro. *J. Agric. Food Chem.* **2004**, *52*, 5032–5039. [CrossRef]
70. Mansour, S.; Djebli, N.; Ozkan, E.E.; Mat, A. In vivo antiinflammatory activity and chemical composition of *Hypericum scabroides*. *Asian Pac. J. Trop. Med.* **2014**, *7*, S514–S520. [CrossRef]
71. Ahn, H.; Lee, G.S. Isorhamnetin and hyperoside derived from water dropwort inhibits inflammasome activation. *Phytomedicine* **2017**, *24*, 77–86. [CrossRef]
72. Li, L.; Zhang, X.H.; Liu, G.R.; Liu, C.; Dong, Y.M. Isoquercitrin suppresses the expression of histamine and pro-inflammatory cytokines by inhibiting the activation of MAP Kinases and NF-κB in human KU812 cells. *Chin. J. Nat. Med.* **2016**, *14*, 407–412. [CrossRef]
73. Chua, L.S. A review on plant-based rutin extraction methods and its pharmacological activities. *J. Ethnopharmacol.* **2013**, *150*, 805–817. [CrossRef]
74. Wang, W.; Sun, C.; Mao, L.; Ma, P.; Liu, F.; Yang, J.; Gao, Y. The biological activities, chemical stability, metabolism and delivery systems of quercetin: A review. *Trends Food Sci. Technol.* **2016**, *56*, 21–38. [CrossRef]
75. Suganthy, N.; Devi, K.P.; Nabavi, S.F.; Braidy, N.; Nabavi, S.M. Bioactive effects of quercetin in the central nervous system: Focusing on the mechanisms of actions. *Biomed. Pharmacother.* **2016**, *84*, 892–908. [CrossRef] [PubMed]
76. Pellegatta, F.; Bertelli, A.A.E.; Staels, B.; Duhem, C.; Fulgenzi, A.; Ferrero, M.E. Different short- and long-term effects of resveratrol on nuclear factor-κB phosphorylation and nuclear appearance in human endothelial cells. *Am. J. Clin. Nutr.* **2003**, *77*, 1220–1228. [CrossRef]
77. Norata, G.D.; Marchesi, P.; Passamonti, S.; Pirillo, A.; Violi, F.; Catapano, A.L. Anti-inflammatory and anti-atherogenic effects of cathechin, caffeic acid and trans-resveratrol in apolipoprotein E deficient mice. *Atherosclerosis* **2007**, *191*, 265–271. [CrossRef]
78. Hou, D.X.; Yanagita, T.; Uto, T.; Masuzaki, S.; Fujii, M. Anthocyanidins inhibit cyclooxygenase-2 expression in LPS-evoked macrophages: Structure-activity relationship and molecular mechanisms involved. *Biochem. Pharmacol.* **2005**, *70*, 417–425. [CrossRef] [PubMed]
79. Pergola, C.; Rossi, A.; Dugo, P.; Cuzzocrea, S.; Sautebin, L. Inhibition of nitric oxide biosynthesis by anthocyanin fraction of blackberry extract. *Nitric Oxide—Biol. Chem.* **2006**, *15*, 30–39. [CrossRef] [PubMed]
80. Karlsen, A.; Retterstøl, L.; Laake, P.; Paur, I.; Bøhn, S.K.; Sandvik, L.; Blomhoff, R. Anthocyanins inhibit nuclear factor-kappaB activation in monocytes and reduce plasma concentrations of pro-inflammatory mediators in healthy adults. *J. Nutr.* **2007**, *137*, 1951–1954. [CrossRef]
81. Kim, J.; Kim, H.; Choi, H.; Jo, A.; Kang, H.; Yun, H.; Im, S.; Choi, C. Anti-inflammatory effects of a *stauntonia hexaphylla* fruit extract in lipopolysaccharide-activated RAW-264.7 macrophages and rats by carrageenan-induced hind paw swelling. *Nutrients* **2018**, *10*, 110. [CrossRef] [PubMed]
82. Schwaiger, S.; Zeller, I.; Pölzelbauer, P.; Frotschnig, S.; Laufer, G.; Messner, B.; Pieri, V.; Stuppner, H.; Bernhard, D. Identification and pharmacological characterization of the anti-inflammatory principal of the leaves of dwarf elder (*Sambucus ebulus* L.). *J. Ethnopharmacol.* **2011**, *133*, 704–709. [CrossRef]
83. van de Stolpe, A.; Caldenhoven, E.; Stade, B.G.; Koenderman, L.; Raaijmakers, J.A.; Johnson, J.P.; van der Saag, P.T. 12-O-tetradecanoylphorbol-13-acetate- and tumor necrosis factor alpha-mediated induction of intercellular adhesion molecule-1 is inhibited by dexamethasone. Functional analysis of the human intercellular adhesion molecular-1 promoter. *J. Biol. Chem.* **1994**, *269*, 6185–6192. [CrossRef]
84. Ueda, A.; Okuda, K.; Ohno, S.; Shirai, A.; Igarashi, T.; Matsunaga, K.; Fukushima, J.; Kawamoto, S.; Ishigatsubo, Y.; Okubo, T. NF-kappa B and Sp1 regulate transcription of the human monocyte chemoattractant protein-1 gene. *J. Immunol.* **1994**, *153*, 2052–2063. [CrossRef]
85. Hsieh, I.-N.; Chang, A.S.-Y.; Teng, C.-M.; Chen, C.-C.; Yang, C.-R. Aciculatin inhibits lipopolysaccharide-mediated inducible nitric oxide synthase and cyclooxygenase-2 expression via suppressing NF-κB and JNK/p38 MAPK activation pathways. *J. Biomed. Sci.* **2011**, *18*, 28. [CrossRef]
86. Tan, K.S.; Qian, L.; Rosado, R.; Flood, P.M.; Cooper, L.F. The role of titanium surface topography on J774A.1 macrophage inflammatory cytokines and nitric oxide production. *Biomaterials* **2006**, *27*, 5170–5177. [CrossRef] [PubMed]
87. Kim, M.S.; Park, S.B.; Suk, K.; Kim, I.K.; Kim, S.Y.; Kim, J.A.; Lee, S.H.; Kim, S.H. Gallotannin isolated from Euphorbia species, 1,2,6-tri-O-galloyl-β-D- allose, decreases nitric oxide production through inhibition of nuclear factor- κB and downstream inducible nitric oxide synthase expression in macrophages. *Biol. Pharm. Bull.* **2009**, *32*, 1053–1056. [CrossRef] [PubMed]

88. Maraslioglu, M.; Oppermann, E.; Blattner, C.; Weber, R.; Henrich, D.; Jobin, C.; Schleucher, E.; Marzi, I.; Lehnert, M. Chronic ethanol feeding modulates inflammatory mediators, activation of nuclear factor- B, and responsiveness to endotoxin in murine kupffer cells and circulating leukocytes. *Mediat. Inflamm.* **2014**, *2014*, 808695. [CrossRef] [PubMed]
89. Zhu, Y.; Zhu, M.; Lance, P. INOS signaling interacts with COX-2 pathway in colonic fibroblasts. *Exp. Cell Res.* **2012**, *318*, 2116–2127. [CrossRef]
90. Barak, V.; Halperin, T.; Kalickman, I. The effect of Sambucol, a black elderberry-based, natural product, on the production of human cytokines: I. Inflammatory cytokines. *Eur. Cytokine Netw.* **2001**, *12*, 290–296.
91. Drummond, G.R.; Selemidis, S.; Griendling, K.K.; Sobey, C.G. Combating oxidative stress in vascular disease: NADPH oxidases as therapeutic targets. *Nat. Rev. Drug Discov.* **2011**, *10*, 453–471. [CrossRef]
92. Blough, N.V.; Zafiriou, O.C. Reaction of Superoxide with Nitric Oxide to Form Peroxonitrite in Alkaline Aqueous Solution. *Inorg. Chem.* **1985**, *24*, 3502–3504. [CrossRef]
93. Gryglewski, R.J.; Palmer, R.M.J.; Moncada, S. Superoxide anion is involved in the breakdown of endothelium-derived vascular relaxing factor. *Nature* **1986**, *320*, 454–456. [CrossRef]
94. Leto, T.L.; Geiszt, M. Role of Nox family NADPH oxidases in host defense. *Antioxidants Redox Signal.* **2006**, *8*, 1549–1561. [CrossRef]
95. Guzik, T.J.; Harrison, D.G. Vascular NADPH oxidases as drug targets for novel antioxidant strategies. *Drug Discov. Today* **2006**, *11*, 524–533. [CrossRef] [PubMed]
96. Abo, A.; Pick, E.; Hall, A.; Totty, N.; Teahan, C.G.; Segal, A.W. Activation of the NADPH oxidase involves the small GTP-binding protein p21rac1. *Nature* **1991**, *353*, 668–670. [CrossRef] [PubMed]
97. Potenza, M.A.; Marasciulo, F.L.; Tarquinio, M.; Tiravanti, E.; Colantuono, G.; Federici, A.; Kim, J.A.; Quon, M.J.; Montagnani, M. EGCG, a green tea polyphenol, improves endothelial function and insulin sensitivity, reduces blood pressure, and protects against myocardial I/R injury in SHR. *Am. J. Physiol.—Endocrinol. Metab.* **2007**, *292*, E1378–E1387. [CrossRef] [PubMed]
98. Galindo, P.; González-Manzano, S.; Zarzuelo, M.J.; Gómez-Guzmán, M.; Quintela, A.M.; González-Paramás, A.; Santos-Buelga, C.; Pérez-Vizcaíno, F.; Duarte, J.; Jiménez, R. Different cardiovascular protective effects of quercetin administered orally or intraperitoneally in spontaneously hypertensive rats. *Food Funct.* **2012**, *3*, 643–650. [CrossRef]
99. Sánchez, M.; Galisteo, M.; Vera, R.; Villar, I.C.; Zarzuelo, A.; Tamargo, J.; Pérez-Vizcaíno, F.; Duarte, J. Quercetin downregulates NADPH oxidase, increases eNOS activity and prevents endothelial dysfunction in spontaneously hypertensive rats. *J. Hypertens.* **2006**, *24*, 75–84. [CrossRef]
100. Sanchez, M.; Lodi, F.; Vera, R.; Villar, I.C.; Cogolludo, A.; Jimenez, R.; Moreno, L.; Romero, M.; Tamargo, J.; Perez-Vizcaino, F.; et al. Quercetin and isorhamnetin prevent endothelial dysfunction, superoxide production, and overexpression of p47phox induced by angiotensin II in rat aorta. *J. Nutr.* **2007**, *137*, 910–915. [CrossRef]
101. Ihm, S.H.; Lee, J.O.; Kim, S.J.; Seung, K.B.; Schini-Kerth, V.B.; Chang, K.; Oak, M.H. Catechin prevents endothelial dysfunction in the prediabetic stage of OLETF rats by reducing vascular NADPH oxidase activity and expression. *Atherosclerosis* **2009**, *206*, 47–53. [CrossRef]
102. Schilder, Y.D.C.; Heiss, E.H.; Schachner, D.; Ziegler, J.; Reznicek, G.; Sorescu, D.; Dirsch, V.M. NADPH oxidases 1 and 4 mediate cellular senescence induced by resveratrol in human endothelial cells. *Free Radic. Biol. Med.* **2009**, *46*, 1598–1606. [CrossRef] [PubMed]
103. Gu, Y.; Xu, Y.C.; Wu, R.F.; Nwariaku, F.E.; Souza, R.F.; Flores, S.C.; Terada, L.S. p47phox participates in activation of RelA in endothelial cells. *J. Biol. Chem.* **2003**, *278*, 17210–17217. [CrossRef]
104. Makowski, L.; Boord, J.B.; Maeda, K.; Babaev, V.R.; Uysal, K.T.; Morgan, M.A.; Parker, R.A.; Suttles, J.; Fazio, S.; Hotamisligil, G.S.; et al. Lack of macrophage fatty-acid-binding protein aP2 protects mice deficient in apolipoprotein E against atherosclerosis. *Nat. Med.* **2001**, *7*, 699–705. [CrossRef] [PubMed]
105. Kazemi, M.R.; McDonald, C.M.; Shigenaga, J.K.; Grunfeld, C.; Feingold, K.R. Adipocyte fatty acid-binding protein expression and lipid accumulation are increased during activation of murine macrophages by toll-like receptor agonists. *Arterioscler. Thromb. Vasc. Biol.* **2005**, *25*, 1220–1224. [CrossRef] [PubMed]
106. Furuhashi, M.; Fucho, R.; Görgün, C.Z.; Tuncman, G.; Cao, H.; Hotamisligil, G.S. Adipocyte/macrophage fatty acid–binding proteins contribute to metabolic deterioration through actions in both macrophages and adipocytes in mice. *J. Clin. Investig.* **2008**, *118*, 2640–2650. [CrossRef] [PubMed]
107. Furuhashi, M.; Tuncman, G.; Görgün, C.Z.; Makowski, L.; Atsumi, G.; Vaillancourt, E.; Kono, K.; Babaev, V.R.; Fazio, S.; Linton, M.F.; et al. Treatment of diabetes and atherosclerosis by inhibiting fatty-acid-binding protein aP2. *Nature* **2007**, *447*, 959–965. [CrossRef]
108. Milne, J.C.; Lambert, P.D.; Schenk, S.; Carney, D.P.; Smith, J.J.; Gagne, D.J.; Jin, L.; Boss, O.; Perni, R.B.; Vu, C.B.; et al. Small molecule activators of SIRT1 as therapeutics for the treatment of type 2 diabetes. *Nature* **2007**, *450*, 712–716. [CrossRef]
109. Csiszar, A.; Labinskyy, N.; Podlutsky, A.; Kaminski, P.M.; Wolin, M.S.; Zhang, C.; Mukhopadhyay, P.; Pacher, P.; Hu, F.; De Cabo, R.; et al. Vasoprotective effects of resveratrol and SIRT1: Attenuation of cigarette smoke-induced oxidative stress and proinflammatory phenotypic alterations. *Am. J. Physiol.—Heart Circ. Physiol.* **2008**, *294*, 2721–2735. [CrossRef]
110. Yoshizaki, T.; Milne, J.C.; Imamura, T.; Schenk, S.; Sonoda, N.; Babendure, J.L.; Lu, J.-C.; Smith, J.J.; Jirousek, M.R.; Olefsky, J.M. SIRT1 Exerts Anti-Inflammatory Effects and Improves Insulin Sensitivity in Adipocytes. *Mol. Cell. Biol.* **2009**, *29*, 1363–1374. [CrossRef]

111. Yu, C.; Chen, Y.; Cline, G.W.; Zhang, D.; Zong, H.; Wang, Y.; Bergeron, R.; Kim, J.K.; Cushman, S.W.; Cooney, G.J.; et al. Mechanism by which fatty acids inhibit insulin activation of insulin receptor substrate-1 (IRS-1)-associated phosphatidylinositol 3-kinase activity in muscle. *J. Biol. Chem.* **2002**, *277*, 50230–50236. [CrossRef]
112. Bettigole, S.E.; Glimcher, L.H. Endoplasmic reticulum stress in immunity. *Annu. Rev. Immunol.* **2015**, *33*, 107–138. [CrossRef]
113. Li, S.; Ye, L.; Yu, X.; Xu, B.; Li, K.; Zhu, X.; Liu, H.; Wu, X.; Kong, L. Hepatitis C virus NS4B induces unfolded protein response and endoplasmic reticulum overload response-dependent NF-κB activation. *Virology* **2009**, *391*, 257–264. [CrossRef]
114. Randjelović, P.; Veljković, S.; Stojiljković, N.; Sokolović, D.; Ilić, I.; Laketić, D.; Randjelović, D.; Randjelović, N. The beneficial biological properties of salicylic acid. *Acta Fac. Med. Naissensis* **2015**, *32*, 259–265. [CrossRef]
115. Yoshiuchi, K.; Kaneto, H.; Matsuoka, T.A.; Kohno, K.; Iwawaki, T.; Nakatani, Y.; Yamasaki, Y.; Hori, M.; Matsuhisa, M. Direct monitoring of in vivo ER stress during the development of insulin resistance with ER stress-activated indicator transgenic mice. *Biochem. Biophys. Res. Commun.* **2008**, *366*, 545–550. [CrossRef]
116. Boden, G.; Duan, X.; Homko, C.; Molina, E.J.; Song, W.; Perez, O.; Cheung, P.; Merali, S. Increase in endoplasmic reticulum stress-related proteins and genes in adipose tissue of obese, insulin-resistant individuals. *Diabetes* **2008**, *57*, 2438–2444. [CrossRef]
117. Sharma, N.K.; Das, S.K.; Mondal, A.K.; Hackney, O.G.; Chu, W.S.; Kern, P.A.; Rasouli, N.; Spencer, H.J.; Yao-Borengasser, A.; Elbein, S.C. Endoplasmic reticulum stress markers are associated with obesity in nondiabetic subjects. *J. Clin. Endocrinol. Metab.* **2008**, *93*, 4532–4541. [CrossRef] [PubMed]
118. Eizirik, D.L.; Cardozo, A.K.; Cnop, M. The role for endoplasmic reticulum stress in diabetes mellitus. *Endocr. Rev.* **2008**, *29*, 42–61. [CrossRef]
119. Lin, J.H.; Walter, P.; Yen, T.S.B. Endoplasmic reticulum stress in disease pathogenesis. *Annu. Rev. Pathol. Mech. Dis.* **2008**, *3*, 399–425. [CrossRef] [PubMed]
120. Ron, D.; Hubbard, S.R. How IRE1 Reacts to ER Stress. *Cell* **2008**, *132*, 24–26. [CrossRef] [PubMed]
121. Ron, D.; Walter, P. Signal integration in the endoplasmic reticulum unfolded protein response. *Nat. Rev. Mol. Cell Biol.* **2007**, *8*, 519–529. [CrossRef]
122. Tsukano, H.; Gotoh, T.; Endo, M.; Miyata, K.; Tazume, H.; Kadomatsu, T.; Yano, M.; Iwawaki, T.; Kohno, K.; Araki, K.; et al. The endoplasmic reticulum stress-C/EBP homologous protein pathway-mediated apoptosis in macrophages contributes to the instability of atherosclerotic plaques. *Arterioscler. Thromb. Vasc. Biol.* **2010**, *30*, 1925–1932. [CrossRef] [PubMed]
123. Li, G.; Scull, C.; Ozcan, L.; Tabas, I. NADPH oxidase links endoplasmic reticulum stress, oxidative stress, and PKR activation to induce apoptosis. *J. Cell Biol.* **2010**, *191*, 1113–1125. [CrossRef]
124. Pahl, H.L.; Baeuerle, P.A. Activation of NF-κB by ER stress requires both ca2+ and reactive oxygen intermediates as messengers. *FEBS Lett.* **1996**, *392*, 129–136. [CrossRef]
125. Deng, J.; Lu, P.D.; Zhang, Y.; Scheuner, D.; Kaufman, R.J.; Sonenberg, N.; Harding, H.P.; Ron, D. Translational Repression Mediates Activation of Nuclear Factor Kappa B by Phosphorylated Translation Initiation Factor 2. *Mol. Cell. Biol.* **2004**, *24*, 10161–10168. [CrossRef]
126. Livak, K.J.; Schmittgen, T.D. Analysis of relative gene expression data using real-time quantitative PCR and the 2-ΔΔCT method. *Methods* **2001**, *25*, 402–408. [CrossRef] [PubMed]
127. Martínez-Sánchez, N.; Seoane-Collazo, P.; Contreras, C.; Varela, L.; Villarroya, J.; Rial-Pensado, E.; Buqué, X.; Aurrekoetxea, I.; Delgado, T.C.; Vázquez-Martínez, R.; et al. Hypothalamic AMPK-ER Stress-JNK1 Axis Mediates the Central Actions of Thyroid Hormones on Energy Balance. *Cell Metab.* **2017**, *26*, 212–229.e12. [CrossRef] [PubMed]
128. González-García, I.; Contreras, C.; Estévez-Salguero, Á.; Ruíz-Pino, F.; Colsh, B.; Pensado, I.; Liñares-Pose, L.; Rial-Pensado, E.; Martínez de Morentin, P.B.; Fernø, J.; et al. Estradiol Regulates Energy Balance by Ameliorating Hypothalamic Ceramide-Induced ER Stress. *Cell Rep.* **2018**, *25*, 413–423.e5. [CrossRef] [PubMed]

plants

MDPI

Article

Jacaranone Derivatives with Antiproliferative Activity from *Crepis pulchra* and Relevance of This Group of Plant Metabolites

Csilla Zsuzsanna Dávid [1], Norbert Kúsz [1], Gyula Pinke [2], Ágnes Kulmány [3], István Zupkó [3], Judit Hohmann [1,4] and Andrea Vasas [1,*]

[1] Department of Pharmacognosy, Interdisciplinary Excellence Centre, University of Szeged, Eötvös u. 6, 6720 Szeged, Hungary; davidzsuzsanna88@gmail.com (C.Z.D.); kusznorbert@gmail.com (N.K.); hohmann.judit@szte.hu (J.H.)
[2] Department of Water and Environmental Sciences, Faculty of Agricultural and Food Sciences, Széchenyi István University, Var Square 2, 9200 Mosonmagyaróvár, Hungary; pinke.gyula@sze.hu
[3] Department of Pharmacodynamics and Biopharmacy, University of Szeged, Eötvös u. 6, 6720 Szeged, Hungary; kulmany.agnes@gmail.com (Á.K.); zupko.istvan@szte.hu (I.Z.)
[4] Interdisciplinary Centre of Natural Products, University of Szeged, Eötvös u. 6, 6720 Szeged, Hungary
* Correspondence: vasas.andrea@szte.hu; Tel.: +36-62546451

Abstract: Jacaranones are a small group of specific plant metabolites with promising biological activities. The occurrence of jacaranones is limited to only a few plant families, with Asteraceae being the most abundant source of these compounds. Therefore, jacaranones can also serve as chemotaxonomic markers. Our phytochemical investigation of *Crepis pulchra* L. (Asteraceae) resulted in three jacaranone derivatives (jacaranone, 2,3-dihydro-2-hydroxyjacaranone, 2,3-dihydro-2-methoxyjacaranone), and (6*R*,9*S*)-3-oxo-α-ionol-β-D-glucopyranoside, fulgidic acid, 12,15-octadecadienoic acid methyl ester, scopoletin and apigenin-7-*O*-β-D-glucoside. This is the first report on the isolation of jacaranones from a species belonging to the Cichorioideae subfamily of Asteraceae. Jacaranone derivatives were subjected to an in vitro antiproliferative assay against a panel of human cancer cell lines (MCF-7, MDA-MB-231, HeLa, and C33A), revealing high or moderate activities, with IC_{50} values ranging from 6.3 to 26.5 μM.

Keywords: jacaranones; *Crepis pulchra*; Cichorioideae; Asteraceae; antiproliferative activity

Citation: Dávid, C.Z.; Kúsz, N.; Pinke, G.; Kulmány, Á.; Zupkó, I.; Hohmann, J.; Vasas, A. Jacaranone Derivatives with Antiproliferative Activity from *Crepis pulchra* and Relevance of This Group of Plant Metabolites. *Plants* **2022**, *11*, 782. https://doi.org/10.3390/plants11060782

Academic Editors: Ivayla Dincheva, Ilian Badjakov and Bistra Galunska

Received: 8 February 2022
Accepted: 14 March 2022
Published: 16 March 2022

1. Introduction

Jacaranone derivatives, bearing an unsaturated cyclohexanone skeleton, occur rarely in the plant kingdom. Jacaranone [methyl 2-(1-hydroxy-4-oxocyclohexa-2,5-dien-1-yl)acetate] (**2**), the methyl ester derivative of quinolacetic acid (**1**), was first isolated from *Jacaranda caucana* (Bignoniaceae) by Ogura et al. in 1976 [1]. Since then, jacaranone derivatives (*n* = 35) have been isolated from other plant species as well, most of them belonging to the family Asteraceae. Almost all of the isolated compounds have been studied for their biological activities (e.g., cytotoxic, antimicrobial, anti-inflammatory, antioxidant, sedative, antiprotozoal, and antifeedant effects), and many of them showed multiple activities [2–8].

The genus *Crepis* (family Asteraceae) includes about 200 annual, biennial, or perennial plant species occurring widely in Eurasia, Africa, and North America [9]. In folk medicine, decoctions prepared from aerial parts of members of the genus *Crepis* are used for the treatment of various diseases, e.g., cough (Uganda) [10]; hepatitis, jaundice, and gallstones (Yemen) [11]; tumors (USA and China) [12]; and cardiovascular diseases, diabetes, cold, catarrh, and eye diseases (Turkey) [13–16]. Some species are traditionally used for their diuretic or laxative properties (Italy) [17,18] or externally for healing wounds, bruises, or inflammation (Spain and Bangladesh) [19,20].

Crepis species are rich sources of guaianolide- and eudesmane-type sesquiterpenes [21–23] and flavonoids [24,25]. In vitro pharmacological assessments revealed that *Crepis* extracts possess hepatoprotective [11], antimicrobial [21], antiviral [26], antioxidant [27], anti-inflammatory [28], and antiproliferative [26,27] activities.

Smallflower hawksbeard (*Crepis pulchra* L.) occurs in different ruderal stands. The species can be found in dry, open habitats, grasslands, pastures, abandoned fields, waste areas, alongside railroads, and roadsides [29,30]. According to the literature data, previously, guaianolide-type sesquiterpenes (8-*epi*-isoamberboin, vernoflexuoside, macrocliniside A, and diaspanoside A) were isolated from the plant by Kisiel et al. [31].

In continuation of our research aiming at the discovery of new bioactive specific metabolites from medicinal plants, the methanol extract of *Crepis pulchra* L. was investigated. The isolated special metabolites are discussed in comparison with the literature dealing with naturally occurring jacaranones.

2. Results

2.1. Isolation of Compounds from C. pulchra

The dried and ground whole plant material (1.25 kg) was extracted with methanol at room temperature. After evaporation, the extract was dissolved in 50% aqueous methanol, and solvent–solvent partition was performed with *n*-hexane, chloroform, and ethyl acetate. The chloroform phase was purified by a combination of different methods, including column chromatography (CC), vacuum liquid chromatography (VLC), thin layer chromatography (TLC), and HPLC to afford eight compounds. The structural determination was carried out by extensive spectroscopic analysis using 1D (^{1}H, JMOD) and 2D NMR (^{1}H-^{1}H COSY, HSQC, HMBC, NOESY) spectroscopy, HRESIMS measurements, and the comparison of the spectral data with the literature values.

The isolated compounds were identified as jacaranone [5], 2,3-dihydro-2-hydroxyjacaranone [5]; 2,3-dihydro-2-methoxyjacaranone [5]; (6*R*,9*S*)-3-oxo-ionol-*β*-D-glucopyranoside [32]; fulgidic acid [33]; 12,15-octadecadienoic acid methyl ester, scopoletin [34]; and apigenin-7-*O*-*β*-D-glucoside [35] (Figure 1). All compounds were isolated for the first time from the plant. Moreover, this was the first time jacaranone derivatives were isolated from a *Crepis* species.

2.2. Antiproliferative Investigation of the Isolated Jacaranones

The jacaranone derivatives, isolated from *C. pulchra*, were subjected to an in vitro cytotoxicity (MTT) assay against human cancer (breast cancer (MCF-7 and MDA-MB-231), and cervical cancer (HeLa and C33A) cell lines (Table 1). Jacaranone proved to be the most active against all four tested cell lines (IC_{50} 6.27–14.61 μM). Its activity was comparable with that of the positive control, cisplatin. 2,3-Dihydro-2-hydroxyjacaranone and 2,3-dihydro-2-methoxyjacaranone differ from jacaranone only in the substitution of C-2 (a hydroxy group in the case of 2,3-dihydro-2-hydroxyjacaranone, and a methoxy group in 2,3-dihydro-2-methoxyjacaranone) and in the saturation of the double bond between C-2–C-3. These modifications resulted in the decrease of the antiproliferative activity in the case of the two jacaranone derivatives. Our results confirm that the presence of an α,β-unsaturated carbonyl group in the molecule is essential for the antiproliferative activity of jacaranones [36].

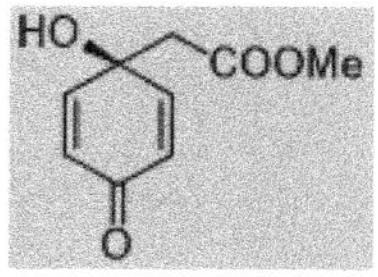

jacaranone

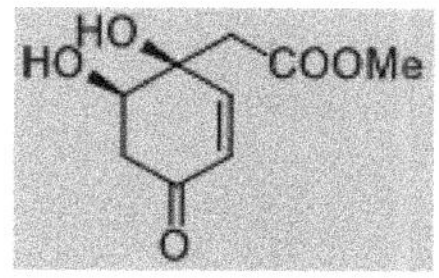

2,3-dihydro-2-hydroxy-jacaranone

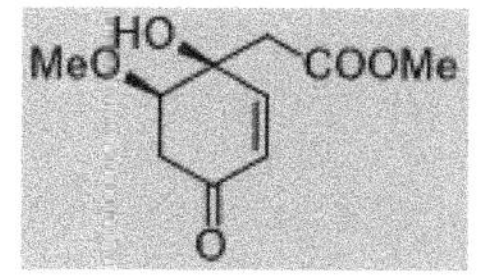

2,3-dihydro-2-methoxy-jacaranone

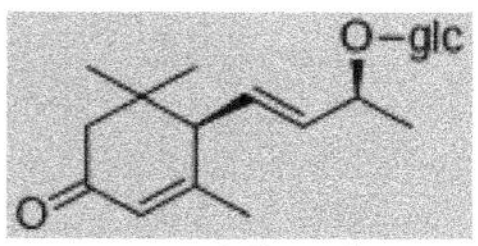

(6*R*,9*S*)-3-oxo-α-ionol-β-D-glucopyranoside

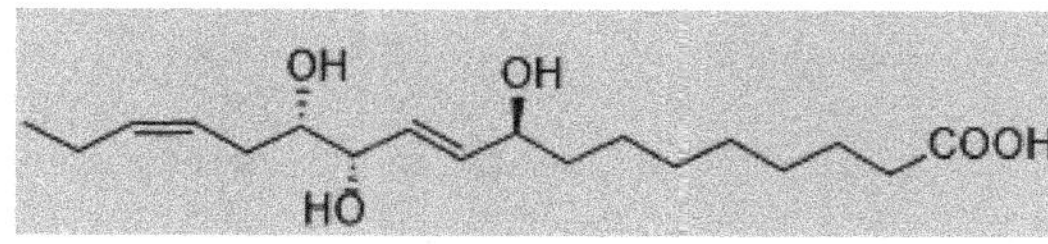

fulgidic acid

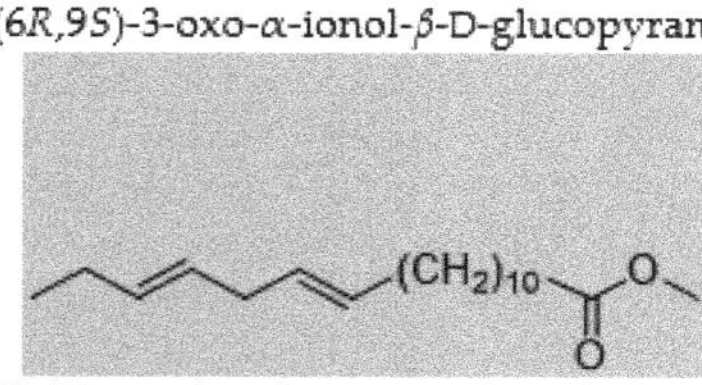

12,15-octadecadienoic acid methyl ester

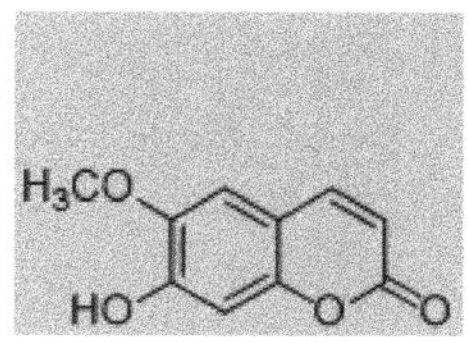

scopoletin

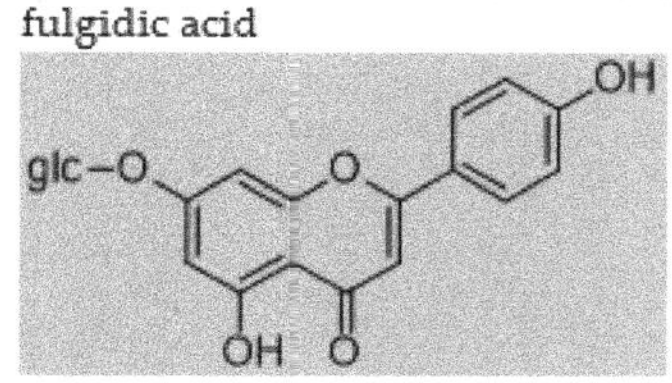

apigenin-7-O-β-D-glucoside

Figure 1. Compounds isolated from *C. pulchra*.

Table 1. Antiproliferative activity of the isolated jacaranone derivatives.

Compound	IC_{50} (μM) [95% Confidence Intervals]			
	MCF-7	**MDA-MB-231**	**HeLa**	**C33A**
jacaranone	10.22 [8.96–11.66]	6.89 [6.29–7.55]	14.61 [13.40–15.93]	6.27 [5.66–6.95]
2,3-dihydro-2-hydroxyjacaranone	>30	26.49 [21.61–32.41]	>30	21.56 [18.86–24.64]
2,3-dihydro-2-methoxyjacaranone	21.30 [19.44–23.35]	17.85 [15.39–20.70]	23.47 [21.36–25.78]	12.52 [11.27–13.92]
cisplatin	6.01 [5.33–6.79]	18.65 [16.67–20.85]	14.02 [12.65–15.56]	3.69 [3.22–3.95]

3. Discussion

3.1. Occurrence of Jacaranones in Nature

Altogether, 35 jacaranones were isolated from 37 plant species; three-quarters of the compounds are monomers (n = 26), and the others are dimers (n = 9) (Figures 2 and 3). Although the best sources of jacaranones are Asteraceae species, certain species of the Acanthaceae [3], Bignoniaceae [1,2,37–41], Delesseriaceae [42], Gesneriaceae [43,44], Oleaceae [45], and Theaceae [4,5] families were also found to be sources of jacaranones (Table 2). Among Asteraceae species, the genus *Senecio* is represented by 20 jacaranone derivative-producing plant species [36,46–62]. Besides the *Bethencourtia* [8], *Packera* [63], and *Pentacalia* [6,7] genera, each is represented by one species. All species belong to the Asteroideae (Tubuliflorae) subfamily of Asteraceae. *Crepis pulchra* is the first representative of the Cichorioideae (Liguliflorae) subfamily. Jacaranone (**2**) is the most common compound, isolated from 28 plant species of the Asteraceae, Bignoniaceae, and Theaceae families.

The basic monomer structure of jacaranones can be modified by different substituents. Regarding the cyclohexanone ring, substitution occurs mainly at C-2, where hydroxy- (**12**), methoxy- (**24, 35**), or ethoxy- (**14, 15**) groups are linked to the ring, or even an epoxide ring can be formed, as in the case of compound **25**. A rare substituent can be found in cases of marinoid F (**10**) and marinoid G (**11**), where a chlorine atom is joined to the cyclohexanone ring at C-3. Most frequently, an ester is formed via the carboxyl group with aliphatic alcohols, e.g., methanol (**2, 16, 18, 21, 22, 24, 25**), ethanol (**3, 13–15, 17, 23**), or butanol (**4**), or with a sugar molecule (**5–8**). A sugar molecule can also be attached to the hydroxy group forming an acetal (**9**, **22**). Compound **26**, isolated from *Senecio giganteus*, is the only compound containing a lactone ring. Jacaranone dimers are formed from two monomers linked through one or two sugar molecules.

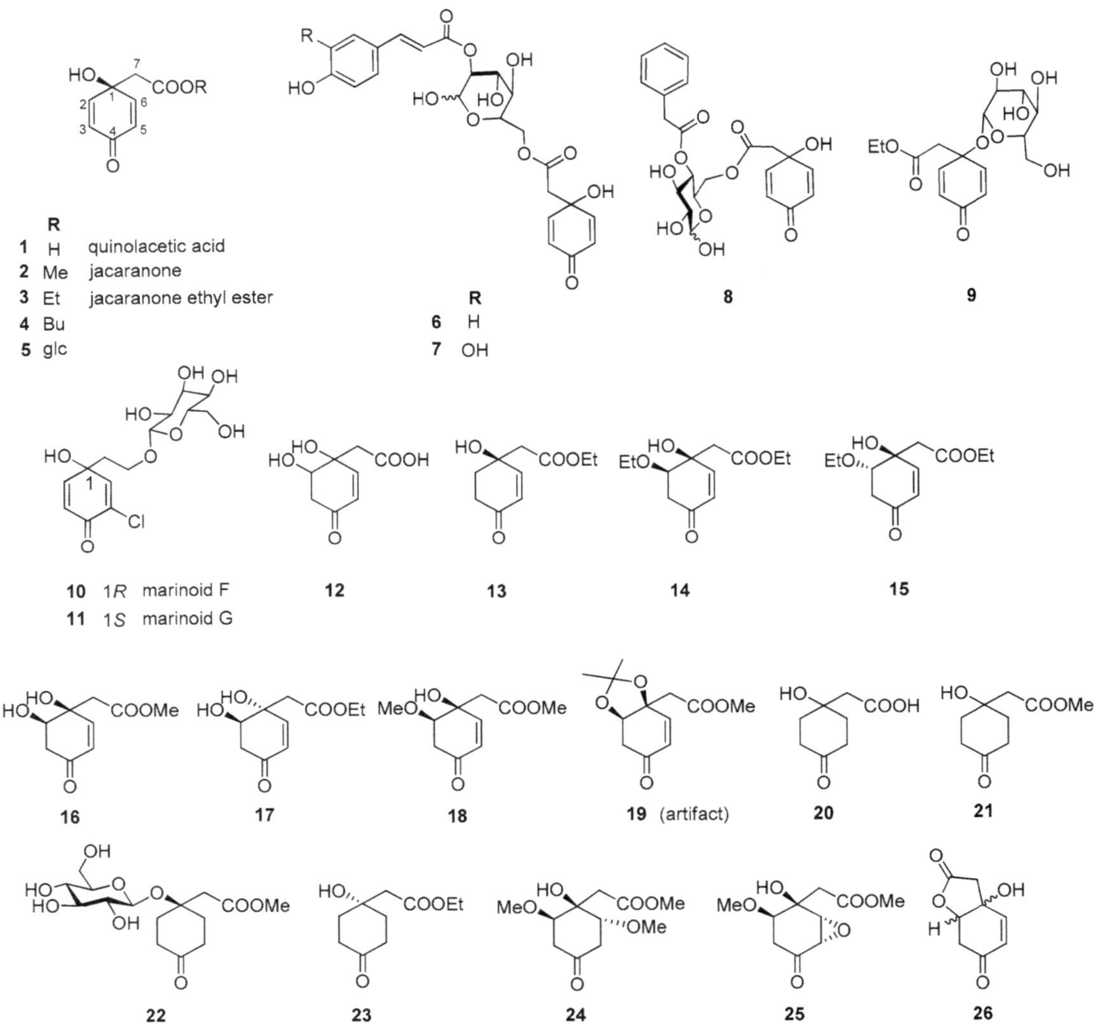

Figure 2. Naturally occurring jacaranone derivatives (monomers) *. * Stereochemistry of the compounds is indicated according to the source literature; Me = methyl, Et = ethyl, Bu = butyl, glc = glucose.

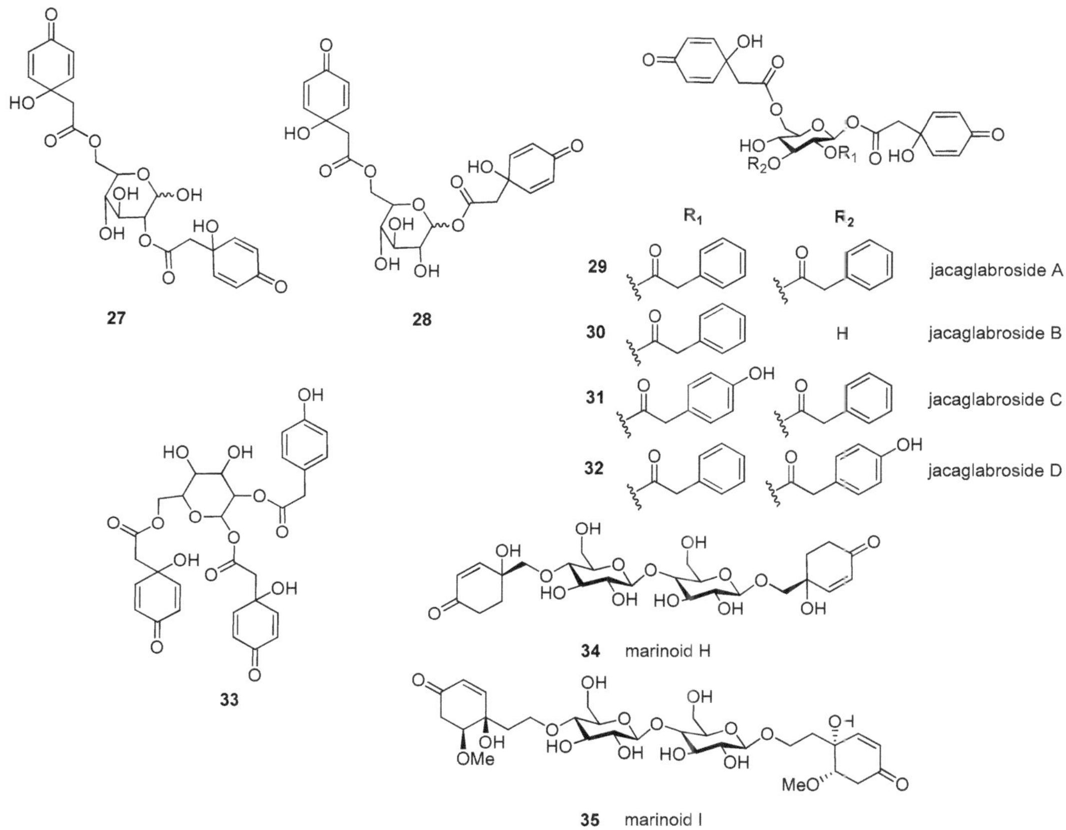

Figure 3. Naturally occurring jacaranone derivatives (dimers) *. * Stereochemistry of each compound is given as it was published.

Table 2. List of plant families and species containing jacaranones.

Family	Species	Compound	Ref.
Asteraceae	*Bethencourtia hermosa*	**3, 17**	[8]
	Jacobaea gigantea = Senecio giganteus	**2, 26**	[46]
	Packera bellidifolia	**2, 21**	[63]
	Pentacalia desiderabilis	**1–3, 21**	[6,7]
	Senecio ambiguous subsp. *ambiguus*	**2**	[47]
	Senecio anonymus	**3**	[48]
	Senecio argunensis	**2, 18, 19, 21, 24, 25**	[36]
	Senecio cannabifolius	**2, 3, 18, 19, 24**	[49,50]
	Senecio cannabifolius var. *integrilifolius*	**21**	[51]
	Senecio carpathicus	**2, 4**	[62]
	Senecio clevelandii	**2, 21**	[52]
	Senecio chrysanoides	**13–15, 23**	[53]
	Senecio erucifolius	**2, 4**	[62]
	Senecio fendleri	**3**	[54]
	Senecio leucanthemifolius	**2**	[55]

Table 2. *Cont.*

Family	Species	Compound	Ref.
	Senecio minutus	**2, 24**	[56]
	Senecio othonnae	**2, 4**	[62]
	Senecio palmatus	**2**	[57]
	Senecio paludosus	**2, 4**	[62]
	Senecio scandens	**1–3, 5–7, 9, 17, 20, 27, 28**	[58–60]
	Senecio scandens var. *incisus*	**6, 7, 12**	[61]
	Senecio subalpinus	**2, 4**	[62]
	Senecio wagneri	**2, 4**	[62]
Acanthaceae	*Avicennia marina*	**10, 11, 34, 35**	[3]
Bignoniaceae	*Jacaranda arborea*	**2, 3**	[2]
	Jacaranda caucana	**2**	[1]
	Jacaranda glabra	**2, 29–32**	[37]
	Jacaranda macrantha	**2**	[38]
	Jacaranda mimosifolia	**2, 33**	[39]
	Jacaranda oxyphylla	**8**	[40]
	Jacaranda puberula	**2**	[41]
Delesseriaceae	*Delesseria sanguineu*	**2**	[42]
Gesneriaceae	*Sinningia mauroana*	**2**	[43]
	Sinningia reitzii	**2**	[44]
Oleaceae	*Forsythia suspensa*	**22**	[45]
Theaceae	*Ternstroemia japonica*	**2, 16, 18**	[5]
	Ternstroemia pringlei	**2**	[4]

3.2. Antiproliferative Activity of Jacaranones

According to the relevant literature, the most promising biological effect of jacaranones is their anticancer activity. The methanolic extract of *Jacaranda caucana* and jacaranone (**2**) were tested against P-388 lymphocytic leukemia and Eagle's 9KB carcinoma cells, and substantial antiproliferative activity was detected [1,64]. The cytotoxicity of **2** was investigated against six tumors (lung large cell carcinoma (COR-L23), colorectal adenocarcinoma (Caco-2), amelanotic melanoma (C32), hepatocellular carcinoma (HepG-2), renal cell adenocarcinoma (ACHN), and hormone dependent prostate carcinoma (LNCaP) and one normal (human fetal lung (MRC-5) cell lines [47,55]. Jacaranone (**2**) showed outstanding action against all tested cell lines (IC_{50} values ranging from 11.31 to 40.57 μM), which was comparable with that of the positive control vinblastine, while **2** did not adversely affect MRC-5 cells. Furthermore, **2** exerted antiproliferative and proapoptotic effects in eight human (A2058, SK-MEL-28, HCT-8, LS160, SiHa, HL-60, SK-BR-3) and one murine (B16F10-Nex2) tumor cell lines in vitro by downregulating Akt and activating p38 MAPK signaling pathways through the generation of reactive oxygen species (ROS). IC_{50} values varied from 9 to 145 μM for human cancer cells and was 17 μM for murine melanoma B16F10-Nex2 cells [65]. Moreover, the protective effect of this quinone (**2**) was also proved in a melanoma syngeneic model in vivo [65]. Jacaranone ethyl ester (**3**) was tested in KB screen and was found to be highly active (ED_{50} value 16.82 μM) [48]. Tian et al. investigated the cytotoxic effect of seven jacaranone analogs against various tumorous cell lines (HCT8, CaEa-17, A2780, HeLa, BEL-7402, KB, PC-3M, A549, BGC-823, and MCF-7) [59]. Jacaranone ethyl ester (**3**) possessed the most potent cytotoxic effect with IC_{50} values at a range of 2.85–4.33 μM. Fluorouracil was applied as a positive control (IC_{50}s 4.15–6.84 μM). Apart from jacaranone ethyl ester (**3**), quinolacetic acid (**1**), jacaranone (**2**), and a new jacaranone-derivative (**17**) also showed relatively high cytotoxic activity. Based on the results of their cytotoxicity assay, the authors found that all the active compounds share the same structural moiety, the α,β-unsaturated carbonyl group, a segment that is known to be of crucial importance for the cytotoxic effect of other compounds as well [59].

Presser et al. synthesized 13 nitrogen-containing jacaranone derivatives from the natural-product-derived cyclohexadienone scaffold and investigated their antiproliferative activity against four human tumorous cell lines (MDA-MB-231 breast cancer, CCRF-CEM leukemia, HCT-116 colon cancer, and U251 glioblastoma), and one non-tumorigenic cell line (MRC-5 lung fibroblasts) at 5 μg/mL and 50 μg/mL concentrations [66]. The positive control vinblastine was applied at 0.01 μg/mL. At 50 μg/mL concentration, almost all derivatives were found to have cytotoxic effect against MDA-MB-231 and CCRF-CEM cells. During their investigations, the authors managed to reveal some structure–activity relationships of jacaranone-based nitrogenous cyclohexadienones as well. It was observed that the most potent compounds shared an α,β-unsaturated imide structural element. In the absence of this structural moiety, no cytotoxic effect could be detected [66].

4. Materials and Methods

4.1. General Experimental Procedures

NMR spectra were recorded in methanol d_4 on a Bruker Avance DRX 500 spectrometer at 500 MHz (^{1}H) and 125 MHz (^{13}C). The signals of the deuterated solvents were chosen as references. The chemical shift values (δ) were given in ppm, and the coupling constants (J) are in Hz. Two-dimensional (2D) experiments were performed with standard Bruker software. In the ^{1}H–^{1}H COSY, HSQC, and HMBC experiments, gradient-enhanced versions were used. Column chromatography (CC) was performed on polyamide (MP Biomedicals Germany GmbH, Eschwege, Germany). Normal and reversed-phase vacuum liquid chromatography (VLC) was carried out on silica gel (Kieselgel 60 GF_{254}, 15 μm, Merck, Darmstadt, Germany) and on reversed phase silica gel [RediSep C-18, 40–60 μm, Teledyne Isco, Lincoln, NE, USA]. Thin-layer chromatography was performed on a Kieselgel 60 RP-18 F_{254} and a Kieselgel 60 F_{254} (Merck, Darmstadt, Germany). Spots on UV active silica gel were detected under UV light (245 nm and 336 nm) and made visible with vanillin sulfuric acid and heating at 105 °C for 2 min. The high-performance liquid chromatographic (HPLC) separation was carried out on a Waters HPLC (Waters 600 controller, Waters 600 pump, and Waters 2998 photodiode array detector), using a normal phase (LiChrospher Si 100 (250 × 4 mm, 5 μm, Merck)) column. The flow rate was 1 mL/min, and the injection volume was 25 μL. The data were acquired and processed with Empower software. All solvents used for CC were of at least analytical grade (VWR Ltd., Szeged, Hungary). Ultrapure water was prepared with a Milli-Q water purification system (Millipore, Molsheim, France).

4.2. Plant Material

The whole plants of *Crepis pulchra* (1.25 kg dried plant material) were collected in the flowering period at Hegyeshalom (Hungary) in July 2019, and were identified by one of the authors, Gyula Pinke (Department of Water and Environmental Sciences, Széchenyi István University). A voucher specimen (No. 895) has been deposited at the Department of Pharmacognosy, University of Szeged, Szeged, Hungary.

4.3. Extraction and Isolation

The dried and ground whole plant of *C. pulchra* (1.25 kg) was percolated with methanol (12.5 L) at room temperature. The crude extract was concentrated in vacuo (152.5 g), redissolved in water, and subjected to solvent–solvent partition with *n*-hexane (5 × 500 mL), chloroform (5 × 500 mL), and ethyl acetate (5 × 400 mL), respectively. The concentrated chloroform-soluble fraction (35.7 g) was further separated by polyamide open-column chromatography with a gradient system of MeOH–H_2O (2:3, 3:2, and 4:1 (600 mL/eluent); each eluent was collected as a fraction (Fractions I–III). Fraction I obtained with 40% MeOH (3.04 g) was subjected to vacuum liquid chromatography (VLC) on silica gel with a gradient system of $CHCl_3$–MeOH (from 99:1 to 7:3 (300 mL/eluent) to yield 10 major fractions (I/1-10). Fraction I/3 (219 mg) was further purified by VLC on reversed phase silica gel using a MeOH–H_2O gradient system (1:9 → 9:1, 100 mL/eluent) to afford 8 subfractions (I/3/1-8). Subfraction I/3/1 (159 mg) was subjected to normal phase VLC (*n*-hexane–

isopropanol, gradient, 95:5 → 6:4, 25 mL/eluent), followed by reversed phase preparative TLC (using MeOH–H_2O 1:1 as an eluent) to yield 4 (**2** (27.3 mg), **16** (4.4 mg), **18** (3.6 mg), and scopoletine (12.9 mg) compounds. Fraction I/9 (1.0 g) was separated by normal phase VLC with a gradient system of $CHCl_3$–MeOH (from 98:2 to 7:3 (50 mL/eluent) to afford 10 subfractions (I/9/1-10). Fraction I/9/6 (345.5 mg) was purified by reverse-phase VLC (gradient system of MeOH–H_2O 1:9 → 8:2, 25 mL/eluent) and gel chromatography on Sephadex LH-20 (using CH_2Cl_2–MeOH 1:1 as an eluent) followed by normal-phase preparative TLC with EtOAc–MeOH–H_2O (100:16:0,5) as a solvent system to yield fulgidic acid (5.19 mg). Fraction I/9/7 (185.0 mg) was separated by gel chromatography on Sephadex LH-20 (using CH_2Cl_2–MeOH 1:1 as an eluent) and then by preparative TLC on reverse-phase silica gel with MeOH–H_2O (1:1) to yield (6*R*,9*S*)-3-oxo-ionol-β-D-glucopyranoside (3.34 mg). Fraction II obtained with 60% MeOH (2.08 g) was subjected to reverse-phase VLC, applying a gradient system of MeOH–H_2O (from 1:9 to 1:0, 200 mL/eluent) to obtain nine main fractions (II/1-9). Separation of fraction II/1 (236.3 mg) by gel chromatography on Sephadex LH-20 afforded five subfractions (II/1/1-5). Subfraction II/1/4 (60.8 mg) was further purified by normal-phase VLC using $CHCl_3$–MeOH as a mobile phase [from pure 1:0 to 6:4 (40 mL/eluent)] followed by preparative TLC on normal-phase silica gel with EtOAc–HCOOH–H_2O (85:10:5) as a solvent system to obtain apigenin-7-*O*-glucoside (7.8 mg). Fraction II/9 (165.1 mg) was subjected to normal-phase VLC, applying a gradient system of cyclohexane–EtOAc–MeOH [from 9:1:0 to 6:3:1 (50 mL/eluent)] to afford ten subfractions (II/9/1-10). Subfraction II/9/2 (4.4 mg) was purified by normal-phase HPLC [isocratic, cyclohexane–EtOAc (8:2)] to yield 12,15-octadecadienoic acid methyl ester (1.5 mg).

4.4. Antiproliferative (MTT) Assay

The growth-inhibition properties of jacaranone derivatives were determined by standard MTT assays on four human malignant gynecological cell lines (breast cancer: MCF-7 and MDA-MB-231, and cervical cancer: HeLa and C33A). All cell lines were maintained in minimal essential medium (MEM) supplemented with 10% fetal bovine serum, 1% nonessential amino acids, and 1% penicillin–streptomycin–amphotericin B mixture in humidified air containing 5% CO_2 at 37 °C. All cell types were seeded into 96-well plates at a density of 5000 with the exception of C33A, which was seeded at a density of 10,000 and treated by increasing concentrations (0.1–30 μM) of the compounds for 72 h under cell culturing conditions. After the incubation, 5 mg/mL MTT [3-(4,5-dimethylthiazol-2-yl)-2,5-diphenyltetrazolium bromide] solution was added to samples for 4 h and precipitated blue formazan crystals were dissolved in DMSO. Absorbance values of the samples were measured at 545 nm using a microplate reader (Stat Fax-2100, Awareness Technologies Inc., Palm City, FL, USA), and untreated cells were used as a control. Normalized sigmoidal concentration−response curves were fitted to the determined data, and the IC_{50} values were calculated by GraphPad Prism 5.01 (GraphPad Software, San Diego, CA, USA). Cisplatin (Ebewe Pharma GmbH, Unterach, Austria) was used as a reference agent in the same concentration range.

5. Conclusions

In our experiment, a combination of different chromatographic techniques resulted in the isolation of eight compounds, among them three jacaranone derivatives from *C. pulchra* for the first time. Jacaranone (**2**) showed the highest antiproliferative activity against MDA-MB-231 (human breast cancer) and C33A (human cervical cancer) cells. Although jacaranones represent a small group of plant special metabolites, they can be interesting either for organic chemists or for pharmacologists because of their promising biological effects. Moreover, they can serve as chemotaxonomic markers. The importance of our results is that smallflower hawksbeard is the first representative of the Cichorioideae subfamily of Asteraceae in which jacaranones have been detected.

Author Contributions: Conceptualization, A.V.; methodology, A.V. and I.Z.; software, investigation, C.Z.D., Á.K. and N.K.; data curation, C.Z.D. and G.P.; writing—original draft preparation, C.Z.D. and A.V.; writing—review and editing, J.H.; project administration, A.V. All authors have read and agreed to the published version of the manuscript.

Funding: This research was funded by the Economic Development and Innovation Operative Program GINOP-2.3.2-15-2016-00012, the UNKP-21-3-SZTE-265 New National Excellence Program, and TKP2021-EGA-32 of the Ministry of Innovation and Technology of Hungary from the source of the National Research, Development and Innovation Fund.

Institutional Review Board Statement: Not applicable.

Informed Consent Statement: Not applicable.

Data Availability Statement: Not applicable.

Conflicts of Interest: The authors declare no conflict of interest.

References

1. Ogura, M.; Cordell, G.A.; Farnsworth, N.R. Potential anticancer agents. III. Jacaranone, a novel phytoquinoid from *Jacaranda caucana*. *J. Nat. Prod.* **1976**, *39*, 255–257.
2. Hirukawa, M.; Zhang, M.; Echenique-Díaz, L.M.; Mizota, K.; Ohdachi, S.; Begué-Quiala, G.; Delgado-Labañino, J.L.; Gámez-Díez, J.; Alvarez-Lemus, J.; Machado, L.G.; et al. Isolation and structure–activity relationship studies of jacaranones: Anti-inflammatory quinoids from the Cuban endemic plant *Jacaranda arborea* (Bignoniaceae). *Tetrahedron Lett.* **2020**, *61*, 152005. [CrossRef]
3. Yi, X.X.; Chen, Y.; Xie, W.P.; Xu, M.B.; Chen, Y.N.; Gao, C.H.; Huang, R.M. Four new jacaranone analogs from the fruits of a Beibu Gulf mangrove *Avicennia marina*. *Mar. Drugs* **2014**, *12*, 2515–2525. [CrossRef]
4. Lozada-Lechuga, J.; Villarreal, M.L.; Fliniaux, M.A.; Bensaddek, L.; Mesnard, F.; del Gutiérrez, M.C.; Cardoso-Taketa, A.T. Isolation of jacaranone, a sedative constituent extracted from the flowers of the Mexican tree *Ternstroemia pringlei*. *J. Ethnopharmacol.* **2010**, *127*, 551–554. [CrossRef] [PubMed]
5. Jo, Y.; Suh, J.; Shin, M.H.; Jung, J.H.; Im, K.S. Jacaranone and related compounds from the fresh fruits of *Ternstroemia japonica* and their antioxidative activity. *Arch. Pharm. Res.* **2005**, *28*, 885–888. [CrossRef]
6. Gomes, K.S.; Tamayose, C.I.; Ferreira, M.J.; Murakami, C.; Young, M.C.; Antar, G.M.; Camilo, F.F.; Sartorelli, P.; Lago, J.H. Isolation of antifungal quinoid derivatives from leaves of *Pentacalia desiderabilis* (Vell.) Cuatre. (Asteraceae) using ionic liquid in the microwave assisted extraction. *Quim. Nova* **2019**, *42*, 156–158. [CrossRef]
7. Morais, T.R.; Romoff, P.; Fávero, O.A.; Reimão, J.Q.; Lourenço, W.C.; Tempone, A.G.; Hristov, A.D.; Di Santi, S.M.; Lago, J.H.; Sartorelli, P.; et al. Anti-malarial, anti-trypanosomal, and anti-leishmanial activities of jacaranone isolated from *Pentacalia desiderabilis* (Vell.) Cuatrec. (Asteraceae). *Parasitol. Res.* **2011**, *110*, 95–101. [CrossRef]
8. Fraga, B.M.; Díaz, C.E.; Amador, L.J.; Reina, M.; Santana, O.; González-Coloma, A. Bioactive compounds from transformed root cultures and aerial parts of *Bethencourtia hermosae*. *Phytochemistry* **2014**, *108*, 220–228. [CrossRef]
9. Eliáš, P.; Turisová, I.; Ťavoda, O. Occurrence of small flower hawksbeard (*Crepis pulchra* L.) in Slovakia. *Thaiszia J. Bot.* **2010**, *20*, 127–135.
10. Namukobe, J.; Kasenene, J.M.; Kiremire, B.T.; Byamukama, R.; Kamatenesi-Mugisha, M.; Krief, S.; Dumontet, V.; Kabasa, J.D. Traditional plants used for medicinal purposes by local communities around the Northern sector of Kibale National Park, Uganda. *J. Ethnopharmacol.* **2011**, *136*, 236–245. [CrossRef]
11. Fleurentin, J.; Hoefler, C.; Lexa, A.; Mortier, F.; Pelt, J.M. Hepatoprotective properties of *Crepis rueppellii* and *Anisotes trisulcus*: Two traditional medicinal plants of Yemen. *J. Ethnopharmacol.* **1986**, *16*, 105–111. [CrossRef]
12. Hartwell, J.L. Plants used against cancer. A survey. *Lloydia* **1968**, *31*, 71–170.
13. Bakar, F.; Acikara, Ö.B.; Ergene, B.; Nebioğlu, S.; Çitoğlu, G.S. Antioxidant activity and phytochemical screening of some Asteraceae plants. *Turk. J. Pharm. Sci.* **2015**, *12*, 36–45. [CrossRef]
14. Dalar, A. Plant taxa used in the treatment of diabetes in Van Province, Turkey. *Int. J. Second. Metab.* **2018**, *5*, 170–184. [CrossRef]
15. Genç, G.E.; Özhatay, N. An ethnobotanical study in Çatalca (European part of İstanbul) II. *Turk. J Pharm. Sci.* **2006**, *3*, 73–89.
16. Kilic, O.; Bagci, E. An ethnobotanical survey of some medicinal plants in Keban (Elazığ-Turkey). *J. Med. Plant Res.* **2013**, *7*, 1675–1684. [CrossRef]
17. Sansanelli, S.; Tassoni, A. Wild food plants traditionally consumed in the area of Bologna (Emilia Romagna region, Italy). *J. Ethnobiol. Ethnomed.* **2014**, *10*, 69. [CrossRef]
18. Guarrera, P.M.; Savo, V. Perceived health properties of wild and cultivated food plants in local and popular traditions of Italy: A review. *J. Ethnopharmacol.* **2013**, *146*, 659–680. [CrossRef]
19. González-Tejero, M.R.; Molero-Mesa, J.; Casares-Porcel, M.; Lirola, M.J.M. New contributions to the ethnopharmacology of Spain. *J. Ethnopharmacol.* **1995**, *45*, 157–165. [CrossRef]
20. Rahman, M. An ethnobotanical investigation on Asteraceae family at Rajshahi, Bangladesh. *Acad. J. Med. Plants* **2013**, *1*, 92–100.

21. Ndom, J.C.; Mbafor, J.T.; Wansi, J.D.; Kamdem, A.W.; Meva'a, L.M.; Vardamides, J.C.; Toukam, F.; Pegyemb, D.; Ngando, T.M.; Laatsch, H.; et al. Sesquiterpene lactones from *Crepis cameroonica* (Asteraceae). *Nat. Prod. Res.* **2006**, *20*, 435–442. [CrossRef]
22. Michalska, K.; Kisiel, W.; Zidorn, C. Sesquiterpene lactones from *Crepis aurea* (*Asteraceae, Cichorieae*). *Biochem. Syst. Ecol.* **2013**, *46*, 1–3. [CrossRef]
23. Ebada, S.S.; El-Kashef, D.H.; Müller, W.E.G.; Proksch, P. Cytotoxic eudesmane sesquiterpenes from *Crepis sancta*. *Phytochem. Lett.* **2019**, *33*, 46–48. [CrossRef]
24. Mañez, S.; Recio, M.C.; Giner, R.M.; Sanz, M.J.; Terencio, M.C.; Peris, J.B.; Stübing, G.; Rios, J.L. A chematoxonomic review of the subtribe Crepidinase based on its phenol constituents. *Biochem. Syst. Ecol.* **1994**, *22*, 297–305. [CrossRef]
25. Zidorn, C.; Schubert, B.; Stuppner, H. Phenolics as chemosystematic markers in and for the genus *Crepis* (*Asteraceae, Cichorieae*). *Sci. Pharm.* **2008**, *76*, 743–750. [CrossRef]
26. Ooi, L.S.; Wang, H.; Luk, C.W.; Ooi, V.E. Anticancer and antiviral activities of *Youngia japonica* (L.) DC (*Asteraceae, Compositae*). *J. Ethnopharmacol.* **2004**, *94*, 117–122. [CrossRef]
27. Zengin, G.; Sarikurkcu, C.; Uyar, P.; Aktumsek, A.; Uysal, S.; Kocak, M.S.; Ceylan, R. *Crepis foetida* L. subsp. *rhoeadifolia* (Bieb.) Celak. as a source of multifunctional agents: Cytotoxic and phytochemical evaluation. *J. Funct. Foods* **2015**, *17*, 698–708. [CrossRef]
28. Barda, C.; Grafakou, M.E.; Kalpoutzakis, E.; Heilmann, J.; Skaltsa, H. Chemical composition of *Crepis foetida* L. and *C. rubra* L. volatile constituents and evaluation of the in vitro anti-inflammatory activity of salicylaldehyde rich volatile fraction. *Biochem. Syst. Ecol.* **2021**, *96*, 104256. [CrossRef]
29. Holub, J. *Crepis pulchra* L. In *Červená Kniha Ohrozených a Vzácnych Druhov Rastlín a Živočíchov SR a ČR*; Čeřovský, J., Feráková, V., Holub, J., Maglocký, Š., Procházka, F., Eds.; Vyššie Rastliny; Príroda a.s.: Bratislava, Slovakia, 1999; Volume 5, p. 355. (In Czech)
30. Bogler, D.J. *Crepis* L. In *Flora of North America: North of Mexico*; Oxford University Press: New York, NY, USA, 2006; Volume 19, pp. 222–239.
31. Kisiel, W.; Gromek, D. Guaianolides from *Crepis pulchra*. *Pol. J. Chem.* **1994**, *68*, 535–538. [CrossRef]
32. Lee, E.H.; Kim, H.J.; Song, Y.S.; Jin, C.; Lee, K.T.; Cho, J.; Lee, Y.S. Constituents of the stems and fruits of *Opuntia ficus-indica* var. *saboten*. *Arch. Pharm. Res.* **2003**, *26*, 1018–1023. [CrossRef]
33. Kurashina, Y.; Miura, A.; Enomoto, M.; Kuwahara, S. Stereoselective synthesis of malyngic acid and fulgidic acid. *Tetrahedron* **2011**, *67*, 1649–1653. [CrossRef]
34. Adfa, M.; Yoshimura, T.; Komura, K.; Koketsu, M. Antitermite activities of coumarin derivatives and scopoletin from *Protium javanicum* Burm. f. *J. Chem. Ecol.* **2010**, *36*, 720–726. [CrossRef]
35. Ersöz, T.; Harput, Ü.Ş.; Saracoğlu, İ.; aliş, İ. Phenolic compounds from *Scutellaria pontica*. *Turk. J. Chem.* **2002**, *26*, 581–588.
36. Tian, Y.Q.; Niu, Y.F.; Shen, T.; Weng, C.W.; Xie, W.D.; Row, K.H. Cyclohexanone derivatives from *Senecio argunensis*. *J. Chem. Res.* **2010**, *34*, 25–27. [CrossRef]
37. Gachet, M.S.; Kunert, O.; Kaiser, M.; Brun, R.; Muñoz, R.A.; Bauer, R.; Schühly, W. Jacaranone-derived glucosidic esters from *Jacaranda glabra* and their activity against *Plasmodium falciparum*. *J. Nat. Prod.* **2010**, *73*, 553–556. [CrossRef]
38. Santos, C.A.; Raslan, D.S.; Chiari, E.; Oliveira, A.B. Bioguided assay of *Jacaranda Macrantha* Cham. (Bignoniaceae). *Acta Hortic.* **1999**, *501*, 151–154. [CrossRef]
39. Rana, A.; Bhangalia, S.; Singh, H.P. A new phenylethanoid glucoside from *Jacaranda mimosifolia*. *Nat. Prod. Res.* **2013**, *27*, 1167–1173. [CrossRef] [PubMed]
40. Pereira, V.V.; Duarte, L.P.; Silva, R.R.; Takahashi, J.A. New jacaranone glucoside from *Jacaranda oxyphylla* leaves. *Nat. Prod. Res.* **2016**, *30*, 2421–2428. [CrossRef] [PubMed]
41. Zanotelli, P.; Locateli, G.; Vecchia, C.D.; Gomes, D.B.; Oliveira, B.M.M.; Lutinski, J.A.; Predebom, A.J.; Miorando, D.; Zanatta, M.E.C.; Steffler, A.M.; et al. Gastroprotective potential of the hydroalcoholic extract from *Jacaranda puberula* in mice. *Rev. Bras. Farmacogn.* **2020**, *30*, 838–843. [CrossRef]
42. Yvin, J.C.; Chevolot, L.; Chevolot-Magueur, A.M.; Cochard, J.C. First isolation of jacaranone from an alga, *Delesseria sanguinea*. A metamorphosis inducer of Pecten larvae. *J. Nat. Prod.* **1985**, *48*, 814–816. [CrossRef]
43. Winiewski, V.; Serain, A.F.; de Sa, E.S.; Salvador, M.J.; Stefanello, M.E.A. Chemical constituents of *Sinningia mauroana* and screening of its extracts for antimicrobial, antioxidant and cytotoxic activities. *Quim. Nova* **2020**, *43*, 181–187. [CrossRef]
44. Silva, A.S.; Amorim, M.S.; Fonseca, M.M.; Salvador, M.J.; de Sa, E.L.; Stefanello, M.E.A. A new cytotoxic naphthoquinone and other chemical constituents of *Sinningia reitzii*. *J. Braz. Chem. Soc.* **2019**, *30*, 2060–2065. [CrossRef]
45. Yan, X.J.; Bai, X.Y.; Liu, Q.B.; Liu, S.; Gao, P.Y.; Li, L.Z.; Song, S.J. Two new glycosides from the fruits of *Forsythia suspense*. *J. Asian Nat. Prod. Res.* **2014**, *16*, 376–382. [CrossRef] [PubMed]
46. Mezache, N.; Derbré, S.; Akkal, S.; Laouer, H.; Séraphin, D.; Richomme, P. Fast counter current chromatography of n-butanolic fraction from *Senecio giganteus* (*Asteraceae*). *Nat. Product Commun.* **2009**, *4*, 1357–1362. [CrossRef]
47. Loizzo, M.R.; Tundis, R.; Statti, G.A.; Menichini, F. Jacaranone: A cytotoxic constituent from *Senecio ambiguus* subsp. *ambiguus* (Biv.) DC. against renal adenocarcinoma ACHN and prostate carcinoma LNCaP cells. *Arch. Pharm. Res.* **2007**, *30*, 701–707. [CrossRef]
48. Gelbaum, L.T.; Zalkow, L.H.; Hamilton, D. Cytotoxic agent from *Senecio anonymus* Wood. *J. Nat. Prod.* **1982**, *45*, 370–372. [CrossRef]
49. Xie, W.D.; Weng, C.W.; Gao, X.; Zhao, H.; Row, K.H. A new farnesene derivative and other constituents from *Senecio Cannabifolius*. *J. Chin. Chem. Soc.* **2010**, *57*, 436–438. [CrossRef]

50. Lajide, L.; Escoubas, P.; Mizutani, J. Cyclohexadienones-insect growth inhibitors from the foliar surface and tissue extracts of *Senecio cannabifolius*. *Experientia* **1996**, *52*, 259–263. [CrossRef]
51. Ma, H.Y.; Yang, L.; Zhang, M.; Wang, C.H.; Wang, Z.T. A new compound from *Senecio cannabifolius* var. *integrilifolius*. *Yao Xue Xue Bao* **2008**, *43*, 626–629.
52. Bohlmann, F.; Zdero, C.; King, R.M.; Robinson, H. The first acetylenic monoterpene and other constituents from *Senecio clevelandii*. *Phytochemistry* **1981**, *20*, 2425–2427. [CrossRef]
53. Wu, C.H.; Zhang, L.; Zhou, P.P.; Sun, M.; Gao, K. Three new jacaranone derivatives from the aerial parts of *Senecio chrysanoides* DC. with their cytotoxic activity. *Phytochem. Lett.* **2015**, *14*, 245–248. [CrossRef]
54. Pettit, G.R.; Einck, J.J.; Brown, P.; Harvey, T.B.; Ode, R.H.; Pase, C.P. Antineoplastic agents. 67. *Senecio fendleri* Gray. *J. Nat. Prod.* **1980**, *43*, 609–616. [CrossRef]
55. Loizzo, M.R.; Tundis, R.; Statti, G.A.; Menichini, F.; Houghton, P.J. In-vitro antiproliferative effects on human tumour cell lines of extracts and jacaranone from *Senecio leucanthemifolius* Poiret. *J. Pharm. Pharmacol.* **2005**, *57*, 897–901. [CrossRef] [PubMed]
56. Torres, P.; Grande, C.; Anaya, J.; Grande, M. Secondary metabolites from *Senecio minutus* and *Senecio boissieri*: A new jacaranone derivative. *Fitoterapia* **2000**, *71*, 91–93. [CrossRef]
57. Xu, H.; Zhang, N.; Casida, J.E. Insecticides in Chinese medicinal plants: Survey leading to jacaranone, a neurotoxicant and glutathione-reactive quinol. *J. Agric. Food Chem.* **2003**, *51*, 2544–2547. [CrossRef]
58. Yang, X.; Yang, L.; Xiong, A.; Li, D.; Wang, Z. Authentication of *Senecio scandens* and *S. vulgaris* based on the comprehensive secondary metabolic patterns gained by UPLC–DAD/ESI-MS. *J. Pharm. Biomed. Anal.* **2011**, *56*, 165–172. [CrossRef]
59. Tian, X.Y.; Wang, Y.H.; Yang, Q.Y.; Yu, S.S.; Fang, W.S. Jacaranone analogs from *Senecio scandens*. *J. Asian Nat. Prod. Res.* **2009**, *11*, 63–68. [CrossRef]
60. Tian, X.Y.; Wang, Y.H.; Yang, Q.Y.; Liu, X.; Fang, W.S.; Yu, S.S. Jacaranone glycosides from *Senecio scandens*. *J. Asian Nat. Prod. Res.* **2006**, *8*, 125–132. [CrossRef]
61. Wang, W.S.; Lu, P.; Duan, C.H.; Feng, J.C. A new jacaranone derivative from *Senecio scandens* var. *incisus*. *Nat. Prod. Res.* **2010**, *24*, 370–374. [CrossRef]
62. Mandić, B.; Gođevac, D.; Vujisić, L.; Trifunović, S.; Tesević, V.; Vajs, V.; Milosavljević, S. Semiquinol and phenol compounds from seven *Senecio* species. *Chem. Pap.* **2011**, *65*, 90–92. [CrossRef]
63. Pérez-Castorena, A.L.; Arciniegas, A.; Martinez, F.; Marquez, C.; Villaseñor, J.L.; Romo de Vivar, A. Chemical constituents of *Packera coahuilensis* and *Packera bellidifolia*. *Biochem. Syst. Ecol.* **2001**, *29*, 203–206. [CrossRef]
64. Ogura, M.; Cordell, G.A.; Farnsworth, N.R. Potential anticancer agents. IV. Constituents of *Jacaranda caucana* Pittier (*Bignoniaceae*). *Lloydia* **1977**, *40*, 157–168. [PubMed]
65. Massaoka, M.H.; Matsuo, A.L.; Figueiredo, C.R.; Farias, C.F.; Girola, N.; Arruda, D.C.; Scutti, J.A.; Romoff, P.; Favero, O.A.; Ferreira, M.J.; et al. Jacaranone induces apoptosis in melanoma cells via ROS-mediated downregulation of Akt and p38 MAPK activation and displays antitumor activity in vivo. *PLoS ONE* **2012**, *7*, e38698. [CrossRef]
66. Presser, A.; Lainer, G.; Kretschmer, N.; Schuehly, W.; Saf, R.; Kaiser, M.; Kalt, M.M. Synthesis of jacaranone-derived nitrogenous cyclohexadienones and their antiproliferative and antiprotozoal activities. *Molecules* **2018**, *23*, 2902. [CrossRef] [PubMed]

plants

MDPI

Article

HS-SPME-GC–MS Volatile Profile Characterization of Peach (*Prunus persica* L. Batsch) Varieties Grown in the Eastern Balkan Peninsula

Dasha Mihaylova [1,*], Aneta Popova [2,*], Radka Vrancheva [3] and Ivayla Dincheva [4]

1 Department of Biotechnology, Technological Faculty, University of Food Technologies, 4002 Plovdiv, Bulgaria
2 Department of Catering and Nutrition, Economics Faculty, University of Food Technologies, 4002 Plovdiv, Bulgaria
3 Department of Analytical Chemistry and Physical Chemistry, Technological Faculty, University of Food Technologies, 4002 Plovdiv, Bulgaria; radka_vrancheva@yahoo.com
4 AgroBioInstitute, Agricultural Academy, 1164 Sofia, Bulgaria; ivadincheva@yahoo.com
* Correspondence: dashamihaylova@yahoo.com (D.M.); popova_aneta@yahoo.com (A.P.)

Abstract: The volatile compounds of eight peach varieties (*Prunus persica* L.)—"Filina", "Gergana", "Ufo-4", "July lady", "Laskava", "Flat Queen", "Evmolpiya", and "Morsiani 90"—growing in Bulgaria were analyzed for the first time. Gas chromatography–mass spectrometry (GC–MS) analysis and the HS-SPME technique revealed the presence of 65 volatile compounds; the main identified components were aldehydes, esters, and fatty acids. According to the provided principal component analysis (PCA) and hierarchical cluster analysis (HCA), the relative quantities of the identified volatile compounds depended on the studied peach variety. The results obtained could be successfully applied for the metabolic chemotaxonomy of peaches.

Keywords: *Prunus persica* L.; gas chromatography–mass spectrometry (GC–MS); volatile compounds; principal component analysis (PCA); hierarchical cluster analysis (HCA); headspace-solid phase micro extraction (HS-SPME)

Citation: Mihaylova, D.; Popova, A.; Vrancheva, R.; Dincheva, I. HS-SPME-GC–MS Volatile Profile Characterization of Peach (*Prunus persica* L. Batsch) Varieties Grown in the Eastern Balkan Peninsula. *Plants* **2022**, *11*, 166. https://doi.org/10.3390/plants11020166

Academic Editor: Barbara Sgorbini

Received: 23 December 2021
Accepted: 6 January 2022
Published: 8 January 2022

1. Introduction

The diversity of volatile compounds is responsible for the unique flavors each food matrix expresses. Aroma is a particularly important and valued feature that illustrates the complex mixture of volatile compounds in foods. The human nose can sense a broad selection of volatile compounds. Although many studies focus on the volatile profiles of various fruit and vegetables, there is a particular need to enhance the available information.

Peaches and nectarines are aroma-dense fruits with a specific, pleasant, and recognizable aroma [1]. Nectarines (*Prunus persica* var. nectarina) may have developed from peach seeds, but their origin is still unknown. The peach (*Prunus persica* L. Batsch), also known as Persian apple, is native to China and Iran. Subsequently, it has spread worldwide. Peaches have a large number of commercial varieties with different shapes, sizes, flesh colors (red, white, or yellow), skin types, seeds, among other variable aspects in relation to this popular fruit, representing a diverse international germplasm [2,3]. The largest producer is China, followed by Italy, Spain, and the United States [2].

The peach is a widely appreciated fruit for consumption, but has not yet been fully studied. The chemical composition of peaches depends on several factors, such as genotype, geographical and climatic conditions, seasonal and meteorological conditions, agronomic practices, stage of maturity, storage conditions, and processing methods [4]. In addition, it has been shown, over the years, that the phytochemicals are not evenly distributed in the fruit tissue; most are concentrated in the rind, particularly in the epidermal and subepidermal layers [5–7].

Volatile compounds, together with sugars and acids, are the main chemical compounds that determine the characteristic aroma and flavor of foods. The peach species holds remarkable characteristics. More than a hundred volatiles have been identified in different peach varieties, with C_6 compounds, esters, benzaldehyde, linalool, C_{13} norisoprenoids, and lactones being the most abundant [8,9]. In general, polyunsaturated fatty acids (PUFAs), such as linoleic acid (18:2) and linolenic acid (18:3), are the main precursors for aroma-related volatiles of peach fruit generated via the lipoxygenase (LOX) pathway or β-oxidation [10]. β-Oxidation leads to the production of the primary aroma in fruits, whereas the LOX system may account for the widest assortment of lipid-derived precursors of aroma compounds in disrupted plant tissues [11]. In addition to their contribution to fruit quality, peach volatiles are also important for the food and fragrance industry, where they are used as flavoring agents. A notable example of a sought-after industrial product with a peach-like aroma is γ-decalactone [12].

Numerous studies demonstrated the large variability in the volatile compound composition of peaches depending on the cultivar, ripening stage, and geographical origin [13–15]. Nectarines generate fewer volatiles than peaches, but have more fruity and floral aroma notes due to greater ester, linalool, and terpinolene production [16]. Authors report changes in volatile aroma-related substances during peach fruit development and ripening after harvest [17–19].

Although the volatile profiles of peaches have been generally widely studied, no information exists on Bulgarian peach varieties. Thus, the goal of this study was to investigate the difference in volatile profiles between eight peach and nectarine varieties in order to provide a tool to evaluate and compare the data on volatiles.

2. Results and Discussion

Peach fruit produces a number of volatile organic compounds (VOCs) [20,21]. The investigation of different peach varieties showed that C_6 compounds, alcohols, aldehydes, and lactones are the most powerful aroma-active compounds [13,14]. Nectarines usually contain C_9 aldehydes, γ-decalactone, and terpenes; peaches contain a majority of $C_{6–10}$ lactones; and flat peaches are dominant in benzaldehyde, γ-decalactone, and δ-dodecalactone [22].

It is no coincidence that volatile compounds are of scientific interest. A quantitative comparison is not always feasible because of the variations in the extraction procedures and quantification procedures used. Different conclusions can be made based on the methods used for collection, the compound concentrations, and the nature of the volatiles produced by fruit. A recent study confirmed significant differences in the number of identified compounds and their quantities exist [23].

The typical taste of most fruits is not present during early formation, but develops after the ripening process [21]. During this period, metabolism changes to catabolism and volatile compounds are formed from the main plant constituents by various biochemical pathways [24]. The climacteric respiration of the fruit aids in the flavor formation during the post-climacteric maturation phase [25]. The richness of the aromatic compounds is of primary importance for consumer acceptance.

Many researchers have confirmed that the same cultivars act differently when subjected to another environment. The varieties "Morsiani 90", "Ufo- 4", "July Lady", and "Flat Queen" are varieties that have been introduced to the Bulgarian geographical region, which means that the orchard management and the ecological factors can result in a different VOCs profile. "Evmolpiya", "Laskava", "Filina", and "Gergana", on the other hand, are local varieties, and their volatile profile is yet to be reported. "Filina" is a result of the breeding selection of "Maycrest" × "July Lady". "Gergana" is created by combining Goldengrand and Aureliogrand varieties. "Laskava" is created by cross-species hybridization, with the participation of the species *Prunus persica* L. Batsch and *Prunus ferganensis* (Kost. and Rjab.) from the parent combination "Hale" × ("Elberta" × "Fergana Yellow"). "Evmolpiya" is a variety obtained by the interspecific hybridization with the participation of *P. persica* var. nucipersica, *P. persca* and *P. davidiana*, from the parent combination "Fantasia" × ("Halle" × *Prunus davidiana*).

2.1. Gas Chromatography–Mass Spectrometry (GC–MS) Profiling of Volatile Compounds of Analyzed Peach Samples

The volatile profiles of eight peach varieties (four local and four introduced) grown in Bulgaria were analyzed by GC–MS. Table 1 is a visual presentation of the results; sixty-five volatile compounds, belonging to seven chemical classes (aldehydes, ketones, alcohols, fatty acids, esters, hydrocarbons, and terpenes), were identified.

Table 1. Identified volatile compounds in peach varieties analyzed by GC–MS. The results are given as % of Total Ion Current *.

Compound	Flavor Contribution **	RI_{lit}	RI_{calc}	"Filina"	"Gergana"	"Ufo-4"	"July Lady"	"Laskava"	"Flat Queen"	"Evmolpiya"	"Morsiani 90"
Aldehydes											
Pentanal	FB	738	741	0.70*	1.17	0.25	1.09	1.14	1.54	0.80	0.99
Hexanal	FFr	800	798	1.95	6.55	3.20	7.13	4.40	2.68	2.24	5.57
(E)-2-Hexenal	N/A	849	850	2.83	4.00	1.35	2.04	7.36	5.30	3.26	6.40
Heptanal	CF	907	909	4.35	1.47	1.58	1.38	1.14	3.95	3.00	1.25
Benzaldehyde	FSw	948	946	0.71	0.49	0.53	0.46	0.48	0.65	0.82	0.42
(E)-2-Heptenal	FW	960	960	0.51	0.35	1.64	1.42	1.49	0.46	0.58	1.30
Octanal	CF	999	1000	1.06	0.73	0.79	0.68	0.71	0.97	1.22	0.62
(E)-2-Octenal	FW	1051	1047	1.43	0.98	0.59	3.12	3.26	1.30	1.65	2.83
2-methyl-Benzaldehyde	FB	1070	1073	0.83	0.57	0.53	0.46	1.71	0.76	0.96	1.49
4-methyl-Benzaldehyde	FB	1084	1085	0.24	0.16	1.44	1.25	1.31	0.21	0.27	1.14
Nonanal	CW	1102	1104	3.89	2.21	2.37	2.06	1.99	3.54	1.48	1.88
(E)-2-Nonenal	CW	1160	1159	1.57	1.07	2.42	2.10	2.20	1.42	1.80	1.91
Decanal	CW	1204	1205	0.39	0.27	0.28	0.25	0.26	0.35	0.44	0.23
(E)-2-Decenal	FW	1250	1253	0.31	0.22	1.50	1.30	1.36	0.29	0.36	1.18
Total aldehydes				*20.77*	*20.24*	*18.47*	*24.74*	*28.81*	*23.42*	*18.88*	*27.21*
Ketones											
3-Octanone	N/A	975	977	0.71	0.49	0.52	0.45	0.25	0.64	0.82	0.41
2-Octanone	NH	991	992	0.56	0.38	0.41	0.36	0.14	0.51	0.64	0.32
γ-hexalactone	FSw	1045	1045	0.29	0.20	0.21	0.19	0.42	0.26	0.33	0.17
2-Nonanone	FW	1090	1088	0.64	0.44	0.47	0.41	0.14	0.58	0.73	0.37
γ-octalactone	SW	1250	1251	1.78	1.22	0.58	2.24	2.80	1.62	1.04	2.04
γ-decalactone	SP	1461	1464	1.11	1.52	1.63	1.42	1.48	1.01	1.28	1.29
γ-dodecalactone	N/A	1673	1675	1.47	2.52	4.00	3.48	3.64	1.34	1.69	3.16
Total ketones				*6.56*	*6.77*	*7.82*	*8.55*	*8.87*	*5.96*	*6.53*	*7.76*
Alcohols											
Pentanol	SW	770	772	1.63	1.12	1.21	1.05	1.10	1.49	1.88	0.95
Hexanol	FFl	851	848	0.49	0.34	0.36	0.32	0.13	0.45	0.57	0.29
Heptanol	NH	920	921	0.74	0.51	0.55	0.47	0.50	0.67	0.85	0.43
Benzyl Alcohol	FFl	1035	1035	0.26	0.18	1.03	0.90	0.94	0.23	0.30	0.81
Nonanol	N/A	1149	1150	1.36	0.93	1.00	0.87	0.39	1.24	1.57	0.79
Total alcohols				*4.48*	*3.08*	*4.15*	*3.61*	*3.06*	*4.08*	*5.17*	*3.27*

Table 1. *Cont.*

Compound	Flavor Contribution **	RI_{lit}	RI_{calc}	"Filina"	"Gergana"	"Ufo-4"	"July Lady"	"Laskava"	"Flat Queen"	"Evmolpiya"	"Morsiani 90"
Fatty Acids											
Butanoic acid	SW	759	760	1.99	1.36	3.31	2.88	3.01	1.81	2.29	2.62
2-methyl-Pentanoic acid	FSw	926	924	1.77	2.72	2.93	2.55	2.66	1.61	2.03	2.31
Hexanoic acid	SS	964	966	2.59	4.80	6.84	5.95	1.22	2.36	2.98	5.41
Octanoic acid	SW	1165	1166	1.74	1.19	2.55	2.21	2.31	1.58	2.00	2.01
Nonanoic acid	NH	1270	1272	2.98	2.05	2.20	1.91	1.20	2.71	3.43	1.74
n-Decanoic acid	CW	1367	1368	2.53	1.74	1.87	1.63	1.07	2.30	2.91	1.48
Dodecanoic acid	FC	1558	1559	3.25	2.23	2.40	2.08	1.18	2.95	3.74	1.89
n-Hexadecanoic acid	FC	1960	1960	1.30	0.89	0.96	0.83	0.49	1.18	1.49	0.76
Total fatty acids				*18.15*	*16.98*	*23.06*	*20.04*	*13.14*	*16.5*	*20.87*	*18.22*
Esters											
Ethyl acetate	FSw	607	610	1.60	1.86	1.99	1.73	1.81	1.46	1.84	1.58
Ethyl pentanoate	FSw	903	905	1.29	1.64	1.76	1.53	1.60	1.17	1.48	1.39
Ethyl tiglate	FFl	940	938	4.76	3.27	3.51	3.06	3.19	4.33	5.48	2.78
Ethyl hexanoate	FSw	998	886	3.99	5.76	1.59	1.38	5.63	3.63	4.59	4.90
Ethyl Heptanoate	FSw	1096	1097	2.08	1.43	1.54	1.34	1.40	1.89	2.40	1.21
Ethyl Benzoate	BD	1170	1173	3.58	2.45	2.64	2.30	2.40	3.25	1.11	2.09
Ethyl Octanoate	FSw	1195	1198	2.04	2.09	2.24	1.95	2.04	2.76	2.35	1.77
Methyl Nonanoate	FC	1226	1225	1.70	2.67	0.88	2.50	2.61	1.54	1.95	2.27
Ethyl oct-(2E)-enoate	FSk	1242	1240	1.55	1.07	1.15	1.00	1.04	1.41	0.79	0.91
1-Octen-3-yl-butanoate	FB	1280	1280	2.19	1.50	0.61	1.40	1.47	1.99	1.52	1.28
Methyl Decanoate	FFl	1320	1322	1.16	0.80	0.72	0.62	0.65	1.06	1.34	0.57
Benzyl butanoate	SP	1344	1345	1.24	2.36	0.76	0.66	0.69	1.13	0.43	0.60
(2E)-Octenyl butanoate	FB	1388	1385	1.63	1.12	1.20	1.05	1.09	1.48	1.87	0.95
Linalool butanoate	FFl	1423	1425	1.19	2.19	1.35	2.04	2.14	2.90	1.36	1.86
2-Phenyl ethyl butanoate	FFl	1435	1436	2.75	1.88	2.03	1.76	1.84	2.50	3.16	1.60
2-Phenyl propyl butanoate	FS	1482	1480	1.86	1.28	3.37	1.19	1.25	1.69	2.14	1.08
Total esters				*34.61*	*33.37*	*27.34*	*25.51*	*30.85*	*34.19*	*33.81*	*26.84*

Table 1. *Cont.*

Compound	Flavor Contribution **	RI_{lit}	RI_{calc}	"Filina"	"Gergana"	"Ufo-4"	"July Lady"	"Laskava"	"Flat Queen"	"Evmolpiya"	"Morsiani 90"
Hydrocarbons											
Undecane	N/A	1098	1095	1.06	0.73	0.79	0.68	0.71	0.97	0.22	0.62
Dodecane	N/A	1200	1202	1.51	1.30	1.40	1.22	1.27	1.37	0.73	1.11
Tridecane	N/A	1302	1304	0.33	1.74	1.87	1.62	0.70	0.30	0.38	1.48
Tetradecane	FW	1400	1401	2.03	2.31	2.05	2.16	1.26	1.85	1.03	1.97
Pentadecane	FW	1497	1495	0.97	0.66	0.71	0.62	0.65	0.88	0.61	0.56
Hexadecane	N/A	1600	1601	0.52	0.36	0.39	0.34	0.35	0.48	0.60	0.31
Heptadecane	N/A	1701	1700	0.85	0.58	0.63	0.54	0.26	0.77	0.97	0.49
Total hydrocarbons				*7.27*	*7.68*	*7.84*	*7.18*	*5.2*	*6.62*	*4.54*	*6.54*
Terpenes											
β-Myrcene	FW	980	985	1.13	2.53	0.96	0.84	2.15	1.02	1.29	2.15
p-Cymene	CF	1018	1020	0.15	1.06	1.14	0.99	0.36	1.40	0.18	0.90
Limonene	CS	1024	1022	0.66	1.96	3.26	2.83	1.49	0.60	0.76	1.67
(Z)-β-Ocimene	NH	1035	1036	0.79	0.54	0.58	0.51	0.25	0.72	0.91	0.46
(E)-β-Ocimene	NH	1042	1041	1.42	0.97	1.04	0.91	0.90	1.29	1.63	0.83
Linalol	FFl	1093	1094	2.22	1.86	2.00	1.74	1.48	1.10	2.75	1.58
(Z)-β-Farnesene	CS	1440	1443	0.26	0.87	0.93	0.81	0.80	1.15	0.30	0.74
(E)-β-Farnesene	N/A	1452	1455	0.60	0.41	0.24	0.39	0.84	0.55	0.69	0.35
Total terpenes				*7.23*	*10.2*	*10.15*	*9.02*	*8.27*	*7.83*	*8.51*	*8.68*

* RI—Kovats retention index; ** FB—fruity, berry; FSw—fruity, sweet; FFr—fruity, fresh; FFl—fruity, floral; CW—citrus, waxy; CF—citrus, fresh; CS—citrus, sweet; SS—sour, sweet; SP—sweet, peachy; SW—sweet, waxy; FC—fatty, coconut; FW—fresh, waxy; BD—bitter, dry; NH—natural, herbal; FSk—fruit skin; N/A—not available.

Aldehydes comprise 21% of the identified compounds, with the dominance of hexanal, (E)-2-hexenal, and nonanal in all peach varieties (Table 1). Some aldehyde compounds are formed in the event of frost damage: octanal, heptanal, and pentanal [26]. The different amounts in the studied samples prove that chilling injuries are variety-dependent, and do not follow the ripening period of the peach.

Aldehydes are flavor-contributing for premature fruit. They bring out a specific fresh-green odor to the fruit. C_6 aldehyde compounds are desired, especially in not fully ripe pears, plums, and apples. Such compounds decrease in quantity during the process of full ripening of the fruit [27]. Following the abovementioned, it can be concluded that in the absence of other unfavorable conditions, the "Laskava" variety, which contains the most aldehydes, can be stored for the longest period, while "Ufo-4" and "Evmolpiya" should reach the market within the shortest time. The melon-like flavor of the "Laskava", "Ufo-4", and "July Lady" varieties could be related to (E)-2-nonenal detection [28]. Hexanal is reported in literature [29] as a major compound in the volatile analysis of nectarines, which is further supported in the currently established results for the "Gergana" and "Morsiani 90" varieties. It is associated with a sweet, fruity taste [30]. Heptanal, 2-hexenal, and octanal, typical for peach varieties, were found to contribute to the fresh odor [31].

Lactones possess high aromatic values in peaches due to their low odor threshold. Lactones, as intramolecular esters of 4- and 5-hydroxy acids, shape the basic peach aroma [22], and have high aroma effects in stone fruits, in general. Among the seven identified ketones, γ-octalactone and γ-dodecalactone, that give peach-like aroma, were in the highest relative concentrations. These lactones act in association with aldehydes, alcohols and terpenoids, which are responsible for the spicy, floral and fruity features in the peach [32]. γ-Decalactone and γ-octalactone are characteristic volatile compounds for peaches. The compound γ-octalactone, which confers a sweet herbaceous, coconut-like odor and taste, was predominant in the "July Lady" and "Laskava" varieties [33]. γ-Decalactone was most abundant in the two nectarine varieties object of analysis, which supports the literature stating that this is the most common compound identified in the pulp of nectarines [13]. It has to be noted that lactone identification is highly dependent on the extraction conditions, which can be identified as a limitation in every study on the subject. The absence or presence of certain lactones can be due to the assessment methodology being used [34,35]. For example, other authors have managed to identify more than ten C_5-C_{10} γ-lactones [8,36]. It has been suggested that lactones in peaches are a result of the β-oxidation pathway of fatty acids [21].

The biosynthesis of fatty acids has been reported to being highly influential on the volatile profile. Eight fatty acids were identified, with hexanoic acid (1.22–6.84% of TIC) being the principle one (Table 1). Nonanoic and dodecanoic acid were the second most abundant of the investigated peach fruits. Fatty acids are important as they serve as carriers for some lipophilic vitamins and bioactive compounds present in fruits, and the presence of essential fatty acids is believed to play an important role in the prevention of cardiovascular diseases [37,38]. Acids most likely contribute little to the aroma profile though, because they normally have high odor detection thresholds [39].

Alcohols represent approximately 8% of the total identified compounds, with pentanol and nonanol predominating. Pentanol, which is responsible for the bouquet and astringent aroma description [40], was the main alcohol in all the studied samples (in the range from 0.95% to 1.88% of TIC). Benzyl alcohol, found in the highest concentrations in "Ufo-4", is described as having a floral aroma [41]. Alcohol dehydrogenase in the fruit mesocarp accumulates throughout ripening [20], and alcohols are usually left undetectable by the consumers. Relatively low alcohol quantities suggest that the fruit is not overripe [42]. Fruit juice pH effectively converts alcohols and aldehydes into flavoring agents [42].

Esters are the main VOCs produced by horticultural crops. Esters, especially straight chain esters, are generally metabolized from fatty acids [43]. The higher the amount of esters, the more pronounced the aroma and the taste of the fruit [44]. Esters, accounting for 25% of the identified compounds, represented the largest group. The composition of esters differed both qualitatively and quantitatively among the peach samples. It has to be noted that the ester distribution is reported to be different within the part of the fruit [36]. The total ester content varied between 25.51% and 34.61% of TIC. Esters contribute to the fruity aroma of peaches [45]. Ethyl hexanoate (in the range from 1.38% to 5.76% of TIC) and ethyl tiglate (between 2.78% and 5.4% of TIC) were present at the highest relative concentrations among the estimated esters. Although the amount of esters was predominant in all eight varieties, the relative TIC was two times smaller than the reported literature average [46]. This is most likely due to the specificity of orchard location, light availability, temperature, and season specificity, as well as ecological location.

The presence of butanoates (B) and hexanoates (H) confirms the ripeness of the studied varieties from the averages of 9.48 (B) and 3.89 (H), 10.5 (B) and 2.61 (H), and 7.19 (B) and 5.33 (H) of the TIC in peaches, flat peaches, and nectarines, respectively.

Seven hydrocarbons were identified, of which tetradecane and tridecane were dominant. The total content of hydrocarbons was in the range from 7.23 to 10.20% of the TIC among the peach varieties. Lu et al. [47] reported the presence of several hydrocarbons in peaches. Tetradecane is considered a creamy descriptor, whereas dodecane is a woody descriptor [48]. The authors reported the presence of (E)-2-nonen-1-ol, 2-methylpropyl acetate, ethyl butanoate, butyl acetate, 3-methyl-1-butyl acetate, ethyl pentanoate, ethyl hexanoate, hexyl acetate, methyl octanoate, and hexyl hexanoate [47]. Other researchers differentiate ethyl acetate as the major ester compound in both peaches and nectarines [49].

Limonene, linalool, and p-cymene are listed as key flavor compounds that form the characteristic aroma profile [27,50]. The predominant compound in the nectarine varieties was β-mycrene, while linalool was the most abundant in the peach varieties, bringing out the floral aroma in them. Terpenes contribute to a floral flavor of fruits [51], and a sweetening taste [20], and are the principal components in plant essential oils. Among the eight identified terpenes, limonene (0.60–3.26% TIC) and linalool (1.1–2.75% TIC) were in the highest relative concentrations. Linalool was also the major terpenoid compound in other peaches and nectarines [13]. Linalool is reported to possess a floral and citrus-like aroma [52]. Mycrene is a precursor of linalool and is reported as an usual representative in peaches and nectarines, characterizing their aroma with a woody note. The currently established results also prove that the more the linalool, the less the myrcene. Other authors also advocate the thesis that myrcene is found in higher amounts in nectarines [35], which is further supported by the current values.

Terpene compounds (i.e., linalool) and alcohols (i.e., 1-hexanol) are reported to be less abundant than aldehydes in apricots and were detected to decrease with ripening [53]. The relative presence of terpenes in the eight studied varieties was lower compared to the aldehydes, in accordance with the results mentioned above.

The distribution of the main chemical families is presented in Figure 1. The volatile compounds were relatively uniformly distributed. The most abundant in all the varieties was esters, followed by aldehydes and fatty acids. The results are in line with those of Ortiz et al. [19], who stated that volatile esters often represent the major contribution in peaches (*Prunus persica* L.). However, fruit flavor is usually a complex mixture of a wide range of compounds [21]. The volatile composition provides significant information about the healthy composition of food, as it is synthesized from essential nutrients [54].

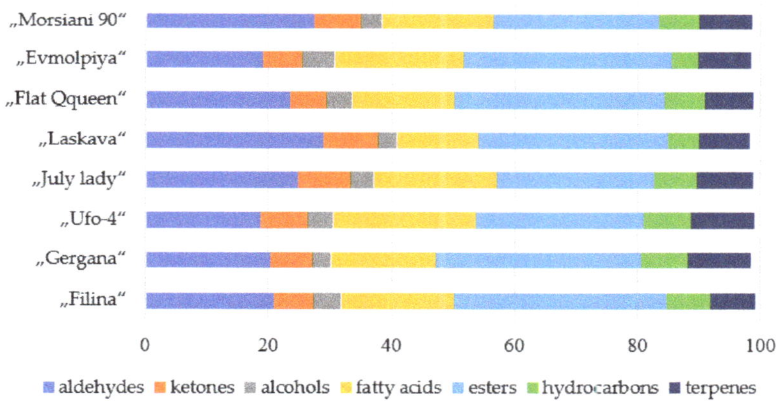

Figure 1. Distribution of volatile compounds according to their chemical families in the studied peach varieties.

The "Laskava" variety had the highest relative total content of aldehydes (28.82% of the TIC) and ketones (8.87% of the TIC), while "Ufo-4" and "Flat Queen" had the lowest content. "Ufo-4" was the variety with the highest total relative quantity of fatty acids (23.05% of TIC) and hydrocarbons (7.83% of TIC). Alcohols and fatty acids were at the lowest total relative concentration in the "Laskava" variety, and the "Evmolpiya" variety had the highest total relative content of alcohols. "Gergana" and "Ufo-4" varieties were with the highest total relative content of terpenes (10.20% of TIC and 10.16% of TIC, respectively).

Based on the odor descriptors, volatile compounds in peaches can be divided into several sensory groups, including green, fruity, and peach-like aromas [13]. To obtain a clear picture of the overall contribution of the identified compounds on the general flavor of the studied peaches, several figures were created. Figure 2 shows the odor/taste distribution [55] in the nectarine varieties based on the VOCs identified in the studied samples.

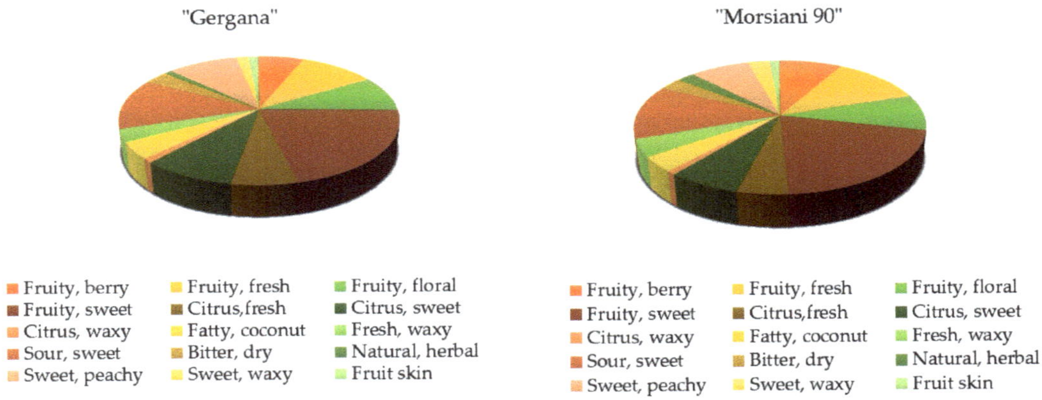

Figure 2. Flavor component distribution (%) in nectarine varieties.

The two nectarine varieties had a dominant fruity, sweet vibe, as well as the sour sweet and sweet peachy descriptors. The "Gergana" variety can be seen as more citrus sweet and the "Morsiani 90" variety is more fruity fresh. Both varieties are considered sweet, but not as floral but fruity.

The results for the two flat peach samples are given in Figure 3. The two flat peach samples are mostly fruity sweet, and floral. The "Ufo-4" variety possesses more sweet citrus scents than the "Flat Queen", which is probably the reason for the more sour sweetness of "Ufo-4" compared to "Flat Queen". This is quite distinctive for the white flesh peaches, as they are usually described as mildly acidic with a distinct sweet taste.

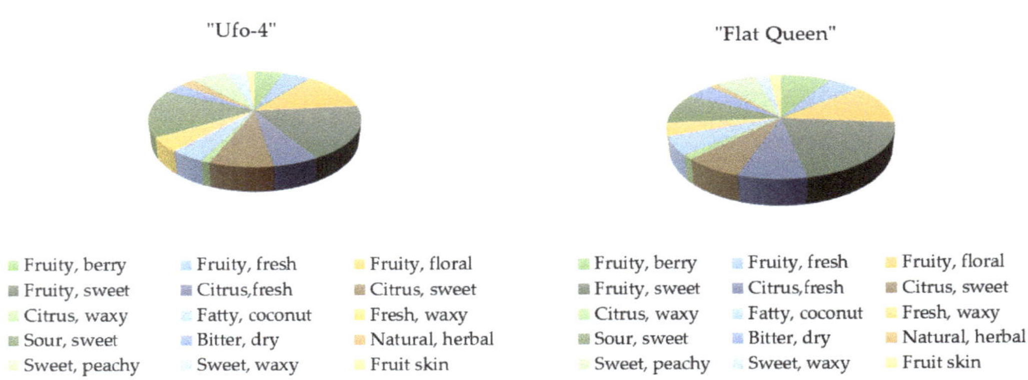

Figure 3. Flavor component distribution (%) in flat peach varieties.

When characterizing the four peach samples (Figure 4), it is evident that they are mainly fruity sweet.

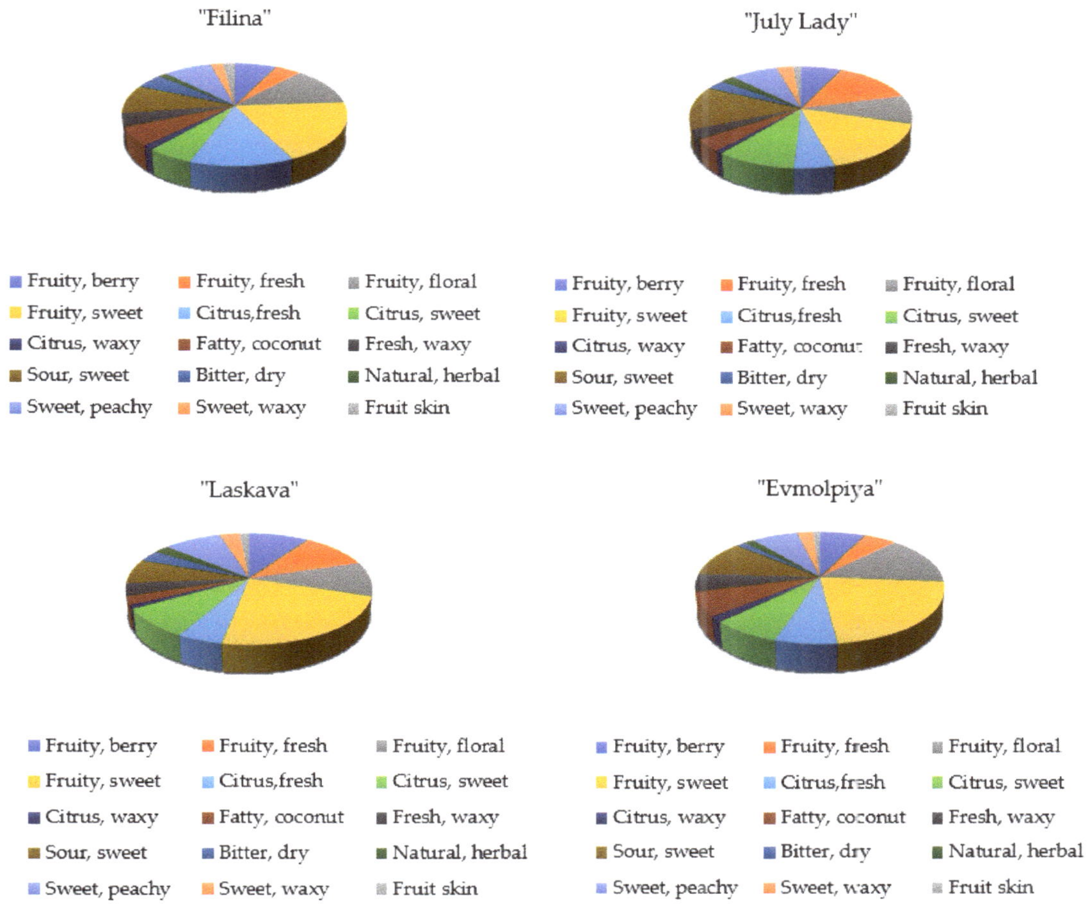

Figure 4. Flavor component distribution (%) in peach varieties.

The "Filina" variety has a moderate citrus fresh scent contributing to its overall flavor, while "July Lady" has a citrus sweet scent. "Evmolpiya" is more sour sweet than "Laskava", while "Laskava" has a specific sweet waxy flavor.

2.2. Principal Component Analysis (PCA) and Hierarchical Cluster Analysis (HCA) of GC–MS Data

PCA is an exploratory technique that is typically used as a descriptive analysis for variable selection in a propensity model. The result quality fluctuates significantly by the number of factors or the factors-to-variables ratio. Considering the variables-to-observations ratio is not a good way to determine the required number of observations [56].

In order to confirm sample differences or similarities, principal component analysis (PCA) and hierarchical cluster analysis (HCA) of the volatile compounds identified were applied. According to the PCA plot obtained, the first two principal components PC1 (47.4%) and PC2 (17.6%) accounted for 65% of the total variance of all identified volatile compounds in the analyzed peach varieties (Figure 5).

(E)-2-Hexanal, pentanal, γ-octalactone, methyl nonanoate, dodecane, ethylbenzoate, hexanal, linalool, 2-methyl-benzaldehyde, β-myrcene, and (Z)-β-farnesene showed high positive loading scores in PC1 that distinguished the "Laskava", "Morsiani 90", and "Gergana" varieties from the other five. Volatile compounds with high negative scores in PC1 were 1-octen-3-yl-butanoate, ethyl hexanoate, (E)-β-farnesene, pentadecane, heptanal, methyl decanoate, nonanal, γ-hexalactone, and ethyl octanoate, which distinguished the "Flat Queen" and "Filina" varieties from the others. The "Evmolpiya" variety appeared clearly different from the other peach varieties, shown by the high negative loading values in PC2 of (E)-2-decenal, n-decanoic acids, 2-nonanone, linalool, and 2-phenyl propyl butanoate. Tetradecane, p-cymene, butanoic acid, 2-methyl-pentanoic acid, ethyl pentanoate, ethyl acetate, hexanoic acid, and limonene clearly differentiated the "July lady" and "Ufo-4" varieties from the other six.

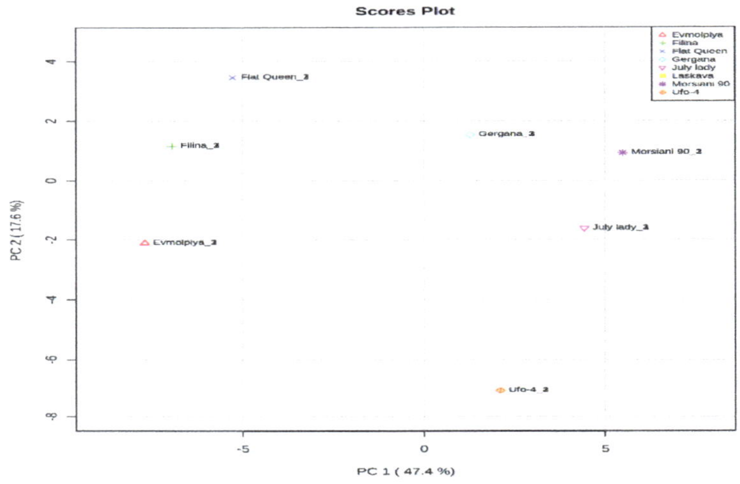

(A)

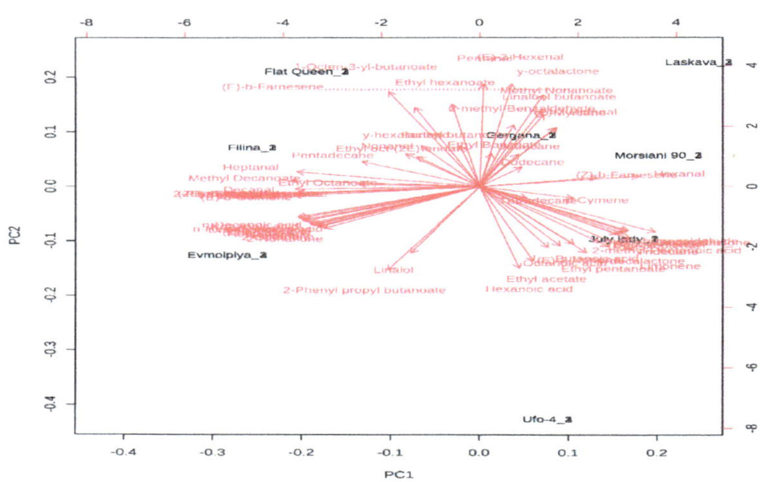

(B)

Figure 5. Principal component analysis (PCA) of GC–MS data for volatile compounds of peach (*Prunus persica* L.) varieties. (**A**) Principal component score plot for the eight peach varieties. (**B**) Eigenvector loading values of compounds identified in the eight peach varieties.

No clear distinguishment between nectarines, peaches, and flat fruits could be stated based on the results. The early season fruits were quite similar to the late season and to the mid-season, and in reverse. Moreover, when clustering the metabolites in polar fractions (phenolic acids, amino acids, organic acids, sugar alcohols, carbohydrates, and saturated and unsaturated fatty acids [57]), different phytochemical similarities were reported.

HCA was performed to understand the relationships between the analyzed varieties. According to the dendrogram and heatmap obtained (Figures 6 and 7), the "Filina" variety had the highest phytochemical similarity to "Flat Queen" and these were grouped in one cluster, with "Morsiani 90" and "July lady" grouped in another cluster. The observed clusters can be explained by the similar quantities of the identified metabolites. HCA also showed the highest diversity among the "Evmolpiya" and "Morsiani 90" varieties because of the significant differences in the quantities of the identified metabolites. According to the results obtained by PCA and HCA, the relative amounts of the identified volatile compounds differed between the studied varieties. In addition, many studies [13,15,51,58,59] also report that the volatile composition of peaches depends on the variety.

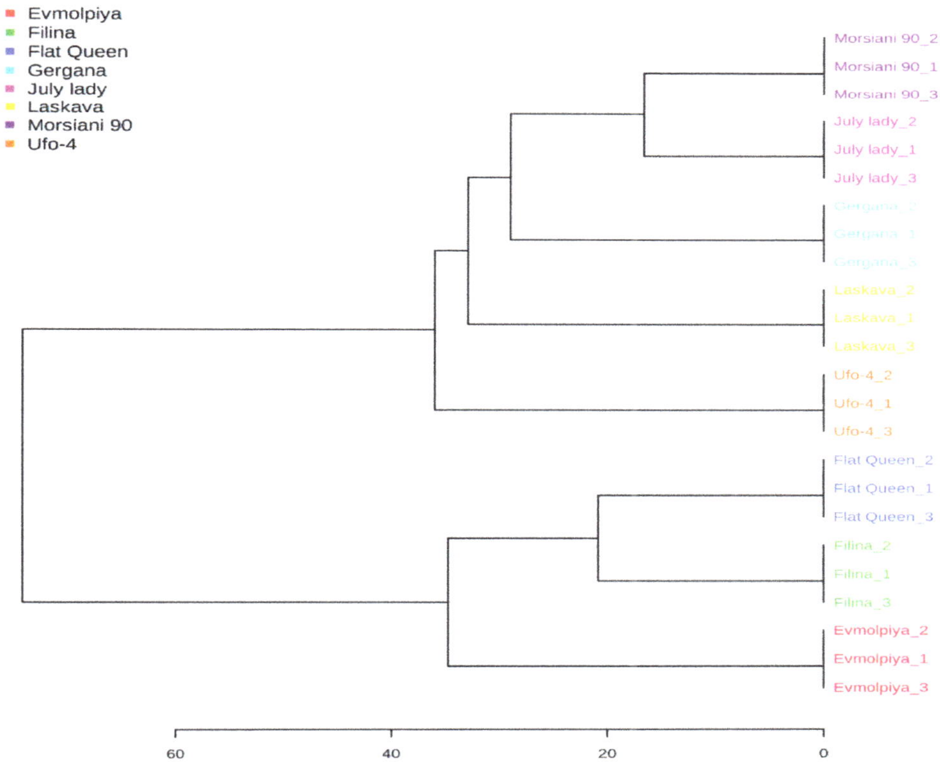

Figure 6. Clustering result of peach varieties shown as a dendrogram (by Euclidean distance measure, and Ward's clustering algorithm).

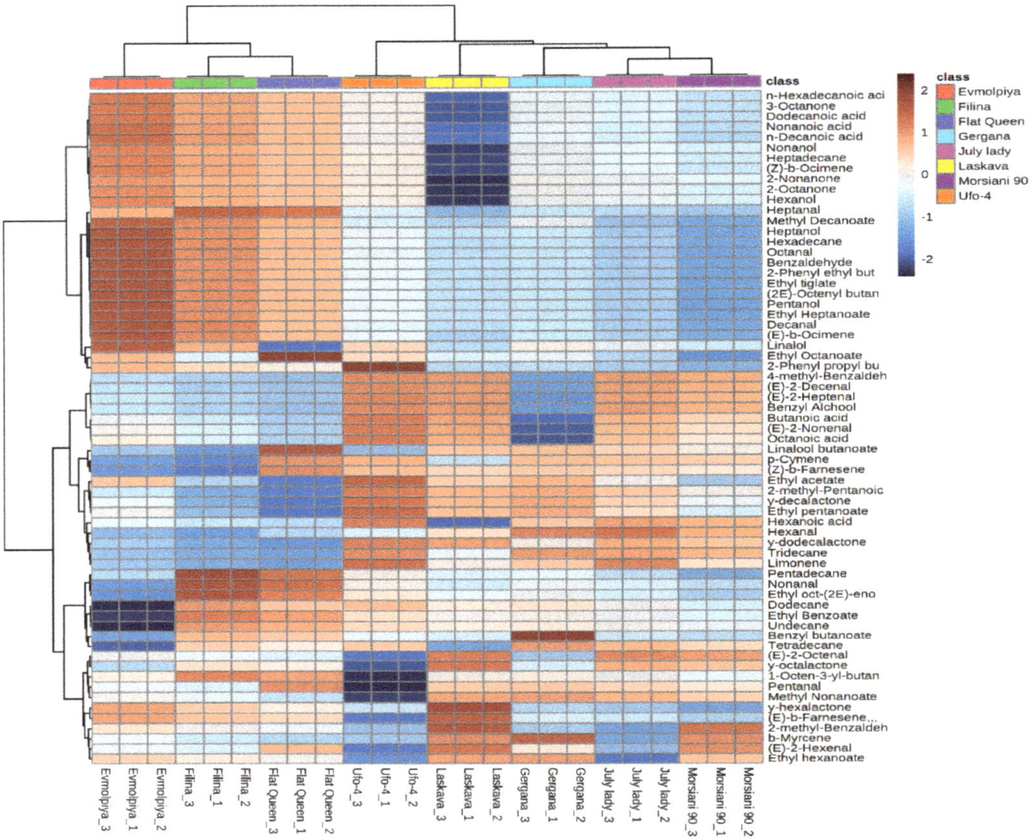

Figure 7. Clustering result of peach varieties shown as a heatmap. The color scale of the heat map ranges from dark brown (value, +2) to dark blue (value, −2). The values were normalized by $\log_{10}$ transformation.

3. Materials and Methods

3.1. Fruit Material

Eight peach and nectarine varieties, from early to late harvesting dates, were the objects of analysis: "Filina" (peach, mid-June), "Ufo-4" (flat peach, white flesh, late June), "Gergana" (nectarine, late June), "Laskava" (peach, early August), "July Lady" (peach, mid-July), "Flat Queen" (flat peach, white flesh, early September), "Evmolpiya" (peach, mid-September), and "Morsiani 90" (nectarine, late September). All were grown on the same plantation in the 2020 season. "Filina", "Gergana", "Laskava", and "Evmolpiya" are Bulgarian varieties created at the Fruit Growing Institute, Plovdiv, through interspecific hybridization. "Flat Queen", "Morsiani 90", "July Lady", and "Ufo-4" are introduced varieties. No bactericides were applied to plants during testing. The undamaged peach, nectarine, and flat fruit were harvested at full ripeness, corresponding to their size, skin color, firmness and total soluble solids values. Fruit on the trees was considered ripe when the growth of the fruit had stopped and the fruit began to soften, exhibited a yellow or orange ground color (which is also specific to each variety), and was easily detached.

3.2. Headspace-Solid Phase Micro Extraction (HS-SPME) and Gas Chromatography–Mass Spectrometry Analysis (GC–MS)

For headspace sampling, a 2 cm SPME fiber assembly Divinylbenzene/Carboxen/Polydimethylsiloxane (DVB/CAR/PDMS, Supelco, Bellefonte, PA, USA) was used.

The HS-SPME technique was used for the extraction of the peach volatile according the procedure described by Uekane et al. [60]. The sampling procedure was automatically performed with a G1888 Network Headspace Sampler that was integrated on-line with the corresponding GC–MS system. An Agilent 7890A GC unit coupled to an Agilent 5975C MSD and a DB-5ms (30 m × 0.25 mm × 0.25 μm) column were used to analyze the volatile compounds in all investigated samples. The oven temperature program was as follows: from 40 °C (hold 1 min) to 250 °C (hold 5 min) at 2 °C/min; carrier gas: helium with flow rate: 1.0 mL/min; transfer line temperature: 270 °C; ion source temperature: 200 °C, EI energy: 70 eV, mass range: 50 to 550 m/z at 1.0 s/decade.

AMDIS software, version 2.64 (Automated Mass Spectral Deconvolution and Identification System, NIST, Gaithersburg, MD, USA) aided in the reading of the obtained mass spectra and the identification of the metabolites. For identification, the separated compounds were compared to the GC–MS spectra and Kovats retention index (RI) of reference compounds in the Golm Metabolome Database (http://csbdb.mpimp-golm.mpg.de/csbdb/gmd/gmd.html, accessed on 25 August 2021) and NIST'08 database (NIST Mass Spectral Database, PC-Version 5.0, 2008 from National Institute of Standards and Technology, Gaithersburg, MD, USA). The 2.64 AMDIS software recorded the RIs of the compounds with a standard n-hydrocarbon calibration mixture (C8–C36, Restek, Teknokroma, Spain).

3.3. Statistical Analysis

PCA and HCA of GC–MS data were conducted using MetaboAnalyst, a web-based platform (www.metaboanalyst.ca, accessed on 17 September 2021) [61]. Three analyses were performed for each of the eight peach varieties. The concentrations of the identified compounds were employed for PCA. All zero values were replaced with a small value (half of the minimum positive values in the original data) assumed to be the detection limit. First, PCA was applied in order to calculate the eigenvector loading values and to identify the major statistically different components among the observations (samples). The GC–MS data were mean-centered and the PCA model was obtained at a confidence level of 95%. The GC–MS data were also subjected to HCA, which produced a dendrogram by Ward's method of hierarchical clustering and Euclidean distance measurement between the analyzed samples. The values were normalized by log10 transformation.

4. Conclusions

Aroma is particularly important in sensory evaluation. The human nose is sophisticated enough to be able to distinct the odor and flavor volatiles but the association between VOCs and aroma/flavor perception is not straightforward. Identifying odor-active compounds is an important stepping stone in the study of volatiles.

No data on VOCs profile for the local ("Filina", "Laskava", "Gergana", "Evmolpiya") or introduced ("Flat Queen", "Morsiani 90", "July Lady", "Ufo-4") peach varieties are available in the scientific literature, which makes this the first report on the subject. In total, sixty-five volatile compounds were identified; aldehydes, esters, and fatty acids were predominant. The overall contribution of the identified compounds on the general flavor of the studied peaches was explored. Some of the main groups represented were sweet, fruity, and floral flavors.

According to the obtained results, the PCA and HCA of volatile compounds can be successfully applied for the metabolic chemotaxonomy of peaches.

Author Contributions: Conceptualization, D.M. and A.P.; methodology, I.D.; software, R.V.; validation, D.M., R.V., A.P. and I.D.; formal analysis, I.D.; investigation, A.P. and D.M.; resources, D.M.; data curation, A.P. and D.M.; writing—original draft preparation, A.P. and D.M.; writing—review and editing, A.P. and D.M.; visualization, A.P. and D.M.; supervision, D.M.; project administration, D.M.; funding acquisition, D.M. All authors have read and agreed to the published version of the manuscript.

Funding: This research and the APC were funded by the Bulgarian National Science Fund, grant number KΠ-06-H37/23.

Institutional Review Board Statement: Not applicable.

Informed Consent Statement: Not applicable.

Data Availability Statement: The data presented in this study are available on request from the corresponding author.

Acknowledgments: This work was supported by the Bulgarian National Science Fund, project no. KΠ-06-H37/23 (granted to Dasha Mihaylova). The authors would like to express their gratitude to Argir Zhivondov from the Fruit Growing Institute, Plovdiv (Bulgaria), and his team for kindly providing the peach samples.

Conflicts of Interest: The authors declare no conflict of interest. The funders had no role in the design of the study; in the collection, analyses, or interpretation of data; in the writing of the manuscript, or in the decision to publish the results.

References

1. Bassi, D.; Mignani, I.; Spinardi, A.; Tura, D. Chapter 23—PEACH (*Prunus persica* (L.) Batsch). In *Nutritional Composition of Fruit Cultivars*; Simmonds, M.S.J., Preedy, V.R., Eds.; Academic Press: San Diego, CA, USA, 2016; pp. 535–571. ISBN 978-0-12-408117-8.
2. Font i Forcada, C.; Gradziel, T.M.; Gogorcena, Y.; Moreno, M.Á. Phenotypic diversity among local Spanish and foreign peach and nectarine [*Prunus persica* (L.) Batsch] accessions. *Euphytica* **2014**, *197*, 261–277. [CrossRef]
3. Davidović, S.M.; Veljović, M.S.; Pantelić, M.M.; Baošić, R.M.; Natić, M.M.; Dabić, D.Č.; Pecić, S.P.; Vukosavljević, P.V. Physicochemical, Antioxidant and Sensory Properties of Peach Wine Made from Redhaven Cultivar. *J. Agric. Food Chem.* **2013**, *61*, 1357–1363. [CrossRef]
4. Liu, H.; Cao, J.; Jiang, W. Evaluation of physiochemical and antioxidant activity changes during fruit on-tree ripening for the potential values of unripe peaches. *Sci. Hortic.* **2015**, *193*, 32–39. [CrossRef]
5. Cantín, C.M.; Moreno, M.A.; Gogorcena, Y. Evaluation of the Antioxidant Capacity, Phenolic Compounds, and Vitamin C Content of Different Peach and Nectarine [*Prunus persica* (L.) Batsch] Breeding Progenies. *J. Agric. Food Chem.* **2009**, *57*, 4586–4592. [CrossRef]
6. Campbell, O.; Padilla-Zakour, O. Phenolic and carotenoid composition of canned peaches (*Prunus persica*) and apricots (Prunus armeniaca) as affected by variety and peeling. *Food Res. Int.* **2013**, *54*, 448–455. [CrossRef]
7. Liu, H.; Cao, J.; Jiang, W. Evaluation and comparison of vitamin C, phenolic compounds, antioxidant properties and metal chelating activity of pulp and peel from selected peach cultivars. *LWT Food Sci. Technol.* **2015**, *63*, 1042–1048. [CrossRef]
8. Eduardo, I.; Chietera, G.; Bassi, D.; Rossini, L.; Vecchietti, A. Identification of key odor volatile compounds in the essential oil of nine peach accessions. *J. Sci. Food Agric.* **2010**, *90*, 1146–1154. [CrossRef]
9. Muto, A.; Müller, C.; Bruno, L.; McGregor, L.; Ferrante, A.; Ada, A.; Chiappetta, C.; Bitonti, M.; Rogers, H.; Spadafora, N. Fruit volatilome profiling through GC × GC-ToF-MS and gene expression analyses reveal differences amongst peach cultivars in their response to cold storage. *Sci. Rep.* **2020**, *10*, 1–16. [CrossRef]
10. Schwab, W.; Davidovich-Rikanati, R.; Lewinsohn, E. Biosynthesis of plant-derived flavor compounds. *Plant J.* **2008**, *54*, 712–732. [CrossRef]
11. Ortiz Catalan, A.; Echeverría, G.; Graell, J.; López, M.L.; Lara, I. Biosynthesis of volatile compounds during on-tree maturation of "Rich Lady" peaches. *Acta Hortic.* **2012**, *962*, 515–522. [CrossRef]
12. Sánchez, G.; Venegas-Calerón, M.; Salas, J.; Monforte, A.; Badenes, M.; Granell, A. An integrative "omics" approach identifies new candidate genes to impact aroma volatiles in peach fruit. *BMC Genom.* **2013**, *14*, 343. [CrossRef]
13. Wang, Y.; Yang, C.; Li, S.; Yang, L.; Wang, Y.; Zhao, J.; Jiang, Q. Volatile characteristics of 50 peaches and nectarines evaluated by HP–SPME with GC–MS. *Food Chem.* **2009**, *116*, 356–364. [CrossRef]
14. Zhu, J.; Xiao, Z. Characterization of the key aroma compounds in peach by gas chromatography–olfactometry, quantitative measurements and sensory analysis. *Eur. Food Res. Technol.* **2019**, *245*, 129–141. [CrossRef]
15. Bavcon-Kralj, M.; Jug, T.; Komel, E.; Fajt, N.; Jarni, K.; Živković, J.; Mujić, I. Aromatic compound in different peach cultivars and effect of preservatives on the final aroma of cooked fruits. *Hem. Ind.* **2014**, *68*, 767–779. [CrossRef]
16. Visai, C.; Vanoli, M. Volatile compound production during growth and ripening of peaches and nectarines. *Sci. Hortic.* **1997**, *70*, 15–24. [CrossRef]

17. Zhang, B.; Xi, W.; Wei, W.; Shen, J.; Ferguson, I.; Chen, K. Changes in aroma-related volatiles and gene expression during low temperature storage and subsequent shelf-life of peach fruit. *Postharvest Biol. Technol.* **2011**, *60*, 7–16. [CrossRef]
18. Zhang, B.; Shen, J.-Y.; Wei, W.-W.; Xi, W.-P.; Xu, C.-J.; Ferguson, I.; Chen, K. Expression of genes associated with aroma formation derived from the fatty acid pathway during peach fruit ripening. *J. Agric. Food Chem.* **2010**, *58*, 6157–6165. [CrossRef]
19. Ortiz, A.; Graell, J.; López, M.L.; Echeverría, G.; Lara, I. Volatile ester-synthesising capacity in 'Tardibelle' peach fruit in response to controlled atmosphere and 1-MCP treatment. *Food Chem.* **2010**, *123*, 698–704. [CrossRef]
20. Seker, M.; Gündoğdu, M.; Ekinci, N.; Gür, E. Recent Developments on Aroma Biochemistry in Fresh Fruits. *Int. J. Innov. Approaches Sci. Res.* **2021**, *5*, 84–103. [CrossRef]
21. El Hadi, M.A.; Zhang, F.-J.; Wu, F.-F.; Zhou, C.-H.; Tao, J. Advances in Fruit Aroma Volatile Research. *Molecules* **2013**, *18*, 8200–8229. [CrossRef]
22. Xi, W.; Zheng, Q.; Lu, J.; Quan, J. Comparative Analysis of Three Types of Peaches: Identification of the Key Individual Characteristic Flavor Compounds by Integrating Consumers' Acceptability with Flavor Quality. *Hortic. Plant J.* **2017**, *3*, 1–12. [CrossRef]
23. Mohammed, J.; Belisle, C.E.; Wang, S.; Itle, R.A.; Adhikari, K.; Chavez, D.J. Volatile Profile Characterization of Commercial Peach (*Prunus persica*) Cultivars Grown in Georgia, USA. *Horticulturae* **2021**, *7*, 516. [CrossRef]
24. Perez, A.G.; Rios, J.J.; Sanz, C.; Olias, J.M. Aroma components and free amino acids in strawberry variety Chandler during ripening. *J. Agric. Food Chem.* **1992**, *40*, 2232–2235. [CrossRef]
25. Reineccius, G. *Flavor Chemistry and Technology*; CRC Press: Boca Raton, FL, USA, 2005; ISBN 020-3485-343.
26. Romero, I.; García-González, D.L.; Aparicio-Ruiz, R.; Morales, M.T. Study of Volatile Compounds of Virgin Olive Oils with "Frostbitten Olives" Sensory Defect. *J. Agric. Food Chem.* **2017**, *65*, 4314–4320. [CrossRef]
27. Gur, E.; Ekinci, N.; Gundogdu, M.A.; Seker, M. Comparison of Fruit Aromatic Compounds of Cardinal Peach, Armking and White Nectarine Varieties. In Proceedings of the International INES Academic Researches Congress, Antalya, Turkey, 18–21 October 2017; pp. 2200–2207.
28. Engel, K.H.; Tressl, R. Studies on the volatile components of two mango varieties. *J. Agric. Food Chem.* **1983**, *31*, 796–801. [CrossRef]
29. Krishna Kumar, S.; Hern, T.; Liscombe, D.; Paliyath, G.; Sullivan, J.A.; Subramanian, J. Changes in the volatile profile of 'Fantasia' nectarines [*Prunus persica* (L.) Batsch, var. nectarina] treated with an enhanced freshness formulation (EFF) containing hexanal. *Hortic. Environ. Biotechnol.* **2020**, *61*, 525–536. [CrossRef]
30. Maul, F.; Sargent, S.A.; Sims, C.A.; Baldwin, E.A.; Balaban, M.O.; Huber, D.J. Tomato Flavor and Aroma Quality as Affected by Storage Temperature. *J. Food Sci.* **2000**, *65*, 1228–1237. [CrossRef]
31. An, K.; Liu, H.; Fu, M.; Qian, M.C.; Yu, Y.; Wu, J.; Xiao, G.; Xu, Y. Identification of the cooked off-flavor in heat-sterilized lychee (*Litchi chinensis* Sonn.) juice by means of molecular sensory science. *Food Chem.* **2019**, *301*, 125282. [CrossRef]
32. Bento, C.; Gonçalves, A.C.; Silva, B.; Silva, L.R. Peach (*Prunus Persica*): Phytochemicals and Health Benefits. *Food Rev. Int.* **2020**, 1–32. [CrossRef]
33. Arctander, S. *Perfume and Flavor Chemicals (Aroma Chemicals)*; Montclair Publishing: Montclair, NJ, USA, 1969.
34. Font i Forcada, C.; Guajardo, V.; Chin-Wo, S.R.; Moreno, M.Á. Association Mapping Analysis for Fruit Quality Traits in *Prunus persica* Using SNP Markers. *Front. Plant Sci.* **2019**, *9*, 2005. [CrossRef]
35. Bianchi, T.; Weesepoel, Y.; Koot, A.; Iglesias, I.; Eduardo, I.; Gratacós-Cubarsí, M.; Guerrero, L.; Hortós, M.; van Ruth, S. Investigation of the aroma of commercial peach (*Prunus persica* L. Batsch) types by Proton Transfer Reaction–Mass Spectrometry (PTR-MS) and sensory analysis. *Food Res. Int.* **2017**, *99*, 133–146. [CrossRef]
36. Aubert, C.; Milhet, C. Distribution of the volatile compounds in the different parts of a white-fleshed peach (*Prunus persica* L. Batsch). *Food Chem.* **2007**, *102*, 375–384. [CrossRef]
37. Duan, Y.; Dong, X.; Liu, B.; Li, P. Relationship of changes in the fatty acid compositions and fruit softening in peach (*Prunus persica* L. Batsch). *Acta Physiol. Plant.* **2013**, *35*, 707–713. [CrossRef]
38. Sánchez-Moreno, C.; Pascual-Teresa, S.D.; Ancos, B.D.; Cano, M.P. Nutritional Values of Fruits. In *Handbook of Fruits and Fruit Processing*; John Wiley & Sons: Hoboken, NJ, USA, 2006; pp. 29–44.
39. Gonçalves, B.; Oliveira, I.; Bacelar, E.; Morais, M.C.; Aires, A.; Cosme, F.; Ventura-Cardoso, J.; Anjos, R.; Pinto, T. Aromas and Flavours of Fruits. In *Generation of Aromas and Flavours*; InTech Open: Rijeka, Croatia, 2018. [CrossRef]
40. Meng, J.; Fang, Y.; Gao, J.; Zhang, A.; Liu, J.; Guo, Z.; Zhang, Z.; Li, H. Changes in aromatic compounds of cabernet sauvignon wines during ageing in stainless steel tanks. *Afr. J. Biotechnol.* **2011**, *10*, 11640–11647. [CrossRef]
41. Mo, E.K.; Sung, C.K. Phenylethyl alcohol (PEA) application slows fungal growth and maintains aroma in strawberry. *Postharvest Biol. Technol.* **2007**, *45*, 234–239. [CrossRef]
42. Li, Y.; Qi, H.; Jin, Y.; Tian, X.; Sui, L.; Qiu, Y. Role of Ethylene in the Biosynthetic Pathway of Related-aroma Volatiles Derived from Fatty Acids in Oriental Sweet Melon. *J. Am. Soc. Hortic. Sci.* **2016**, *141*, 327–338. [CrossRef]
43. Jennings, W.G.; Tressl, R. Production of volatile compounds in the ripening of 'Bartlett' pear. *Chem. Mikrobiol. Technol. Leb.* **1974**, *3*, 52–55.
44. Şeker, M.; Kaçan, A.; Gür, E.; Ekİncİ, N.; Gündoğdu, M.A. Investigation of aromatic compounds of peach and nectarine varieties grown in Canakkale ecological conditions. *TABAD Tarım Bilim. Araştırma Derg.* **2013**, *6*, 62–67.

45. Wang, Y.; Yang, C.; Liu, C.; Xu, M.; Li, S.; Yang, L.; Wang, Y. Effects of Bagging on Volatiles and Polyphenols in "Wanmi" Peaches during Endocarp Hardening and Final Fruit Rapid Growth Stages. *J. Food Sci.* **2010**, *75*, S455–S460. [CrossRef]
46. Espino-Díaz, M.; Sepúlveda, D.R.; González-Aguilar, G.; Olivas, G.I. Biochemistry of Apple Aroma: A Review. *Food Technol. Biotechnol.* **2016**, *54*, 375–397. [CrossRef]
47. Lu, P.F.; Qiao, H.L.; Xu, Z.C.; Cheng, J.; Zong, S.X.; Luo, Y.Q. Comparative analysis of peach and pear fruit volatiles attractive to the oriental fruit moth, *Cydia molesta*. *J. Plant Interact.* **2013**, *9*, 388–395. [CrossRef]
48. Yahia, E.M. Postharvest physiology and biochemistry of fruits and vegetables. *Postharvest Physiol. Biochem. Fruits Veg.* **2018**, 1–476. [CrossRef]
49. Rizzolo, A.; Bianchi, G.; Vanoli, M.; Lurie, S.; Spinelli, L.; Torricelli, A. Electronic nose to detect volatile compound profile and quality changes in "spring Belle" peach (*Prunus persica* L.) during cold storage in relation to fruit optical properties measured by time-resolved reflectance spectroscopy. *J. Agric. Food Chem.* **2013**, *61*, 1671–1685. [CrossRef]
50. Gündoğdu, M.A.; Ekinci, N.; Kaleci, N.; Şeker, M. Determination of Aromatic Compounds of Some Promising Pomegranate Genotypes. *Ecol. Life Sci.* **2018**, *13*, 142–150. [CrossRef]
51. Engel, K.H.; Flath, R.A.; Buttery, R.G.; Mon, T.R.; Teranishi, R.; Ramming, D.W. Investigation of volatile constituents in nectarines. 1. Analytical and sensory characterization of aroma components in some nectarine cultivars. *J. Agric. Food Chem.* **2002**, *36*, 549–553. [CrossRef]
52. Feng, S.; Huang, M.; Crane, J.H.; Wang, Y. Characterization of key aroma-active compounds in lychee (*Litchi chinensis* Sonn.). *J. Food Drug Anal.* **2018**, *26*, 497–503. [CrossRef]
53. González-Agüero, M.; Troncoso, S.; Gudenschwager, O.; Campos-Vargas, R.; Moya-León, M.A.; Defilippi, B.G. Differential expression levels of aroma-related genes during ripening of apricot (Prunus armeniaca L.). *Plant Physiol. Biochem.* **2009**, *47*, 435–440. [CrossRef]
54. Goff, S.A.; Klee, H.J. Plant volatile compounds: Sensory cues for health and nutritional value? *Science* **2006**, *311*, 815–819. [CrossRef]
55. The Good Scents Company Information System. Available online: http://www.thegoodscentscompany.com/ (accessed on 21 December 2021).
56. Mundfrom, D.J.; Shaw, D.G.; Ke, T.L. Minimum Sample Size Recommendations for Conducting Factor Analyses. *Int. J. Test.* **2005**, *5*, 159–168. [CrossRef]
57. Mihaylova, D.; Desseva, I.; Popova, A.; Dincheva, I.; Vrancheva, R.; Lante, A.; Krastanov, A. GC-MS Metabolic Profile and α-Glucosidase-, α-Amylase-, Lipase-, and Acetylcholinesterase-Inhibitory Activities of Eight Peach Varieties. *Molecules* **2021**, *26*, 4183. [CrossRef]
58. Amine, E.K.; Baba, N.H.; Belhadj, M.; Deurenberg-Yap, M.; Djazayery, A.; Forrestre, T.; Galuska, D.A.; Herman, S.; James, W.P.T.; M'Buyamba Kabangu, J.R.; et al. Diet, nutrition and the prevention of chronic diseases. *Am. J. Clin. Nutr.* **2003**, *60*, 644–645. [CrossRef]
59. Montero-Prado, P.; Bentayeb, K.; Nerín, C. Pattern recognition of peach cultivars (*Prunus persica* L.) from their volatile components. *Food Chem.* **2013**, *138*, 724–731. [CrossRef]
60. Uekane, T.M.; Nicolotti, L.; Griglione, A.; Bizzo, H.R.; Rubiolo, P.; Bicchi, C.; Rocha-Leão, M.H.M.; Rezende, C.M. Studies on the volatile fraction composition of three native Amazonian-Brazilian fruits: Murici (*Byrsonima crassifolia* L., Malpighiaceae), bacuri (*Platonia insignis* M., Clusiaceae), and sapodilla (*Manilkara sapota* L., Sapotaceae). *Food Chem.* **2017**, *219*, 13–22. [CrossRef]
61. Chong, J.; Wishart, D.S.; Xia, J. Using MetaboAnalyst 4.0 for Comprehensive and Integrative Metabolomics Data Analysis. *Curr. Protoc. Bioinform.* **2019**, *68*, e86. [CrossRef]

plants

MDPI

Article

Studies Regarding the Pharmaceutical Potential of Derivative Products from *Plantain*

Marilena-Gabriela Olteanu Zaharie [1], Nicoleta Radu [1,2,*], Lucia Pirvu [3], Marinela Bostan [4,5], Mariana Voicescu [6], Mihaela Begea [7,*], Mariana Constantin [2,8], Catalina Voaides [1], Narcisa Babeanu [1] and Viviana Roman [4]

1 Faculty of Biotechnology, University of Agronomic Sciences and Veterinary Medicine of Bucharest, 59 Marasti Boulevard, 011464 Bucharest, Romania; mzaharie@yahoo.com (M.-G.O.Z.); catalina.voaides@biotehnologii.usamv.ro (C.V.); narcisa.babeanu@biotehnologii.usamv.ro (N.B.)
2 Biotechnology Department, National Institute of Chemistry and Petrochemistry R & D of Bucharest, 202 Splaiul Independentei Street, 060021 Bucharest, Romania; marriconstantin@yahoo.com
3 Biotechnology Department, National Institute of Chemical Pharmaceutical R & D, 112 Vitan Road, 031299 Bucharest, Romania; lucia.pirvu@yahoo.com
4 Institute of Virology Stefan S. Nicolau, Center of Immunology, 285 Mihai Bravu Avenue, 030304 Bucharest, Romania; marinela.bostan@yahoo.com (M.B.); rviviana50@yahoo.com (V.R.)
5 Immunology Department, Victor Babes National Institute of Pathology, 050096 Bucharest, Romania
6 Institute of Physical Chemistry Ilie Murgulescu, 202 Splaiul Independentei, 060021 Bucharest, Romania; voicescu@icf.ro
7 Faculty of Biotechnical Systems Engineering, Politehnica University of Bucharest, 313 Splaiul Independentei, 060026 Bucharest, Romania
8 Faculty of Pharmacy, University Titu Maiorescu, 178 Calea Vacaresti, 040051 Bucharest, Romania
* Correspondence: nicoleta.radu@biotehnologii.usamv.ro (N.R.); ela_begea@yahoo.com (M.B.)

Citation: Zaharie, M.-G.O.; Radu, N.; Pirvu, L.; Bostan, M.; Voicescu, M.; Begea, M.; Constantin, M.; Voaides, C.; Babeanu, N.; Roman, V. Studies Regarding the Pharmaceutical Potential of Derivative Products from *Plantain*. *Plants* **2022**, *11*, 1827. https://doi.org/10.3390/plants11141827

Academic Editor: Ivayla Dincheva

Received: 7 June 2022
Accepted: 6 July 2022
Published: 12 July 2022

Publisher's Note: MDPI stays neutral with regard to jurisdictional claims in published maps and institutional affiliations.

Abstract: In this study, three types of extracts isolated from leaves of Plantain (*Plantago lanceolata*) were tested for their chemical content and biological activities. The three bioproducts are combinations of polysaccharides and polyphenols (flavonoids and iridoidic compounds), and they were tested for antioxidant, antifungal, antitumor, and prebiotic activity (particularly for polysaccharides fraction). Briefly, the iridoid-enriched fraction has revealed a pro-oxidant activity, while the flavonoid-enriched fraction had a high antioxidant potency; the polysaccharide fraction also indicated a pro-oxidant activity, explained by the co-presence of iridoid glycosides. All three bioproducts demonstrated moderate antifungal effects against *Aspergillus* sp., *Penicillium* sp., and dermatophytes, too. Studies in vitro proved inhibitory activity of the three fractions on the leukemic tumor cell line THP-1, the main mechanism being apoptosis stimulation, while the polysaccharide fraction indicated a clear prebiotic activity, in the concentration range between 1 and 1000 μg/mL, evaluated as higher than that of the reference products used, inulin and dextrose, respectively.

Keywords: *Plantago lanceolata*; derivative products; antitumor activity; prebiotic activity

1. Introduction

Plantain (*Plantago lanceolata*) has been used since ancient times in traditional medicine, either as such or in the form of tinctures, decoctions, or infusions. At the base of the medicinal properties are the active principles [1–5], with antioxidant, and antitumor properties [6–9] as polysaccharides [10–12], polyphenols [3,5,8] alkaloids, terpenoids (ursolic acid, oleanolic acid), caffeic acid derivatives (plantamajoside, acteoside or verbascoside), iridoid glycosides (aucubin, catalpol), fatty acids (palmitic acid, linolenic acid, linoleic acid, myristic acid), vitamins [1–3], macroelements and microelements [13]. Analytical studies performed on the essential oil obtained by hydrodistillation from the leaves of *Plantain* showed that it contains a wide range of organic compounds, with fatty acids representing about 41% of the oils fraction [2]. According to the reported data, *Plantago lanceolata* contains up to 1 g of total phenolic compounds per 100 g of vegetal material, regardless of whether the plant is obtained from controlled crops or spontaneous flora [5–7]. The main phenolic compounds are caffeic acid and luteolin and apigenin derivates [5,7,8,11].

Caffeic acid derivatives found in *Plantaginaceaea* are the plantamajoside or acteoside type, the last one is also known as verbascoside [7]. The most prominent flavonoid derivates are luteolin 7-O-glucoside, hispidulin 7-O-glucuronide, luteolin 7-O-diglucoside, apigenin 7-O-glucoside, nepetin 7-O-glucoside, luteolin 6-hydroxy 4′-methoxy 7-galactoside, and homoplantaginin, baicalein, hispidulin, plantaginin, and scutallarein [9]. Studies also evidenced the presence of alkaloid compounds, for example, indicain and plantagonin [1]. The terpenoid compounds isolated from the leaves and leaf wax of *Plantago sp.* were loliolid, ursolic acid, oleanolic acid, and sitosterol acid 18 β-glycyrrhetinic. Another important class in *Plantaginaceae* is that of iridoid glycosides; aucubin was noticed as the main iridoid glycoside in *Plantago major*—this species may also contain small quantities of asperuloside, majoroside, 10-hydroxymajoroside, 10-acetoxymajoroside, catapol, gardoside, geniposidic acid, melittoside [1,8]. The iridoid glycosides content from *Plantago lanceolata* attain 2–3% (w/w); the main compounds found here are acucubin, catalpol, asperuloside, globularin, and desacetylasperuloside-acid methyl ester [11,14]. Generally, iridoids are oxygenated monoterpenoids, derived from geraniol [15]; these compounds play a major role in plant defense against herbivores, insects [16], ambient stress (temperature, drought) [17], and are respectively in control of interactions between plant roots and mycorrhiza [15]. Other important compounds in *Plantago* sp. are soluble polysaccharides, mucilage-type polymers based on sugars, and uronic acids, respectively [10–12]. The main sugars in polysaccharide fraction hydrolysate were arabinose, rhamnose, galactose, and galacturonic acid [12]. Particularly, the seeds of *Plantago* sp., namely Psyllium, contain xylose, arabinose, galacturonic acid, galactose, glucuronic acid, rhamnose, galactose, glucose, and a highly esterified pectin polysaccharide, acidic arabinogalactan, composed of arabinose, galactose, rhamnose, and galacturonic acid [1]. Furthermore, the crude mucilage also revealed the copresence of protein compounds, following essential amino acids (the evidenced amino acids were lysine, histidine, arginine, aspartic acid, threonine, serine, glutamic acid, proline, glycine, alanine, valine, methionine, isoleucine, leucine, tyrosine, and phenylalanine) [10]. The main fatty acids identified in *Plantago sp.* are lignoceric acid, palmitic acid, stearic acid, oleic acid, linoleic acid, linolenic acid, myristic acid, 9-hydroxy-cis-11-octadecenoic acid, arachidic acid, and behenic acid [1,2]. Literature data also indicates *Plantago* sp. as containing important amounts of minerals and microelements, for example, K, P, Cl, Na, Mg, Fe, Mn, Sr, Sc, Ti, V, Cr, Co, Ni, Ga, As, Br, Rb, Mo, Ag, Au, and Sb essential and trace elements [13,18–20].

Regarding the biological activities [18–23], the alcoholic extracts of *Plantaginaceae* have been proven to exhibit antimicrobial [18–20], antiviral [21], antioxidant [23,24], and antitumor activities [24–27]. Antimicrobial behavior was manifested against: bacteria (such as *Staphylococcus aureus*, *Bacillus subtilis*, *Escherichia coli*, and *Salmonella typhimuium*); fungi (such as *Candida albicans*, *Candida galabrata*, *Candida krusei* [28]); as well as micromycetes (such as *Alternaria alternata*, *Pyrenophora teres*, *Pyrenophora tritici-repentis* [29], *Fusarium solani* [30], and *Ustilago scitaminea* [31]). Antioxidant activity [3,7,23] was attributed to contained secondary metabolites, mainly polyphenolics content as well as antitumor potency, proved on numerous human cancer cell lines (e.g., human gastric cancer cells, human solid tumor cell lines, human non–small-cell lung cancer, human colorectal cancer, and human renal cell carcinoma) [24–27]. Research performed on *Plantain* from Isparta, Turkey, by Bahadori et al. [7] reveals the existence of compounds such as verbascoside, phenylethanoid glycosides, phenolic acids (chlorogenic acid, rosmarinic acid), as well flavonoid glycosides (i.e., hesperidin, hyperoside) in the alcoholic extract, analyzed by LC–ESI–MS/MS [7]. Sanna and collab [8], in the studies performed on alcoholic extracts obtained from *Plantain* developed from germplasm of Sardinia, Italy, found that in the case of plants cultivated on three types of soils, the content of polyphenolic compounds attains a maximum level in leaves in January and April. Researchers found in the alcoholic extract of *Plantain* leaves compounds such as chlorogenic acid, neo chlorogenic acid, cryptochlorogenic acid, and verbascoside. Luteolin was found only in an ethanolic extract from *Plantain* leaves harvested in July.

Luteolin derivates were proved able to reduce gastric acid synthesis, via smooth muscle inhibition, while caffeic acid normalized gall bladder function. The mucilage fraction from *Plantago lanceolata* has major therapeutic importance because it helps the regeneration of tissue damaged by forming a protective layer, under which the diseased tissue is regenerated [32–34].

Based on the scientific data presented in the literature, the present work aimed to study the biological activities of three extracts obtained from Plantain. In this regard, three fractions derived from *Plantain*, the polysaccharidic fraction (PP), the flavonoidic fraction (PF), and the iridoidic fraction (PI), respectively, were used. For these fractions, studies "in vitro" to determine their antioxidant, antifungal (using common micromycetes and dermatophytes), and antitumor properties were performed. The polysaccharides fraction was used in studies regarding the prebiotic properties.

2. Materials and Methods

2.1. Plant Bioproducts

Three types of solid products were obtained from dry leaves of *Plantago lanceolata* L. (source: Romanian indigenous plants, cultivated at Orastie, Romania, and marketed by Fares Company as dry plants, voucher specimen deposited in ICCF *Plant Material Storing Room*). These plants, ground and sifted (mesh diameter: 1 mm), were used to obtain solid bioproducts, according to the methodology presented below, established by the National Institute for Chemical-Pharmaceutical R&D (ICCF Bucharest), Romania [20,35].

(a) A quantity of 500 g of plantain powder firstly was extracted with 5000 mL of distilled water, by boiling under reflux. The aqueous extract was concentrated at 250 mL at 50 °C, using a rotary evaporator (Heidolph, Schwabach, Germany); then, the concentrate was treated with 1250 mL of 96% methanol solvent. The resulting precipitate was then dissolved into 250 mL distilled water and then treated with 1250 mL methanol (the operation was repeated twice). The final precipitate was dried in the oven (Gallenkamp, UK) at 45 °C and resulted in a fine gray powder which was considered as the polysaccharidic fraction (PP).

(b) The solid vegetal material from the first extraction was re-extracted with the methanolic solution resulting from the polysaccharides precipitation (approx. 4500 mL total volume) for 1 h, at boiling temperature under reflux, in a continuous stirring system operated at 300 rpm. The resulting methanolic extract was concentrated at solid residue at 50 °C using a rotary evaporator after that, the residue was dissolved into 250 mL of distilled water. The aqueous solution was further extracted (three times) with 250 mL ethyl acetate, at boiling, under reflux, and the combined ethyl acetate fractions were concentrated in a rotary evaporator at 50 °C The residue obtained was dissolved into ethanol and precipitated with ethyl ether. The final precipitate was considered the flavonoidic fraction (PF).

(c) The three aqueous fractions resulting from (a) and (b) were combined and then filtrated on the active charcoal. The resulting aqueous extract was concentrated in a rotary evaporator at 50 °C and the obtained solid residue was considered the iridoidic fraction (PI).

The three *Plantain* extracts PP, PF, and PI were qualitatively analyzed by high-performance thin-layer chromatography (HPTLC) and X-ray diffraction (XRD). The biological properties of these extracts were assessed in vitro by studies regarding antioxidant, antifungal, and antitumor activity. Additionally, for the extract enriched in polysaccharides (PP), studies regarding the probiotic activity were performed.

2.2. Regents Used in HPTLC Analysis

Chemical solvents (i.e., methanol, ethanol, ethyl acetate, formic acid, acetic acid) as well as the reference products (i.e., hyperoside (>97%), rutin (min. 95%), kaempferol (95%), cosmosiin (97%), vitexin (>96%), cynaroside (95%), chlorogenic acid (>95%), caffeic acid (99%), gallic acid (98%), and kaempferol (97%)) were purchased from Fluka and Sigma-Aldrich Co (Bucharest, Romania); reference compounds were prepared as 10^{-3}M solution in 70% ethanol solution.

2.3. Qualitative Determination

High-performance thin-layer chromatography studies were performed using specific solvent systems (ethyl acetate-glacial acetic acid-formic acid-water, 100:12:12:26) [11] in order to assess the polyphenols distribution in the three extracts (PP, PF, and PI).

Crystallographic properties of the three solid extracts were assigned with X-ray diffraction analysis (XRD), using a Rigaku SmartLab 9 kW diffractometer (Rigaku Corp., Tokyo, Japan), operated at 45 kV and 200 mA, $Cu_{K\alpha}$ radiation 1.54059 Å, in 2θ/θ scanning mode, between 2 and 90° (2θ). The obtained diffractograms were analyzed using the dedicated software PDXL and were processed for graphical representation using the graphing software Origin 2016 Pro (OriginLab Corporation, Northampton, MA, USA).

2.4. Antioxidant Activity Estimation

The antioxidant activity (AA) was assessed by the chemiluminescence (CL) method with luminol, using a GLOMAX 20/20 luminometer, single reagent injector, Model E 5321-PROMEGA, operated at wavelength λ = 430 nm. For this purpose, a quantity of 0.01 g from each solid extract was dissolved into 3 mL dimethyl sulfoxide (DMSO) in an ultrasonic bath, obtaining three stock solutions that contain 3.33 mg/mL for each extract. These solutions were stored in the refrigerator at 5 °C until use. From these solutions (samples for analysis), aliquots of 10–100 μL were used in the chemiluminescence studies. Reagents: solution LH2 (5-amino-2,3-dihydro-1,4-ftalazindione), with concentration c = 2.5×10^{-5}M, in DMSO; buffer solution TRIS HCl with c = 50 mM with pH = 8.5; H_2O_2 with c = 30 mM. Samples for analysis. Reagent blank (witness): 200 μL LH2 + 750 μL buffer solution + 50 μL H_2O_2. The testing of each sample was made by adding the reagents in the following order: 200 μL LH2 + 700 μL buffer solution + 50 μL sample + 50 μL H_2O_2. The volume of the added sample was subtracted from the volume of buffer solution so that the final volume of all reagents (with the analyzed sample) was 1000 μL.

Antioxidant activity (%), by means of the oxygen free radicals scavenging, of used extracts was calculated using Relation (1):

$$\text{A.A.} = \frac{(\text{Iw} - \text{Is}) \times 100}{\text{Iw}} \tag{1}$$

with Iw and Is as the CL intensity measured 5 s from the beginning of the CL reaction for the witness (Iw) and for each plant extract containing the sample (Is), respectively.

All measurements were made in triplicate; the average values of them in all representations were used, with standard deviations.

2.5. Antifungal Activities

2.5.1. Strains Used in Antifungal Studies

(a) micromycetes: *Aspergillus niger* ATTC 1015, (*A. niger*), *Aspergillus terreus* ATTC 1012 (*A. terreus*), *Penicillium citrinum* ATTC 10105 (*P. citrinum*), *Penicillium digitatum* ATTC 9849 (*P. digitatum*), *Penicillium* sp. 1, and *Penicillium* sp. 2 (the last two species were isolated from infected wood).

(b) dermatophytes: *Microsporum canis* ATTC 10214 (*M. canis*), *Trichophyton mentagrophytes* ATTC 18748 (*T. mentagrophytes*), *Microsporum gypseum* ATTC 24102 (*M. gypseum*), and *Scopulariopsis brevicaulis* ATTC 1102 (*S. brevicaulis*).

2.5.2. Methodology Used in Antifungal Studies

All the fungus strains were cultivated on Petri plates with PDA (potato dextrose agar) by spore inoculation; after 7 days, the spores of each strain were inoculated in a Petri plate

(90 mm diameter) on PDA. After 15 min, in each Petri plate, three discs of cellulosic sterile paper of 5 mm diameter were placed at equal distance, impregnated with 20 μL from each extract prepared at the point 2.4. Separately, the solvent effect (DMSO) was evaluated by placing three discs of cellulosic papers impregnated with DMSO in Petri plates. The Petri plates used in quantification of the DMSO effect were previously inoculated with each studied fungal strain. All the plates were incubated for 48 h at 25 °C; after that, the inhibition diameters were measured. All results were presented as the average of three measurements, with corresponding standard deviation.

2.6. *Antitumor Activities*

2.6.1. Cell Viability Assay by Flow Cytometry Analysis

The cell viability was assessed by the ability of Propidium Iodide (PI) to pass through the membrane of dead cells interacting with the DNA of the nuclei and emitting red fluorescence light. THP1 cells (2×10^3 cell/well) were seeded in 96-microwell plates and treated with different concentrations of each of the PP, PF, and PI extracts at different concentrations (i.e., 0.05 mg/mL; 0.1 mg/mL; 0.2 mg/mL; 0.5 mg/mL; 1 mg/mL) for 24 h. Each solid extract was dissolved in a culture medium with 0.01% DMSO. The cells without PI staining were used as a negative control. Tumor cells THP-1 either untreated or treated with *Plantago* bioproducts for 24 h were incubated for 10 min with a 2 μg/mL solution of PI on ice in the dark. A total of 10,000 cells/sample were acquired using a BD Canto II flow cytometer. The flow cytometric analysis was performed using DIVA 6.2 software (Becton Dickinson Immunocytometry Systems, San Jose, CA, USA) to discriminate viable cells (FITC $-$ PI$^-$) from necrotic cells (FITC + PI).

2.6.2. Cell Cultures

A leukemic THP-1 cell line (ATCC TIB 202) was used in experiments regarding the antitumor properties. This cell line was grown in RPMI 1640 medium, supplemented with the following reagents: 10% heat-inactivated fetal bovine serum (FBS); 2 mM glutamine; 100 U/mL penicillin, and 100 μg/mL streptomycin, at 37 °C in a humidified incubator with 5% CO_2 atmosphere. Cultures were maintained by the addition of fresh medium or the replacement of medium every 2 to 3 days. Before treatment, the cell cultures were centrifugated and resuspended at 4×10^5 viable cells/mL.

2.6.3. Apoptosis Assay—Annexin V-FITC/PI Double Staining

The apoptosis assay was carried out with the Annexin V-FITC kit using the manufacturer's protocol [36–38]. Cells were washed with cold PBS three times and resuspended in 1 mL PBS. To determine the rate of apoptosis, the treated and untreated 1×10^5 cells/mL were resuspended in cold binding buffer and stained simultaneously with 5 μL FITC-Annexin V (green fluorescence) and 5 μL propidium iodide (PI) in the dark at room temperature for 15 min. Then, 400 μL of Annexin V binding buffer was added and 10,000 cells/per sample were acquired using a BD Canto II flow cytometer. The analysis was performed using the DIVA 6.2 software to discriminate viable cells (FITC $-$ PI$^-$) from necrotic cells (FITC $-$ PI$^+$) and early apoptosis (FITC + PI$^-$) from late apoptosis (FITC + PI).

2.7. *Prebiotic Activities*

2.7.1. Microorganisms

In prebiotic activities, we use four types of strains, respectively: *Lactobacillus plantarum* ATTC 8014 (*L. plantarum*), *Lactobacillus casei* ATTC 393 (*L. casei*), *Lactobacillus reuteri* ATTC 55730 (*L. reuteri*), and *Saccharomyces cerevisiae* (*S. cerevisiae*) (isolated from commercially Dr. Oetker bakery yeast). *Lactobacillus* species were incubated for 24 h on a liquid MRS medium without agar at 37 °C. *Saccharomyces cerevisiae* was grown on the specific media YPG.

2.7.2. Methodology Used in Prebiotic Activities

For screening activities, 24-well microplates were used; for *Lactobacillus* sp. we used liquid MRS medium with or without dextrose; for *Saccharomyces* sp. we used YPG medium with or without dextrose. Into each well of the plate, the same volume of MRS or YPG medium with or without dextrose was distributed. In each well, a serial dilution was made with a volume of 0.25 mL sterile water (in the case of culture medium with dextrose) or with sterile stock solution, which contains PP or inulin. The solutions which contain polysaccharides, inulin, or dextrose were made in sterile water. Each row of the plate had the following concentrations of prebiotics (μg/mL): 5000; 1000; 200; 40; 8; and 1.6. In each well, the inoculum of the studied probiotic strain was evenly distributed, at the rate of inoculation of 1:10, from each probiotic strain, which contains 4×10^8 CFU/mL. As a reference, the culture mediums MRS (for *Lactobacillus* sp.) or YPG (for *Saccharomyces* sp.) were used. The influence of PP, inulin, and dextrose on each probiotic microorganism was studied by replacing dextrose from MRS or YPG with PP or inulin, using serial dilutions. All tests were made in three replications, and each measurement was performed three times. The OD (optical density) of the well plate was measured after 24 h and 48 h at 620 nm using a DYNEX plate reader (DYNEX Technologies, MRS, VA, USA).

3. Results

3.1. Plant Bioproduct Characterization

Quantification of polyphenolic compounds in the three extracts (PP, PF, and PI) by HPTLC (Figure 1) reveals that the polyphenols were distributed in all the three fractions. This analysis indicates some differences regarding the quality distributions of polyphenolic compounds between the three bioproducts; therefore, in PP and PI, the bands obtained appear to be almost identical, and most probably the PP extract contains iridoidic compounds, too. Regarding the PF extract, this fraction contains luteolin and hyperoside derivatives compounds (s1, s2), rutin (s3), chlorogenic acid (s4) hyperoside (s5), cynaroside (s6), caffeic acid (s8), and kaempferol (s9).

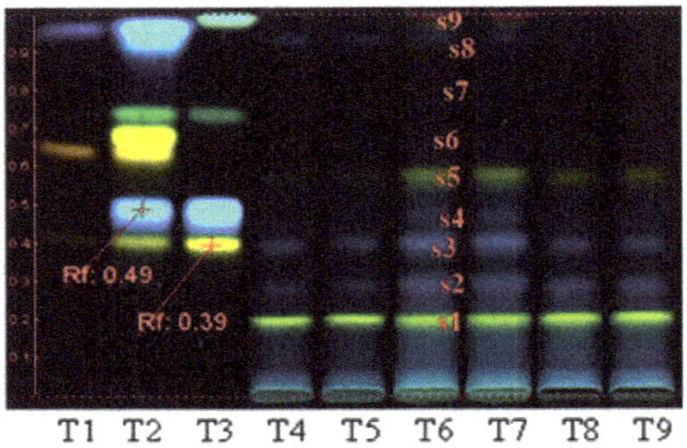

Figure 1. Qualitative assessment of polyphenolic compounds distribution in the three extracts obtained from *Plantago lanceolata*, by HPTLC analysis. T1: quercetin-3-O-rutinoside/rutin, quercetin-3-O-galactoside/hyperoside and gallic acid; T2: quercetin-3-O-rutinoside/rutin, chlorogenic acid, quercetin-3-O-galactoside/hyperoside, luteolin-7-O-glucoside/cynaroside, apigenin-8-C-glucoside/vitexin, caffeic acid; T3: rutin, chlorogenic acid, apigenin-7-O-glucoside/cosmosiin and kaempferol; T4–T5: polysaccharides fraction PP; T6–T7: polyphenols fraction PF; T8–T9: iridoids fraction PI.

The PI fraction may contain luteolin and hyperoside derivatives (s1, s2), rutin (s3), traces of hyperoside (s5), and chlorogenic acid (s4). The quantitative analyses performed on the same plant (i.e., plant source: Fares, Orastie, Romania), and already reported [4,6,11,35], showed that the content of polyphenolic compounds in the bioproducts separated from aqueous media (polysaccharidic fraction) can reach 0.89% GAE/g dry plant [4]. Regarding the bioproducts separated from alcoholic media, from the same source of raw materials, the reported data showed that the crude extract may contain 0.89% total polyphenols [6]. The polyphenolcarboxilic acids represent 0.7% and flavonoids 0.63% [6]. The analysis performed on the bioproducts obtained from the alcoholic fractions by electrospray ionization mass spectrometry indicated the presence of apigenin, luteolin, aucubin, chlorogenic acid, catalpol, and luteolin 7-O-β glucosides as polyphenolic compounds. Regarding iridoidic compounds, Plantain may contain iridoid glycosides such as aucubin and catalpol in quantities of a maximum 3%/plant [11].

Results obtained from X-ray diffraction analysis (Figure 2a,b) revealed the fact that the three extracts obtained are different: the PF fraction has a classical diffractogram, specific to amorphous substances (without crystalline structure) (Figure 2a, black line). Regarding the PI extract, the obtained diffractogram indicates a crystalline structure, with two high-intensity diffraction peaks, located at 14.178 and 20.2693 theta and at 29.2959 theta, respectively (Figure 2a red line and details in Figure 2b), and interplanar distances ranging between 1.5742 to 18.0627 Å (Table 1). As for the PP fraction, it has a weakly crystalline structure, indicating a low-intensity diffraction peak located at 15.625 theta (Figure 2a, blue line), and other very low diffraction peaks situated at 7.5, 9.75, and 12.5 theta.

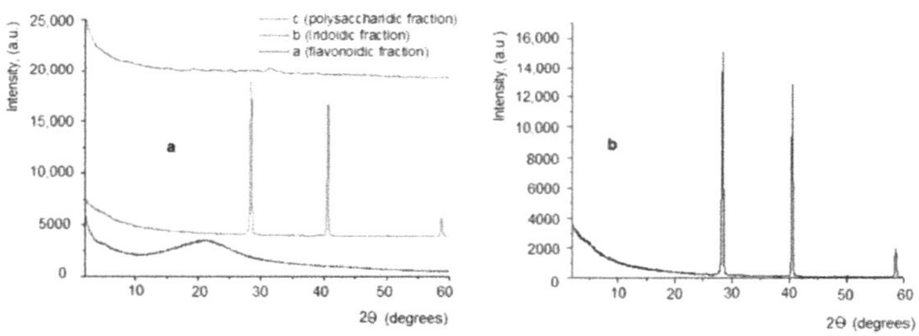

Figure 2. X-ray diffraction analysis. (**a**) Diffractograms obtained for extracts of *Plantago lanceolata*: a = PP; b = PI; c = PF; (**b**) Diffractogram details from the sample PI (iridoidic fraction).

Table 1. Interplanar distances obtained for the sample PI.

Peak	2θ (°)	θ (°)	d, Å	Peak Intensity
1	4.8883	2.4442	18.0626	456.9
2	25.5078	12.7539	3.4892	51.02
3	27.2186	13.6093	3.2736	79.41
4	27.9222	13.9611	3.1927	321.18
5	28.3559	14.1780	3.1449	21501.72
6	40.5385	20.2693	2.2235	14142.32
7	47.7775	23.8888	1.9021	40.15
8	50.1652	25.0826	1.8170	162.86
9	58.5917	29.2959	1.5742	2821.06

3.2. Antioxidant Properties

Due to the high sensitivity and rapidity, the addition of compounds into a chemiluminescence (CL) system leads to the reduction of the CL emission, thus scavenging the reactive oxygen species. Experimental data on PP/PF/PI fractions obtained from CL studies revealed a strong antioxidant effect for the PF fraction (Figure 3b inset) for all tested doses (10–100 µL), whereas, for the PP (tested doses, 10–40 µL) and PI (tested doses, 10–50 µL) samples, a strong prooxidant effect was observed (Figure 3a inset, Figure 3c inset). Moreover, the prooxidant effect of the PP fraction (Figure 3a inset) is more pronounced than that of the PI fraction (Figure 3c inset).

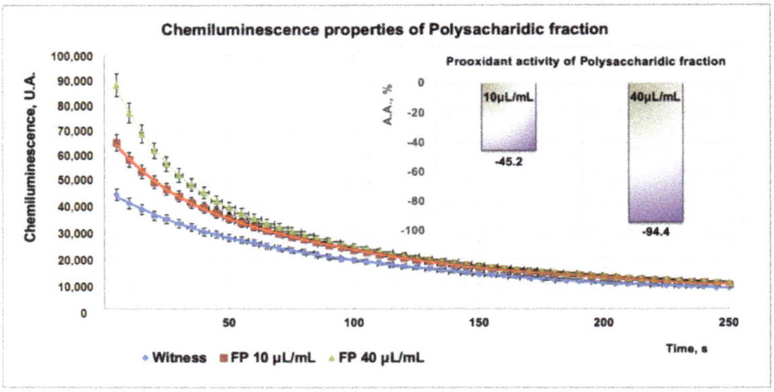

(a)

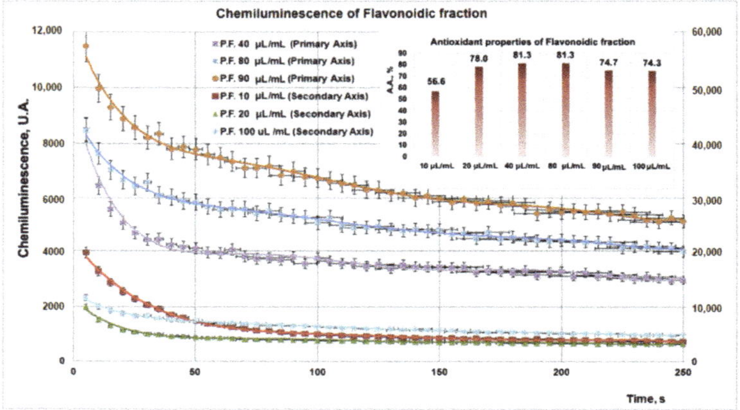

(b)

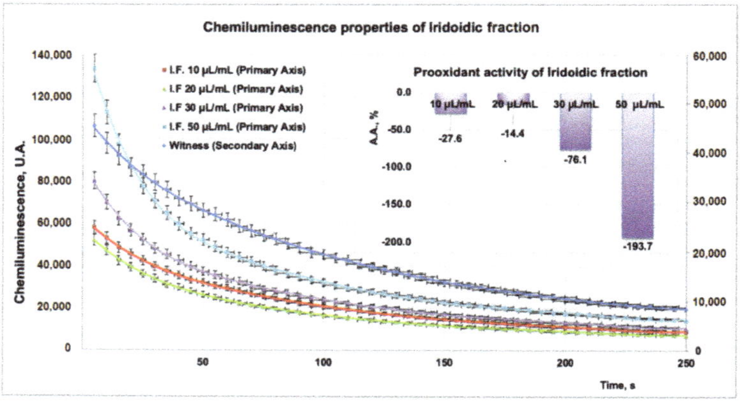

(c)

Figure 3. (**a**). Studies regarding chemiluminescence/prooxidant properties of PP fraction. (**b**) Studies regarding chemiluminescence/antioxidant properties of PF fraction (with the same witness as in (**a**)). (**c**) Studies regarding chemiluminescence/prooxidant properties of PI fraction.

3.3. Antifungal Properties of Plantago Bioproducts

Studies performed with DMSO reveal that after 48 h the aprotic solvent did not have an inhibitory effect on fungal strains involved in this study. Regarding the action of the PP bioproduct on fungal dermatophytes (Figure 4a), the moderate action is obtained at 48 h of exposure for *M. canis* (8.4 mm inhibition diameter), followed by *T. mentagrophytes* (7.8 mm inhibition diameter), *M. gypseum* (7.4 mm inhibition diameter), and *S. brevicaulis* (inhibition diameter 5.7 mm). In the case of micromycetes, the PP fraction has a moderate effect on *Penicillium* sp. 2 (inhibition diameter of 8.8 mm) and *Penicillium* sp. 1 (inhibition diameter of 6.4 mm) (Figure 4a). The flavonoidic fraction acts similarly on dermatophytes (Figure 4b), hence moderate activity is obtained for *M. canis* (inhibition diameter 8.4 mm), followed by *T. mentagrophytes* (6.8 mm inhibition diameter), *S. brevicaulis* (6.7 mm diameter of inhibition), and *M. gypseum* (6.2 mm inhibition diameter).

Moderate effects of the PF fraction were also observed for micromycetes (Figure 4b) such as *A. terreus* (inhibition diameter 8.8 mm) followed by *P. citrinum* (inhibition diameter 8 mm), *P. digitatum* (inhibition diameter 7.9 mm), *A. niger* (inhibition diameter 6.9 mm), *Penicillium* sp. 2 (inhibition diameter 6.8 mm), and *Penicillium* sp. 1 (inhibition diameter 6.1 mm).

The studies regarding the effect of PI fraction on fungal dermatophytes (Figure 4c) revealed a moderate action on *M. canis* (inhibition diameter 8.2 mm), followed by *S. brevicaulis* (inhibition diameter of 7.9 mm), *T. mentagrophytes* (7.4 mm inhibition diameter), and *M. gypseum* (diameter of inhibition of 6.8 mm). In the case of micromycetes, the PI fraction (Figure 4c) showed a moderate effect on *A. terreus* (inhibition diameter 7.1 mm), followed by *Penicillium* sp. 2 (inhibition diameter of 6.9 mm), *P. citrinum*, (diameter of inhibition of 6.1 mm), *P. digitatum* (inhibition diameter of 5.8 mm), *A. niger* (inhibition diameter of 5.3 mm) and *Penicillium* sp. 1 (inhibition diameter of 5.1 mm).

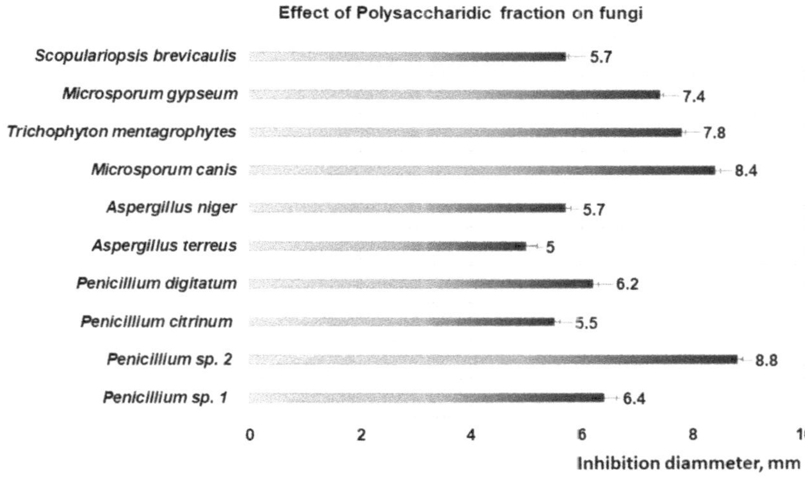

(**a**)

Figure 4. *Cont.*

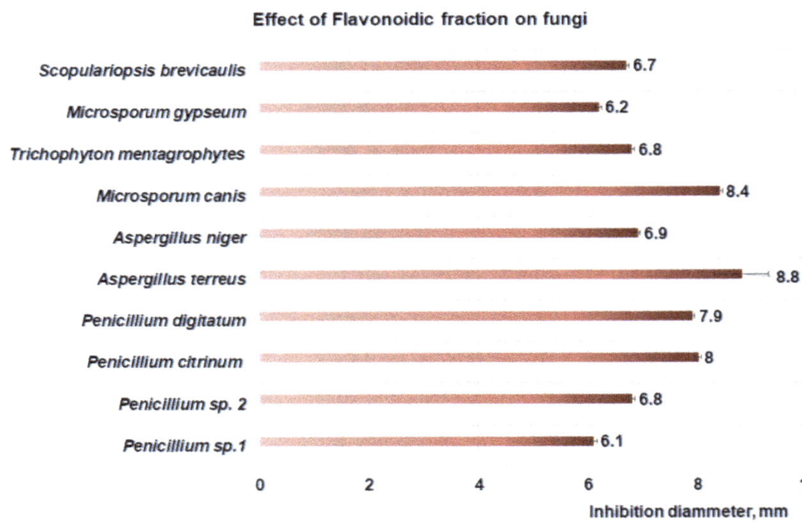

(b)

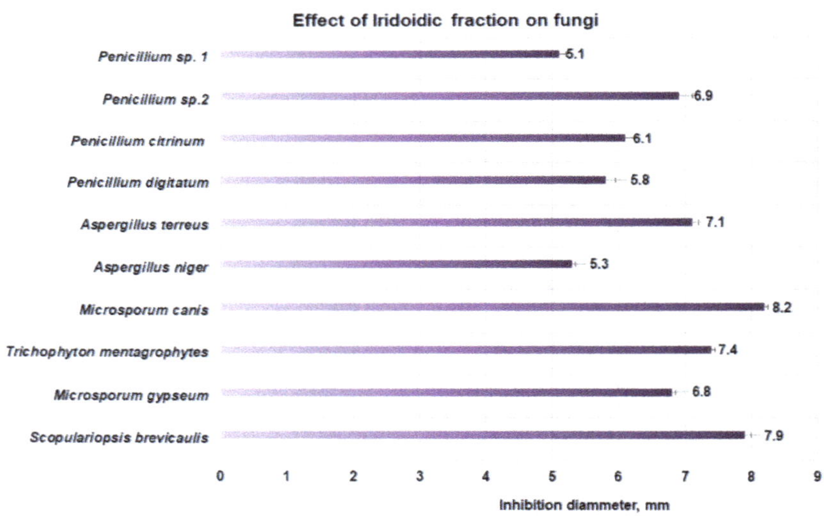

(c)

Figure 4. (**a**) The effect of polysaccharidic fraction on different fungal strains. (**b**) Effect of flavonoidic fraction on different fungal strains. (**c**) Effect of iridoidic fraction on different fungal strains.

3.4. Antitumor Properties of Plantago Bioproducts

The study performed on the THP-1 cell line regarding the action of PP (Figure 5a–c) showed that under the action of this bioproduct, the viability of the cells decreased with increasing PP concentration in the culture medium (Figure 5a), followed by a decreasing power function so that at 1 mg PP/mL the viability of the cells became 65%. The apoptosis values obtained for the same PP concentrations in culture media indicated that up to

0.1 mgPP/mL, the apoptotic process increased up to 35% and after this concentration the level of apoptotic cells decreased, even if the PP concentration in culture media increased. The apoptotic process of THP-1 cells that have been exposed to the action of PP bioproduct can be approximated with a polynomial function of degree three, with a good correlation coefficient (0.9171) (Figure 5b). The necrosis process appears to attain maximal value (20%) at 0.05 mg PP/mL (Figure 5c), and this fact suggests that most of the cells that are in the apoptotic state pass into the necrosis stage. By matching all experimental data obtained for the PP bioproduct, it can be estimated that the mechanism of destroying the THP-1 cell line is complex and the apoptosis process affects it to a large extent. The antitumor effect of PP bioproduct can be due to some polyphenolic compounds that pass in the PP fraction when it is separated from the *Plantain* leaves.

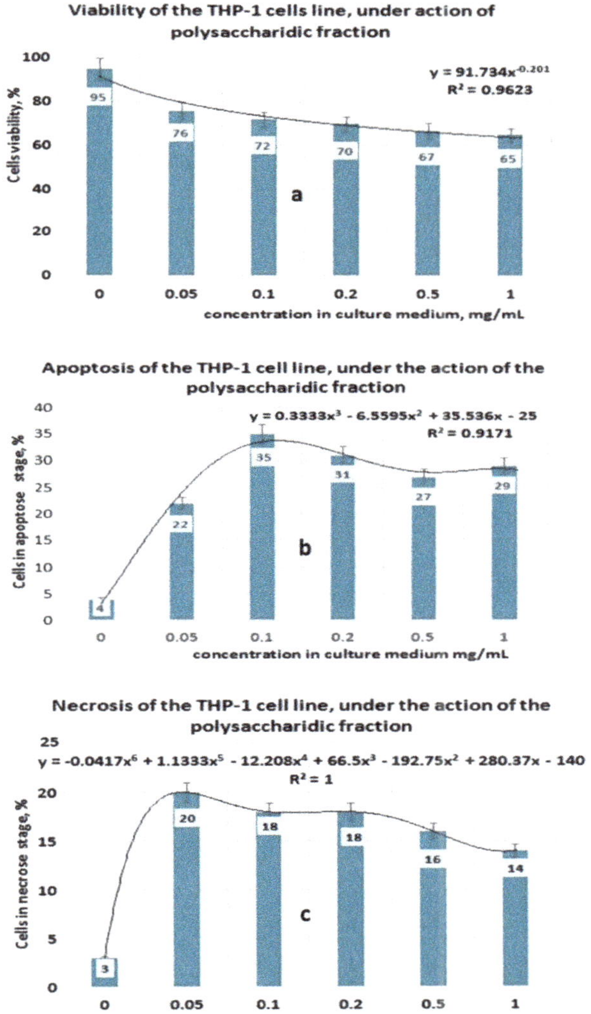

Figure 5. Effect of exposure at PP fraction on THP–1 cell lines. (**a**) Viability of THP–1 cell line under action of PP fraction; (**b**) Apoptosis of THP–1 cell line under action of PP fraction; (**c**) Necrosis of THP–1 cell line under action of PP fraction.

Investigations performed with the PF fraction (Figure 6a–c) indicated that for the increasing concentration of PF in the culture medium, the cell viability decreased, and reached 63% at 1 mg PF/mL. This process can be mathematically characterized by a decreasing power function (Figure 6a), as in the case of the PP fraction as well. At the same time, by increasing the PF concentration in the culture medium, the cells found in the apoptosis and necrosis stages increased (Figure 6b,c) and both processes can be mathematically characterized by an increasing power function (Figure 6a). Based on a mathematical model established in this study, it can be observed that the necrosis process is relatively faster in comparison with apoptosis. Thus, at 1 mg PF/mL, 21% of THP-1 cells were in the apoptosis stage and 23% in the necrosis stage. The mechanisms appear to be complex, involving the apoptotic and necrosis processes. Interesting for this type of bioproduct is the fact that all investigated parameters (i.e., viability, apoptosis, necrosis) can be mathematically characterized by the same type of function (i.e., a decreasing or increasing power function).

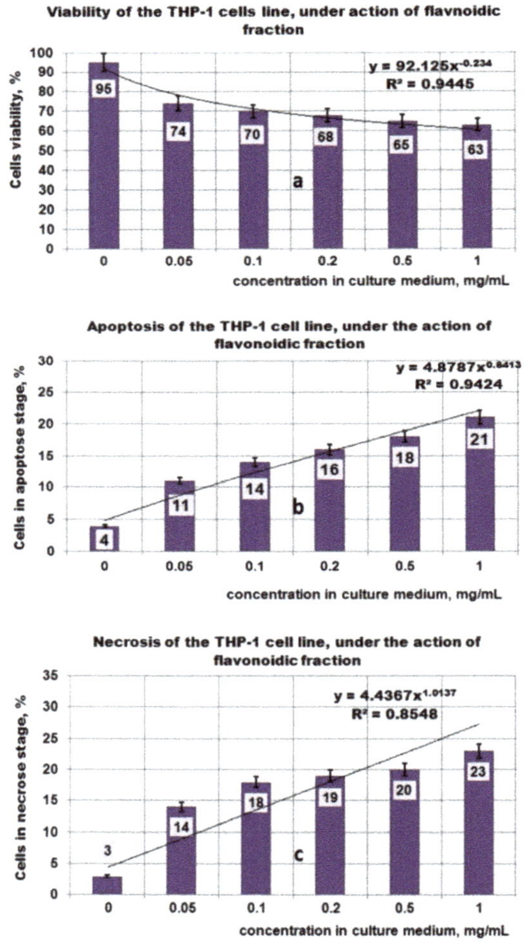

Figure 6. Effect of exposure at PF fraction on THP–1 cell lines. (**a**) Viability of THP–1 cell line under action of PF fraction; (**b**) Apoptosis of THP–1 cell line under action of PF fraction; (**c**) Necrosis of THP–1 cell line under action of PF fraction.

The results obtained in the case of the fraction which contains iridoidic compounds (Figure 7a–c) are much more spectacular. Under the action of PI bioproduct in the range of (0–1) mg/mL, the viability of the THP-1 cell line decreased from 95% to 42%; this process can be mathematically approximated with a decreasing power function (Figure 7a). Regarding the influence of the PI fraction on apoptosis, the results indicated a maximum effect at 0.05 mg/mL when 26% of the THP-1 cell line found it in the apoptotic stage (Figure 7b). At a concentration greater than 0.05 mg PI/mL, the apoptotic process decreased until 14% for 1 mg PI/mL. This apoptotic process due to the presence of PI bioproduct in the studied concentration range can be well described by a polynomial function of degree six (Figure 7b). The necrosis process of THP-1 cells due to the presence of PI in the culture medium is characterized by an increasing power function, and ranged from 3% for 0 mg PI/mL to 27% at 1 mg PI/mL (Figure 7c).

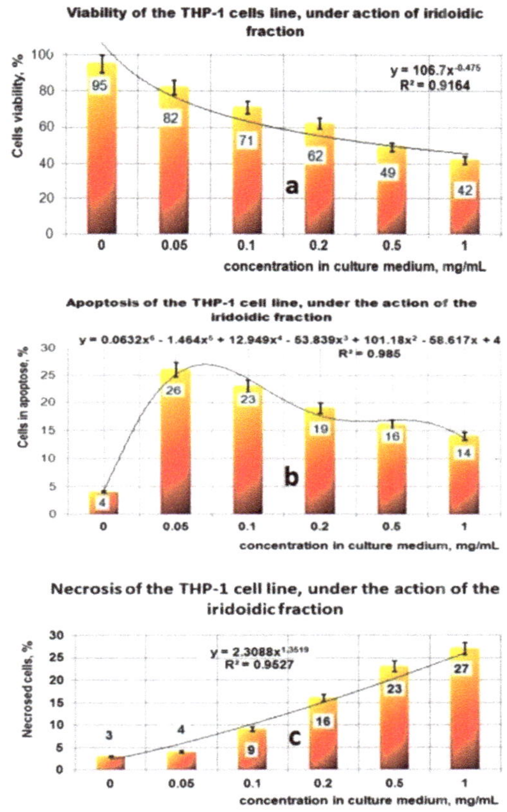

Figure 7. Effect of exposure at PI fraction on THP–1 cell lines. (**a**) Viability of THP–1 cell line under action of PI fraction; (**b**) Apoptosis of THP–1 cell line under action of PI fraction; (**c**) Necrosis of THP–1 cell line under action of PI fraction.

3.5. Prebiotic Activity of Plantain Bioproducts

Experiments regarding prebiotic activity were performed on the PP extract only, because both PF and PI extracts contain large quantities of polyphenolic compounds without prebiotic effects. The growth of *L. plantarum* after 24 h of exposure appears to be well influenced by the presence of PP fraction, in comparison with inulin (small growth) and dextrose (no growth) (Figure 8a). At 24 h, the best results are obtained at a concentration of 5 mgPP/mL (OD = 0.98) for *L. plantarum*.

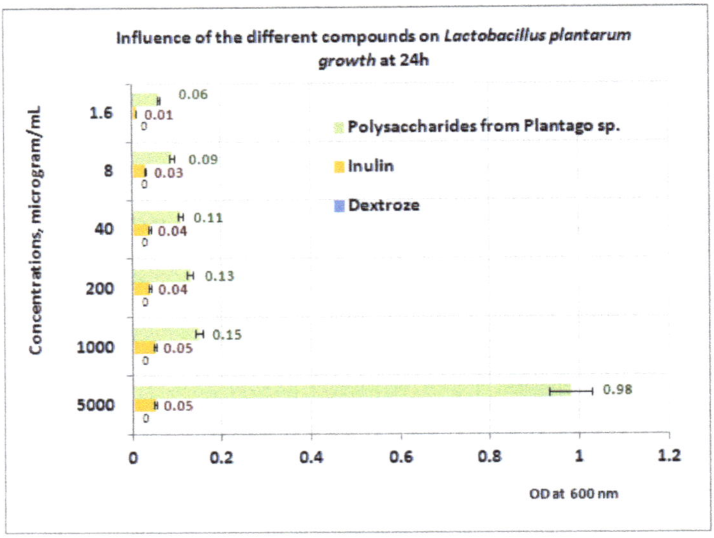

(**a**)

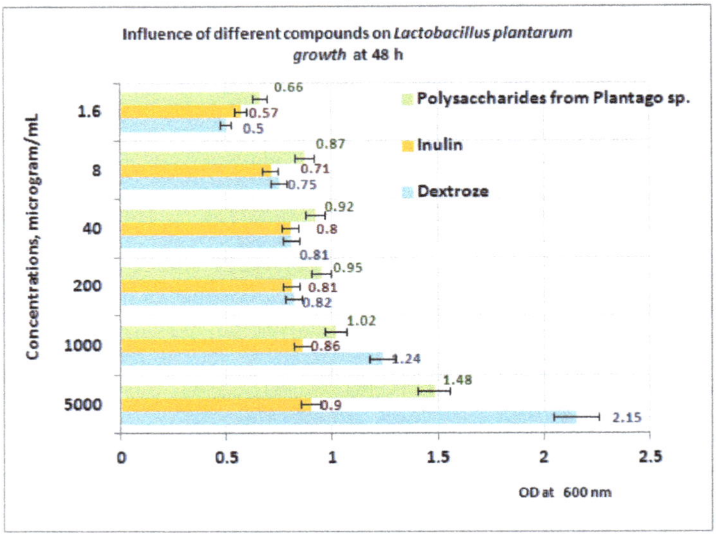

(**b**)

Figure 8. (**a**) Influence of PP on *Lactobacillus plantarum* growth at 24 h. (**b**) Influence of PP on *L. plantarum* growth at 48 h.

At 48 h of exposure, the measurements performed reveal that in the concentration range of 1–5 mg/mL the *L. plantarum* growth is best influenced in the presence of dextrose, followed by PP bioproduct (Figure 8b). For a concentration range situated between (1.6 ÷ 200) µg/mL, the best results regarding the *L. plantarum* growth were obtained for PP bioproduct followed by dextrose and inulin. The obtained results are probably due to the intensification of the depolymerization processes of the PP bioproduct, under the action of metabolic compounds resulting from *L. plantarum* growth, processes followed by the appearance of monosaccharides in culture medium, which are more accessible for probiotic strain.

In the case of *L. reuteri* (Figure 9a), the best results appear to be obtained after 24 h in favor of the PP fraction, in the PP concentration range of (40–5000) µg/mL. After 48 h of exposure to PP (Figure 9b), the measurements performed indicate that in the concentration range of (1 ÷ 5) mg PP/mL, the growth is best influenced by the presence of inulin, followed by dextrose and PP (Figure 9b). At 200 µg PP/mL, the best results are obtained for inulin, followed by PP and dextrose. If the concentration range of the PP fraction is between (1.6 ÷ 40) µg/mL, the best results regarding the *L. reuteri* growth are obtained in the presence of PP followed by inulin and dextrose. These results are probably due to the absence of metabolic compounds of the *L. reuteri*, without influence or with small influence regarding the depolymerization processes of the PP bioproduct.

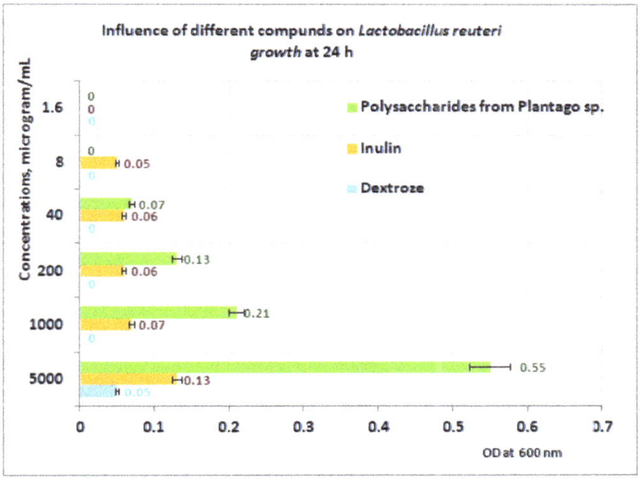

(a)

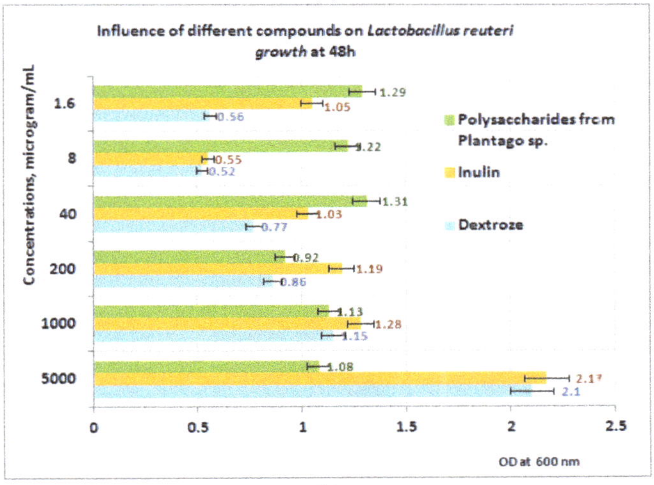

(b)

Figure 9. (**a**) Influence of PP on *L. reuteri* growth at 24 h. (**b**) Influence of PP on *L. reuteri* growth at 48 h.

In the case of *L. casei* (Figure 10a,b), the growth after 24 h of exposure to PP appears to be well influenced by the presence of the PP fraction compared to inulin and dextrose (Figure 10a). At 24 h, the best results are obtained at a concentration of 5 mg PP/mL (OD = 0.86). After 48 h of exposure, the measurements performed indicate that in the concentration range of (1 ÷ 5) mg PP/mL the growth of the *L. casei* is best influenced in the presence of dextrose, followed by PP and inulin (Figure 10b). In the concentration range situated between (1.6 ÷ 40) μg/mL, the best results are obtained for PP bioproduct, followed by inulin and dextrose.

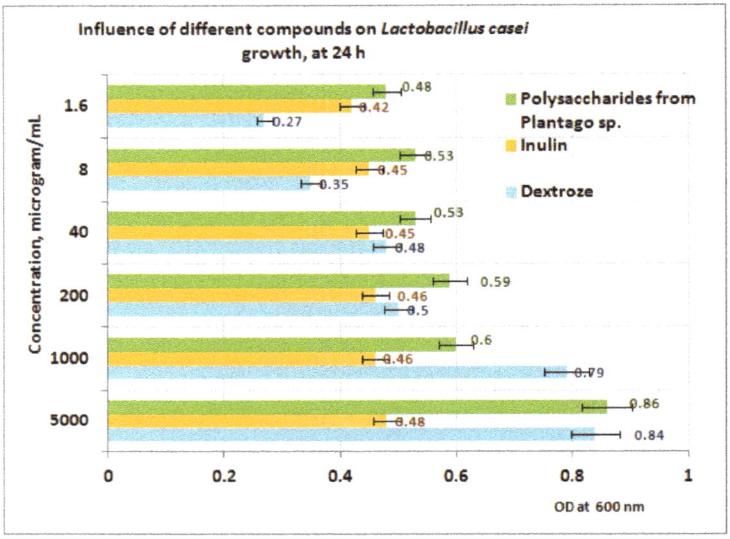

(**a**)

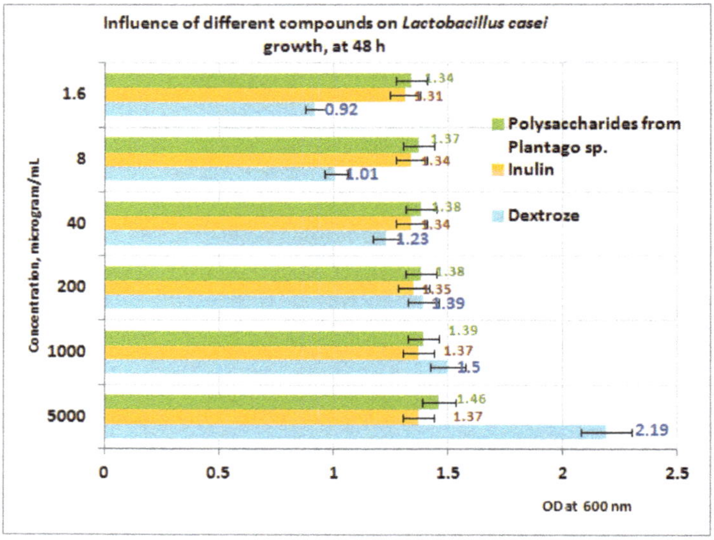

(**b**)

Figure 10. (**a**) Influence of PP on *L. casei* growth at 24 h. (**b**) Influence PP on *L. casei* growth at 48 h.

Studies performed on *S. cerevisiae* (Figure 11a,b) after 24 h of exposure to PP revealed a positive influence of PP fraction regarding its growth in the concentration range of (8 ÷ 40) μg/mL, when the best results are obtained for the PP fraction followed by inulin and dextrose (Figure 11a). The data obtained after 48 h of exposure indicate a slow growth of *S. cerevisiae* at the same concentration range of PP (Figure 11b).

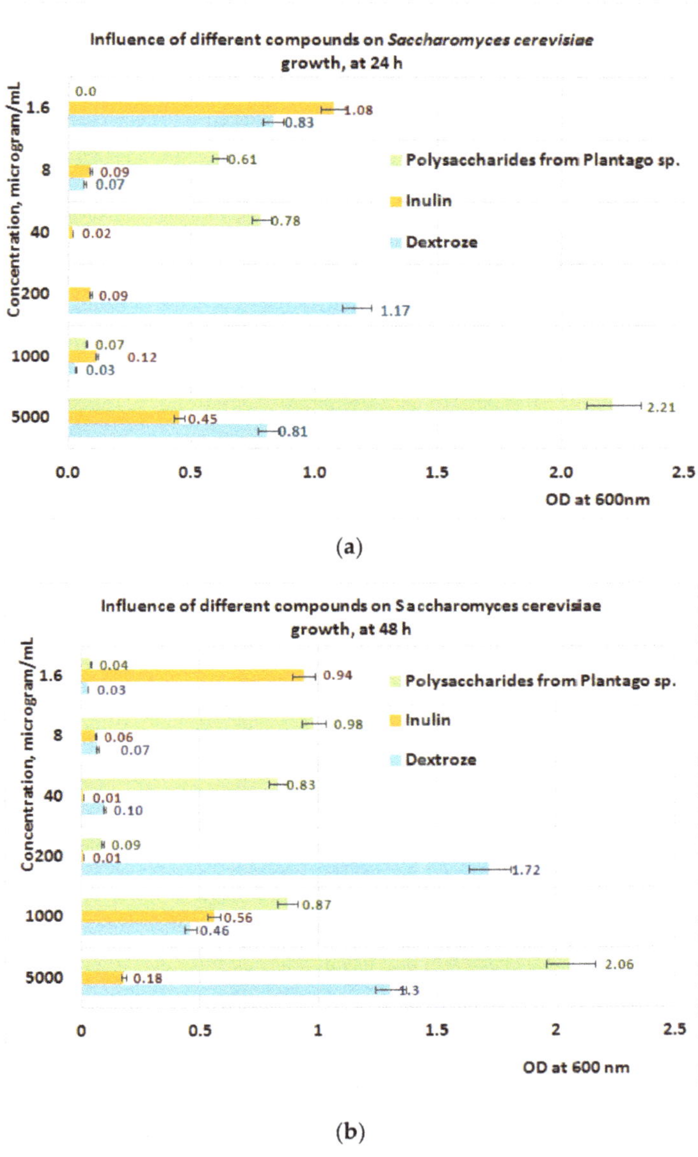

Figure 11. (**a**) Influence of PP on *S. cerevisiae* growth at 24 h. (**b**) Influence of PP on *S. cerevisiae* growth at 48 h.

4. Discussion

Analyses performed with the HPTLC technique show that the three extracts, codified PP (polysaccharidic fraction), PF (flavonoidic fraction), and PI (iridoidic fraction), contain variable amounts of the main polyphenolic compounds in *Plantaginaceae*, caffeic acid derivates and luteolin derivates. X-ray diffraction analysis on the three bioproducts also indicated an amorphous structure for the PF fraction, but a crystalline structure for the PI and PP fractions (specifically with two high-intensity diffraction peaks, located at 14.178, 20.2693, and 29.2959 theta in the case of the PI fraction, and a lower intensity diffraction peak located at 15.625, 7.5, 9.75, and 12.5 theta in the case of the PP fraction).

The antioxidant activity of the PF fraction is due to the presence of compounds such as luteolin and hyperoside derivatives compounds, rutin, chlorogenic acid, hyperoside, cynaroside, caffeic acid, and kaempferol [39] all evidenced by HPTLC analysis. These results are in agreement with results obtained by Abate and collab [40] who reported that the methanolic extract of *Plantain* leaves may contain 4-dihydroxyphenylacetic acid, (+)-catechin, pyrocatechol, vanillin, verbascoside, epicatechin, taxifolin, hesperidin, rosmarinic acid, pinoresinol, eriodictyol, and kaempferol. The same authors report that in the alcoholic extract there exists significant levels of kaempferol (43.64 mg/g), luteolin (5.35 mg/g), apigenin (8.27 mg/g), p-hydroxybenzoic acid (149.46 mg/g), 2,5-dihydroxybenzoic acid (16.20 mg/g), proto-catechuic acid (103.48 mg/g), vanillic acid (411.52 mg/g), gallic acid (212.01 mg/g), apigenin (184.38 mg/g), (cymaroside) luteolin-7-O-glucoside (119.15 mg/g), and quercetin-3-O-glucoside (34.67 mg/g) [40] Regarding the flavonoidic compounds, in the same study the existence of compounds such as luteolin-7-O-glucuronide, luteolin, apigenin, luteolin-7-O-glucoside, and quercetin-3-O-D-galactopyranoside, 3, 5, 7, 4-tetrahydroxyflavonol, apigenin-6,8-di-C-glucoside, luteolin-7-O-glucoside, and 7-O-glucuronide-3′-glucoside, as well as quercetin-3-rutinoside, 7-O-glucuronide, and apigenin-7-O-glucoside was reported. Other studies [41] reported the presence of compounds such as 3,4-dihydroxyphenylacetic acid, catechin, pyrocatechol, vanillin, verbascoside, epicatechin, and taxifolin, hesperidin, rosmarinic acid, pinoresinol, eriodictyol, and kaempferol. Regarding the flavonoidic compounds, the same scientists found the existence of compounds such as luteolin-7-O-glucuronide, luteolin, apigenin, luteolin-7-O-glucoside, and quercetin-3-O-D-galactopyranoside. The pro-oxidant activity is due likely to the low level of polyphenolic compounds in the PP and PI fractions, and for this reason, these fractions are not able to act as ion scavengers.

However, these two fractions are not the same as we can see from the XRD analysis. The co-presence of polyphenolic compounds in the three types of bioproducts from *P. lanceolata* can also explain the antimicrobial and antitumor activity. Regarding the probiotic activity, this can be due to the good hydrolyzation of *Plantago* polysaccharides which are present in the units of galacturonic acid (Gal A), glucose (Glc), arabinose (Ara), and rhamnose (Rh). This fact can be responsible for the good metabolization of probiotic bacteria, in comparison with inulin. Regarding the elemental unit contents of the PP fraction, Lukova and collab [12] in studies performed on polysaccharidic fractions obtained from *Plantago lanceolata*, found that the fractions soluble in water contained as major units GalA (70.58%), Ara (29.42%), and traces of Rha. Another study performed by Kardosova in 1992 [40] indicates that the mucilages obtained from dry leaves of *Plantago lanceolata* contain D-Glc (21.9%), D-Gal (35.8%), L-Ara (26%), uronic acid (6.9%), D mannose (D-Man) (4.6%), and L-Rha (4.6%). Zhang and collab [42] found that the polysaccharides obtained by extraction with hot water at 80 °C from *Plantago* sp. contain units of Gal A (64.88%) and Ara (29.42%); these types of polysaccharides can stimulate the activity and growth of the gut microbiome. Other studies were performed by Lukova and collab. in 2020 [43]. on probiotic microorganisms such as *L. acidophilus*, *L. sakei*, and *L. brevis*, and revealed the probiotic activities of the polysaccharidic fraction obtained from *Plantago major*. This fraction contains GalA (55.38%), Glc (21.5%), Ara (9.88%), Gal (8.02), Rha (3.17%), and xylose (Xy) (2.05%). Bioproducts with polysaccharides obtained from *Plantago* species also contain macro- and microelements and did not show antimicrobial activities against *E. coli* and *S. aureus* [18]. Regarding the biological activities of this bioproduct, moderate activity against microfungi and the leukemic cell line can be due to iridoid glycosides [44]. The content of the bioproduct enriched in flavonoids found in this study (PF) is in agreement with other studies [5,7,45]. Regarding the antioxidant activity, similar results were reported by Lukova and collab. on alcoholic extract obtained from *P. lanceolata;* this extract contains polyphenolic compounds (17.37 mg GAE/g) that have antioxidant properties, evidenced by tests performed with DPPH (AA = 59%) [39].

Antimicrobial activities of different bioproducts obtained in various solvents are reported for microorganisms such as *E. coli*, *S. aureus*, *P. aeruginosa*, *K. pneumoniae*, *P. mirabilis*, *C. albicans*, and various *Streptoccosus* species as *S. aglactiae*, *S. pneumoniae*, *S. bovis*, *S. mutans*, *S. sobrinus*, *S. parasanguinis*, and *S. viridans* [46]. Ziarno and collab [47] in studies performed on aqueous extract obtained from *Plantago lanceolata,* which contains 41.84 mg GAE/g, did not find an inhibitory effect of the polyphenolic content on the bacterial population of *S. thermophilus* or *L. delbruecki* spp. bulgaricus in the field of concentrations ranging between 0.2–1.4% during 4 h of fermentation. If the concentration of aqueous extract increased to 3% in the culture media, the bacterial population decreased slightly with (0.5–0.6) logarithmic units for each microorganism involved in the study [47]. Beara and collab. in studies performed in vitro, put into evidence the antitumor properties of the alcoholic extract obtained from *P. lancelolata*; according to these scientists, cytotoxicity was observed on MRC-5 (IC560 = 551.69 μg/mL), HeLa (172.32 μg/mL), MCF-7 (142.78 μg/mL), and HT-9 (IC50 = 405.5 μg/mL) [5]. Chiang and collab. performed studies on leukemic cell line types HL-60, K562, CCRF-CEM, and P3HRI and revealed that the aucubin, ferulic acid, p-coumaric acid, and vanillic acid show low antileukemic activities (IC50 = (26–56 μg/mL)), whereas luteolin has a strong antileukemic activity (IC50 < 18 μg/mL) [48]. Weber et al. found that the caffeic acid derivates named plantamajoside, which are found in leaves of *Plantago lanceolata*, exhibit antitumor activities against HL-60 and P338 cell lines [49]. Experimental studies performed "in vitro" by Yang and collab. with these compounds (obtained by preparative chromatography) on HL-60 and P338 cell lines indicated a value of IC 50 for HL-60 and P338 greater than 100 μM [50]. Regarding the mechanism of action, other studies indicate that under the action of flavonoids (i.e., kaempferol) the activation of p53 is stimulated, and the level of proapoptotic proteins such as Bax and Bcl-2 is upregulated, favoring the beginning of apoptosis processes in leukemic cells [51, 52]. The biological activity of PI products is due probably to iridoid glycosides such as aucubin, catalpol, or its derivatives, for which antimicrobial activities were reported against bacteria and fungi [20,40,46]. Other studies performed on catalpol and aucubin reported antimicrobial activities of these two compounds against *E. coli*, *E. faecalis* (MIC 512 μg/mL), *P. aeruginosa*, *S. aureus* (MIC = 256 μg/mL), *C. albicans* (MIC = 128 μg/mL and 256 μg/mL, respectively), *C. krusei* and *C. parapsilopsis* (MIC = 256μg/mL) [40,46]. These compounds exhibit cytotoxicity in K562 cell lines by inhibiting proliferation due to phase G1 from the cell cycle. Iridoid glycosides inhibit tumor proliferation by upregulation of p53 or p21 genes, which stops the cell cycle, as well by cell accumulation in the phase G0/G1 [44].

The analysis of the results obtained by the in vitro experiments performed on the THP-1 tumor cell line in this study led to the following aspects:

- the PP fraction reduced the cell viability of the THP-1 tumor cell line in a concentration-dependent manner. This PP fraction has antitumor properties which are sustained by its ability to induce apoptosis or necrosis of THP-1 tumor cells;
- in the case of the PF fraction, the antitumor effect seems to be higher, this being supported by the mathematical analysis of the results which shows a significant increase in both the apoptotic process and the necrosis that correlates with the decrease in viability;
- the fraction containing iridoidic compounds (PI) has a strong antitumor effect on THP-1 cells, which is demonstrated by the significant decrease of cell viability depending on the increase in concentration and by the amplification of apoptosis and necrosis processes leading to tumor cell death.

The increase in the necrosis level of tumor cells comparatively with apoptosis level suggests that the level of the PI concentration used in in vitro experiments is too high for this type of cell line.

Prebiotic activities of the polysaccharides fraction are due to extracellular secretion of hydrolases of lactobacilli, which in this way are able to metabolize galactose, sucrose, and arabinoxylan taking into account that the L-arabinose, D-galactose, and D-galacturonic acid are the main components in crude polysaccharides fractions [53–55].

The different behavior of *S. cerevisiae* can be explained by the fact that the yeast is not able to metabolize the main components of the polysaccharidic fraction from Plantain (like arabinose regarding galactose, *S. cerevisiae* can use this substrate, but with much less yield in comparison with dextrose [56,57].

5. Conclusions

In this study, three types of extracts obtained from leaves of *P. lanceolata*, which contain polysaccharides (PP), flavonoids (PF), and iridoids (PI), were studied in terms of biological activities of high interest to the pharmaceutical industry.

The antioxidant activity of the PF fraction and the HPTLC results represent proof that PF contains flavonoids such as luteolin, rutin (luteolin-7-O-glucoside), kaempferol, and cymaroside, as well as the hydroxycinnamic acids (caffeic acid). In terms of prooxidant activities, we supposed that the fractions PP and PI do not possess polyphenolic compounds in a sufficient concentration to act as radical scavengers, and most probably this is a reason for which these two fractions isolated from aqueous media act as prooxidants.

The main conclusions and potential applicability were as follows: the three extracts were studied (PP, PI, PF), and all indicated moderate antifungal activity against dermatophytes such as *M. canis*, *T. mentagropythes*, *M. gypseum*, and *S. brevicaulis* and micromycetes type *Penicillium* sp. and *Aspergillus* sp.; in vitro studies on the leukemic cell line type THP-1 indicated moderate to augmented antitumor activity, the PF and PI fractions suggesting an inhibitory mechanism based on the stimulation of the apoptosis process. Specifically, the antitumor activity for the three bioproducts decreased in the order PI > PF > PP; the PP extract confirmed the prebiotic potential of some important microorganisms found in human microbiota, in particular on *L. plantarum*, *L. reuteri*, and *L. casei*. In conclusion, in vitro experiments revealed the antitumor effect of the analyzed fractions obtained from *P. lanceolata*, which encourages the extension of studies on several types of normal cells and tumor cells.

Author Contributions: Conceptualization: N.R. and N.B. Methodology: N.R., M.B. (Marinela Bostan), V.R., L.P. and M.V. Probiotic activity determination: M.-G.O.Z., N.R. and M.C. Antifungal determinations: M.-G.O.Z., M.C., M.B. (Mihaela Begea) and C.V. Antitumor activities: M.B. (Marinela Bostan) and V.R. Antioxidant activities: M.V., N.R. and M.-G.O.Z. Writing—original draft preparation: N.R. Writing—review and editing, N.R., L.P., M.B. (Mihaela Begea), M.V. and V.R. Supervision: N.R., N.B. and V.R. All authors have read and agreed to the published version of the manuscript.

Funding: This research received no external funding.

Institutional Review Board Statement: Not applicable.

Informed Consent Statement: Not applicable.

Conflicts of Interest: The authors declare no conflict of interest.

Abbreviations

MRS	cultivation media for Lactobacillus species
YPG	cultivation media that contain yeast extract, peptone, and glucose
DPPH	2,2-diphenyl-1-picrylhydrazyl (i.e., 2,2-Diphenyl-1-(2,4,6-trinitrophenyl)hydrazin-1-yl)
GAE	Galic Acid equivalents
MRC-5	diploid cell culture line composed of fibroblasts
HeLa	cervical cancer cell line
MCF-7	human breast cancer cell line
HT-9	colorectal tumor cell line
HL-60	promyelocytic leukemia cell line
K562	human chronic myeloid leukemia cell line
CCRF-CEM	acute lymphoblastic leukemia cell line
P3HRI	Burkitt lymphoma cells
P338	murine leukemia cell line
Bax	proapoptotic protein; apoptosis regulator
Bcl-2	proapoptotic protein apoptosis regulator
p53	tumor suppressor gene
p21	protein which regulates cell proliferation by inhibiting the cell cycle through the cyclin kinase pathway
G1	cell phase in which the cell grows physically larger copies of organelles and makes the molecular building blocks it will need in later steps
G0	cell phase also known as the resting phase is the phase of the cell cycle during which a cell is neither dividing nor preparing to divide

References

1. Adom, M.B.; Taher, M.; Mutalabisin, M.F.; Amri, M.S.; Kudos, M.B.A.; Sulaiman, M.W.A.W.; Sengupta, P.; Susanti, D. Chemical constituents and medical benefits of *Plantago major*. *Biomed. Pharmacother.* **2017**, *96*, 348–360. [CrossRef] [PubMed]
2. Bajer, T.; Janda, V.; Bajerová, P.; Kremr, D.; Eisner, A.; Ventura, K. Chemical composition of essential oils from *Plantago lanceolata* L. leaves extracted by hydrodistillation. *J. Food Sci. Technol.* **2016**, *53*, 1576–1586. [CrossRef]
3. Varban, R.; Varban, D. Comparative study of the active ingredient content *Plantago lanceolata* L. *ProEnvironment* **2012**, *5*, 248–250.
4. Pirvu, L.C.; Nita, S.; Rusu, N.; Bazdoaca, C.; Neagu, G.; Bubueanu, C.; Udrea, M.; Udrea, R.; Enache, A. Effects of Laser Irradiation at 488, 514, 532, 552, 660, and 785 nm on the Aqueous Extracts of *Plantago lanceolata* L.: A Comparison on Chemical Content, Antioxidant Activity and $Caco_2$ Viability. *Appl. Sci.* **2022**, *12*, 5517. [CrossRef]
5. Beara, I.N.; Lesjak, M.M.; Orčić, D.Z.; Simin, N.Đ.; Četojević-Simin, D.D.; Božin, B.N.; Mimica-Dukić, N.M. Comparative analysis of phenolic profile, antioxidant, anti-inflammatory, and cytotoxic activity of two closely related Plantago species: *Plantago altissima* L. and *Plantago lanceolata* L. *LWT Food Sci. Technol.* **2012**, *47*, 64–70. [CrossRef]
6. Nichita, C.; Neagu, G.; Cucu, A.; Virginia Vulturescu, V.; Berteşteanu, S.V.G. Antioxidative Properties of *Plantago Lanceolata* L. Extracts Evaluated by Chemiluminescence Method. *AgroLife Sci. J.* **2016**, *5*, 95–102.
7. Bahadori, M.B.; Sarikurkcu, C.; Kocak, M.S.; Calapoglu, M.; Uren, M.C.; Ceylan, O. Plantago lanceolata as a source of health beneficial phytochemicals: Phenolic profile and antioxidant capacity. *Food Biosci.* **2020**, *34*, 100536. [CrossRef]
8. Sanna, F.; Piluzza, G.; Campesi, G.; Molinu, M.G.; Re, G.A.; Sulas, L. Antioxidant Contents in a Mediterranean Population of *Plantago lanceolata* L. Exploited for Quarry Reclamation Interventions. *Plants* **2022**, *11*, 791. [CrossRef]
9. Karadeniz, A.; Alexie, G.; Greten, H.J.; Andersch, K.; Efferth, T. Cytotoxicity of medicinal plants of the west Canadian which in Native Americans towards sensitive and multidrug-resistant cancer cells. *J. Ethnopharmacol.* **2015**, *168*, 191–200. [CrossRef]
10. Kardosova, A. Polysaccharides from the Leaves of *Plantago lanceolata* L., var. LIBOR: An a-D-Glucan. *Chem. Pap.* **1992**, *46*, 127–130.
11. Grigore, A.; Bubueanu, C.; Pirvu, L.; Ionita, L.; Toba, G. *Plantago lanceolata* L. Crops—Source of valuable raw materials for various industrial applications. *Sci. Pap. Ser. A Agron.* **2015**, *57*, 207–214.
12. Lukova, P.K.; Karcheva-Barkevanka, D.P.; Nikolova, M.M.; Iliev, I.I.; Mladenov, R.D. Comparison of structure and antioxidant activity of polysaccharides from the leaves of *Plantago major* L., *P. media* and *P. lanceolata*. *Bulg. Chem. Commun.* **2017**, *49*, 282–288.
13. Irmgard, U.; John-Deesbach, W. Essential and trace element content of medicinal plants and their infusions. *Botanica* **1984**, *58*, 255–266.
14. Gurib-Fakim, A. *Plantago lanceolata L. Record from Protabase*; Schmelzer, G.H., Gurib-Fakim, A., Eds.; PROTA (Plant Resources of Tropical Africa/Ressources Végétales de l'Afrique Tropicale): Wageningen, The Netherlands, 2006; Available online: http://www.prota4u.org (accessed on 18 May 2022).
15. Szucs, I.; Escobar, M.; Grodzinski, B. *Emerging Roles for Plant Terpenoids. Agriculture and Related Biotechnologies*; Elsevier: Amsterdam, The Netherlands, 2011; Volume 4, pp. 273–286.

16. Pankoke, H.; René Gehring, R.; Müller, C. Impact of the dual defence system of *Plantago lanceolata* (Plantaginaceae) on performance, nutrient utilisation, and feeding choice behavior of *Amata mogadorensis larvae* (Lepidoptera, Erebidae). *J. Insect Physiol.* **2015**, *82*, 99–108. [CrossRef]
17. Orians, C.M.; Schweiger, R.; Dukes, J.S. Combined impacts of prolonged drought and warming on plant size and foliar chemistry. *Ann. Bot.* **2019**, *124*, 41–52. [CrossRef]
18. Radu, N.; Ghita, I.; Rau, I. Therapeutic Effect of Polysaccharides from Plantago Species. *Mol. Cryst. Liq. Cryst.* **2010**, *523*, 236–246.
19. Radu, N.; Ghita, I.; Coman, O.; Rau, I. Therapeutic Effect of Flavonoids Derived from Plantago Species. *Mol. Cryst. Liq. Cryst.* **2010**, *523*, 273–281.
20. Radu, N.; Ghita, I.; Rau, I. Therapeutic Effect of Irridoidic Compounds from Plantago Species. *Mol. Cryst. Liq. Cryst.* **2010**, *523*, 289–296.
21. Sahakyan, N.Z.; Ginovyan, M.M.; Petrosyan, M.T.; Trchounian, A.H. Antibacterial and anti-phage activity of *Plantago major* L. *Proc. Yerevian State Univ.* **2019**, *53*, 59–64.
22. Yang, J.; Lee, H.; Sung, J.; Kim, Y.; Jeong, H.S.; Lee, J. Conversion of Rutin to Quercetin by Acid Treatment in Relation to Biological Activities. *Prev. Nutr. Food Sci.* **2019**, *24*, 313–320. [CrossRef]
23. Farcaș, A.D.; Zăgrean-Tuza, C.; Vlase, L.; Gheldiu, A.-M.; Pârvu, M.; Moț, A.C. EPR fingerprinting and antioxidant response of four selected *Plantago* species. *Stud. Ubb Chem.* **2020**, *65*, 209–220. [CrossRef]
24. Yang, C.; Shi, Z.; You, L.; Du, Y.; Ni, J.; Yan, D. Neuroprotective Effect of Catalpol via Anti-Oxidative, Anti-Inflammatory, and Anti-Apoptotic Mechanisms. *Front. Pharmacol.* **2020**, *11*, 690. [CrossRef]
25. Bhattamisra, S.K.; Yap, K.H.; Rao, V.; Choudhury, H. Multiple Biological Effects of an Iridoid Glucoside, Catalpol, and Its Underlying Molecular Mechanisms. *Biomolecules* **2020**, *10*, 32. [CrossRef]
26. Mirza-Aghazadeh-Attari, M.; Ekrami, E.M.; Aghdas, S.A.M.; Mihanfar, A.; Hallaj, S.; Yousefi, B.; Safa, A.; Majidinia, M. Targeting PI3K/Akt/mTOR signaling pathway by polyphenols: Implication for cancer therapy. *Life Sci.* **2020**, *255*, 117481. [CrossRef]
27. Caparica, R.; Júlio, A.; Araújo, M.E.M. Anticancer Activity of Rutin and Its Combination with Ionic Liquids on Renal Cells. *Biomolecules* **2020**, *10*, 233. [CrossRef]
28. Vandana, J.; Gupta, A.K.; Mukerjee, A. Pharmacological Activities of Miraculous Plant *Plantago Major* L.: A Review. *IJCPS* **2017**, *6*, 26–37.
29. Sanchez-Maldonado, A.F.; Schieber, A.; Ganzle, M.G. Antifungal activity of secondary plant metabolites from potatoes (*Solanum tuberosum* L.): Glycoalkaloids and phenolic acids show synergistic effects. *J. Appl. Microbiol.* **2016**, *120*, 955–965. [CrossRef]
30. Martínez, G.; Regente, M.; Jacobi, S.; Del Rio, M.; Pinedo, M.; de la Canal, L. Chlorogenic acid is a fungicide active against phytopathogenic fungi. *Pestic. Biochem. Physiol.* **2017**, *140*, 30–35. [CrossRef]
31. Santiago, R.; de Armas, R.; Blanch, M.; Vicente, C.; Legaz, M.E. In vitro effects of caffeic acid upon the growth of the fungi *Sporisorium scitamineum*. *J. Plant Interact.* **2010**, *5*, 233–240. [CrossRef]
32. Bone, K.; Mills, S. (Eds.) *Principles and Practice of Phytotherapy. Modern Herbal Medicine*; Elsevier: Amsterdam, The Netherlands, 2013; pp. 17–82.
33. De Lira Mota, K.S.; Dias, G.E.N.; Pinto, M.E.F.; Luiz-Ferreira, Â.; Monteiro Souza-Brito, A.R.; Hiruma-Lima, C.A.; Barbosa-Filho, J.M.; Batista, L.M. Flavonoids with gastroprotective activity. *Molecules* **2009**, *14*, 979–1012. [CrossRef]
34. Pirvu, L.; Panteli, M.; Bubueanu, C.; Coprean, D. *Centaurea cyanus* L. Polysaccharides and Polyphenols Cooperation in Achieving Strong Rat Gastric Ulcer Protection. *Cent. Eur. J. Chem.* **2015**, *13*, 910–921. [CrossRef]
35. Pintilie, G.; Paraschiv, I.; Manaila, N.; Ocnaru, D.; Armatu, A.; Colceru, S.; Pirvu, L.; Rughinis, D.; Nita, S. Cascade of bioactive compounds from *Plantago lanceolata* L. cultivated in Romania. *Planta Med.* **2007**, *73*, 265. [CrossRef]
36. Bostan, M.; Mihaila, G.; Petrica-Matei, G.; Radu, N.; Hainarosie, R.; Stefanescu, C.-D.; Roman, V.; Diaconu, C.C. Resveratrol Modulation of Apoptosis and Cell Cycle Response to Cisplatin in Head and Neck Cancer Cell Lines. *Int. J. Mol. Sci.* **2021**, *22*, 6322. [CrossRef] [PubMed]
37. Bostan, M.; Petrica-Matei, G.; Radu, N.; Hainarosie, R.; Stefansescu, C.D.; Diaconu, C.; Roman, V. The Effect of Resveratrol or Curcumin on Head and Neck Cancer Cells Sensitivity to the Cytotoxic Effects of Cisplatin. *Nutrients* **2020**, *12*, 2596. [CrossRef]
38. Neagu, M.; Constantin, C.; Bostan, M.; Caruntu, C.; Ignat, S.R.; Costache, M. Proteomics technologies "lens" for epithelial-mesenchymal transition process identification in oncology. *Anal. Cell. Pathol.* **2019**, *2019*, 3565970. [CrossRef]
39. Lukova, P.; Karcheva-Bahchevanska, D.; Dimitrova-Dyulgerova, I.; Katsarov, P.; Mladenov, R.; Iliev, I. A Comparative Pharmacognostic Study and Assessment of Antioxidant Capacity of Three Species From Plantago Genus. *Farmacia* **2018**, *66*, 609–614. [CrossRef]
40. Abate, L.; Bachheti, R.K.; Tadesse, M.G.; Bachheti, A. Ethnobotanical Uses, Chemical Constituents, and Application of *Plantago lanceolata* L. *J. Chem.* **2022**, *2022*, 1532031. [CrossRef]
41. Ferrazzano, G.F.; Cantile, T.; Roberto, L.; Ingenito, A.; Catania, M.R.; Roscetto, E.; Palumbo, G.; Zarrelli, A.; Pollio, A. Determination of the in vitro and in vivo antimicrobial activity on salivary Streptococci and Lactobacilli and chemical characterization of the phenolic content of a *Plantago lanceolata* infusion. *BioMed Res. Int.* **2015**, *2015*, 286817. [CrossRef]
42. Zhang, S.; Hu, J.; Sun, Y.; Tan, H.; Yin, J.; Geng, F.; Nie, S. Review of structure and bioactivity of *Plantago* (Plantaginaceae) polysaccharides. *Food Chem.* **2021**, *12*, 100158. [CrossRef]
43. Lukova, P.; Nikolova, M.; Petit, E.; Elboutachfaiti, R.; Vasileva, T.; Katsarov, P.; Manev, H.; Gardarin, C.; Pierre, G.; Michaud, P.; et al. Prebiotic Activity of Poly- and Oligosaccharides Obtained from *Plantago major* L. Leaves. *Appl. Sci.* **2020**, *10*, 2648. [CrossRef]

44. Kim, C.W.; Choi, K.C. Potential Roles of Iridoid Glycosides and Their Underlying Mechanisms against Diverse Cancer Growth and Metastasis: Do They Have an Inhibitory Effect on Cancer Progression? *Nutrients* **2021**, *13*, 2974. [CrossRef]
45. Janković, T.; Zdunić, G.; Beara, I.; Balog, K.; Pljevljakušić, D.; Stešević, D.; Šavikin, K. Comparative study of some polyphenols in Plantago species. *Biochem. Syst. Ecol.* **2012**, *42*, 69–74. [CrossRef]
46. Kahraman, Ç.; Tatli, İ.İ.; Kart, D.; Ekizoğlu, M.; Akdemir, Z.Ş. Structure Elucidation and Antimicrobial Activities of Secondary Metabolites from the Flowery Parts of *Verbascum mucronatum* Lam. *Turk. J. Pharm. Sci.* **2018**, *15*, 231–237. [CrossRef]
47. Ziarno, M.; Kozłowska, M.; Scibisz, I.; Kowalczyk, M.; Pawelec, S.; Stochmal, A.; Szleszynski, B. The Effect of Selected Herbal Extracts on Lactic Acid Bacteria Activity. *Appl. Sci.* **2021**, *11*, 3898. [CrossRef]
48. Chiang, L.C.; Chiang, W.; Chang, M.Y.; Ng, L.T.; Lin, C.C. Antileukemic activity of selected natural products in Taiwan. *Am. J. Chin. Med.* **2003**, *31*, 37–46. [CrossRef]
49. Weber, R.H.; Mondolot, L.; Kelly, M.T.; Lykke, A.M. Plantamajoside—A current review. *Phytochem. Lett.* **2015**, *12*, 42–53.
50. Yang, X.; Yuan, J.; Wan, J. Cytotoxic phenolic glycosides from *Boschniakia himalaica*. *Chem. Nat. Compd.* **2012**, *48*, 555–558. [CrossRef]
51. Saraei, R.; Marofi, F.; Naimi, A.; Talebi, M.; Ghaebi, M.; Javan, N.; Salimi, O.; Hassanzadeh, A. Leukemia therapy by flavonoids: Future and involved mechanisms. *J. Cell. Physiol.* **2019**, *234*, 8203–8220. [CrossRef]
52. Cotoraci, C.; Ciceu, A.; Sasu, A.; Miutescu, E.; Hermenean, A. The Anti-Leukemic Activity of Natural Compounds. *Molecules* **2021**, *26*, 2709. [CrossRef]
53. Wang, Y.; Wu, J.; Lv, M.; Shao, Z.; Hungwe, M.; Wang, J.; Bai, X.; Xie, J.; Wang, Y.; Geng, W. Metabolism Characteristics of Lactic Acid Bacteria and the Expanding Applications in Food Industry. *Front. Bioeng. Biotechnol.* **2021**, *9*, 612285. [CrossRef]
54. Brautigam, M.; Franz, G.G. Structural Features of Plantago lanceolata Mucilage. *Planta Med.* **1985**, *51*, 293–297. [CrossRef] [PubMed]
55. Ahmad, T.A.; Ibrahem, M.M.; Chloob, Z.M. Identification and estimation of carbohydrate and amino acids contents of mucilage from callus cultures of *Plantago lanceolata* L. *Plant Arch.* **2020**, *20*, 1875–1879.
56. Endalur Gopinarayanan, V.; Nair, N.U. Pentose metabolism in Saccharomyces cerevisiae: The Need to Engineer Global Regulatory Systems. *Biotechnol. J.* **2019**, *14*, e1800364. [CrossRef] [PubMed]
57. Lee, K.S.; Hong, M.E.; Jung, S.C.; Ha, S.J.; Yu, B.J.; Koo, H.M.; Park, S.M.; Seo, J.-H.; Kweon, D.H.; Park, J.C.; et al. Improved galactose fermentation of *Saccharomyces cerevisiae* through inverse metabolic engineering. *Biotechnol. Bioeng.* **2011**, *108*, 621–631. [CrossRef]

plants

Article

Comparative Phytochemical Analysis of *Aronia melanocarpa* L. Fruit Juices on Bulgarian Market

Oskan Tasinov [1], Ivayla Dincheva [2], Ilian Badjakov [2], Christina Grupcheva [3] and Bistra Galunska [1,*]

[1] Department of Biochemistry, Molecular Medicine and Nutrigenomics, Faculty of Pharmacy, Medical University of Varna, 84B Tzar Osvoboditel Blvd., 9002 Varna, Bulgaria; oskan.tasinov@gmail.com
[2] AgroBioInstitute, Agricultural Academy, 8 Dr. Tsankov Blvd., 1164 Sofia, Bulgaria; ivadincheva@yahoo.com (I.D.); ibadjakov@gmail.com (I.B.)
[3] Department of Ophthalmology and Visual Sciences, Faculty of Medicine, Medical University of Varna, 15 Doyran Street, 9000 Varna, Bulgaria; hristina.grupcheva@mu-varna.bg
* Correspondence: bistra.galunska@gmail.com

Abstract: *Aronia melanocarpa* L. (black chokeberry), belonging to the Rosaceae family, contains high amounts of polyphenolics and therefore exhibits one of the highest antioxidant and anti-inflammatory activities among berry fruits. Chokeberries are used in the food industry for juice, nectar, and wine production and as colorants. We aimed to compare the phytochemical composition of three chokeberry juices commercially available in the local market as sources of beneficial phytochemicals. Using GC–MS and LC–MS/MS, we performed the identification and quantitation of polar compounds and polyphenolics. The concentrations of 13 amino acids, including 6 essential amino acids, 10 organic acids, 20 sugar alcohols and derivatives, 14 saccharides, 12 fatty acids and esters, and 38 polyphenols, were estimated. One of the analyzed juices had the highest polyphenolic content (5273.87 ± 63.16 μg/mL), possibly due to 2.9 times higher anthocyanin concentration compared to anthocyanins in other tested juices. This study provides new data concerning phytochemical composition in terms of amino acids, organic acids, sugar acids, fatty acids and their esters, and polyphenols as phytocomponents of commercially available chokeberry juices. Results show that after all processing techniques and possibly different plant growth conditions, chokeberry juices are a valuable source of health-promoting phytochemicals such as phenolic acids, pro-anthocyanins, and anthocyanins, thus considering them as functional foods. We demonstrated a diversity of the active substances in bioactive foods marketed as "same"; therefore, the standardized therapeutic effect could be expected only by the utilization of food supplements with guaranteed constant content.

Keywords: *Aronia melanocarpa*; fruit juice; phenolic acids; polyphenols; functional foods

Citation: Tasinov, O.; Dincheva, I.; Badjakov, I.; Grupcheva, C.; Galunska, B. Comparative Phytochemical Analysis of *Aronia melanocarpa* L. Fruit Juices on Bulgarian Market. *Plants* **2022**, *11*, 1655. https://doi.org/10.3390/plants11131655

Academic Editor: Maria José U. Ferreira

Received: 9 June 2022
Accepted: 20 June 2022
Published: 22 June 2022

1. Introduction

Aronia melanocarpa L. (AM, black chokeberry) belongs to the Rosaceae family, subfamily Maloideae, and is native to the eastern parts of North America shrub, growing to a height of 2–3 m and forming purplish to black berries. It was transferred to Europe at the beginning of the 20th century and had recently been cultivated mainly in eastern European countries and Germany [1]. Chokeberry fruits are the most used part of the plant [2]. They are commonly used in the European food industry for the production of syrups, juices, jellies, nectars, wines, fruit teas, and dietary supplements [3–5]. In addition, the high content of anthocyanins in chokeberries suggests their usage as natural food colorants [6–8].

Chokeberries are used in North American traditional medicine as astringents and as a remedy for cold treatment. In Russia and Eastern European countries, they are mostly known as natural antihypertensive and anti-atherosclerotic remedies for the treatment of achlorhydria, avitaminoses, convalescence, and hemorrhoids [1,5,9–11].

A. melanocarpa L. fruits' are rich in anthocyanins, flavonols, flavanols, proanthocyanidins, and phenolic acids [12]. Chokeberries are especially high in cyanidin glycosides, proanthocyanidins mono-, di-, and trimers, and hydroxycinnamic acids such as chlorogenic acids and quercetin glycosides [13]. The high polyphenol content of fruits is related to their strong in vitro [14,15] and in vivo antioxidant activity, including modulation of antioxidant enzymes [16–18]. In addition, they exert anti-inflammatory activity by decreasing inflammatory cytokine production and improving lipid profile by reducing chylomicron, LDL, and triglyceride levels and increasing HDL [19–22]. Regular consumption of chokeberry juice reduces blood pressure in individuals with metabolic syndrome and mild hypercholesterolemia [21,23]. The health benefits of chokeberry juice are supported by scientific studies pointing to its hepatoprotective, gastroprotective, antidiabetic, and anticancer activities [24–28]. AM juice exerts in vitro bacteriostatic activity against *Staphylococcus aureus* and *Escherichia coli*, antiviral activity against influenza type A virus [9], and reduces urinary tract infections [29].

The nutritional contribution and main biological effects of chokeberry fruits as a potential functional food are associated predominantly with their phytochemical content and especially with the presence of chlorogenic acids, cyanidin glycosides, and quercetin derivatives [30]. Since the periods of ripening and the time of harvesting of ripe fruits may vary, it is expected that fruit juices harvested from different regions and produced by different companies may vary in their polyphenolic, sugar, organic, and sugar acids content and, as a consequence, in their biological activity [4].

In recent years, the interest in locally produced functional foods, including chokeberry fruit juices, has been growing. Therefore, it would be useful to provide the consumers with a scientific assessment and an overview of the content of bioactive phytocompounds from different chokeberry juices available on the local market.

Recently we aimed to perform a comparative phytochemical analysis of chokeberry juices produced locally and available on the Bulgarian food market. Thus, assessing their quality as a source of bioactive compounds and functional food with known beneficial health effects, a selection among the tested AM fruit juices would be made and recommend the juice with the highest quality in dietary guidelines.

2. Results

2.1. Phytochemical Content and Composition

Detailed phytochemical analyses of selected *Aronia melanocarpa* L. fruit juices revealed the presence of 13 amino acids (AA), 10 organic acids (OA), 20 sugar alcohols and derivatives, 14 saccharides, 12 saturated and unsaturated acids and esters, and 38 polyphenols including anthocyanins, proanthocyanidins, stilbenes, cyclohexanecarboxylic acid, hydroxycinnamic acids, and flavonol glycosides.

2.1.1. Polar Compounds

Analyses of the AA content revealed 41.8% essential AA for all tested AM juices. Juice 3 stands out with the highest AA content (106.88 ± 1.56 μg/mL) vs. juice 1 ($p < 0.01$) and 2 ($p < 0.001$) (Table 1). In all analyzed AM drinks, the highest content was determined for L-proline (22% of AA), L-aspartic acid (17.94% of AA), and L-phenylalanine (11.23% of AA). L-phenylalanine represents 26.9% of all detected essential AA in all analyzed AM drinks (Table 1).

Among the polar OAs highest concentrations in all AM drinks, pyroglutamic acid (5-oxoproline) (30.39% of OA) and isocitric acid (16.37% of OA) were found (Table 1). The highest concentrations of total OA (129.96 ± 2.01 μg/mL) and each individual OA were found in juice 3, followed by juice 2 (124.91 ± 1.93 μg/mL), and juice 1 (118.55 ± 2.34 μg/mL).

The dominating sugar alcohols in all analyzed AM drinks are sorbitol and its derivative sorbitol 6-phosphate (34.9% of alcohols), followed by glycerol and glycerol 3-phosphate (20.3% of alcohols), and then arabinitol (13% of alcohols). The sugar content was highest for juice 3, followed by juice 2 and 1.

Table 1. Polar phytochemicals identified in fraction A of *Aronia melanocarpa* L. fruit juice using GC-MS. The concentration is given in µg/mL. Results are presented as mean ± standard deviation.

Compound	AM Juice 1 Content, µg/mL	AM Juice 2 Content, µg/mL	AM Juice 3 Content, µg/mL
Amino Acids			
L-Valine	3.24 ± 0.06	3.41 ± 0.05	3.62 ± 0.08
L-Leucine	8.63 ± 0.17	9.10 ± 0.14	9.46 ± 0.15
L-Isoleucine	9.08 ± 0.18	9.57 ± 0.15	9.96 ± 0.15
L-Threonine	4.16 ± 0.08	4.39 ± 0.07	4.56 ± 0.07
L-Phenylalanine	10.98 ± 0.22	11.57 ± 0.18	12.04 ± 0.19
L-Lysine	4.68 ± 0.09	4.93 ± 0.08	5.13 ± 0.08
L-Proline	21.43 ± 0.42	22.59 ± 0.35	23.50 ± 0.36
Glycine	4.05 ± 0.08	4.27 ± 0.07	4.44 ± 0.07
Serine	2.77 ± 0.06	2.92 ± 0.05	3.04 ± 0.05
L-Aspartic acid	17.49 ± 0.34	18.42 ± 0.28	19.17 ± 0.29
L-Asparagine	6.63 ± 0.13	6.98 ± 0.11	7.27 ± 0.11
L-Glutamic acid	1.44 ± 0.03	1.52 ± 0.02	1.58 ± 0.02
L-Tyrosine	2.85 ± 0.06	3.00 ± 0.05	3.12 ± 0.05
Total essential AAs	40.78 ± 0.79	42.97 ± 0.67	44.76 ± 0.61
Total non-essential AAs	56.66 ± 1.12	59.71 ± 0.92	62.12 ± 0.96
Total AAs	97.44 ± 1.90	102.67 ± 1.58	106.88 ± 1.56
Organic Acids			
Succinic acid	13.54 ± 0.27	14.26 ± 0.22	14.84 ± 0.23
Fumaric acid	7.08 ± 0.14	7.46 ± 0.11	7.76 ± 0.12
Malic acid	9.88 ± 0.20	10.41 ± 0.16	10.83 ± 0.17
Pyroglutamic acid (5-oxoproline)	36.03 ± 0.71	37.96 ± 0.59	39.50 ± 0.61
4-Aminobutyric acid	6.10 ± 0.12	6.42 ± 0.10	6.68 ± 0.11
2-Hydroxyglutaric acid	4.36 ± 0.09	4.59 ± 0.07	4.78 ± 0.07
2-Ketoglutaric acid	8.60 ± 0.17	9.06 ± 0.14	9.42 ± 0.15
Phenylpyruvic acid	2.33 ± 0.05	2.46 ± 0.04	2.56 ± 0.04
2,3-Dihydroxybutanedioic acid	11.23 ± 0.23	11.84 ± 0.19	12.32 ± 0.19
Isocitric acid	19.41 ± 0.38	20.45 ± 0.31	21.28 ± 0.33
Total organic acids	118.55 ± 2.34	124.91 ± 1.93	129.96 ± 2.01
Sugar Alcohols			
Glycerol	38.69 ± 0.76	40.78 ± 0.63	42.42 ± 0.66
Digalactosylglycerol	7.48 ± 0.15	7.88 ± 0.12	8.20 ± 0.13
Threitol	8.20 ± 0.17	8.65 ± 0.13	8.99 ± 0.14
Erythreol	2.23 ± 0.05	2.35 ± 0.04	2.45 ± 0.04
Xylitol	4.50 ± 0.09	4.75 ± 0.08	4.94 ± 0.08
Arabinitol	37.12 ± 0.73	39.11 ± 0.61	40.69 ± 0.63
L-Glycerol-3-phosphate	18.99 ± 0.37	20.00 ± 0.31	20.82 ± 0.33
Manitol	3.19 ± 0.06	3.36 ± 0.05	3.50 ± 0.06
Sorbitol	52.77 ± 1.04	55.61 ± 0.86	57.86 ± 0.90
Galactitol	2.05 ± 0.04	2.16 ± 0.04	2.25 ± 0.03
Myo-inositol	7.19 ± 0.14	7.57 ± 0.12	7.88 ± 0.12
Galactosylglycerol	25.92 ± 1.10	28.93 ± 0.91	31.31 ± 0.95
Sorbitol-6-phosphate	46.41 ± 0.91	48.90 ± 0.76	50.87 ± 0.79
myo-Inositol-1-phosphate isomer	6.04 ± 0.12	6.37 ± 0.10	6.62 ± 0.10
myo-Inositol-2-phosphate isomer	7.96 ± 0.16	8.39 ± 0.13	8.73 ± 0.13
myo-Inositol-1-phosphate isomer	3.54 ± 0.07	3.73 ± 0.06	3.88 ± 0.06
myo-Inositol-2-phosphate isomer	7.36 ± 0.15	7.76 ± 0.12	8.07 ± 0.13
Maltitol; alpha-D-Glc-(1,4)-D-sorbitol	5.25 ± 0.11	5.54 ± 0.09	5.76 ± 0.09
Galactinol isomer; alpha-D-Gal-(1,3)-myo-Inositol	0.74 ± 0.02	0.78 ± 0.02	0.81 ± 0.02
Galactinol isomer; alpha-D-Gal-(1,3)-myo-Inositol	3.93 ± 0.08	4.14 ± 0.07	4.31 ± 0.07
Total sugar alcohols	282.09 ± 6.12	298.87 ± 5.10	312.14 ± 5.30
Sugar acids			
Glyceric acid	18.27 ± 0.36	19.25 ± 0.30	20.02 ± 0.31
Erithreonic acid	2.84 ± 0.06	2.99 ± 0.05	3.11 ± 0.05
Threonic acid	9.00 ± 0.18	9.49 ± 0.15	9.87 ± 0.16
Pentonic acid	8.24 ± 0.17	8.69 ± 0.13	9.04 ± 0.14
Ribonic acid	5.10 ± 0.10	5.37 ± 0.08	5.59 ± 0.09
Glucuronic acid isomer	9.09 ± 0.18	9.58 ± 0.15	9.96 ± 0.15

Table 1. *Cont.*

Compound	AM Juice 1 Content, μg/mL	AM Juice 2 Content, μg/mL	AM Juice 3 Content, μg/mL
Galacturonic acid isomer	17.05 ± 0.34	17.97 ± 0.28	18.69 ± 0.29
Glucuronic acid isomer	13.96 ± 0.27	14.71 ± 0.22	15.30 ± 0.24
Gluconic acid isomer	1.91 ± 0.04	2.02 ± 0.03	2.10 ± 0.03
Galacturonic acid isomer	3.09 ± 0.06	3.26 ± 0.05	3.39 ± 0.06
Glucuronic acid isomer	4.15 ± 0.08	4.37 ± 0.07	4.55 ± 0.07
Galactonic acid	6.78 ± 0.14	7.15 ± 0.11	7.43 ± 0.12
Gluconic acid isomer	3.98 ± 0.08	4.19 ± 0.07	4.36 ± 0.07
Glucaric acid	16.81 ± 0.92	22.66 ± 5.21	31.32 ± 0.79
Galactaric acid	3.63 ± 0.07	3.82 ± 0.06	3.97 ± 0.06
Gluconic acid-6-phosphate	1.64 ± 0.04	1.73 ± 0.03	1.80 ± 0.03
Total sugar acids	125.54 ± 3.06	137.24 ± 4.16	150.51 ± 2.65
Saccharides (mono-, di-, and tri-)			
Xylose methoxyamine	6.36 ± 0.13	6.52 ± 0.21	6.97 ± 0.11
Arabinose methoxyamine	13.37 ± 0.15	14.28 ± 0.22	14.86 ± 0.23
Fructose isomer	15.33 ± 0.30	16.16 ± 0.25	16.81 ± 0.26
Fructose isomer	20.24 ± 0.40	21.33 ± 0.33	22.19 ± 0.34
Sorbose isomer	30.12 ± 0.60	31.74 ± 0.49	33.02 ± 0.51
Sorbose isomer	22.86 ± 0.45	24.10 ± 0.37	25.07 ± 0.39
Galactose isomer	37.70 ± 0.74	39.72 ± 0.61	41.33 ± 0.64
Galactose isomer	14.85 ± 0.30	15.65 ± 0.24	16.28 ± 0.25
Glucose isomer	18.58 ± 0.37	19.58 ± 0.31	20.37 ± 0.32
Glucose isomer	14.56 ± 0.29	15.34 ± 0.24	15.96 ± 0.25
Fructose-6-phosphate isomer	17.35 ± 0.34	18.28 ± 0.28	19.02 ± 0.29
Mannose-6-phosphate isomer	3.71 ± 0.08	3.91 ± 0.06	4.07 ± 0.06
Galactose-6-phosphate isomer	20.12 ± 0.40	21.21 ± 0.33	22.06 ± 0.34
Glucose-6-phosphate isomer	32.42 ± 0.64	34.17 ± 0.53	35.54 ± 0.55
Fructose-6-phosphate isomer	6.22 ± 0.13	6.56 ± 0.10	6.82 ± 0.11
Galactose-6-phosphate isomer	3.56 ± 0.07	3.75 ± 0.06	3.90 ± 0.06
Glucose-6-phosphate isomer	4.84 ± 0.10	5.10 ± 0.08	5.31 ± 0.08
Sucrose isomer; alpha-D-Glc-(1,2)-beta-D-Fru isomer	26.57 ± 0.52	28.00 ± 0.43	29.13 ± 0.45
Trehalose; alpha-D-Glc-(1,1)-alpha-D-Glc isomer	10.82 ± 0.21	11.40 ± 0.18	11.86 ± 0.18
Melibiose isomer; alpha-D-Gal-(1,6)-D-Glc isomer	19.91 ± 0.40	20.98 ± 0.32	21.83 ± 0.34
Melibiose isomer; alpha-D-Gal-(1,6)-D-Glc isomer	20.14 ± 0.40	21.23 ± 0.33	22.09 ± 0.34
Sucrose isomer; alpha-D-Glc-(1,2)-beta-D-Fru isomer	22.02 ± 0.44	23.20 ± 0.36	24.14 ± 0.38
Trehalose; alpha-D-Glc-(1,1)-alpha-D-Glc isomer	17.28 ± 0.34	18.20 ± 0.28	18.94 ± 0.29
Raffinose; alpha-D-Gal-(1,6)-alpha-D-Glc-(1,2)-beta-D-Fru isomer	13.83 ± 0.27	14.57 ± 0.23	15.16 ± 0.24
Raffinose; alpha-D-Gal-(1,6)-alpha-D-Glc-(1,2)-beta-D-Fru isomer	27.44 ± 0.54	28.91 ± 0.45	30.08 ± 0.47
Total saccharides	440.21 ± 8.38	463.89 ± 6.90	482.82 ± 7.48
Saturated, unsaturated acids and esters			
9-(E)-Hexadecenoic acid	9.13 ± 0.18	9.61 ± 0.15	10.00 ± 0.15
9-(Z)-Hexadecenoic acid	7.03 ± 0.14	7.41 ± 0.12	7.71 ± 0.12
Heptadecanoic acid	8.10 ± 0.16	8.54 ± 0.13	8.88 ± 0.14
Hexadecatrienoic acid	5.19 ± 0.11	5.47 ± 0.08	5.70 ± 0.09
Hexadecanoic acid (Palmitic acid)	7.03 ± 0.14	7.41 ± 0.11	7.71 ± 0.12
Heptadecanoic acid	6.49 ± 0.13	6.83 ± 0.11	7.11 ± 0.11
9,12-(Z,Z)-Octadecadienoic acid (Linoleic acid)	10.38 ± 0.21	10.94 ± 0.17	11.39 ± 0.18
9,12,15-(Z,Z,Z)-Octadecatrienoic acid (Linolenic acid)	9.01 ± 0.18	9.50 ± 0.15	9.88 ± 0.15
Nonadecanoic acid	2.50 ± 0.05	2.64 ± 0.04	2.75 ± 0.04
Octadecanoic acid (Stearic acid)	11.91 ± 0.23	12.55 ± 0.19	13.05 ± 0.20
(2E,4E)-2,4-Octadecadienoic acid	16.76 ± 0.33	17.66 ± 0.27	18.38 ± 0.29
1-Monopalmitin	14.79 ± 0.29	15.58 ± 0.24	16.21 ± 0.25
Monooctadecanoylglycerol	9.24 ± 0.18	9.73 ± 0.15	10.13 ± 0.16
beta-Sitosterol	16.31 ± 0.32	17.18 ± 0.26	17.88 ± 0.28
Total saturated, unsaturated acids and esters	141.34 ± 2.77	148.95 ± 2.31	154.96 ± 2.40

AM—*Aronia melanocarpa* L.; AAs—amino acids. All metabolites are trimethylsilyl derivatives. Essential AAs are given in *italic*. Additional data regarding chromatographic parameters and total ion chromatogram of tested polar compounds are given in Table S1 and Figure S1, respectively.

Dominating monosaccharides are galactose and its 6-phosphate derivative (17.3% of all analyzed saccharides), followed by glucose and its 6-phosphate derivative (15.99% of saccharides), and fructose and its 6-phosphate derivative (13.42% of saccharides) (Table 1). Sucrose (11.03% of saccharides) was the main disaccharide, and raffinose (9.37% of saccharides) was the main trisaccharide. Juice 3 was leading in saccharide content as well.

Octadecadienoic acid (11.86% of lipids) was with the highest content among all detected fatty acids, followed by octadecanoic acid (8.42% of lipids), while beta-sitosterol is the sterol, representing 11.54% of detected lipids (Table 1).

2.1.2. Polyphenolic Content

It is well known that fruit juices of *A. melanocarpa* L. are rich in polyphenolic compounds, but data on the quantitative and qualitative composition of the polyphenols they contain are very limited. Highest concentration of polyphenolics was detected in juice 3 (5273.87 ± 63.16 μg/mL), followed by juice 2 (4460.53 ± 136.67 μg/mL), and juice 1 (4351.83 ± 75.38 μg/mL) (Table 2).

The most abundant anthocyanin was cyanidin-3-O-galactoside, comprising a minimum of 70% of anthocyanins in all analyzed samples. The content of cyanidin-3-O-galactoside was about 2.6 times higher in juice 3 compared to juice 2 and 1. Epicatechine, the major proanthocyanidin in all tested samples, exhibited the highest concentration in juice 2 (269.47 ± 17.35 μg/mL) (Table 2). The leading in proanthocyanidin polymer content was juice 3, containing 552.05 ± 5.49 μg/mL epicatechin dimers and 732.42 ± 7.29 μg/mL trimers. The only detected stilbene in the tested juices was trans-resveratrol-3-O-glucoside (from 44.37 ± 1.79 μg/mL in juice 3 to 39.80 ± 1.63 μg/mL in juice 1). Quinic acid was the only detected cyclohexanecarboxylic acid (from 84.95 ± 0.42 μg/mL to 81.74 ± 2.27 μg/mL, in juice 3 and 1, respectively). Dominating hydroxycinnamic acids in the tested AM fruit juices were neochlorogenic and chlorogenic acid, followed by 3-O-p-coumaroylquinic acid. The total amount of hydroxycinnamic acids in the analyzed juices represents 51.52%, 51.38%, and 45.35% for juices 1, 2, and 3, respectively (Table 2). Hyperoside was found to be the major flavonol (14.93% of flavonols) detected in all analyzed samples.

We compared the portions of different polyphenol classes within each of the analyzed juices (Figure 1). When calculated, the percentage of detected hydroxycinnamic acids varies from 45.3% to 51.5%, followed by the proanthocyanidins from 29.3% to 34.4%, anthocyanins from 8.1% to 19.9%, and flavonols from 3% to 3.3% as the lowest portion of identified polyphenols.

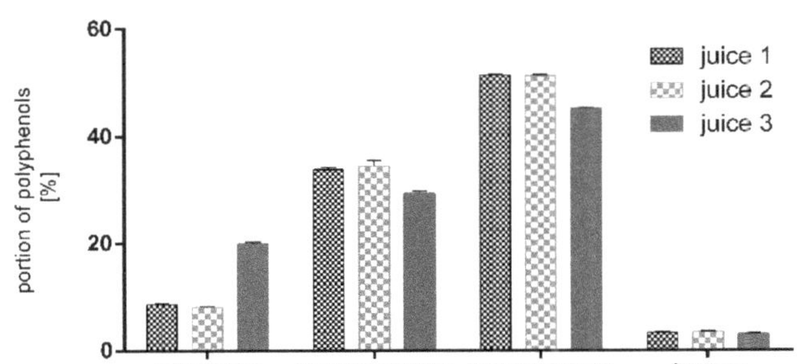

Figure 1. Content of anthocyanins, proanthocyanidins, hydroxycinnamic acids, and flavanols in the tested chokberry juices. Data are presented as % ± SD of the total polyphenolic content.

Table 2. Polyphenolics identified in fractions B and C of the SE FAE using LC-PDA-ESI-MS/MS. The concentrations are given in μg/mL. Results are presented as mean ± standard deviation.

Compound	AM Juice 1 μg/mL	AM Juice 2 μg/mL	AM Juice 3 μg/mL
	Anthocyanins		
Cyanidin-3-O-galactoside (idaein)	278.91 ± 6.56	262.69 ± 8.78	735.80 ± 17.41
Cyanidin-3-O-glucoside (chrysanthemin)	23.04 ± 0.61	25.62 ± 0.99	55.34 ± 1.84
Cyanidin-3-O-arabinoside	60.98 ± 2.81	60.69 ± 1.82	228.72 ± 7.53
Cyanidin-3-O-xyloside	10.91 ± 0.39	10.79 ± 0.52	30.88 ± 2.1
Total anthocyanins	373.84 ± 9.73	359.80 ± 10.46	1050.75 ± 24.86
	Proanthocyanidin monomers		
Catechin	29.64 ± 2.42	34.60 ± 3.28	23.37 ± 1.02
Epicatechin	239.17 ± 3.05	269.47 ± 17.35	237.40 ± 3.75
Total proanthocyanidin monomers	268.81 ± 2.89	304.07 ± 14.69	260.77 ± 3.01
	Proanthocyanidin dimers		
EC→EC(1)	128.33 ± 2.71	131.72 ± 4.09	137.45 ± 1.37
EC→EC(2)	126.71 ± 2.68	130.03 ± 4.00	135.72 ± 1.35
EC→EC(3)	142.15 ± 3.01	145.89 ± 4.50	152.25 ± 1.52
EC→EC(4)	118.22 ± 2.50	121.35 ± 3.76	126.63 ± 1.26
Total proanthocyanidin dimers	515.41 ± 10.90	529.00 ± 16.35	552.05 ± 5.49
	Proanthocyanidin trimers		
EC→EC→EC (1)	168.32 ± 3.34	172.63 ± 5.35	180.13 ± 1.80
EC→EC→EC (2)	181.06 ± 3.59	185.69 ± 5.76	193.76 ± 1.93
EC→EC→EC (3)	148.67 ± 2.95	152.46 ± 4.73	159.09 ± 1.59
EC→EC→EC (4)	186.36 ± 3.69	191.12 ± 5.93	199.43 ± 1.99
Total proanthocyanidin trimers	684.41 ± 13.56	701.90 ± 21.76	732.42 ± 7.29
	Stilbenes		
trans-Resveratrol-3-O-glucoside	39.80 ± 1.63	40.71 ± 1.42	44.37 ± 1.79
	Cyclohexanecarboxylic acid		
Quinic acid	81.74 ± 2.27	82.66 ± 1.55	84.95 ± 0.42
	Hydroxycinnamic acids		
3-O-Caffeoylquinic acid (chlorogenic acid)	423.08 ± 7.35	432.49 ± 13.41	451.29 ± 4.49
Caffeic acid-O-galactoside	73.66 ± 1.28	75.29 ± 2.34	78.57 ± 0.78
Caffeic acid-O-glucoside	55.71 ± 0.97	56.94 ± 1.76	59.42 ± 0.59
5-O-Caffeoylquinic acid (neochlorogenic acid)	676.03 ± 11.75	691.05 ± 21.43	721.10 ± 7.18
p-Coumaric acid-O-glucoside	176.36 ± 3.07	180.27 ± 5.59	188.11 ± 1.88
3-O-p-Coumaroylquinic acid	298.04 ± 5.18	304.67 ± 9.45	317.92 ± 3.17
Feruloylquinic acid	185.72 ± 3.23	189.85 ± 5.89	198.11 ± 1.97
4-O-p-Coumaroylquinic acid	164.01 ± 2.85	167.66 ± 5.20	174.95 ± 1.74
Ferulic acid-O-galactoside	98.23 ± 1.71	100.41 ± 3.11	104.78 ± 1.04
Ferulic acid-O-glucoside	91.22 ± 1.58	93.25 ± 2.89	97.30 ± 0.97
Total hydroxycinnamic acids	2242.06 ± 38.98	2291.88 ± 71.06	2391.55 ± 23.80
	Flavonol glycosides		
Quercetin-3-O-rhamnosyl-galactoside	19.07 ± 0.33	19.70 ± 0.61	20.55 ± 0.20
Quercetin-3-O-galactoside (hyperoside)	21.77 ± 0.38	22.48 ± 0.70	23.45 ± 0.23
Kaempferol-3-O-galactoside	8.32 ± 0.15	8.59 ± 0.27	8.96 ± 0.09
Quercetin-3-O-rhamnosyl-glucoside	15.19 ± 0.27	15.69 ± 0.49	16.36 ± 0.16
Quercetin-3-O-glucoside (isoquercetin)	17.02 ± 0.30	17.57 ± 0.55	18.33 ± 0.18
Kaempferol-3-O-glucoside (astragalin)	7.42 ± 0.13	7.66 ± 0.24	7.99 ± 0.08
Quercetin-3-O-arabinoside (guaiaverin)	12.51 ± 0.22	12.93 ± 0.40	13.48 ± 0.13
Quercetin-3-O-xyloside	10.42 ± 0.18	10.76 ± 0.33	11.23 ± 0.11
Kaempferol-3-O-rhamnosyl-galactoside	9.34 ± 0.17	9.65 ± 0.30	10.06 ± 0.10
Kaempferol-3-O-rhamnosyl-glucoside	6.83 ± 0.12	7.05 ± 0.22	7.36 ± 0.07
Kaempferol-3-O-arabinoside	8.32 ± 0.14	8.59 ± 0.26	8.96 ± 0.09
Kaempferol-3-O-xyloside	9.55 ± 0.17	9.86 ± 0.31	10.29 ± 0.10
Total flavonol glycosides	145.75 ± 2.53	150.52 ± 4.67	157.02 ± 1.53
Total analyzed polyphenols	4351.83 ± 75.38	4460.53 ± 136.67	5273.87 ± 63.16

EC—epicatechin; AM—*Aronia melanocarpa* L. Additional data regarding precursor ion and fragment ion mass-to-charge ratios (m/z) of the analyzed polyphenols are given in Table S2. Representative LC-PDA-ESI-MS/MS chromatograms of detected polyphenols are given in Figure S2 (anthocyanins), Figure S3 (proanthocyanidin monomers), Figure S4 (proanthocyanidin di- and trimers), Figure S5 (stilbenes), Figure S6 (hydroxycinnamic acids), Figure S7 (hydroxycinnamic acids), Figures S8–S11 (flavonols).

3. Discussion

As there are high variations in environmental and climate conditions, soil characteristics, maturation level, harvesting period, etc., the phytochemical composition and respective health benefits of chokeberries vary as well [31–33]. Recently, we have analyzed and compared the phytochemical composition and content of selected bioactive compounds in three *A. melanocarpa* L. fruit juices available on the Bulgarian market, produced by different local companies, and harvested from different regions. We have provided for the first time new qualitative and quantitative data regarding the phytochemical composition (AAs, Oas, sugar acids and alcohols, fatty acids and esters, and polyphenols) of commercially available chokeberry juices produced locally.

3.1. Amino Acids

Aas, especially essential ones, are valuable food components for living organisms. Literature data concerning the AA content of chokeberry juices are very scarce. There are data that AM fruits are rich in dry matter [34,35]; however, the amount of total protein is low [36]. The main amino acids, including essentials, found in chokeberry fruit pomace are glutamic and aspartic acid, arginine, tyrosine, histidine, leucine, lysine, cysteine, alanine, serine, and threonine [37]. We performed for the first time a comparative analysis of the amino acid content of three selected chokeberry juices from different local producers. The highest amino acid content, including essential AA, was found in juice 3 (106.88 ± 1.56 μg/mL). Thus, chokeberry fruit juice may be considered a natural source of AA.

3.2. Organic Acids

The OAs reported in chokeberries are tartaric, citric, isocitric, malic, succinic, fumaric, ascorbic, shikimic, and oxalic acids [33,38,39]. Literature data have shown the presence of citric, malic, oxalic, and tartaric acids in commercially available chokeberry juices [40]. In addition, we also detected succinic, isocitric, fumaric, pyroglutamic, 4-aminobutyric, 2-hydroxyglutaric, 2-ketoglutaric, phenylpyruvic, and 2.3-dihydroxybutanedioic acids in the tested chokeberry juices commercially available. It might be presumed that the content of pyroglutamic acid (5-oxoproline) in tested chokeberry juices may be due to the high content of its keto-derivative proline.

3.3. Saccharides, Sugar Acids and Alcohols

There are wide variations in the content of carbohydrates in chokeberry fresh fruits [12]. Monosaccharides reported by other authors in chokeberry juice are glucose and fructose [40]. In addition to glucose and fructose detected in all tested AM juices, we also found sugars such as galactose, sorbose, arabinose, and xylose. The main detected disaccharide in chokeberry juice was sucrose, as reported by other authors as well [29]. In our study, we also found the presence of trehalose and melibiose and raffinose, the main trisaccharide found for the first time in chokeberry juice.

There is only one study reporting the presence of galacturonic acid in chokeberry fruit pomace [41]. There are almost no data concerning the detailed sugar acid content in chokeberry fruit juices. We recently found the presence of glyceric, eritnreonic, threonic, pentonic, ribonic, glucuronic, galacturonic, gluconic, galactonic, glucaric, and galactaric acid.

There are data that fruits [38] and juice [40] are considerably rich in sugar alcohols, mainly sorbitol. This finding was confirmed by our results as well, revealing that sorbitol is the main sugar alcohol in the tested samples (52.77 ± 1.04 μg/mL in juice 1; 57.86 ± 0.90 μg/mL in juice 3). In addition, we identified a few more sugar alcohols in chokeberry juice, including glycerol and arabinitol in higher concentrations and threitol, erythreol, xylitol, manitol, inositol, and galactitol in lower concentrations.

3.4. Fatty Acids and Esters

It was shown that *A. melanocarpa* fruits are rich in phospholipids, sterols, and α-tocopherols [42]. There are data that polyunsaturated fatty acids and especially linoleic acid are the main portions of fatty acids found in dried pomace and seeds of chokeberry [37,42]. The same study reports that β-sitosterol is the main sterol in chokeberry seeds, followed by campesterol and δ-avenasterol [42]. We have also detected β-sitosterol in chokeberry juice as well (Table 1). In our study, we reported for the first time the presence of fatty acids such as hexadecenoic, heptadecanoic, hexadecatrienoic, hexadecanoic (palmitic acid), octadecadienoic (linoleic acid), octadecanoic (stearic acid) acids, and esters as 1-monopalmitin, monooctadecanoyl glycerol (Table 1). The dominating fatty acids in the tested juice samples were the nonessential stearic acid and the essentials linoleic and linolenic acids. Health beneficial effects of β-sitosterol and essential fatty acids as vascular protectors and cholesterol-lowering agents are widely reported [43–46]. Considering recent findings regarding the fatty acid profile of chokeberry fruit juice, we may add new data explaining the potential of AM fruits and fruit juice in lipid profile improvement [19,21,22].

3.5. Phenolic Compounds

Plant-derived polyphenols are among the main bioactive compounds in our diet and comprise the main portion of antioxidants we intake. Their wide range of health benefits makes the proper intake of naturally derived foods and drinks rich in polyphenols an essential part of a healthy diet. Chokeberries have the highest polyphenolic content among different berries [47]. They are rich mostly in proanthocyanidins, anthocyanins, and phenolic acids but low in flavonols [15]. This observation concerning the phenolic content was also confirmed by our results (Table 2).

Application of different fruit processing and juice extraction techniques on the one hand, the time and period of harvesting and level of ripening on the other, may cause variations in the phenolic content of chokeberry fruits and fruit products, including pomace and juice [4,39,48–50]. However, chokeberry products, including juices, remain very rich in phenolic compounds and high in antioxidant activity [51]. An important fact is that warm and dry weather correlates with higher content of phenolic compounds [52]. As was suggested already by other authors, the warm climate in Bulgaria may positively affect the polyphenolic content of local chokeberry fruits and juices [39]. In our study, all assessed AM juices available on the Bulgarian food market revealed a high content of anthocyanins, proanthocyanidins, and hydroxycinnamic acids. We found the highest anthocyanin content compared to other tested samples in juice 3. This might be explained by the period of harvesting, as the anthocyanins double after the fifth week of harvesting [4], or by the technology of juice production [49].

In our comparative study, we found cyaniding-3-O-galactoside to be the main anthocyanin, followed by cyaniding-3-O-arabinoside. A similar observation regarding the presence of cyanidin-3-O-glucoside and cyanidin-3-O-xyloside in chokeberry fruits and juices was reported by other authors [2,15]. In accordance with our previous results [53], we found that the commercially available juices contain mainly epicatechine and ten times less catechin. There are data for a notably high degree of proanthocyanidin polymerization in chokeberry juices and extracts [54], which may explain the higher levels of their mono- and oligomeric forms [15,39]. In the tested juices, the content of anthocyanins and proanthocyanidins varied from 8.1% to 19.9% and from 33.7% to 29.3%, respectively (Figure 1). The observed high variation in anthocyanin content between tested juices might be due to the higher content of anthocyanins in juice 3 and especially of cyaniding-3-O-galactoside and arabinoside.

Most of the literature concerns the content of *trans*-resveratrol in chokeberry wine and not resveratrol and resveratrol-3-glucoside levels in chokeberry fruit juice [55]. Recently, we found a low concentration of trans-resveratrol-3-O-glucoside in the tested juices.

Herrmann et al. [56] reported that hydroxycinnamic acids are the most abundant phenolic acids in plants. They represent about 50% of all detected polyphenols in our samples and vary from 45.3% (juice 3) to 51.4% (juice 2) and to 51.5% (juice 1). Neochlorogenic and chlorogenic acids are dominating among all detected phenolic acids in our samples. Chlorogenic acid and 3-O-p-coumaroylquinic acid were previously reported in chokeberry fruit juice [53]; coumaric and caffeic acid glucosides were found in chokeberry fruits [57] and were also found in our samples. We have newly reported the presence of caffeic acid-O-galactoside and 4-O-p-coumaroylquinic acid in chokeberry juice. Quinic and ferulic acid were also found in chokeberry juices [39]. We recently reported the presence of ferulic acid glucoside and galactoside and feruloylquinic acid.

Other authors showed that flavonol content in chokeberry fruit juice and extracts is relatively low and represents only 1.3% of phenolic content [15]. In agreement with these data, in our samples, we found flavonol content between 3% and 3.3% of all analyzed polyphenols, mainly presented by quercetin glycosides. Flavonol glycosides such as quercetin-3-O-rhamnosyl-galactoside and glucoside, quercetin-3-O-galactoside and glucoside [58], quercetin-3-O-arabinoside and 3-O-xyloside [59,60], kaempferol-3-O-glucoside [59], and 3-O-galactoside [60] found in our samples, were reported in AM fruits also by other authors. In addition, we established the presence of kaempferol-3-O-rhamnosyl-galactoside and glucoside, kaempferol-3-O-arabinoside and xyloside in our commercially available chokeberry juices.

Phenolic content may vary in chokeberry fruit juice even within a year of harvesting and production [4,39]. Even standardized plant extracts need periodical detailed phytochemical analysis. Therefore, the variations in chokeberry fruit juices' phytochemical composition might be a prerequisite for the differences in their biological effects. Thus, updating the information for the types of bioactive ingredients, their quantity and quality may be useful for the evaluation of the health benefits of chokeberry juices as functional foods.

The novelty of our study is that, for the first time, a detailed comparative analysis was performed regarding the phytochemical composition of locally produced and commercially available AM fruit juices. Moreover, some of the detected phytocompounds were reported for the first time. The obtained phytochemical data could be used as a basis for the evaluation of Bulgarian AM fruit juices as functional foods.

A limitation of the current study is that it provides a snapshot regarding the phytochemical composition of locally produced AM fruit juices. It would be useful to seek relationships between juice phytochemical composition and some growth conditions, such as climate changes during the years and soil characteristics. The future objective would be a longitudinal study on the composition and quality of fruit juices produced by the same companies.

4. Materials and Methods

4.1. Plant Material

In the recent study, we analyzed samples from three selected juices from *Aronia melanocarpa* L. fresh fruits harvested locally, produced by three different Bulgarian companies, and available on the food market. All tested juices were produced from cultivated Aronia melanocarpa L. plants. There is information for the grown conditions only for the plants used for the production of juices 2 and 3. There is no available information regarding soil agrochemical characteristics.

Information regarding the analyzed samples and the technology of their production is presented in Table 3.

4.2. Phytochemical Analysis

Phytochemical analysis was described in more detail in our previous study [61].

For each juice, we used six bottles/bags and tested five parallel samples of each.

4.2.1. Extraction

In brief, the sample preparation includes solid phase extraction (SPE) on Discovery® DSC-18 column (5 g, 20 mL, Sigma-Aldrich Co. LLC, St. Louis, MO, USA). Filtered (0.45 µm PTFE filter, Waters, Milford, MA, USA) juice samples were loaded onto the SPE columns, and the anthocyanin fraction (C) was eluted 12 mL acetonitrile containing 0.1% (v/v) formic acid; the fraction containing phenolic acids, flavanols, and flavonols (B) was eluted with

12 mL ethyl acetate. The polar fraction (A) was eluted with 12 mL of water containing 0.2% (*v*/*v*) formic acid. The dry residues of all eluates were obtained after evaporation under reduced pressure at a temperature below 40 °C.

Table 3. Characteristics of analyzed samples from 3 juices made by *Aronia melanocarpa* L. fresh fruits.

Characteristic	Juice 1	Juice 2	Juice 3
Organic fruits/Bio product label	N/A	yes	yes
Source of fruits/region in Bulgaria	N/A	Central Stara Planina mountain, near Kapinovski Monastery temperate-continental climate with pronounced mountain influence; latitude: 42.978677; longitude: 25.747620; altitude 203 m.	Central Stara Planina mountain, near city of Troyan climate: temperate-continental climate; latitude: 42.883, longitude: 24.717; altitude 446 m.
Processing	• immediately after harvest • cold press • pasteurization	• immediately after harvest • cold press • pasteurization	• immediately after harvest • cold press • pasteurization
Additives	NO sugars, NO preservatives, NO additives	NO sugars, NO preservatives, NO additives	NO sugars, NO preservatives, NO additives
Package	250 mL glass bottle	1.5 L bag in box	270 mL glass bottle

4.2.2. GC-MS Analysis of Fraction A

A total of 0.2 mL of fraction A was lyophilized (6 h, −20 °C). The derivatization step was performed using methoxyamine hydrochloride (300.0 µL, 20.0 mg/mL in pyridine) on Thermo-Shaker TS-100 (1 h/70 °C/300 rpm). A total of 100.0 µL N,O-Bis (trimethylsilyl)trifluoroacetamide (BSTFA) were added to the mixture under heating (40 min/70 °C 300 rpm; Thermoshaker, Analytik Jena AG, Jena, Germany) and 1.0 µL of the solution was subjected to GC-MS analysis (Agilent GC 7890, Agilent MD 5975; column HP-5: length 30 m, diameter 0.32 mm, film thickness 0.25 µm). A temperature gradient was used for optimal separation: initial 100 °C for 2 min; ramp up to 180 °C with 15 °C/min for 1 min; ramp up to 300 °C with 5 °C/min for 10 min. Injector and detector temperatures were 250 °C. Helium was used as a carrier gas, with a rate of 1.0 mL/min. The MS scanning range was 50–550 m/z.

4.2.3. LC-MS/MS Analysis of Fractions B and C

Fractions B and C were analyzed by LC-PDA-ESI-MS/in negative ESI mode for fraction B and in positive ESI mode for fraction C, as previously described [61].

The dry residues of fractions B and C were dissolved in 200 µL methanol:formic acid, (99:1 *v*/*v*), and 2 µL of the filtered solution (0.22 µm PTFE filter) were subjected to LC-PDA-ESI-MS/MS analysis.

Mass-spectrometric analysis was done on LTQ Orbitrap mass spectrometer (Thermo Scientific, Hemel Hempstead, UK) equipped with an ESI source. Operation parameters: source voltage—4 kV; sheath, auxiliary, and sweep gas—20, 10, and 2 arbitrary units, respectively; capillary temperature—275 °C. The analysis was done in full scan mode, the resolution 30,000 at m/z 400, and data-dependent MS/MS events were acquired at a resolving power of 15,000. Ions with lower intensity were analyzed in MS2 mode, resolution power of 15,000 at m/z 400, isolation width 100 amu. Precursor fragmentation was performed at collision energy 30 V, activation time 10 ms. The mass range in FTMS mode was from m/z 100 to 1000. XCalibur software v2.0.7 (Thermo Fisher Scientific, Hemel Hempstead, UK) was used for data analyses.

Chromatographic analysis was done on Accela chromatograph (Thermo Scientific, Waltham, MA, USA). Optimal separation was achieved on Kinetex C18 column (100 Å, 2.6 µm, 150 × 2.1 mm, Phenomenex Inc, Torrance, CA, USA) in a gradient elution mode: A—water/0.1% formic acid; B—acetonitrile; 0 min, 10% B; 1 min, 10% B; 15 min, 30% B; 22 min, 50% B; 28 min, 100% B; 34 min, 100% B, 36 min, 10% B; flow rate 0.3 mL/min.

4.2.4. Qualitative and Quantitative Analyses

The identification of compounds in fraction A was carried out by two approaches: (1) by comparison of the retention times and Kovach indexes (RI) with the same parameters of corresponding pure standards; and (2) by using Golm Metabolome Database libraries (http://csbdb.mpimp-golm.mpg.de/csbdb/gmd/gmd.html, accessed on 30 August 2021) and NIST'08 (National Institute of Standards and Technology, Gaithersburg, MD, USA). Using the first approach, we confirmed the presence of fifteen phenolic compounds in our samples. The second approach was used for the identification of the remaining compounds.

Phenolics in fractions B and C were quantified by the external standard method as previously described [62].

4.3. Statistical Analysis

For statistical data analysis, we used GraphPad Prism v7.0 software (GraphPad Software, Inc.; La Jolla, CA, USA). The values of $p < 0.05$ were considered as significant. Data were presented as mean ± SD. All analyses were performed in triplicates.

5. Conclusions

This study provides for the first time a detailed comparative phytochemical analysis of local commercially available chokeberry juices and reports newly detected amino acids, organic acids, sugar acids, fatty acids and esters, and polyphenols. Considering the results, we may conclude that the juice acquisition techniques, pasteurization, and expected differences in plant growth conditions may cause differences in the phytochemical composition of chokeberry juices. We demonstrated a diversity of the active substances in bioactive foods marketed as "same"; therefore, the standardized therapeutic effect could be expected only by the utilization of food supplements with guaranteed constant content. Having in mind the well-known health effects of the detected bioactive compound, we may suggest that commercial black chokeberry juices are a valuable source of health-promoting phenolic acids, proanthocyanidins, and anthocyanins, thus characterizing them as a functional food.

Supplementary Materials: The following supporting information can be downloaded at: https://www.mdpi.com/article/10.3390/plants11131655/s1, Figure S1: Representative chromatogram of analyzed polar compounds (fraction A) by GC-MS technique, Figure S2: Representative LC-PDA-ESI-MS/MS chromatogram of *Aronia melanocarpa* L. fruit anthocyanins (1-Cyanidin-3-O-Galactoside, 2-Cyanidin-3-O-Glucoside, 3-Cyanidin-3-O-Arabinoside, 4-Cyanidin-3-O-Xyloside), Figure S3: Representative LC-PDA-ESI-MS/MS chromatogram of *Aronia melanocarpa* L. fruit proanthocyanidin monomers, Figure S4: Representative LC-PDA-ESI-MS/MS chromatogram of *Aronia melanocarpa* L. fruit proanthocyanidin di- and trimers, Figure S5: Representative LC-PDA-ESI-MS/MS chromatogram of *Aronia melanocarpa* L. fruit stilbenes, Figure S6: Representative LC-PDA-ESI-MS/MS chromatogram of *Aronia melanocarpa* L. fruit hydroxycinnamic acids (1–3-O-Caffeoylquinic acid, 2–Caffeic ac-id-O-galactoside, 3–Caffeic acid-O-glucoside, 4–5-O-Caffeoylquinic acid, 5–p-Coumaric ac-id-O-glucoside, 6–3-O-p-Coumaroylquinic acid, 7–Feruloylquinic acid; 8–4 -O-p-Coumaroylquinic acid; 9–Ferulic acid-O-galactoside; 10–Ferulic acid-O-glucoside), Figure S7: Representative LC-PDA-ESI-MS/MS chromatogram of *Aronia melanocarpa* L. fruit flavonols (1-Kaempferol-3-O-arabinoside, 2-Kaempferol-3-O-xyloside), Figure S8: Representative LC-PDA-ESI-MS/MS chromatogram of *Aronia melanocarpa* L. fruit flavonols (1-Quercetin-3-O-galactoside, 2-Quercetin-3-O-glucoside), Figure S9: Representative LC-PDA-ESI-MS/MS chromatogram of *Aronia melanocarpa* L. fruit flavonols (1-Kaempferol-3-O-rhamnosyl-galactoside, 2-Kaempferol-3-O-rhamnosyl-glucoside), Figure S10: Representative LC-PDA-ESI-MS/MS chromatogram of *Aronia melanocarpa* L. fruit flavonols (1-Quercetin-3-O-rhamnosyl-galactoside, 2-Quercetin-3-O-rhamnosyl-glucoside), Figure S11: Representative LC-PDA-ESI-MS/MS chromatogram of *Aronia melanocarpa* L. fruit flavonols (1-Kaempferol-3-O-galactoside, 2-Kaempferol-3-O-glucoside), Table S1: Relative Kovat's reten-tion index (RI) of analyzed polar compounds (fraction A) presented in Table 1, using GC-MS technique, Table S2: Precursor ion and fragment ion mass-to-charge ratios (m/z) of the analyzed polyphenols using the LC-PDA-ESI-MS/MS technique.

Author Contributions: Writing—original draft, methodology, review and editing O.T.; writing—methodology, resources and formal analysis, I.D. and I.B.; supervision, conceptualization review and editing, C.G.; writing—review and editing, methodology, resources, supervision, conceptualization B.G. All authors have read and agreed to the published version of the manuscript.

Funding: This research was funded by Fund "Medical Science", at the Medical University of Varna, grant number FMN-87/21.12.2015.

Data Availability Statement: The data presented in this study are available in the article.

Conflicts of Interest: The authors declare no conflict of interest.

References

1. Kulling, S.E.; Rawel, H.M. Chokeberry (*Aronia melanocarpa*)—A review on the characteristic components and potential health effects. *Planta Med.* **2008**, *74*, 1625–1634. [CrossRef] [PubMed]
2. Ochmian, I.; Grajkowski, J.; Smolik, M. Comparison of some morphological features, quality and chemical content of four cultivars of chokeberry fruits (*Aronia melanocarpa*). *Not. Bot. Horti Agrobot. Cluj-Napoca* **2012**, *40*, 253–260. [CrossRef]
3. Kitrytė, V.; Kraujalienė, V.; Šulniūtė, V.; Pukalskas, A.; Venskutonis, P.R. Chokeberry pomace valorization into food ingredients by enzyme-assisted extraction: Process optimization and product characterization. *Food Bioprod. Process.* **2017**, *105*, 36–50. [CrossRef]
4. Bolling, B.W.; Taheri, R.; Pei, R.; Kranz, S.; Yu, M.; Durocher, S.N.; Brand, M.H. Harvest date affects aronia juice polyphenols, sugars, and antioxidant activity, but not anthocyanin stability. *Food Chem.* **2015**, *187*, 189–196. [CrossRef]
5. Kokotkiewicz, A.; Jaremicz, Z.; Luczkiewicz, M. Aronia plants: A review of traditional use, biological activities, and perspectives for modern medicine. *J. Med. Food* **2010**, *13*, 255–269. [CrossRef]
6. Kapci, B.; Neradova, E.; Čizkova, H.; Voldrich, M.; Rajchl, A.; Capanoglu, E. Investigating the antioxidant potential of chokeberry (*Aronia melanocarpa*) products. *J. Food Nutr. Res.* **2013**, *52*, 219–229.
7. Vagiri, M.; Jensen, M. Influence of juice processing factors on quality of black chokeberry pomace as a future resource for colour extraction. *Food Chem.* **2017**, *217*, 409–417. [CrossRef]
8. Carle, R.; Schweiggert, R.M. *Handbook on Natural Pigments in Food and Beverages: Industrial Applications for Improving Food Color*; Woodhead Publishing Ltd.: Sawston, UK, 2016; ISBN 9780081003923.
9. Valcheva-Kuzmanova, S.V.; Belcheva, A. Current knowledge of Aronia melanocarpa as a medicinal plant. *Folia Med.* **2006**, *48*, 11–17.
10. Wawer, I. *The Power of Nature: Aronia Melanocarpa*, 1st ed.; Nature's Print Ltd.: London, UK, 2006.
11. Domarew, C.A.; Holt, R.R.; Goodman-Snitkoff, G. A Study of Russian Phytomedicine and Commonly Used Herbal Remedies. *J. Herb. Pharmacother.* **2002**, *2*, 31–48. [CrossRef]
12. Sidor, A.; Gramza-Michałowska, A. Black Chokeberry *Aronia melanocarpa* L.—A Qualitative Composition, Phenolic Profile and Antioxidant Potential. *Molecules* **2019**, *24*, 3710. [CrossRef]
13. Taheri, R.; Connolly, B.A.; Brand, M.H.; Bolling, B.W. Underutilized chokeberry (*Aronia melanocarpa, Aronia arbutifolia, Aronia prunifolia*) accessions are rich sources of anthocyanins, flavonoids, hydroxycinnamic acids, and proanthocyanidins. *J. Agric. Food Chem.* **2013**, *61*, 8581–8588. [CrossRef] [PubMed]
14. Jakobek, L.; Šeruga, M.; Krivak, P. The influence of interactions among phenolic compounds on the antiradical activity of chokeberries (*Aronia melanocarpa*). *Int. J. Food Sci. Nutr.* **2011**, *62*, 345–352. [CrossRef] [PubMed]
15. Oszmiański, J.; Wojdylo, A. Aronia melanocarpa phenolics and their antioxidant activity. *Eur. Food Res. Technol.* **2005**, *221*, 809–813. [CrossRef]
16. Kardum, N.; Takić, M.; Šavikin, K.; Zec, M.; Zdunić, G.; Spasić, S.; Konić-Ristić, A. Effects of polyphenol-rich chokeberry juice on cellular antioxidant enzymes and membrane lipid status in healthy women. *J. Funct. Foods* **2014**, *9*, 89–97. [CrossRef]
17. Kardum, N.; Petrović-Oggiano, G.; Takic, M.; Glibetić, N.; Zec, M.; Debeljak-Martacic, J.; Konić-Ristić, A. Effects of glucomannan-enriched, aronia juice-based supplement on cellular antioxidant enzymes and membrane lipid status in subjects with abdominal obesity. *Sci. World J.* **2014**, *2014*, 869250. [CrossRef]
18. Rugină, D.; Diaconeasa, Z.; Coman, C.; Bunea, A.; Socaciu, C.; Pintea, A. Chokeberry anthocyanin extract as pancreatic β-cell protectors in two models of induced oxidative stress. *Oxid. Med. Cell. Longev.* **2015**, *2015*, 429075. [CrossRef]
19. Kim, B.; Park, Y.; Wegner, C.J.; Bolling, B.W.; Lee, J. Polyphenol-rich black chokeberry (*Aronia melanocarpa*) extract regulates the expression of genes critical for intestinal cholesterol flux in Caco-2 cells. *J. Nutr. Biochem.* **2013**, *24*, 1564–1570. [CrossRef]
20. Martin, D.A.; Taheri, R.; Brand, M.H.; Draghi, A.; Sylvester, F.A.; Bolling, B.W. Anti-inflammatory activity of aronia berry extracts in murine splenocytes. *J. Funct. Foods* **2014**, *8*, 68–75. [CrossRef]
21. Skoczyńska, A.; Jedrychowska, I.; Poreba, R.; Affelska-Jercha, A.; Turczyn, B.; Wojakowska, A.; Andrzejak, R. Influence of chokeberry juice on arterial blood pressure and lipid parameters in men with mild hypercholesterolemia. *Pharmacol. Rep.* **2007**, *59*, 177–182.
22. Daskalova, E.; Delchev, S.; Vladimirova-Kitova, L.; Kitov, S.; Denev, P. Black chokeberry (*Aronia melanocarpa*) functional beverages increase hdl-cholesterol levels in aging rats. *Foods* **2021**, *10*, 1641. [CrossRef]
23. Sikora, J.; Broncel, M.; Mikiciuk-Olasik, E. Aronia melanocarpa elliot reduces the activity of angiotensin I-converting enzyme—In vitro and EX vivo studies. *Oxid. Med. Cell. Longev.* **2014**, *2014*, 739721. [CrossRef] [PubMed]

24. Valcheva-Kuzmanova, S.; Borisova, P.; Galunska, B.; Krasnaliev, I.; Belcheva, A. Hepatoprotective effect of the natural fruit juice from Aronia melanocarpa on carbon tetrachloride-induced acute liver damage in rats. *Exp. Toxicol. Pathol.* **2004**, *56*, 195–201. [CrossRef] [PubMed]
25. Valcheva-Kuzmanova, S.; Marazova, K.; Krasnaliev, I.; Galunska, B.; Borisova, P.; Belcheva, A. Effect of Aronia melanocarpa fruit juice on indomethacin-induced gastric mucosal damage and oxidative stress in rats. *Exp. Toxicol. Pathol.* **2005**, *56*, 385–392. [CrossRef] [PubMed]
26. Simeonov, S.B.; Botushanov, N.P.; Karahanian, E.B.; Pavlova, M.B.; Husianitis, H.K.; Troev, D.M. Effects of Aronia melanocarpa juice as part of the dietary regimen in patients with diabetes mellitus. *Folia Med.* **2002**, *44*, 20–23.
27. Qin, B.; Anderson, R.A. An extract of chokeberry attenuates weight gain and modulates insulin, adipogenic and inflammatory signalling pathways in epididymal adipose tissue of rats fed a fructose-rich diet. *Br. J. Nutr.* **2012**, *108*, 581–587. [CrossRef]
28. Sharif, T.; Alhosin, M.; Auger, C.; Minker, C.; Kim, J.H.; Etienne-Selloum, N.; Bories, P.; Gronemeyer, H.; Lobstein, A.; Bronner, C.; et al. Aronia melanocarpa juice induces a redox-sensitive p73-related caspase 3-dependent apoptosis in human leukemia cells. *PLoS ONE* **2012**, *7*, e32526. [CrossRef]
29. Handeland, M.; Grude, N.; Torp, T.; Slimestad, R. Black chokeberry juice (*Aronia melanocarpa*) reduces incidences of urinary tract infection among nursing home residents in the long term-a pilot study. *Nutr. Res.* **2014**, *34*, 518–525. [CrossRef]
30. Jurendić, T.; Ščetar, M. Aronia melanocarpa products and by-products for health and nutrition: A review. *Antioxidants* **2021**, *10*, 1052. [CrossRef]
31. Sidor, A.; Drożdżyńska, A.; Gramza-Michałowska, A. Black chokeberry (*Aronia melanocarpa*) and its products as potential health-promoting factors—An overview. *Trends Food Sci. Technol.* **2019**, *89*, 45–60. [CrossRef]
32. Kader, A.; Barrett, D. Classification, Composition of Fruits, and Postharvest Maintenance of Quality. In *Processing Fruits*; Barrett, D.M., Somogyi, L., Ramaswamy, H.S., Eds.; CRC Press: Boca Raton, FL, USA, 2004; pp. 3–22.
33. Šnebergrová, J.; Cížková, H.; Neradová, E.; Kapci, B.; Rajchl, A.; Voldrich, M. Variability of characteristic components of aronia. *Czech J. Food Sci.* **2014**, *32*, 25–30. [CrossRef]
34. Ochmian, I.; Oszmiański, J.; Skupień, K. Chemical composition, phenolics, and firmness of small black fruits. *J. Appl. Bot. Food Qual.* **2009**, *83*, 64–69.
35. Skupień, K.; Oszmiański, J. The effect of mineral fertilization on nutritive value and biological activity of chokeberry fruit. *Agric. Food Sci.* **2007**, *16*, 46–55. [CrossRef]
36. Červenka, L. Moisture adsorption characteristics of black currant (*Ribes nigrum* L.), black elderberry (*Sambucus nigra* L.) and Chokeberry (*Aronia melanocarpa*, [MINCHX.] ELL.) samples at different temperatures. *J. Food Process Eng.* **2011**, *34*, 1419–1434. [CrossRef]
37. Pieszka, M.; Gogol, P.; Pietras, M.; Pieszka, M. Valuable components of dried pomaces of chokeberry, black currant, strawberry, apple and carrot as a source of natural antioxidants and nutraceuticals in the animal diet. *Ann. Anim. Sci.* **2015**, *15*, 475–491. [CrossRef]
38. Djuric, M.; Brkovic, D.; Miloševic, D.; Pavlovic, M.; Curčic, S. Chemical characterisation of the fruit of black chokeberry grown on different types of soil. *Rev. Chim.* **2015**, *66*, 178–181.
39. Denev, P.; Kratchanova, M.; Petrova, I.; Klisurova, D.; Georgiev, Y.; Ognyanov, M.; Yanakieva, I. Black chokeberry (*Aronia melanocarpa* (Michx.) Elliot) fruits and functional drinks differ significantly in their chemical composition and antioxidant activity. *J. Chem.* **2018**, *2018*, 9574587. [CrossRef]
40. Sosnowska, D.; Podsędek, A.; Kucharska, A.Z.; Redzynia, M.; Opęchowska, M.; Koziołkiewicz, M. Comparison of in vitro anti-lipase and antioxidant activities, and composition of commercial chokeberry juices. *Eur. Food Res. Technol.* **2016**, *242*, 505–515. [CrossRef]
41. Sójka, M.; Kołodziejczyk, K.; Milala, J. Polyphenolic and basic chemical composition of black chokeberry industrial by-products. *Ind. Crops Prod.* **2013**, *51*, 77–86. [CrossRef]
42. Zlatanov, M.D. Lipid composition of Bulgarian chokeberry, black currant and rose hip seed oils. *J. Sci. Food Agric.* **1999**, *79*, 1620–1624. [CrossRef]
43. Best, M.M.; Duncan, C.H.; van Loon, E.J.; Wathen, J.D. Lowering of serum cholesterol by the administration of a plant sterol. *Circulation* **1954**, *10*, 201–206. [CrossRef]
44. Kassis, A.N.; Vanstone, C.A.; AbuMweis, S.S.; Jones, P.J.H. Efficacy of plant sterols is not influenced by dietary cholesterol intake in hypercholesterolemic individuals. *Metabolism* **2008**, *57*, 339–346. [CrossRef] [PubMed]
45. Chan, J.K.; Bruce, V.M.; McDonald, B.E. Dietary α-linolenic acid is as effective as oleic acid and linoleic acid in lowering blood cholesterol in normolipidemic men. *Am. J. Clin. Nutr.* **1991**, *53*, 1230–1234. [CrossRef] [PubMed]
46. Marangoni, F.; Agostoni, C.; Borghi, C.; Catapano, A.L.; Cena, H.; Ghiselli, A.; La Vecchia, C.; Lercker, G.; Manzato, E.; Pirillo, A.; et al. Dietary linoleic acid and human health: Focus on cardiovascular and cardiometabolic effects. *Atherosclerosis* **2020**, *292*, 90–98. [CrossRef] [PubMed]
47. Jakobek, L.; Šeruga, M.; Medvidović-Kosanović, M.; Novak, I. Antioxidant activity and polyphenols of Aronia in comparison to other berry species. *Agric. Conspec. Sci.* **2007**, *72*, 301–306.
48. Mayer-Miebach, E.; Adamiuk, M.; Behsnilian, D. Stability of chokeberry bioactive polyphenols during juice processing and stabilization of a polyphenol-rich material from the by-product. *Agriculture* **2012**, *2*, 244–258. [CrossRef]

49. Kobus, Z.; Nadulski, R.; Wilczyński, K.; Kozak, M.; Guz, T.; Rydzak, L. Effect of the black chokeberry (*Aronia melanocarpa* (Michx.) Elliott) juice acquisition method on the content of polyphenols and antioxidant activity. *PLoS ONE* **2019**, *14*, e0219585. [CrossRef]
50. Gralec, M.; Wawer, I.; Zawada, K. Aronia melanocarpa berries: Phenolics composition and antioxidant properties changes during fruit development and ripening. *Emir. J. Food Agric.* **2019**, *31*, 214–221. [CrossRef]
51. Tolić, M.T.; Jurčević, I.L.; Krbavčić, I.P.; Marković, K.; Vahčić, N. Phenolic content, antioxidant capacity and quality of chokeberry (*Aronia melanocarpa*) products. *Food Technol. Biotechnol.* **2015**, *53*, 171–179. [CrossRef]
52. Tolić, M.T.; Krbavčić, I.P.; Vujević, P.; Milinović, B.; Jurčević, I.L.; Vahčić, N. Effects of Weather Conditions on Phenolic Content and Antioxidant Capacity in Juice of Chokeberries (*Aronia melanocarpa* L.). *Polish J. Food Nutr. Sci.* **2017**, *67*, 67–74. [CrossRef]
53. Oszmiański, J.; Lachowicz, S. Effect of the production of dried fruits and juice from chokeberry (*Aronia melanocarpa* L.) on the content and antioxidative activity of bioactive compounds. *Molecules* **2016**, *21*, 1098. [CrossRef]
54. Hellström, J.K.; Shikov, A.N.; Makarova, M.N.; Pihlanto, A.M.; Pozharitskaya, O.N.; Ryhänen, E.L.; Kivijärvi, P.; Makarov, V.G.; Mattila, P.H. Blood pressure-lowering properties of chokeberry (*Aronia mitchurinii*, var. Viking). *J. Funct. Foods* **2010**, *2*, 163–169. [CrossRef]
55. Gumienna, M.; Lasik, M.; Czarnecki, Z. Bioconversion of grape and chokeberry wine polyphenols during simulated gastrointestinal in vitro digestion. *Int. J. Food Sci. Nutr.* **2011**, *62*, 226–233. [CrossRef] [PubMed]
56. Herrmann, K. Occurrence and content of hydroxycinnamic and hydroxybenzoic acid compounds in foods. *Crit. Rev. Food Sci. Nutr.* **1989**, *28*, 315–347. [CrossRef] [PubMed]
57. Dudonné, S.; Dubé, P.; Anhê, F.F.; Pilon, G.; Marette, A.; Lemire, M.; Harris, C.; Dewailly, E.; Desjardins, Y. Comprehensive analysis of phenolic compounds and abscisic acid profiles of twelve native Canadian berries. *J. Food Compos. Anal.* **2015**, *44*, 214–224. [CrossRef]
58. Slimestad, R.; Torskangerpoll, K.; Nateland, H.S.; Johannessen, T.; Giske, N.H. Flavonoids from black chokeberries, *Aronia melanocarpa*. *J. Food Compos. Anal.* **2005**, *18*, 61–68. [CrossRef]
59. Tian, Y.; Liimatainen, J.; Alanne, A.L.; Lindstedt, A.; Liu, P.; Sinkkonen, J.; Kallio, H.; Yang, B. Phenolic compounds extracted by acidic aqueous ethanol from berries and leaves of different berry plants. *Food Chem.* **2017**, *220*, 266–281. [CrossRef]
60. Mikulic-Petkovsek, M.; Slatnar, A.; Stampar, F.; Veberic, R. HPLC-MS n identification and quantification of flavonol glycosides in 28 wild and cultivated berry species. *Food Chem.* **2012**, *135*, 2138–2146. [CrossRef]
61. Tasinov, O.; Dincheva, I.; Badjakov, I.; Kiselova-Kaneva, Y.; Galunska, B.; Nogueiras, R.; Ivanova, D. Phytochemical Composition, Anti-Inflammatory and ER Stress-Reducing Potential of Sambucus ebulus L. Fruit Extract. *Plants* **2021**, *10*, 2446. [CrossRef]
62. Kiselova-Kaneva, Y.; Galunska, B.; Nikolova, M.; Dincheva, I.; Badjakov, I. High resolution LC-MS/MS characterization of polyphenolic composition and evaluation of antioxidant activity of Sambucus ebulus fruit tea traditionally used in Bulgaria as a functional food. *Food Chem.* **2022**, *367*, 130759. [CrossRef]

plants

MDPI

Article

Phytochemical Constitution, Anti-Inflammation, Anti-Androgen, and Hair Growth-Promoting Potential of Shallot (*Allium ascalonicum* L.) Extract

Warintorn Ruksiriwanich [1,2,3,*], Chiranan Khantham [1], Anurak Muangsanguan [1], Chuda Chittasupho [1,2], Pornchai Rachtanapun [3,4], Kittisak Jantanasakulwong [3,4], Yuthana Phimolsiripol [3,4], Sarana Rose Sommano [2,3,5], Korawan Sringarm [2,3,6], Emilia Ferrer [7] and Francisco J. Barba [7]

1 Department of Pharmaceutical Sciences, Faculty of Pharmacy, Chiang Mai University, Chiang Mai 50200, Thailand; ckhantham@gmail.com (C.K.); anurak_m@cmu.ac.th (A.M.); chuda.c@cmu.ac.th (C.C.)
2 Cluster of Research and Development of Pharmaceutical and Natural Products Innovation for Human or Animal, Chiang Mai University, Chiang Mai 50200, Thailand; sarana.s@cmu.ac.th (S.R.S.); korawan.s@cmu.ac.th (K.S.)
3 Cluster of Agro Bio-Circular-Green Industry, Faculty of Agro-Industry, Chiang Mai University, Chiang Mai 50100, Thailand; pornchai.r@cmu.ac.th (P.R.); kittisak.jan@cmu.ac.th (K.J.); yuthana.p@cmu.ac.th (Y.P.)
4 School of Agro-Industry, Faculty of Agro-Industry, Chiang Mai University, Chiang Mai 50100, Thailand
5 Department of Plant and Soil Sciences, Faculty of Agriculture, Chiang Mai University, Chiang Mai 50200, Thailand
6 Department of Animal and Aquatic Sciences, Faculty of Agriculture, Chiang Mai University, Chiang Mai 50200, Thailand
7 Department of Preventive Medicine and Public Health, Food Science, Toxicology and Forensic Medicine, Faculty of Pharmacy, University of Valencia, 46100 Valencia, Spain; emilia.ferrer@uv.es (E.F.); francisco.barba@uv.es (F.J.B.)
* Correspondence: warintorn.ruksiri@cmu.ac.th

Citation: Ruksiriwanich, W.; Khantham, C.; Muangsanguan, A.; Chittasupho, C.; Rachtanapun, P.; Jantanasakulwong, K.; Phimolsiripol, Y.; Sommano, S.R.; Sringarm, K.; Ferrer, E.; et al. Phytochemical Constitution, Anti-Inflammation, Anti-Androgen, and Hair Growth-Promoting Potential of Shallot (*Allium ascalonicum* L.) Extract. *Plants* **2022**, *11*, 1499. https://doi.org/10.3390/plants11111499

Academic Editors: Ivayla Dincheva, Ilian Badjakov and Bistra Galunska

Received: 12 May 2022
Accepted: 1 June 2022
Published: 2 June 2022

Abstract: In Thai folklore wisdom, shallot (*Allium ascalonicum* L.) was applied as a traditional herbal medicine for hair growth promotion with no scientific evidence. Androgenetic alopecia (AGA) is a progressive hair loss caused by multiple factors, including androgen hormones, inflammation, and oxidative stress. Conventional medicines (finasteride, dutasteride, corticosteroids, and minoxidil) have been used with limited therapeutic efficacy and unpleasant side effects. In this study, we aimed to give the first estimation of bioactive compounds in shallot extract and evaluate the hair growth-promoting activities regarding anti-inflammatory and gene expression modulation involving androgen, Wnt/β-catenin, sonic hedgehog, and angiogenesis pathways. The results reveal that phenolic compounds (quercetin, rosmarinic, and *p*-coumaric acids) are the major constituents of the methanolic shallot extract. Compared with the lipopolysaccharide-stimulated control group (2.68 ± 0.13 μM), nitric oxide production was remarkably diminished by shallot extract (0.55 ± 0.06 μM). Shallot extract improves hair growth promotion activity, as reflected by the downregulation of the androgen gene expression (*SRD5A1* and *SRD5A2*) and the upregulation of the genes associated with Wnt/β-catenin (*CTNNB1*), sonic hedgehog (*SHH*, *SMO*, and *GIL1*), and angiogenesis (*VEGF*) pathways. These findings disclose the new insights of shallot extract on hair growth promotions. Shallot extract could be further developed as nutraceutical, nutricosmetic, and cosmeceutical preparations for AGA treatment.

Keywords: androgenetic alopecia; anti-hair loss; hair growth promotion; shallot; *Allium ascalonicum*; anti-inflammatory; 5α-reductase; *SRD5A2*; Wnt/β-catenin

1. Introduction

Shallot (*Allium ascalonicum* L.) from the Alliaceae family is a valuable horticultural spice that originated in Southeast Asia. Shallot bulbs have been widely utilized as a major component in Asian diets and traditional herbal medicines in Thailand and India [1]. This plant possesses various properties, including anti-cancer, anti-diabetes, antimicrobial, anti-inflammatory, and antioxidant activities [1–3]. In addition, several health benefits of shallot have been reported, including wound healing and maintaining healthy skin and hair [1,4]. The most abundant components of shallots are phenolic compounds, saponins, and carbohydrates [5,6]. In addition, fresh shallot bulbs have been used in traditional Thai folklore to treat bacterial skin infections, tinea capitis, and hair loss.

Androgenetic alopecia (AGA) is a chronic hair loss, characterized by hair follicle miniaturization and perifollicular inflammation [7]. The pathogenesis of AGA is extensively influenced by genetic factors and androgens [8]. Androgen-mediated follicular miniaturization is the most elucidated pathogenesis of AGA [7,8]. Testosterone is metabolized to dihydrotestosterone (DHT)—a potent androgen—by steroid 5α-reductases [8]. Additionally, a histological evaluation of AGA's scalp found perifollicular inflammation [9]. Another study proposed that nitric oxide-mediated perifollicular inflammation arose in response to DHT level [10]. Antiandrogenic medicines (finasteride, dutasteride, and spironolactone), anti-inflammatory corticosteroids, and the anti-seborrheic tretinoin have been used to attenuate AGA progression [11]. However, those have shown limited efficacy and unsatisfactory side effects [11,12]. Presumably, other factors exist in the gradual process of hair follicle miniaturization besides the androgenic effect [9].

The hair growth cycle consists of four phases: anagen (growth phase), catagen (regression phase), telogen (resting phase), and exogen (shedding phase) [13,14]. Multiple signaling pathways involve the transition between these phases, including Wnt/β-catenin, sonic hedgehog, and angiogenesis signaling pathways [9,13]. The Wnt/β-catenin and sonic hedgehog signaling pathways involve the differentiation and proliferation of hair follicle cells [15,16]. Vascular endothelial growth factor (VEGF) has been reported to be implicated in the angiogenesis in the anagen phase and facilitate the supply of oxygen and nutrients to hair follicles [13]. Minoxidil has been widely used to stimulate hair growth in males and females with AGA due to its ability to dilate the blood vessels in the follicle [11]. The synthetic medicines targeting Wnt/β-catenin and sonic hedgehog signaling pathways have not been implemented for the treatment of hair loss [13,15].

Hair follicle inflammation has been confirmed as a possible factor in the pathogenesis of AGA [17]. The nitric oxide (NO) level, which mediates inflammatory reaction in hair follicles, was increased in response to the DHT level [10]. In addition, it has been revealed that serum samples of AGA contained a higher level of NO than the control group [18]. The ultraviolet radiation stimulated NO production in keratinocytes of hair follicles. Consequently, the proinflammatory cytokines were released, and the recruitment of immune cells was facilitated, resulting in the damage of hair roots [19].

Conventional medicines for AGA have been shown to have several side effects, such as scalp dryness, skin irritation, erectile dysfunction, and testicular pain [12]. These limitations contribute to the reduction in individual compliance with hair loss treatment. Since AGA requires long-term treatment, alternative treatments and natural herbal medicines have gained attention due to their advantages, including fewer side effects, a broad spectrum of hair growth-promoting activities, and affordable prices [13,20].

Currently, no strongly scientific evidence that supports the beneficial effects of shallot on AGA has been established. Furthermore, the effects of shallot on hair growth regulation at the cellular level have not been elucidated. The conversion of fresh shallots to concentrate extract might create a new perspective for developing nutraceuticals, nutricosmetics, and cosmeceuticals containing shallot extract for anti-hair loss. Considering all above reasons, with this first study, we aimed to estimate the bioactive compounds of shallot extract and investigate the anti-inflammatory activity and gene expression regulation involving androgen, Wnt/β-catenin, sonic hedgehog, and angiogenesis signaling pathways for AGA treatment.

2. Results

2.1. Extraction Yield and Bioactive Compound Estimation

The physical appearance of shallot extract was a non-greasy paste with pink color. The extraction yield was 9.11 ± 0.23% *w*/*w* based on dry material. As shown in Table 1 and Figure 1, the most abundant bioactive compound in shallot extract was total phenolic content (4.96 ± 0.42 GAE/g), followed by contents of proteins, polysaccharides, and flavonoids. *p*-Coumaric acid, rosmarinic acid, and quercetin were the major phenolic compounds in shallot extract.

Table 1. Content of bioactive compounds of shallot extract.

Bioactive Compounds	Content	
Total polysaccharide content	0.90 ± 0.06	mg D-glucose/g
Total protein content	1.01 ± 0.04	mg BSAE/g
Total phenolic content	4.69 ± 0.42	mg GAE/g
Total flavonoid content	<0.003	mg EGCGE/g
Phenolic compounds		
p-Coumaric acid	1.091 ± 0.011	mg/g
Quercetin	0.029 ± 0.002	mg/g
Rosmarinic acid	0.234 ± 0.007	mg/g

Note: Milligrams of gallic acid equivalents per gram of extract (mg GAE/g extract); milligrams of epigallocatechin gallate equivalents per gram of extract (mg EGCGE/g extract); milligrams of D-glucose equivalents per gram of extract (mg D-glucose/g extract); milligrams of bovine serum albumin equivalents per gram of extract (mg BSAE/g extract).

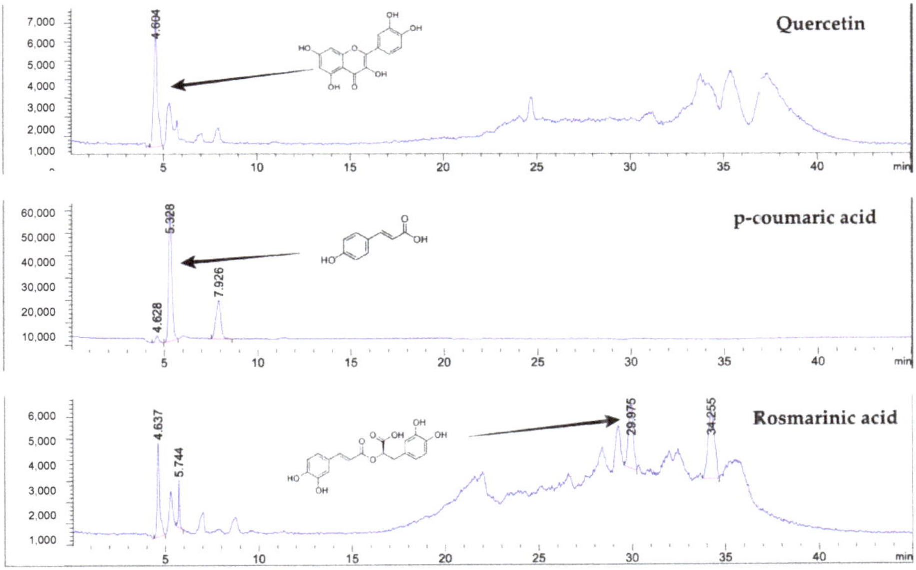

Figure 1. Chromatogram of bioactive compounds in shallot extract analyzed by liquid chromatography–mass spectrometry (LC-MS).

2.2. Cell Viability

In order to evaluate the anti-inflammation and effects on gene expression profiling of shallot extract, the viability of cells used in this study, namely DU-145 and hHFDPC, was assessed by the sulforhodamine B (SRB) assay [21]. In Figure 2, shallot extract at the concentration above 0.5 mg/mL showed cytotoxicity and significantly decreased viability of all types of cells compared to corresponding solvent-treated control groups. The highest concentration of shallot extract (0.1 mg/mL) that gave the viability of RAW 264.7 cells above 80% was classified as a non-toxicity concentration and selected for all experiments [22].

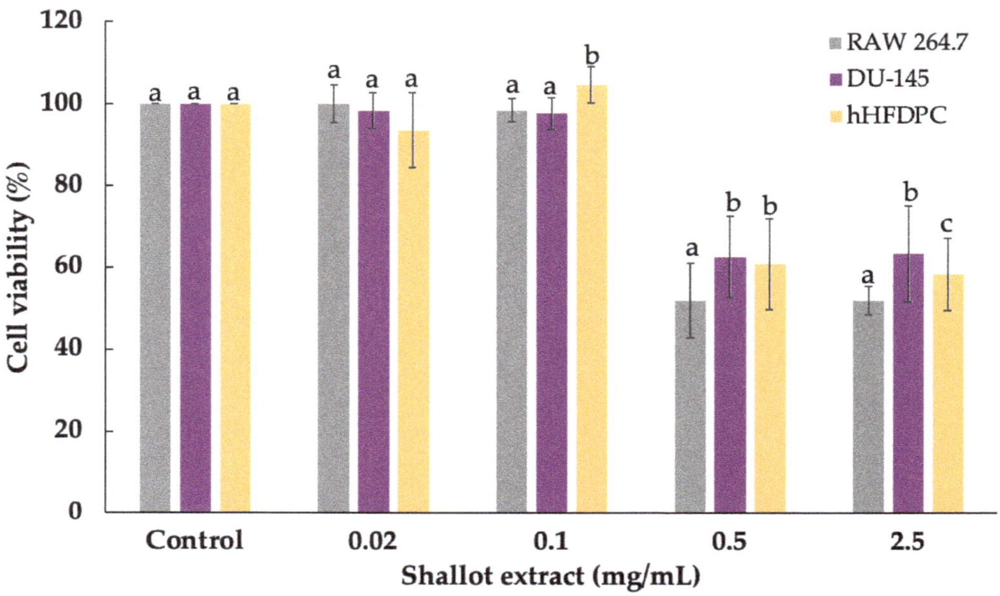

Figure 2. Cell viability of RAW 264.7 macrophage cells (RAW 264.7), DU-145 human prostate cancer cells (DU-145), and human hair follicle dermal papilla cells (hHFDPC) after shallot extract treatment for 24 h with different concentrations (0.02 to 2.5 mg/mL) was determined by sulforhodamine B (SRB) assay. Different letters (a, b, and c) indicate statistical differences (p-value < 0.05) in the cell viability of each concentration.

2.3. *Anti-Inflammatory Activity*

Therefore, diclofenac sodium (DF) at the same concentration (0.1 mg/mL) with no toxicity was selected to compare the inhibitory effect on NO production. The level of accumulated nitrite, which is the stable metabolite of NO, was indirectly quantified. In Figure 3, the concentration of nitrite in the lipopolysaccharide (LPS)-stimulated control group (2.68 $\pm$ 0.13 μM) explicitly increased compared with a solvent-pretreated group (0.43 $\pm$ 0.06 μM). The pretreatment with shallot extract significantly decreased the nitrite production to 0.55 $\pm$ 0.06 μM compared with the LPS-stimulated group ($p < 0.05$). Moreover, the NO inhibition of the DF-pretreated group (0.36 $\pm$ 0.01 μM) was significantly comparable to shallot extract.

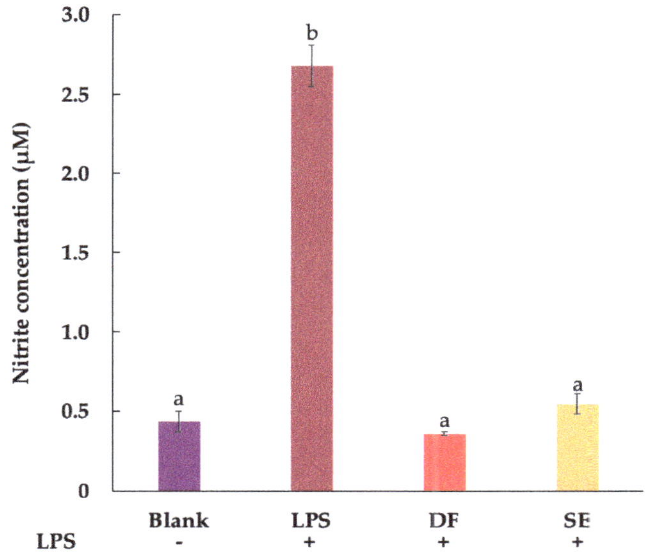

Figure 3. Effects of shallot extract (SE) and diclofenac sodium (DF) at the same concentration of 0.1 mg/mL on nitrite production in the lipopolysaccharide (LPS)-stimulated RAW 264.7 murine macrophages for 24 h compared to solvent-treated control without LPS (blank) and LPS-stimulated control (+LPS). Different letters (a and b) indicate statistical significance ($p < 0.05$) in comparison to +LPS and DF.

2.4. Effect of Shallot Extract on Expressions of Genes Associated with Androgenetic Alopecia

In this study, we evaluated the regulatory effect of shallot extract (0.1 mg/mL) on mRNA expressions of genes associated with the pathogenesis of AGA, including androgen pathway (*SRD5A1*, *SRD5A2*, and *SRD5A3*), sonic hedgehog pathway (*SHH*, *SMO* and *GIL1*), Wnt/β-catenin pathway (*CTNNB1*), and VEGF signaling pathway (*VEGF*). The reference standard compounds, including finasteride, dutasteride, purmorphamine, and minoxidil, were used at the same concentration of 0.1 mg/mL in all experiments. The results are illustrated in Figure 4.

Steroid 5α-reductase types 1, 2, and 3 were encoded by *SRD5A1*, *SRD5A2*, and *SRD5A3*. Remarkably, the expressions of *SRD5A1* (fold change of 0.73 ± 0.14) and *SRD5A2* (fold change of 0.25 ± 0.16) were significantly suppressed in the shallot extract group compared with the control group (Figure 4a,b). There were no significant differences between shallot extract and the standard drugs (finasteride and dutasteride) regarding *SRD5A2* suppression (Figure 4b). However, the reduction in *SRD5A3* expression was not observed in shallot extract.

The molecules in sonic hedgehog pathways, which are sonic hedgehog (shh), smoothened (SMO), and GLI family zinc finger 1 (GIL1), were encoded by *SHH*, *SMO*, and *GIL1*, respectively. Compared with the control group, treatment with shallot extract upregulated the expressions of *SHH* (Figure 4d), *SMO* (Figure 4e), and *GIL1* (Figure 4f) in hHFDPC. In addition, the expression of *GIL1* markedly increased with a fold change of 1.53 ± 0.29, compared with the purmorphamine and control groups.

The gene encoding β-catenin is *CTNNB1*. In Figure 4g, the mRNA level of *CTNNB1* in the treatment of shallot extract in hHFDPC distinctly elevated with a fold change of 3.51 ± 0.41, compared with the minoxidil (1.15 ± 0.04) and control groups. On the other hand, minoxidil induced the upregulation of *VEGF* (fold change of 8.79 ± 0.96), as is evident in Figure 4h. *VEGF* expression was moderately upregulated with a fold change of 2.25 ± 0.82 in hHFDPC treated with shallot extract.

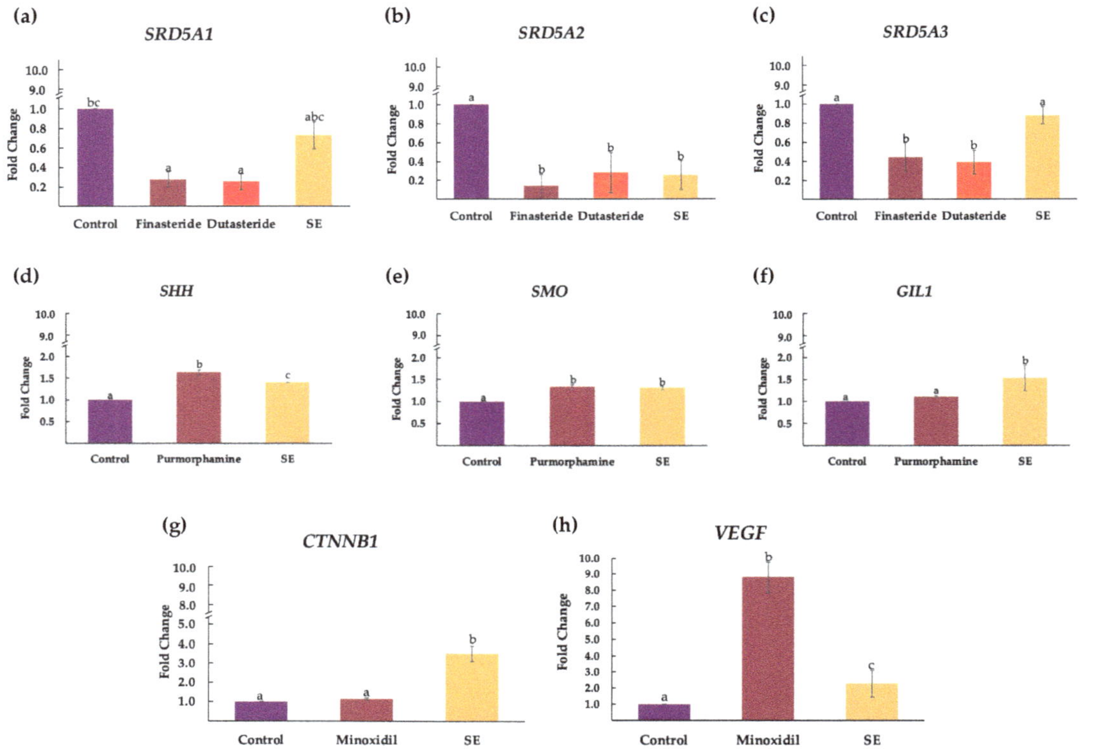

Figure 4. Effect of shallot extract (SE, 0.1 mg/mL) on expressions of genes associated with androgenetic alopecia: (**a**) *SRD5A1*; (**b**) *SRD5A2*; (**c**) *SRD5A3*; (**d**) *SHH*; (**e**) *SMO*; (**f**) *GIL1*; (**g**) *CTNNB1*; (**h**) *VEGF*. DU-145 human prostate cancer cells (DU-145) were used to observe the expressions of genes in the androgen pathway (*SRD5A* genes), whereas human hair follicle dermal papilla cells (hHFDPC) were used to study the remaining pathways. Different letters (a, b, and c) indicate statistical significance ($p < 0.05$) in comparison to control, finasteride (0.1 mg/mL), dutasteride (0.1 mg/mL), purmorphamine (0.1 mg/mL), and minoxidil (0.1 mg/mL).

3. Discussion

In this present study, phenolic compounds have been identified as the major compounds of methanolic shallot extract. Additionally, *p*-coumaric acid, quercetin, and rosmarinic acid were detected. Shallot extract diminished the NO production and secretion, contributing to anti-inflammatory activity. Shallot extract suppressed the expressions of *SRD5A1* and *SRD5A2* in DU-145 cell lines, whereas the expressions of genes associated with hair growth activation (*SHH*, *SMO*, *GIL1*, *CTNNB1*, and *VEGF*) were upregulated in hHFDPC. These findings demonstrate that shallot extract possesses hair growth-promoting effects through inhibiting inflammatory and androgen pathways. Wnt/β-catenin, sonic hedgehog, and VEGF pathways were also activated.

The disturbance of hair cycles leads to the elongation of resting phase, which contributes to AGA. Furthermore, it has been verified that progressive hair follicle miniaturization involves anagen shortening and premature catagen entry [9,23]. Since the anagen phase governs hair length, terminal hair eventually transforms to vellus hair in AGA [23].

Specialized mesenchymal cells, hHFDPC, play a vital role in hair follicle development and hair growth by supporting multi-potent stem cells, cytokines, growth factors, and nutrients [24]. Synchronized intercellular signaling cascades of hHFDPC and other adjacent cells implicate the hair follicle's formation, maintenance, and homeostasis [7,23,25].

The bioactive compound estimations in our study revealed the presence of phenolic compounds, especially *p*-coumaric acid, quercetin, and rosmarinic acid, as the major components. According to current literature, shallot bulbs contain several phenolic compounds, including apigenin, eriodictyol, gallic acid, quercetin, isoquercetin, kaempferol, catechin, and tannic acid [26]. Additionally, it has been confirmed that flavonoids and their glycosides, including quercetin and isorhamnetin, were detected in the methanolic extract of shallot bulbs [5]. Two novel furostanol saponins, named ascalonicoside A1/A2 (1a/1b) and ascalonicoside B(4), have been found in the methanolic shallot extract [1,5]. The bioactive compounds estimation of shallot extract indicated the other constituents that have not been identified in the current experimental system. Therefore, other unknown compounds in the extract required further analysis.

Plant phenolics compounds are known to possess anti-inflammatory potential, which can interact with free radicals and impede cyclooxygenase (COX), lipoxygenase, and inducible nitric oxide synthase (iNOS) [27]. The inflammation in hair follicles is triggered by oxidative stress and androgens [27,28]. Wolf et al. reported that the excessive NO production and the expression of iNOS in hHFDPC were induced by DHT [10]. It has been reported that *p*-coumaric acid possessed the anti-inflammation activity due to suppression of the nuclear factor kappa B in LPS-stimulated RAW 264.7 macrophage cells [29]. The protein expression of the pro-inflammatory enzymes (COX and iNOS) and NO production were inhibited by rosmarinic acid and quercetin [30–32]. Shallot extract in this study that contained those compounds exhibited anti-inflammatory potential by attenuating NO production and reducing inflammatory-induced perifollicular damage in AGA.

Androgens affect the function of human skin, including wound healing, development of sebaceous glands, and hair growth [8]. Testosterone can be catalyzed into DHT by steroid 5α-reductases [12]. In hair follicles of AGA, DHT binds the androgen receptors, leading to hair follicle miniaturization and diminishing the period of the anagen phase [14,23]. The activities of steroid 5α-reductase type 1 and 2 in balding hair follicles were higher than in non-balding hair follicles [33]. Furthermore, the expressions of gene encoding steroid 5α-reductases (*SRD5A1*, *SRD5A2*, and *SRD5A3*) were found to upregulate in the androgen-sensitive hair follicles of AGA [34,35]. Our results postulate that shallot extract significantly attenuates the mRNA expression of *SRD5A1* and *SRD5A2*, and slightly suppresses the *SRD5A3* expression. The bioactive compounds in shallot extract may affect the regulatory elements for the expressions of *SRD5A* genes differently [36], which was not discovered in this study. So far, there is no study about the effects of bioactive compounds on the transcriptional regulatory elements associated with *SRD5A* genes and AGA. Quercetin has been reported to possess anti-inflammatory activity and anti-androgen activity through the inhibition of steroid 5α-reductases and the downregulation of androgen receptors [37–40]. Quercetin-rich extracts including *Ginkgo biloba*, *Camelia sinensis*, and *Cuscuta reflexa* exhibited promising hair growth-promoting activities via the attenuation of steroid 5α-reductases [41]. In addition, rosmarinic acid and ursolic acid, the major components of *Rosmarinus officinalis*, showed inhibitory activity towards steroid 5α-reductases [42]. It is suggested that shallot extract could reverse androgen-induced alopecia by suppressing the expressions of *SRD5A* and contributing to reducing its translation.

The Wnt/β-catenin signaling pathway is a dominant pathway that involves the development of hair follicles and sebaceous glands [11,43]. Moreover, this pathway mediates the initiation and maintenance of the anagen phase [9]. It has been proposed that there might be crosstalk between the Wnt/β-catenin and androgen pathways [44]. The differentiation of hair follicle stem cells was abolished by DHT [45]. Dickkopf 1 (DKK-1), a Wnt antagonist, promoted the premature onset of catagen and cell apoptosis [46]. A previous study demonstrated that hHFDPC secreted DKK-1 in response to DHT [47].

Furthermore, DHT induced the downregulation of β-catenin in hHFDPC, suggesting the androgen-induced inhibition of the Wnt/β-catenin signaling pathway [45]. β-catenin, encoded by *CTNNB1*, is known to induce the transition from telogen to anagen, leading to hair regrowth and a new hair cycle [48]. The disruption of *CTNNB1* expression contributed to abnormal hair growth in mice [49]. Recent studies have reported that quercetin increased the expression of Wnt and β-catenin [50,51]. Our results indicate that shallot extract in this study notably upregulated the expression of *CTNNB1*, leading to the elevation of the translation of β-catenin. The accumulation of β-catenin may prolong the growing phases of the anagen hair cycle, providing the promotion of hair growth [15].

Likewise, the sonic hedgehog signaling pathway regulates hair growth and hair follicle development [44,52]. It has been presumed that sonic hedgehog signaling is the downstream pathway of the Wnt/β-catenin signaling to regulate hair follicle induction [53]. The sonic hedgehog signaling is initiated by the interaction of Shh to its receptor called Patched. Consequently, Smo is dissociated from the inhibition of Patched and activates the downstream transcription factors of Gli1 [54]. This signal contributes to hair growth activation by inducing telogen-to-anagen transition during the hair follicle cycle [48]. The absence of Shh and Smo implicated the impairments of growth and morphogenesis of hair follicles [16]. It has been demonstrated that a small molecule agonist of the sonic hedgehog pathway enhanced hair growth and promoted the anagen phase in mice through the activation of gene expressions of *SHH* and *GIL1* [55]. Furthermore, the retardation of *SHH* influenced hair follicle morphogenesis and hair growth [56]. Previous studies indicated that *Polygonum multiflorum* and *Thujae occidentalis*, which are the sources of quercetin and coumarins, induced hair growth through the upregulation of Shh and β-catenin [48,57–59]. Moreover, our findings revealed that shallot extract activated the expressions of *SHH*, *SMO*, and *GIL1* genes in hHFDPC. These effects may lead to hair growth-promoting activity of shallot extract and the induction of telogen to the anagen phase in hair follicles.

VEGF is the important mediator that regulates blood vessel formation, wound healing, and hair growth [15,60]. Perifollicular vascularization is extensively active in the anagen phase and correlated with the upregulation of VEGF in follicular keratinocytes, leading to the acceleration of hair regrowth [61]. The size of hair follicles and the diameter of the hair shaft were also increased due to VEGF [62]. In addition, the mechanisms of minoxidil involved the upregulation of *VEGF* and its receptor in hHFDPC, leading to the promotion of angiogenesis in the anagen phase [13]. The expression levels of *VEGF* in both male and AGA were significantly lower than control without AGA [63]. Rosmarinic acid in shallot extract could enhance the protein expression of VEGF [64]. Recently, quercetin has exhibited wound healing potential by enhancing the VEGF level [50]. In our study, the expression of *VEGF* was slightly upregulated by shallot extract. This may help to promote the angiogenesis around hair follicles and stimulate hair growth.

The major constituents of shallot extract were phenolic compounds, especially quercetin, *p*-coumaric acid, and rosmarinic acid. Our findings also found that shallot extract exhibited the anti-inflammation and the modulation of genes associated with androgen, Wnt/β-catenin, sonic hedgehog, and VEGF signaling pathways. Targeting these biochemical signaling pathways of hair growth regulation would benefit AGA, a multifactorial disorder. Synergistic activities of phenolic components might contribute to shallot extract's hair growth-promoting activities. However, further studies are required to elucidate the other bioactive compounds in shallot extract and their effects on specific signaling pathways. In the present study, an exploration of the hair growth-promoting activities of shallot extract was undertaken. Shallot extract could be applied for the development of nutraceuticals, nutricosmetics, and cosmeceuticals for AGA.

4. Materials and Methods

4.1. Preparation of Extract

Shallot (*Allium ascalonicum* L.) was purchased from the local market (Chiang Mai, Thailand) on 10 February 2021, and authenticated by the Pharmaceutical and Natural

Products Research and Development Unit, Faculty of Pharmacy, Chiang Mai University (reference specimens no. PNPRDU63027). Two kilograms of shallot bulbs were blended by a food blender into a paste and macerated in methanol (ratio of solid/solvent: 1:2) for 24 h [4,5]. Then, the extract solution was filtered through Whatman filter paper no. 4 and no. 1. The clear solution was concentrated and evaporated at 50 °C by an evaporator (Hei-VAP value, Heidolph, Schwabach, Germany) until it was completely dried. Samples were kept at 4 °C for further analysis.

4.2. Phytochemical Estimations

4.2.1. Total Phenolic Content

The Folin–Ciocalteu colorimetric method was used to determine total phenolic content. The reaction consists of the Folin–Ciocalteu reagent, sodium carbonate, and the phenolic compounds in the extract. The concentrations of standard gallic acid in the range of 0.01 to 0.2 mg/mL and their absorbances were used to plot standard curves. The results are expressed as milligrams of gallic acid equivalents per gram of extract (mg GAE/g extract) [65].

4.2.2. Total Flavonoid Content

The aluminum chloride colorimetric method was employed to estimate the total flavonoid content in the sample. Aluminum chloride reacts with the C-4 keto group and either the C-3 or C-5 hydroxyl group of flavones and flavonols, resulting in a stable-colored complex. Different concentrations of (−)-epigallocatechin gallate (EGCG) in the range of 0.01 to 0.3 mg/mL and their absorbances were plotted to create the standard curve. The results are expressed in terms of milligrams of EGCG equivalents per gram of extract (mg EGCGE/g extract) [65].

4.2.3. Total Polysaccharide Content

The anthrone-sulfuric acid method was conducted to quantify total polysaccharide content. The reaction of anthrone and the extract in acidic conditions was performed at 100 °C, resulting in blue-green solutions. The absorbances of D-glucose in various concentrations from 0.01 to 0.6 mg/mL were used to generate the calibration curve. The results were milligrams of D-glucose equivalents per gram of extract (mg D-glucose/g extract) [32].

4.2.4. Total Protein Content

Total protein content was estimated by the Lowry method. The Folin–Ciocalteu reagent was used to interact with the cuprous ions and the side chains of tyrosine, tryptophan, and cysteine in the sample, and afterward, a blue-green color was produced. The standard protein was bovine serum albumin (BSA) with the concentration range of 0.01 to 2 mg/mL. The results are expressed as milligrams of BSA equivalents per gram of extract (mg BSA/g) [32].

4.3. Determination of Phenolic Compounds by Liquid Chromatography–Mass Spectrometry (LC-MS)

In accordance with the procedure of Arjin et al. [66], the samples were dissolved in 0.01% formic acid and ethanol (1:1, *v/v*) to achieve a final concentration of 1 mg/mL, and then purified with the QuEChERS dispersive SPE kit, fat + pigments (Agilent Technology, Santa Clara, CA, USA) prior to being filtered through a 0.22 μm membrane. The phenolic compounds of extract were quantified and analyzed in liquid chromatography (Agilent 1260 Infinity II series), equipped with an electrospray ion quadrupole mass spectrometer 6130 (Agilent Tech., Santa Clara, CA, USA), according to the reported method [39,67]. Solvent A was 5% formic acid. Solvent B was 5% formic acid in 10% water and 85% acetonitrile. The gradient elution was programmed as follows: 80% A at 0–8 min, 80% to 25% A at 8–24 min, 25% A at 24–28 min, 25% to 70% A at 28–34 min, 70% to 80% A at

34–36 min, and 80% A at 36–45 min. The injection volume was 5 μL. For chromatographic separation, a Restek Ultra C18 reversed-phase column (250 × 4.6 mm, 5 μm, Restek Corporation, Bellefonte, PA, USA) was used. The column oven temperature and flow rates were 30 °C and 0.5 mL/min. For mass spectrometry, the negative selected ion monitoring was implemented. Nitrogen gas was used as desolvation gas with a flow rate of 12 L/min and nebulizer pressure of 60 psi. Other parameters were programmed: a capillary voltage of −3 kV, a gas temperature °C, a fragmentation voltage of 70 V, and the full scan spectra from 100 to 1200 m/z with an acquisition rate of 250 ms/spectrum. Data acquisition and integration were processed with OpenLab software (Agilent Tech., Santa Clara, CA, USA).

4.4. *Cell Viability Assay*

Human hair follicle dermal papilla cells (hHFDPC: Promo Cell GmbH, Heidelberg, Germany) were grown in Follicle Dermal Papilla Cell Growth Medium Kit (cat no. C-26501) supplemented with 1% antibiotic-antimycotic 100× solution (Gibco™, cat no. 15240062). RAW 264.7 macrophage cells and DU-145 human prostate cancer cells were obtained from the American Type Culture Collection (Rockville, MD, USA). DU-145 cells were cultured in Roswell Park Memorial Institute 1640 medium (RPMI-1640; cat no. 31800022) containing 10% fetal bovine serum (FBS; cat no. 16000044) and 1% antibiotic-antimycotic 100× solution. RAW 264.7 macrophage cells were grown in Dulbecco's Modified Eagle Medium (DMEM; cat no. 31600083) supplemented with 10% FBS and 1% antibiotic-antimycotic 100× solution. Cells were incubated at 37 °C in a 5% CO_2 humidified atmosphere.

The sulforhodamine B (SRB) assay was used to determine the cytotoxic potential of shallot extract and standard reference compounds (diclofenac sodium (DF), finasteride, dutasteride, purmorphamine, and minoxidil) in a concentration range from 0.02 to 2.5 mg/mL [21]. Briefly, cells were seeded in 96-well plates (10^4 cells/well) and incubated for 24 h. The monolayer cells were washed and treated with the tested samples (0.02–1 mg/mL). After 30 h of incubation, cultured cells were fixed on plates, and washed, dried, and stained with the SRB solution (Sigma Chemical, St. Louis, MO, USA). Tris-EDTA buffer was added to solubilize the dye extracted from stained cells. The optical density (OD) was acquired by a microplate reader (EZ Read 400, Biochrom, Cambridge, UK) at 515 nm. The highest concentration providing the percentages of cell viability above 80% was considered as non-cytotoxicity and was selected for further experiments. The percentage of cell viability was calculated by Equation (1):

$$\text{Cell viability } (\%) = \left(\frac{OD_{sample} - OD_{blank}}{OD_{control} - OD_{blank}}\right) \times 100 \quad (1)$$

4.5. *Anti-Inflammatory Activity*

The Griess reaction colorimetric assay kit (Invitrogen, Thermo Fisher Scientific, Inc., Eugene, OR, USA) was used to determine the nitric oxide (NO) level in the culture medium [22]. The quantification of NO was indirectly estimated by measuring nitrite, which is the final inert product of NO. Briefly, RAW 264.7 macrophage cells were seeded into 96-well plates (10^4 cells/well) and incubated for 24 h. The cells were pretreated with 0.1 mg/mL of diclofenac sodium (DF), 0.1 mg/mL of shallot extract, and solvent (blank). After pretreatment for 2 h, cells were incubated with and without lipopolysaccharides (LPS: Sigma Chemical, St. Louis, MO, USA). After incubation for 24 h, 150 μL of each supernatant solution was reacted with 20 μL of Griess reagent mixture and incubated for 30 min at room temperature. Then, the absorbance was read at 570 nm. The standard curve equation of reference standard sodium nitrite was used to calculate the nitrite concentration.

4.6. *Semi-Quantitative Reverse Transcription and Polymerase Chain Reaction*

DU-145 cells were used to study the expressions of genes in the androgen pathway. hHFDPC were utilized for the other remaining experiments [68,69]. Shallot extract was compared to the reference standard compounds (finasteride, dutasteride, purmorphamine, and minoxidil) at the same concentration of 0.1 mg/mL. Total RNA was isolated from cells using the E.Z.N.A.® Total RNA Kit I (Omega Bio-Tek, Norcross, GA, USA), according to the manufacturer's instructions. Qubit™ 4 fluorometer (Invitrogen, Carlsbad, CA, USA) and Qubit™ RNA HS Assay Kit (Invitrogen, Carlsbad, CA, USA) were used to determine the concentration of the purified RNA. The RNA solution was maintained at −20 °C until it was used. Gene expression levels were carried out by the semi-quantitative RT-PCR [70]. Complementary DNA was synthesized using the MyTaq™ One-Step RT-PCR Kit (Bioline, Memphis, TN, USA). Primer sequences used are as follows: *SRD5A1*: AGCCATTGTGCAGTGTATGC and AGCCTCCCCTTGGTATTTTG; *SRD5A2*: TGAATACCCTGATGGGTGG and CAAGCCACCTTGTGGAATC; *SRD5A3*: TCCTTCTTTGCCCAAACATC and TCCTTCTTTGCCCAAACATC; *SHH*: AAAAGCTGACCCCTTTAGCC and GCTCCGGTGTTTTCTTCATC; *SMO*: GAAGTGCCCTTGGTTCGGACA and CCGCCAGTCAGCCACGAAT; *GIL1*: GCAGGGAGTGCAGCCAATACAG and GAGCGGCGGCTGACAGTATA; *CTNNB1*: CCCACTAATGTCCAGCGTTT and AACCAAGCATTTTCACCAGG; *VEGF*: CTACCTCCACCATGCCAAGT and GCGAGTCTGTGTTTTTGCAG; *GAPDH*: GGAAGGTGAAGGTCGGAGTC and CTCAGCCTTGACGGTGCCATG.

Agarose gel electrophoresis was performed to detect the RT-PCR products [70]. The gel images and band intensity were acquired by the Gel Doc™ EZ System (Version 3.0; Bio-Rad) and Image Lab™ software (Bio-Rad). The expression of target genes was normalized by the *GAPDH* expression value and expressed as the relative expression value. Each sample was analyzed in triplicate.

4.7. Statistical Analysis

All the tests were conducted in triplicates. Results are expressed as a mean ± standard error of the mean. Statistical comparisons were performed using the Jamovi version 1.6.23 (The Jamovi Project, Sydney, Australia). One-way analysis of variance followed by Tukey's test was used to determine the statistical differences between the mean of pairs. p-value < 0.05 was considered significant.

5. Conclusions

Shallot is one of the most essential horticultural ingredients in Asian cuisine and is used as a traditional medicine for hair loss. Our findings show that shallot extract contains phenolic compounds, namely, quercetin, rosmarinic, and *p*-coumaric acids, contributing to its anti-inflammation potential via NO inhibition. Interestingly, the gene expressions of *SRD5A2* were downregulated by shallot extract and comparable to standard drugs (finasteride and dutasteride), leading to the reduction in androgenic effects on androgen-sensitive hair follicles. On the other hand, shallot extract enhanced the expressions of *CTNNB1*, *SHH*, *SMO*, *GIL1*, and *VEGF*, providing hair growth promotion effects via the maintenance of the anagen phase and the improvement of blood flow in hair follicles. In summary, shallot extract could promote hair growth by anti-inflammation and regulations of genes in androgen, sonic hedgehog, Wnt/β-catenin, and angiogenesis signaling pathways. The results of this study provide a sufficient basis for the utilization of shallot extract, which could be further developed as nutraceuticals, nutricosmetics, and cosmeceuticals for promoting hair growth.

Author Contributions: Conceptualization, W.R. and C.K.; methodology, W.R. and C.K.; software, K.S.; validation, C.K. and S.R.S.; formal analysis, E.F.; investigation, C.K. and A.M.; resources, W.R. and C.K.; data curation, C.K. and K.J.; writing—original draft preparation, W.R. and C.K.; writing—review and editing, W.R., C.K., C.C., P.R., Y.P., and F.J.B.; visualization, A.M.; supervision, W.R.; project administration, W.R.; funding acquisition, W.R. All authors have read and agreed to the published version of the manuscript.

Funding: This research work was partially supported by Chiang Mai University, grant number 19/2565.

Data Availability Statement: Not applicable.

Acknowledgments: The authors are grateful to Chiang Mai University (grant number 19/2565) and the Faculty of Pharmacy, Chiang Mai University, for supporting research facilities.

Conflicts of Interest: The authors declare no conflict of interest.

References

1. Sun, W.; Shahrajabian, M.H.; Cheng, Q. The insight and survey on medicinal properties and nutritive components of shallot. *J. Med. Plant Res.* **2019**, *13*, 452–457.
2. Leelarungrayub, N.; Chanarat, N.; Rattanapanone, V. Potential activity of Thai shallot (*Allium ascalonicum* L.) extract on the prevention of hemolysis and glutathione depletion in human erythrocyte from oxidative stress. *Chiang Mai J. Sci.* **2004**, *3*, 225–234.
3. Raeisi, S.; Sharifi-Rad, M.; Quek, S.Y.; Shabanpour, B.; Sharifi-Rad, J. Evaluation of antioxidant and antimicrobial effects of shallot (*Allium ascalonicum* L.) fruit and ajwain (*Trachyspermum ammi* (L.) Sprague) seed extracts in semi-fried coated rainbow trout (*Oncorhynchus mykiss*) fillets for shelf-life extension. *LWT-Food Sci. Technol.* **2016**, *65*, 112–121. [CrossRef]
4. Owoyele, B.; Abioye, A.; Afinowi, N.; Jimoh, S.; Soladoye, A. Analgesic and anti-inflammatory effects of *Allium ascalonicum*. *Trop. J. Health Sci.* **2006**, *13*, 28–32. [CrossRef]
5. Fattorusso, E.; Iorizzi, M.; Lanzotti, V.; Taglialatela-Scafati, O. Chemical composition of shallot (*Allium ascalonicum* Hort.). *J. Agric. Food Chem.* **2002**, *50*, 5686–5690. [CrossRef] [PubMed]
6. Leelarungrayub, N.; Rattanapanone, V.; Chanarat, N.; Gebicki, J.M. Quantitative evaluation of the antioxidant properties of garlic and shallot preparations. *Nutrition* **2006**, *22*, 266–274. [CrossRef]
7. Katzer, T.; Leite Junior, A.; Beck, R.; da Silva, C. Physiopathology and current treatments of androgenetic alopecia: Going beyond androgens and anti-androgens. *Dermatol. Ther.* **2019**, *32*, e13059. [CrossRef] [PubMed]
8. Ceruti, J.M.; Leirós, G.J.; Balañá, M.E. Androgens and androgen receptor action in skin and hair follicles. *Mol. Cell. Endocrinol.* **2018**, *465*, 122–133. [CrossRef]
9. Anastassakis, K. Hormonal and genetic etiology of male androgenetic alopecia. In *Androgenetic Alopecia From A to Z: Vol.1 Basic Science, Diagnosis, Etiology, and Related Disorders*; Anastassakis, K., Ed.; Springer International Publishing: Cham, Switzerland, 2022; pp. 135–180. [CrossRef]
10. Wolf, R.; Schönfelder, G.; Paul, M.; Blume-Peytavi, U. Nitric oxide in the human hair follicle: Constitutive and dihydrotestosterone-induced nitric oxide synthase expression and NO production in dermal papilla cells. *J. Mol. Med.* **2003**, *81*, 110–117. [CrossRef]
11. Zgonc Škulj, A.; Poljšak, N.; Kočevar Glavač, N.; Kreft, S. Herbal preparations for the treatment of hair loss. *Arch. Dermatol.* **2020**, *312*, 395–406. [CrossRef]
12. Dhariwala, M.Y.; Ravikumar, P. An overview of herbal alternatives in androgenetic alopecia. *J. Cosmet. Dermatol.* **2019**, *18*, 966–975. [CrossRef] [PubMed]
13. Herman, A.; Herman, A.P. Mechanism of action of herbs and their active constituents used in hair loss treatment. *Fitoterapia* **2016**, *114*, 18–25. [CrossRef]
14. Manosroi, A.; Ruksiriwanich, W.; Manosroi, W.; Abe, M.; Manosroi, J. In vivo hair growth promotion activity of gel containing niosomes loaded with the *Oryza sativa* bran fraction (OSF3). *Adv. Sci. Lett.* **2012**, *16*, 222–228. [CrossRef]
15. Choi, B.Y. Targeting Wnt/β-catenin pathway for developing therapies for hair loss. *Int. J. Mol. Sci.* **2020**, *21*, 4915. [CrossRef] [PubMed]
16. Zhang, H.; Nan, W.; Wang, S.; Song, X.; Si, H.; Li, T.; Li, G. Epigallocatechin-3-gallate promotes the growth of mink hair follicles through sonic hedgehog and protein kinase B signaling pathways. *Front. Pharmacol.* **2018**, *9*, 674. [CrossRef]
17. Trüeb, R.M. Molecular mechanisms of androgenetic alopecia. *Exp. Gerontol.* **2002**, *37*, 981–990. [CrossRef]
18. Balık, A.R.; Balık, Z.B.; Aktaş, A.; Neşelioğlu, S.; Karabulut, E.; Karabulut, A.B. Examination of androgenetic alopecia with serum biomarkers. *J. Cosmet. Dermatol.* **2021**, *20*, 1855–1859. [CrossRef]
19. Sasaki, M.; Shinozaki, S.; Morinaga, H.; Kaneki, M.; Nishimura, E.; Shimokado, K. iNOS inhibits hair regeneration in obese diabetic (*ob/ob*) mice. *Biochem. Biophys. Res. Commun.* **2018**, *501*, 893–897. [CrossRef]
20. Manosroi, A.; Ruksiriwanich, W.; Abe, M.; Sakai, H.; Aburai, K.; Manosroi, W.; Manosroi, J. Physico-chemical properties of cationic niosomes loaded with fraction of rice (*Oryza sativa*) bran extract. *J. Nanosci. Nanotechnol.* **2012**, *12*, 7339–7345. [CrossRef]
21. Orellana, E.A.; Kasinski, A.L. Sulforhodamine B (SRB) assay in cell culture to investigate cell proliferation. *Bio. Protoc.* **2016**, *6*, e1984. [CrossRef]
22. Nazir, Y.; Linsaenkart, P.; Khantham, C.; Chaitep, T.; Jantrawut, P.; Chittasupho, C.; Rachtanapun, P.; Jantanasakulwong, K.; Phimolsiripol, Y.; Sommano, S.R.; et al. High efficiency in vitro wound healing of *Dictyophora indusiata* extracts via anti-inflammatory and collagen stimulating (MMP-2 inhibition) mechanisms. *J. Fungus* **2021**, *7*, 1100. [CrossRef] [PubMed]
23. Lolli, F.; Pallotti, F.; Rossi, A.; Fortuna, M.C.; Caro, G.; Lenzi, A.; Sansone, A.; Lombardo, F. Androgenetic alopecia: A review. *Endocrine* **2017**, *57*, 9–17. [CrossRef] [PubMed]
24. Madaan, A.; Verma, R.; Singh, A.T.; Jaggi, M. Review of hair follicle dermal papilla cells as *in vitro* screening model for hair growth. *Int. J. Cosmet. Sci.* **2018**, *40*, 429–450. [CrossRef] [PubMed]
25. Li, S.; Chen, J.; Chen, F.; Wang, C.; Guo, X.; Wang, C.; Fan, Y.; Wang, Y.; Peng, Y.; Li, W. Liposomal honokiol promotes hair growth via activating Wnt3a/β-catenin signaling pathway and down regulating TGF-β1 in C57BL/6N mice. *Biomed. Pharmacother.* **2021**, *141*, 111793. [CrossRef]

26. Sittisart, P.; Yossan, S.; Prasertsan, P. Antifungal property of chili, shallot and garlic extracts against pathogenic fungi, Phomopsis spp., isolated from infected leaves of para rubber (*Hevea brasiliensis* Muell. Arg.). *Agric. Nat. Resour.* **2017**, *51*, 485–491. [CrossRef]
27. Hussain, T.; Tan, B.; Yin, Y.; Blachier, F.; Tossou, M.C.B.; Rahu, N. Oxidative stress and inflammation: What polyphenols can do for us? *Oxid. Med. Cell. Longev.* **2016**, *2016*, 7432797. [CrossRef]
28. English Jr, R.S. A hypothetical pathogenesis model for androgenic alopecia: Clarifying the dihydrotestosterone paradox and rate-limiting recovery factors. *Med. Hypotheses* **2018**, *111*, 73–81. [CrossRef]
29. Lee, M.; Rho, H.S.; Choi, K. Anti-inflammatory effects of a *P*-coumaric acid and kojic acid derivative in LPS-stimulated RAW264.7 macrophage cells. *Biotechnol. Bioprocess. Eng.* **2019**, *24*, 653–657. [CrossRef]
30. Huang, N.; Hauck, C.; Yum, M.Y.; Rizshsky, L.; Widrlechner, M.P.; McCoy, J.A.; Murphy, P.A.; Dixon, P.M.; Nikolau, B.J.; Birt, D.F. Rosmarinic acid in *Prunella vulgaris* ethanol extract inhibits lipopolysaccharide-induced prostaglandin E2 and nitric oxide in RAW 264.7 mouse macrophages. *J. Agric. Food Chem.* **2009**, *57*, 10579–10589. [CrossRef]
31. Hämäläinen, M.; Nieminen, R.; Vuorela, P.; Heinonen, M.; Moilanen, E. Anti-inflammatory effects of flavonoids: Genistein, kaempferol, quercetin, and daidzein inhibit STAT-1 and NF-κB activations, whereas flavone, isorhamnetin, naringenin, and pelargonidin inhibit only NF-κB activation along with their inhibitory effect on iNOS expression and NO production in activated macrophages. *Mediat. Inflamm.* **2007**, *2007*, 45673.
32. Khantham, C.; Linsaenkart, P.; Chaitep, T.; Jantrawut, P.; Chittasupho, C.; Rachtanapun, P.; Jantanasakulwong, K.; Phimolsiripol, Y.; Sommano, S.R.; Prom-u-thai, C.; et al. Antioxidation, anti-inflammation, and regulation of *SRD5A* gene expression of *Oryza sativa* cv. Bue Bang 3 CMU husk and bran extracts as androgenetic alopecia molecular treatment substances. *Plants* **2022**, *11*, 330. [CrossRef] [PubMed]
33. Sawaya, M.E.; Price, V.H. Different levels of 5α-reductase type I and II, aromatase, and androgen receptor in hair follicles of women and men with androgenetic alopecia. *J. Invest. Dermatol.* **1997**, *109*, 296–300. [CrossRef]
34. Sánchez, P.; Serrano-Falcón, C.; Torres, J.; Serrano, S.; Ortega, E. 5α-Reductase isozymes and aromatase mRNA levels in plucked hair from young women with female pattern hair loss. *Arch. Dermatol.* **2018**, *310*, 77–83. [CrossRef] [PubMed]
35. Asada, Y.; Sonoda, T.; Ojiro, M.; Kurata, S.; Sato, T.; Ezaki, T.; Takayasu, S. 5α-reductase type 2 is constitutively expressed in the dermal papilla and connective tissue sheath of the hair follicle in vivo but not during culture in vitro. *J. Clin. Endocrinol. Metab.* **2001**, *86*, 2875–2880. [PubMed]
36. Li, J.; Ding, Z.; Wang, Z.; Lu, J.-F.; Maity, S.N.; Navone, N.M.; Logothetis, C.J.; Mills, G.B.; Kim, J. Androgen regulation of 5α-reductase isoenzymes in prostate cancer: Implications for prostate cancer prevention. *PLoS ONE* **2011**, *6*, e28840. [CrossRef] [PubMed]
37. Yang, F.; Song, L.; Wang, H.; Wang, J.; Xu, Z.; Xing, N. Quercetin in prostate cancer: Chemotherapeutic and chemopreventive effects, mechanisms and clinical application potential. *Oncol. Rep.* **2015**, *33*, 2659–2668. [CrossRef]
38. Hiipakka, R.A.; Zhang, H.-Z.; Dai, W.; Dai, Q.; Liao, S. Structure–activity relationships for inhibition of human 5α-reductases by polyphenols. *Biochem. Pharmacol.* **2002**, *63*, 1165–1176. [CrossRef]
39. Ruksiriwanich, W.; Khantham, C.; Linsaenkart, P.; Chaitep, T.; Jantrawut, P.; Chittasupho, C.; Rachtanapun, P.; Jantanasakulwong, K.; Phimolsiripol, Y.; Sommano, S.R. In vitro and in vivo regulation of *SRD5A* mRNA expression of supercritical carbon dioxide extract from *Asparagus racemosus* Willd. Root as anti-sebum and pore-minimizing active ingredients. *Molecules* **2022**, *27*, 1535. [CrossRef]
40. Chittasupho, C.; Manthaisong, A.; Okonogi, S.; Tadtong, S.; Samee, W. Effects of quercetin and curcumin combination on antibacterial, antioxidant, in vitro wound healing and migration of human dermal fibroblast cells. *Int. J. Mol. Sci.* **2022**, *23*, 142. [CrossRef]
41. Rondanelli, M.; Perna, S.; Peroni, G.; Guido, D. A bibliometric study of scientific literature in Scopus on botanicals for treatment of androgenetic alopecia. *J. Cosmet. Dermatol.* **2016**, *15*, 120–130. [CrossRef]
42. Murata, K.; Noguchi, K.; Kondo, M.; Onishi, M.; Watanabe, N.; Okamura, K.; Matsuda, H. Promotion of hair growth by *Rosmarinus officinalis* leaf extract. *Phytother. Res.* **2013**, *27*, 212–217. [CrossRef] [PubMed]
43. Kishimoto, J.; Burgeson, R.E.; Morgan, B.A. Wnt signaling maintains the hair-inducing activity of the dermal papilla. *Genes Dev.* **2000**, *14*, 1181–1185. [CrossRef] [PubMed]
44. Zhang, R.; Li, Y.; Jia, K.; Xu, X.; Li, Y.; Zhao, Y.; Zhang, X.; Zhang, J.; Liu, G.; Deng, S.; et al. Crosstalk between androgen and Wnt/β-catenin leads to changes of wool density in FGF5-knockout sheep. *Cell Death Dis.* **2020**, *11*, 407. [CrossRef] [PubMed]
45. Leiros, G.J.; Attorresi, A.I.; Balaña, M.E. Hair follicle stem cell differentiation is inhibited through cross-talk between Wnt/β-catenin and androgen signalling in dermal papilla cells from patients with androgenetic alopecia. *Br. J. Dermatol. Suppl.* **2012**, *166*, 1035–1042. [CrossRef] [PubMed]
46. Kwack, M.H.; Ahn, J.S.; Kim, M.K.; Kim, J.C.; Sung, Y.K. Preventable effect of L-threonate, an ascorbate metabolite, on androgen-driven balding via repression of dihydrotestosteroneinduced dickkopf-1 expression in human hair dermal papilla cells. *BMB Rep.* **2010**, *43*, 688–692. [CrossRef]
47. Kwack, M.H.; Sung, Y.K.; Chung, E.J.; Im, S.U.; Ahn, J.S.; Kim, M.K.; Kim, J.C. Dihydrotestosterone-inducible Dickkopf 1 from balding dermal papilla cells causes apoptosis in follicular keratinocytes. *J. Invest. Dermatol.* **2008**, *128*, 262–269. [CrossRef]
48. Zhang, N.n.; Park, D.K.; Park, H.J. Hair growth-promoting activity of hot water extract of *Thuja orientalis*. *BMC Complement. Altern. Med.* **2013**, *13*, 9. [CrossRef]

49. Huelsken, J.; Vogel, R.; Erdmann, B.; Cotsarelis, G.; Birchmeier, W. beta-Catenin controls hair follicle morphogenesis and stem cell differentiation in the skin. *Cell* **2001**, *105*, 533–545. [CrossRef]
50. Mi, Y.; Zhong, L.; Lu, S.; Hu, P.; Pan, Y.; Ma, X.; Yan, B.; Wei, Z.; Yang, G. Quercetin promotes cutaneous wound healing in mice through Wnt/β-catenin signaling pathway. *J. Ethnopharmacol.* **2022**, *290*, 115066. [CrossRef]
51. Jin, Z.; Ke, J.; Guo, P.; Wang, Y.; Wu, H. Quercetin improves blood-brain barrier dysfunction in rats with cerebral ischemia reperfusion via Wnt signaling pathway. *Am. J. Transl. Res.* **2019**, *11*, 4683–4695.
52. Gritli-Linde, A.; Hallberg, K.; Harfe, B.D.; Reyahi, A.; Kannius-Janson, M.; Nilsson, J.; Cobourne, M.T.; Sharpe, P.T.; McMahon, A.P.; Linde, A. Abnormal hair development and apparent follicular transformation to mammary gland in the absence of hedgehog signaling. *Dev. Cell* **2007**, *12*, 99–112. [CrossRef] [PubMed]
53. Veltri, A.; Lang, C.; Lien, W.H. Concise review: Wnt signaling pathways in skin development and epidermal stem cells. *Stem Cells* **2018**, *36*, 22–35. [CrossRef] [PubMed]
54. Yu, S.H.; Kim, Y.; Jung, N.; Hwang, J.W.; Kim, N.; Ha, J.C.; Kim, M.J.; Lee, Y.; Choi, Y.S.; Han, K.; et al. Hair growth-promoting effect of recombinant human sonic hedgehog proteins. *Biomed. Dermatol.* **2019**, *3*, 7. [CrossRef]
55. Paladini, R.D.; Saleh, J.; Qian, C.; Xu, G.X.; Rubin, L.L. Modulation of hair growth with small molecule agonists of the hedgehog signaling pathway. *J. Invest. Dermatol.* **2005**, *125*, 638–646. [CrossRef]
56. St-Jacques, B.; Dassule, H.R.; Karavanova, I.; Botchkarev, V.A.; Li, J.; Danielian, P.S.; McMahon, J.A.; Lewis, P.M.; Paus, R.; McMahon, A.P. Sonic hedgehog signaling is essential for hair development. *Curr. Biol.* **1998**, *8*, 1058–1068. [CrossRef]
57. Naser, B.; Bodinet, C.; Tegtmeier, M.; Lindequist, U. *Thuja occidentalis* (*Arbor vitae*): A review of its pharmaceutical, pharmacological and clinical properties. *Evid. Based Complement. Altern.* **2005**, *2*, 69–78. [CrossRef]
58. Lin, L.; Ni, B.; Lin, H.; Zhang, M.; Li, X.; Yin, X.; Qu, C.; Ni, J. Traditional usages, botany, phytochemistry, pharmacology and toxicology of *Polygonum multiflorum* Thunb.: A review. *J. Ethnopharmacol.* **2015**, *159*, 158–183. [CrossRef] [PubMed]
59. Park, H.J.; Zhang, N.; Park, D.K. Topical application of *Polygonum multiflorum* extract induces hair growth of resting hair follicles through upregulating Shh and β-catenin expression in C57BL/6 mice. *J. Ethnopharmacol.* **2011**, *135*, 369–375. [CrossRef]
60. Cebe-Suarez, S.; Zehnder-Fjällman, A.; Ballmer-Hofer, K. The role of VEGF receptors in angiogenesis; complex partnerships. *Cell. Mol. Life Sci.* **2006**, *63*, 601–615. [CrossRef]
61. Yano, K.; Brown, L.F.; Detmar, M. Control of hair growth and follicle size by VEGF-mediated angiogenesis. *J. Clin. Investig.* **2001**, *107*, 409–417. [CrossRef]
62. Lee, C.Y.; Su, C.H.; Chiang, C.Y.; Wu, C.N.; Kuan, Y.H. Observation of the expression of vascular endothelial growth factor and the potential effect of promoting hair growth treated with chinese herbal BeauTop. *Evid. Based Complement. Altern.* **2021**, *2021*, 6667011. [CrossRef] [PubMed]
63. Kubanov, A.; Gallyamova, Y.A.; Korableva, O. The study of growth factors in patients with androgenic alopecia. *Biomed. Pharmacol. J.* **2017**, *10*, 1219–1228. [CrossRef]
64. Formiga, R.d.O.; Alves Júnior, E.B.; Vasconcelos, R.C.; Araújo, A.A.; de Carvalho, T.G.; de Araújo Junior, R.F.; Guerra, G.B.C.; Vieira, G.C.; de Oliveira, K.M.; Diniz, M.d.F.F.M.; et al. Effect of *p*-cymene and rosmarinic acid on gastric ulcer healing–involvement of multiple endogenous curative mechanisms. *Phytomedicine* **2021**, *86*, 153497. [CrossRef] [PubMed]
65. Leksawasdi, N.; Taesuwan, S.; Prommajak, T.; Techapun, C.; Khonchaisri, R.; Sittilop, N.; Halee, A.; Jantanasakulwong, K.; Phongthai, S.; Nunta, R.; et al. Ultrasonic extraction of bioactive compounds from green soybean pods and application in green soybean milk antioxidants fortification. *Foods* **2022**, *11*, 588. [CrossRef]
66. Arjin, C.; Hongsibsong, S.; Pringproa, K.; Seel-Audom, M.; Ruksiriwanich, W.; Sutan, K.; Sommano, S.R.; Sringarm, K. Effect of ethanolic *Caesalpinia sappan* fraction on in vitro antiviral activity against porcine reproductive and respiratory syndrome virus. *Vet. Sci.* **2021**, *8*, 106. [CrossRef] [PubMed]
67. Wisetkomolmat, J.; Arjin, C.; Satsook, A.; Seel-Audom, M.; Ruksiriwanich, W.; Prom-u-Thai, C.; Sringarm, K. Comparative analysis of nutritional components and phytochemical attributes of selected Thai rice bran. *Front. Nutr.* **2022**, *9*. [CrossRef]
68. Manosroi, A.; Chankhampan, C.; Kietthanakorn, B.O.; Ruksiriwanich, W.; Chaikul, P.; Boonpisuttinant, K.; Sainakham, M.; Manosroi, W.; Tangjai, T.; Manosroi, J. Pharmaceutical and cosmeceutical biological activities of hemp (*Cannabis sativa* L. var. sativa) leaf and seed extracts. *Chiang Mai J. Sci.* **2019**, *46*, 180–195.
69. Khantham, C.; Yooin, W.; Sringarm, K.; Sommano, S.R.; Jiranusornkul, S.; Carmona, F.D.; Nimlamool, W.; Jantrawut, P.; Rachtanapun, P.; Ruksiriwanich, W. Effects on steroid 5-alpha reductase gene expression of Thai rice bran extracts and molecular dynamics study on SRD5A2. *Biology* **2021**, *10*, 319. [CrossRef]
70. Wang, F. Semi-Quantitative RT-PCR: An Effective Method to Explore The Regulation of Gene Transcription Level Affected by Environmental Pollutants. In *Environmental Toxicology and Toxicogenomics: Principles, Methods, and Applications*; Pan, X., Zhang, B., Eds.; Springer: New York, NY, USA, 2021; pp. 95–103.

plants

MDPI

Article

Computation Screening of Multi-Target Antidiabetic Properties of Phytochemicals in Common Edible Mediterranean Plants

Vlasios Goulas [1,*], Antonio J. Banegas-Luna [2], Athena Constantinou [1], Horacio Pérez-Sánchez [2] and Alexandra Barbouti [3]

[1] Department of Agricultural Sciences, Biotechnology and Food Science, Cyprus University of Technology, Lemesos 3603, Cyprus; ai.konstantinou@edu.cut.ac.cy

[2] Structural Bioinformatics and High Performance Computing (BIO-HPC) Research Group, UCAM Universidad Católica de Murcia, 30107 Guadalupe, Spain; ajbanegas@ucam.edu (A.J.B.-L.); hperez@ucam.edu (H.P.-S.)

[3] Department of Anatomy-Histology-Embryology, Faculty of Medicine, School of Health Sciences, University of Ioannina, 45110 Ioannina, Greece; abarbout@uoi.gr

* Correspondence: vlasios.goulas@cut.ac.cy; Tel.: +357-2500-2141

Abstract: Diabetes mellitus is a metabolic disease and one of the leading causes of deaths worldwide. Numerous studies support that the Mediterranean diet has preventive and treatment effects on diabetes. These effects have been attributed to the special bioactive composition of Mediterranean foods. The objective of this work was to decipher the antidiabetic activity of Mediterranean edible plant materials using the DIA-DB inverse virtual screening web server. A literature review on the antidiabetic potential of Mediterranean plants was performed and twenty plants were selected for further examination. Subsequently, the most abundant flavonoids, phenolic acids, and terpenes in plant materials were studied to predict their antidiabetic activity. Results showed that flavonoids are the most active phytochemicals as they modulate the function of 17 protein-targets and present high structural similarity with antidiabetic drugs. Their antidiabetic effects are linked with three mechanisms of action, namely (i) regulation of insulin secretion/sensitivity, (ii) regulation of glucose metabolism, and (iii) regulation of lipid metabolism. Overall, the findings can be utilized to understand the antidiabetic activity of edible Mediterranean plants pinpointing the most active phytoconstituents.

Keywords: antidiabetic activity; diabetes; DIA-DB web server; flavonoids; in silico study; mechanism of action; phenolic acids; terpenes

Citation: Goulas, V.; Banegas-Luna, A.J.; Constantinou, A.; Pérez-Sánchez, H.; Barbouti, A. Computation Screening of Multi-Target Antidiabetic Properties of Phytochemicals in Common Edible Mediterranean Plants. *Plants* **2022**, *11*, 1637. https://doi.org/10.3390/plants11131637

Academic Editors: Ivayla Dincheva, Ilian Badjakov and Bistra Galunska

Received: 30 May 2022
Accepted: 16 June 2022
Published: 21 June 2022

1. Introduction

Diabetes mellitus (DM), or merely diabetes, is a complex chronic disease, which requires continues medical care, reducing the patient's quality of life and raising the medical cost if not treated properly. According to the World Health Organization (WHO), DM has entered the top 10 leading causes of deaths worldwide, following a significant percentage increase of 70% from 2000 to 2030 [1]. From a medical point of view, it is a chronic noncommunicable disease arising from impaired insulin secretion and insulin resistance, leading to its defining feature of hyperglycemia. It can affect different organ systems in the body and, over time, causes several complications including neuropathy, nephropathy, retinopathy, cardiovascular disease, stroke, and peripheral artery disease [2]. The more common types of DM are classified into type 1 (diabetes mellitus (T1DM)), type 2, and gestational diabetes. Type 1 DM is characterized by absolute insulin deficiency associated with pancreatic β cells destruction [3], while type 2 DM, which accounts for nearly 95% of individuals, is mainly due to insulin resistance (IR) and deficiency in insulin secretion. Gestational diabetes, on the other hand, develops during pregnancy and usually disappears after giving birth. [3].

Recently, Newman and Cragg (2019) stated and classified all approved therapeutic agents for DM, from the 1st of January 1981 to the 30th of September 2019. According to their report, over 54% of registered drugs for DM are natural products or mimic natural products. It is obvious that natural products are an attractive reservoir of antidiabetic compounds [4]. In addition, numerous studies demonstrate an antidiabetic potential of natural products [5]. The antidiabetic effects of many phytochemicals including polyphenols, terpenes, alkaloids, saponins, and quinones have been well-documented [6,7]. Furthermore, clinical trials with medicinal plants and natural products have been conducted, whereas some of them have been used for the development of herbal formulations controlling DM. Regarding the mechanism of action of natural products, (i) the inhibition of α-glucosidase and α-amylase in the digestive tract, (ii) the boost of insulin secretion and pancreatic β cell proliferation, (iii) the regulation of glucose uptake and glucose transporters, (iv) the inhibition of protein tyrosine phosphatase 1B activity, and (v) the reduction in the generation of oxidative stress are the main modes of action of pure phytochemicals and crude extracts. [3].

Epidemiological studies also highlight that the adoption of a healthy dietary pattern such as the Mediterranean diet contributes to the prevention and treatment of Type 2 DM [8,9]. The beneficial effects of the Mediterranean diet are correlated with weight control and the consumption of foods rich in nutrients with various health benefits [10]. Fruit and vegetables are essential components of the Mediterranean diet and contain many bioactive phytochemicals. Taking into consideration a current study where researchers fished new antidiabetic compounds from various natural products, the search for new antidiabetic agents in Mediterranean edible plants stands as a challenge [5].

In the present study, we strived to clarify the antidiabetic potential of edible Mediterranean plant materials that contain common phytochemicals, with the employment of in silico virtual screening methodologies. Plants are complex mixtures of several primary and secondary metabolites and the classic approach including the isolation and the evaluation of the antidiabetic activity is time-consuming and complicated. Thus, the employment of the DIA-DB inverse virtual screening web server allows the rapid evaluation of several compounds for antidiabetic activity and predicts the possible mode of action of antidiabetic compounds.

2. Results and Discussion

2.1. Plant and Phytochemical Selection

At first, a list of plants that are widely distributed in Mediterranean flora and/or the Mediterranean diet was prepared. The in vitro and in vivo antidiabetic activity of these plants were checked with the implementation of an extensive literature review. Twenty plants were selected for further examination as they exert both in vitro and in vivo antidiabetic activity (Table 1). All selected plants are edible and/or can be consumed after cooking or processing. Regarding the plant taxonomy, they belong to 11 families. Five plants come from the Lamiaceae and four from the Apiaceae family. Both Amaryllidaceae and Asteraceae families includes two plant materials for further study. Other plant materials belong to Brassicaceae, Ranunculaceae, Plantaginaceae, Rutaceae, Rosaceae, Oleaceaceae, and Vitaceae families.

A literature review showed that the in vitro antidiabetic activity of Mediterranean plants is mostly correlated with their inhibitory effects on enzymes related to DM. About 85% of the selected plant material have α-glucosidase and/or α-amylase inhibitory activity. Both enzymes are considered as carbohydrate-hydrolyzing enzymes and are linked with postprandial hyperglycaemia as they regulate the absorption of glucose [11]. Studies also demonstrated the stimulation of insulin secretion using different pancreatic β-cells [12,13]. Coriander and black mustard seeds, thyme, and summer savory also enhance in vitro the glucose uptake in cell lines or rat muscle pieces [12–14]. Furthermore, in vitro studies indicated that Mediterranean plants promote the proliferation of pancreatic β-cells [13], inhibit dipeptidyl peptidase-4 inhibition [15], prevent the formation of advanced glycation end-products [16], and decrease the fat accumulation in *Caenorhabditis elegans* [14].

Table 1. A comprehensive summary of in vitro and in vivo antidiabetic properties of edible Mediterranean plants.

Common Name	Scientific Name	Plant Part	In vitro Antidiabetic Effects	In vivo Antidiabetic Effects	References
Black cumin	*Nigella sativa* L.	Seeds	▪ Increase in insulin secretion ▪ Induction in proliferation of pancreatic β-cells ▪ Stimulation of glucose uptake	▪ Reduction in blood glucose ▪ Antihyperlipidemic effects	[13,17–19]
Black mustard	*Brassica nigra* L.	Aerial plants, seeds	▪ α-glucosidase inhibition	▪ Reduction in blood glucose ▪ Decrease in glycosylated hemoglobin (HbA1c) ▪ Antihyperlipidemic effects ▪ Insulinotropic effect	[20–22]
Broadleaf plantain	*Plantago major* L.	Leaves	▪ α-amylase inhibition	▪ Reduction in blood glucose ▪ Stimulation of insulin secretion	[23–25]
Citrus fruits	Citrus spp.	Peel	▪ α-glucosidase inhibition ▪ α-amylase inhibition	▪ Reduction in blood glucose ▪ Antihyperlipidemic effects ▪ Stimulation of insulin secretion ▪ Protective effects for tissue damage	[26–30]
Coriander	*Coriandrum sativum* L.	Seeds	▪ Stimulation of insulin secretion ▪ Enhancement of glucose uptake	▪ Stimulation of insulin secretion ▪ Enhancement of glucose uptake ▪ Reduction in blood glucose ▪ Antihyperlipidemic effects	[12]
Cumin	*Cuminum cyminum* L.	Seeds	▪ α-glucosidase inhibition ▪ α-amylase inhibition	▪ Reduction in blood glucose ▪ Antihyperlipidemic effects ▪ Decrease of HbA1c	[31–34]
Dill	*Anethum graveolens* L.	Seeds	▪ α-glucosidase inhibition ▪ α-amylase inhibition	▪ Reduction in blood glucose ▪ Antihyperlipidemic effects ▪ Control of colonic motility disorder	[35–37]
Garlic	*Allium sativum* L	Bulb	▪ dipeptidyl peptidase-4 inhibition ▪ α-glucosidase inhibition	▪ Reduction in blood glucose ▪ Antihyperlipidemic effects ▪ Decrease of insulin resistance	[15,38–40]
Grapes	*Vitis vinifera* L.	Seeds, skins	α-glucosidase inhibition α-amylase inhibition	▪ Reduction in blood glucose ▪ Antihyperlipidemic effects ▪ Decrease in HbA1c	[41,42]
Marjoram	*Origanum majorana* L.	Aerial parts	▪ α-glucosidase inhibition ▪ Inhibition of advanced glycation end-product (AGE) formation	▪ Inhibition of (AGE) formation ▪ Reduction in blood glucose ▪ Antihyperlipidemic effects	[16,43]
Olive	*Olea europaea* L.	Leaves	▪ α-glucosidase inhibition ▪ α-amylase inhibition	▪ Reduction in blood glucose ▪ Stimulation of insulin secretion ▪ Antihyperlipidemic effects ▪ Decrease in histopathological changes ▪ α-amylase inhibition	[44–48]
Onion	*Allium sepa* L.	Bulbs, skins	▪ α-glucosidase inhibition	▪ Reduction in blood glucose ▪ Stimulation of insulin secretion	[49–52]
Parsley	*Petroselinum sativum/crispum*	Leaves	▪ α-glucosidase inhibition ▪ α-amylase inhibition	▪ Reduction in blood glucose ▪ Stimulation of insulin secretion ▪ Antihyperlipidemic effects	[53–55]
Rosemary	*Rosmarinus officinalis* L.	Aerial parts	▪ α-glucosidase inhibition	▪ Stimulation of insulin secretion	[56]
Sage	*Salvia officinalis* L.	Aerial parts	▪ α-glucosidase inhibition ▪ α-amylase inhibition	▪ Reduction in blood glucose ▪ Stimulation of insulin secretion ▪ Antihyperlipidemic effects	[57–59]
Sow thistle	*Sonchus oleraceus* L.		▪ α-glucosidase inhibition ▪ α-amylase inhibition	▪ Reduction in blood glucose ▪ Stimulation of insulin secretion ▪ Antihyperlipidemic effects ▪ Protective effects for tissue damage	[60–63]
Strawberry	*Fragaria* spp.	Leaves, fruits	▪ α-glucosidase inhibition ▪ α-amylase inhibition	▪ Reduction in blood glucose ▪ Antihyperlipidemic effects Improvement of liver functions	[64–67]

Table 1. *Cont.*

Common Name	Scientific Name	Plant Part	In vitro Antidiabetic Effects	In vivo Antidiabetic Effects	References
Summer savory	*Satureja hortensis* L.	Aerial parts	▪ Stimulation of insulin-dependent glucose uptake	▪ Reduction in blood glucose	[68]
Thyme	*Thymus vulgaris* L.	Aerial parts	▪ α-glucosidase inhibition ▪ α-amylase inhibition ▪ Stimulation of insulin-dependent glucose uptake ▪ Decrease of fat accumulation in *Caenorhabditis elegans*	▪ Reduction in blood glucose ▪ Antihyperlipidemic effects	[14,69–71]
Yarrow	*Achillea santolina* L.	Aerial parts	▪ α-amylase inhibition	▪ Reduction in blood glucose ▪ Antihyperlipidemic effects	[72–74]

Several works also document the in vivo antidiabetic effects of selected Mediterranean plants (Table 1). This activity is mainly established using animal models such as healthy and streptozotocin or alloxan-induced diabetic rats or mice. There are also clinical trials that report their antidiabetic activity in patients with type 2 DM and overweight adults. The reduction in glucose in blood, the stimulation of insulin secretion, and the antihyperlipidemic properties are the main antidiabetic effects of selected Mediterranean plants according to the literature review. In addition, the decrease in glycosylated hemoglobin (HbA1c), the protective effects from tissue damage, and the improvement of liver functions have been also mentioned. It is noteworthy that all selected plants exert antidiabetic effects through multiple mechanisms. It is attributed to the synergism of two or more phytochemicals as plants are complex mixtures of bioactive phytochemicals.

In the next step, the phytochemical composition of active Mediterranean plants was elucidated. The constituents were classified into three groups, namely phenolic acids, flavonoids, and terpenes. The most abundant phytochemicals from each group were selected to further investigate their antidiabetic potential using the DIA-DB inverse virtual screening web server. Table 2 summarizes the phytochemicals and their distribution in the Mediterranean plants, which is as follows: quercetin is present in 19 plants, caffeic acid in 16, ferulic acid in 15, and luteolin in 15 plants. These are the most common phytochemicals in the Mediterranean plants in the present study. On the other hand, chrysoeriol, ellagic acid, and hesperidin are found less frequently in Mediterranean plants (in 5, 6, and 6 plants, respectively). All structures of the studied phytoconstituents are presented in Figure 1.

Table 2. Occurrence of phytochemicals in Mediterranean plant materials and their SMILES notations.

Compound	SMILES Notation	Plants	References
	Phenolic acids		
Caffeic acid	C1=CC(=C(C=C1C=CC(=O)O)O)O	black mustard, broadleaf plantain, citrus peel, coriander, cumin, dill, garlic, grape skin, marjoram, olive leaf, parsley, rosemary, sage, santolina, thyme	[43,75–85]
Cinnamic Acid	C1=CC=C(C=C1)C=CC(=O)O	dill, grape skin, marjoram, olive leaf, sage, thyme	[43,81–83,86,87]
Ellagic acid	C1=C2C3=C(C(=C1O)O)OC(=O)C4=CC (=C(C(=C43)OC2=O)O)O	black mustard, broadleaf plantain, citrus fruits, grape skin, parsley, strawberry	[75,88–92]
Ferulic acid	COC1=C(C=CC(=C1)C=CC(=O)O)O	black cumin, citrus peel, coriander, cumin, dill, garlic, marjoram, olive leaf, onion, rosemary, sage, santolina, thyme, strawberry	[43,77–79,82,85,87,93–96]
Gallic acid	C1=C(C=C(C(=C1O)O)O)C(=O)O	black mustard, black cumin, citrus peel, cumin, dill, garlic, grape skin and seeds, marjoram, olive leaf, onion, strawberry,	[43,76,77,82,86,93,94,97–101]
p-Coumaric acid	C1=CC(=CC=C1C=CC(=O)O)O	black cumin, broadleaf plantain, citrus peel, coriander, cumin, dill, garlic, grape skin, marjoram, olive leaf, parsley, sage, santolina, thyme	[43,76–82,84,85,87,93,99,102]
Rosmarinic acid	C1=CC(=C(C=C1CC(C(=O)O)OC(=O)C=CC2=CC (=C(C=C2)O)O)O)O	cumin, dill marjoram, olive leaf, rosemary, sage, santolina, summer savory, thyme	[43,82,83,97,102–104]

Table 2. *Cont.*

Compound	SMILES Notation	Plants	References
		Flavonoids	
Apigenin	C1=CC(=CC=C1C2=CC(=O)C3=C(C=C(C=C3O2)O)O)O	black mustard, black cumin, broadleaf plantain, citrus peel, cumin, garlic, marjoram, olive leaf, parsley, santolina, sage, sow thistle, summer savory, thyme,	[43,75,76,80,82,85,87,93,97,104–107]
Catechin	C1C(C(OC2=CC(=CC(=C21)O)O)C3=CC(=C(C=C3)O)O)O	black mustard, black cumin, broadleaf plantain, citrus peel, coriander, cumin grape skin, olive leaf, onion, strawberry	[77,78,82,93,100,108–111]
Chrysoeriol	COC1=C(C=CC(=C1)C2=CC(=O)C3=C(C=C(C=C3O2)O)O)O	broadleaf plantain, citrus peel, coriander, olive leaf, parsley,	[112–116]
Hesperidin	CC1C(C(C(C(O1)OCC2CC(CC(O2)OC3=CC(=C4C(=O)CC(OC4=C3)C5=CC(=C(C=C5)OΨ)O)O)O)O)O)O)O)O	citrus peels, marjoram, olive leaf, rosemary, strawberry, summer savory	[43,77,82,117–119]
Kaempferol	C1=CC(=CC=C1C2=C(C(=O)C3=C(C=C(C=C3O2)O)O)O)O	black mustard, citrus peel, coriander, cumin, grape skin, onion, parsley, rosemary, santolina, strawberry, summer savory, thyme	[75,77,78,85,94,99,101,120–123]
Luteolin	C1=CC(=C(C=C1C2=CC(=O)C3=C(C=C(C=C3O2)O)O)O)O	broadleaf plantain, citrus peel, coriander, cumin, garlic, marjoram, olive leaf, parsley, rosemary, sage, santolina, sow thistle, strawberry, summer savory, thyme	[123–127]
Quercetin	C1=CC(=C(C=C1C2=C(C(=O)C3=C(C=C(C=C3O2)O)O)O)O)O	black mustard, broadleaf plantain, citrus peel, coriander, cumin, dill, garlic, grape skin, marjoram, olive leaf, onion, parsley, rosemary, sage, sow thistle, summer savory, strawberry, thyme	[85,92,93,95–98,102,109,110,117,122,124–129]
Rutin	CC1C(C(C(C(O1)OCC2C(C(C(C(O2)OC3=C(OC4=CC(=CC(=C4C3=O)O)O)C5=CC(=C(C=C5)O)O)O)O)O)O)O)O	black mustard, broadleaf plantain, citrus peel, dill, garlic, marjoram, olive leaf, rosemary, sage, santolina, strawberry, summer savory,	[90–92,97,98,101,102,111,114,122,129,130]
		Terpenes	
alpha-pinene	CC1=CCC2CC1C2(C)C	black mustard, citrus peel, coriander, cumin, dill, grape skins, marjoram, olive leaf, rosemary, sage, santolina, summer savory, thyme	[43,105,131–139]
alpha-terpinene	CC1=CC=C(CC1)C(C)C	cumin, marjoram, parsley, rosemary, sage, santolina, summer savory, thyme	[43,140–146]
beta-pinene	CC1(C2CCC(=C)C1C2)C	citrus peel, coriander, cumin, grape skins, marjoram, olive leaf, rosemary, sage, summer savory, thyme	[43,77,133,134,136–139]
camphene	CC1(C2CCC(C2)C1=C)C	coriander, marjoram, rosemary, sage, santolina, summer savory, thyme,	[43,74,133,138,139]
gamma-terpinene	CC1=CCC(=CC1)C(C)C	citrus peel, coriander, dill, marjoram, parsley, rosemary, sage, santolina, summer savory, thyme	[43,77,135,139,141–147]
limonene	CC1=CCC(CC1)C(=C)C	black mustard, citrus peel, coriander, dill, marjoram, onion, rosemary, santolina, summer savory, thyme	[43,77,132,133,135,139,142,145,146]
beta-myrcene	CC(=CCCC(=C)C=C)C	citrus peel, coriander, dill, grape skins, marjoram, olive leaf, parsley, rosemary, sage, summer savory, thyme	[43,105,133,135–139,141,145]
sabinene	CC(C)C12CCC(=C)C1C2	citrus peel, coriander, marjoram, olive leaf, parsley, sage, santolina, summer savory, thyme	[43,74,105,133,137,139,141,143,145]

2.2. *Estimation of Antidiabetic Activity of Phytochemicals using Virtual Screening*

The phytochemicals were screened for antidiabetic activity against 17 diabetes targets using the DIA-DB inverse virtual screening web server. The docking scores of the crystallized ligands ranged from -10.4 to -1.5 kcal mol^{-1}. A docking cutoff score of -8 kcal mol^{-1} was set, as it was deemed a reasonable average docking score that covered the most active compounds for each protein target (Table 3). Results revealed interesting findings for the antidiabetic potential of phytochemicals. More specifically, all terpenes presented docking scores bigger than -8 kcal mol^{-1}; thus, the predicted antidiabetic activities of all terpenes are negligible or weak. The studied terpenes cannot be considered promising antidiabetic compounds for further investigation. In addition, the antidiabetic effects of edible plant materials cannot be correlated with the presence of these terpenes.

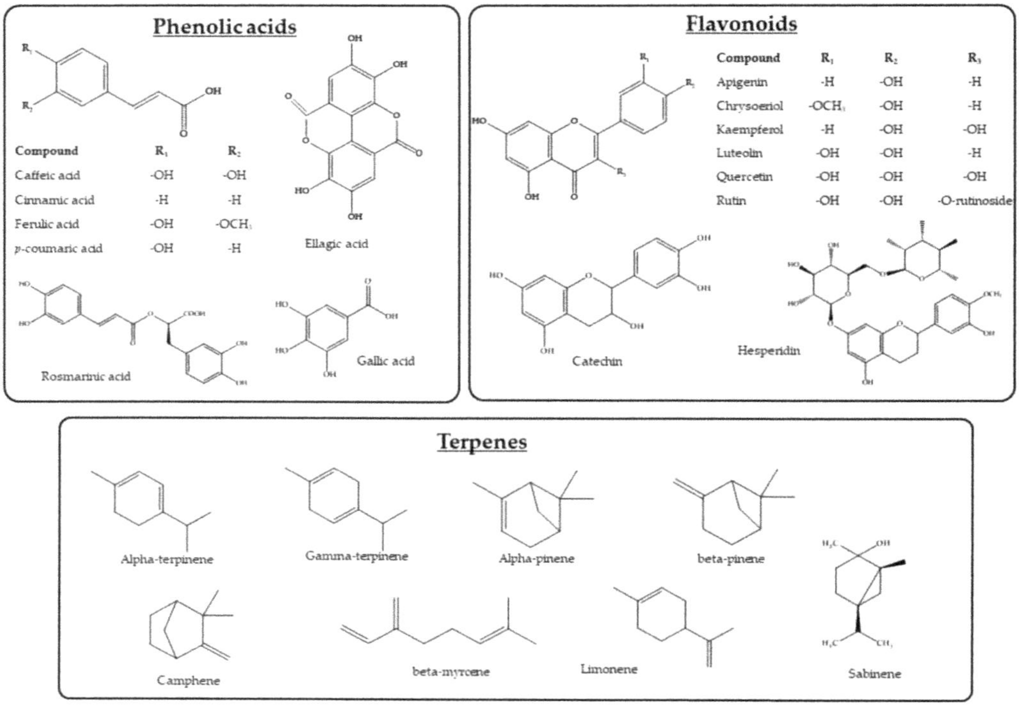

Figure 1. Structures of the studied phytochemicals present in Mediterranean plants.

Table 3. The docking cut-off score of active phytochemicals for each protein target. The score is given in parentheses and expressed as kcal mol^{-1}.

Protein Target	PDB Code	Function	Phytochemicals
Regulation of insulin secretion and/or sensitivity			
DPP4	4A5S	Stimulation of insulin secretion from pancreas degrading and inactivating glucagon-like peptide-1 [148]	Apigenin (−8.2), catechin (−8.3), chrysoeriol (−8.1), ellagic acid (−8.3), hesperidin (−10.4), kaempferol (−8.4), quercetin (−8.3), rutin (−9.1)
FFAR1	4PHU	Binding of free fatty acids to receptor results in increase in glucose-stimulated insulin secretion [149]	Apigenin (−8.4), caffeic acid (−8.0), ferulic acid (−8.0), hesperidin (−8.7), luteolin (−8.2)
HSD11B1	4K1L	Activates the synthesis of active glucocorticoids [150]	Apigenin (−9.0), catechin (−9.0), chrysoeriol (−9.1), ellagic acid (−8.6), hesperidin (−9.9), kaempferol (−9.2), luteolin (−9.4), quercetin (−9.8), rutin (−9.7)
INSR	3EKN	Regulates glucose uptake and synthesis of glycogen, lipid, and protein [151]	Ellagic acid (−8.2), hesperidin (−9.3), rutin (−8.4)
PTPN9	4GE6	Reduces insulin sensitivity, dephosphorylating the insulin receptor [152]	Ellagic acid (−8.0), hesperidin (−8.8), rutin (−8.6)
RBP4	2WR6	Reduces insulin signaling and promotes gluconeogenesis [153]	Apigenin (−9.9), catechin (−9.0), chrysoeriol (−9.6), ellagic acid (−8.7), kaempferol (−9.5), luteolin (−9.9), quercetin (−9.6)

Table 3. *Cont.*

Protein Target	PDB Code	Function	Phytochemicals
		Regulation of glucose metabolism	
AKR1B1	3G5E	Catalyzes the reduction of glucose to sorbitol [154]	Apigenin (−9.1), catechin (−9.1), chrysoeriol (−9.0), ellagic acid (−8.8), hesperidin (−8.1), kaempferol (−8.6), luteolin (−9.1), quercetin (−8.8)
AMY2A	4GQR	Regulates the digestion of starch to glucose [155]	Apigenin (−8.3), catechin (−8.4), chrysoeriol (−8.5), hesperidin (−8.9), kaempferol (−8.0), luteolin (−9.4), quercetin (−9.8), rutin (−9.0)
GCK	3IMX	Phosphorylates glucose for glycolysis or synthesis of glycogen [156]	Hesperidin (−10.2), rutin (−8.6)
MGAM	3L4Y	Regulates the digestion of starch to glucose [157]	Hesperidin (−8.2), rutin (−8.3)
PDK2	4MPC	Regulates glucose oxidation through the inactivation of pyruvate dehydrogenase complex [158]	Hesperidin (−9.1), rutin (−8.1)
PYGL	3DDS	Regulates phosphorolysis of glycogen in glycogenesis [159]	Hesperidin (−8.6), rutin (−8.6)
		Regulation of lipid metabolism	
NR5A2	4DOR	Regulates the expression of genes involved in the synthesis of bile acid and cholesterol, and steroidogenesis [160]	Hesperidin (−8.4)
PPARA	3FEI	Regulates the expression of genes involved in lipid metabolism [161]	Rutin (−8.4)
PPARD	3PEQ	Regulates the expression of genes involved in fatty acid catabolism [162]	Chrysoeriol (−8.0), hesperidin (−9.1), rutin (−8.9)
PPARG	2FVJ	Regulates the expression of genes involved in adipogenesis and lipid oxidation [163]	Apigenin (−8.1), catechin (−8.2), chrysoeriol (−8.4), ellagic acid (−8.3), hesperidin (−1035), kaempferol (−8.5), luteolin (−8.2), quercetin (−8.4), rutin (−9.8)
RXPRA	1FM9	Heterodimerizes with PPARs [164]	Apigenin (−9.2), chrysoeriol (−9.3), hesperidin (−8.0), luteolin (−9.1)

Ellagic, caffeic, and ferulic acids were active phenolic acids (Table 3). The antidiabetic potential of ellagic acid is linked with five protein targets, whereas the regulation, secretion, and/or sensitivity of insulin is the main mechanism of action of ellagic acid. In vitro and in vivo studies also reported the antidiabetic effect of ellagic acid through the stimulation of insulin [165,166]. Regarding hydroxycinnamic acids, caffeic and ferulic acids modulate the functions of only one protein target. Previous studies also manifest their antidiabetic effects by modulating insulin-signaling molecules [167,168]. A quantitative structure–activity relationship analysis for hydroxycinnamic acids will be interesting as the other acids such cinnamic acid, coumaric acid, and rosmarinic acid did not bInd with protein targets.

Indisputably, results demonstrated that flavonoids are the most promising group of compounds. All studied flavonoids bInd. with protein targets related with DM. The antidiabetic potential of flavonoids is correlated with three possible mechanisms, namely the (i) regulation of insulin secretion and/or sensitivity, (ii) regulation of glucose metabolism, and (iii) regulation of lipid metabolism. DPP4, HSD11B1, AKR1B1, AMY2A, and PPARG protein targets interacted with the majority of flavonoids. Figure 2 shows hesperidin and rutin binds with 15 and 12 protein targets, respectively, whereas the other flavonoids modulate the function of 6 to 8 protein targets. It is noteworthy that both hesperidin and rutin are glycosylated flavonoids. Furthermore, a comparison between rutin and its aglycone form (quercetin) demonstrates the superiority of the glycosylated form. Jadhav and Puchchakayala (2012) also stated that rutin had a higher antidiabetic and hypoglycemic activity than quercetin in normoglycemic and streptozotocin -nicotinamide-induced diabetic rats performing blood glucose and serum lipid profile measurements [169]. Hesperidin is less distributed than rutin but is also a strong multi-target antidiabetic agent according to the computational measurements. These findings are in line with the literature; hesperidin is considered to have preventive and/or therapeutic effects against DM as it regulates glucose and lipid metabolism [170,171]. Overall, flavonoids are an interesting pool of antidiabetic compounds and further study is needed to probe the structural characteristics of flavonoids that are mainly linked with antidiabetic activity.

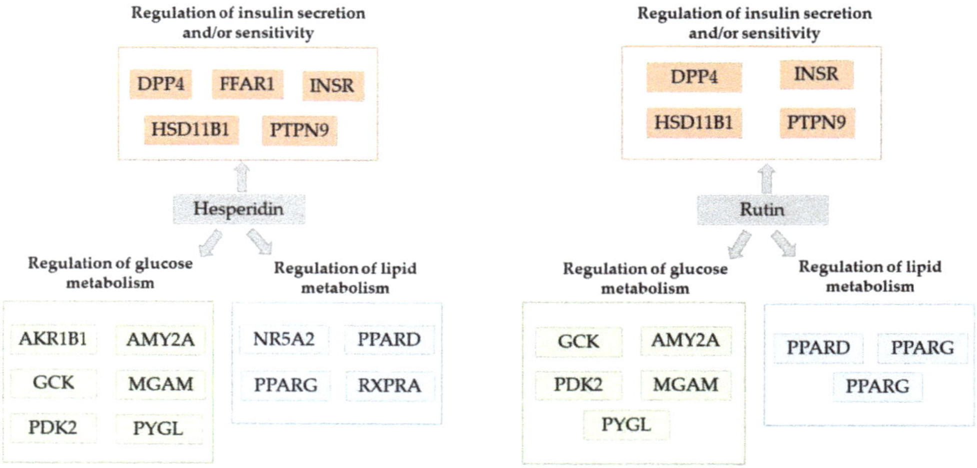

Figure 2. Interaction of hesperidin and rutin with protein targets related with diabetes mellitus as predicted with the employment of DIA-DB inverse virtual screening web server.

2.3. Evaluation of Molecular Similarity of Predicted Active Phytochemicals and Known/Experimental Antidiabetic Drugs

In the next step, the Tanimoto similarity index was used to quantify the similarity of the studied phytochemicals with 190 known or experimental antidiabetic drugs. A Tanimoto score of 0.7 or greater indicated a robust molecular similarity. According to our results, all tested phytochemicals have structural similarities with antidiabetic drugs. Although these findings are contrary to their predicted antidiabetic activity, the in-depth interpretation of results confirmed the previous results. More specifically, the highest molecular similarity was found for flavonoids. A previous study also correlated flavonoids with the antidiabetic activity of herbs and spices [2]. A strong molecular similarity with 132 and 131 antidiabetic drugs was detected for hesperidin (69.5%) and rutin (68.9%). Both flavonoids also bInd. with the highest number of protein targets according to our results. Furthermore, chrysoeriol (57.4%), luteolin (51.1%), and apigenin (50.5%) also showed some structural similarity with antidiabetic drugs. Similarly to the predicted antidiabetic activity, the glycosylated flavonoids had more structural similarities with antidiabetic drugs than aglycone flavonoids. From a chemical point of view, the most active flavonoids belong to the group of flavanones, flavanols, and flavones. In summary, the molecular similarity test also confirmed the potent antidiabetic effects of flavonoids.

3. Materials and Methods

3.1. Literature Review

The literature review on the antidiabetic potential of Mediterranean plants was performed using the databases of Scopus [172] and Google Scholar [173]. More specifically, the search included the terms "antidiabetic activity", "diabetes", "hyperglycemia", "hypoglycemic activity", "alpha-glucosidase", and "alpha-amylase" in combination with "scientific plant name" and "common plant name" of Mediterranean plants. The antidiabetic potency of several plants was searched according to a list, which was prepared based on their abundance in the Mediterranean flora and/or diet.

Subsequently, data for the phytochemical composition of plants were collected using the same databases. The terms "scientific plant name", "common plant name", "phytochemicals", "bioactive compounds", "Liquid chromatography", "Nuclear Magnetic Resonance (NMR)", and "Liquid Chromatography-Mass Spectroscopy (LC-MS)" were used to build the compound library.

3.2. Determination of Anti-Diabetic Activity Using DIA-DB Inverse Virtual Screening Web Server

At first, a simplified molecular-input line-entry system (SMILES) notation was created for each compound, which was obtained from PubChem. [174]. All SMILES notations are demonstrated in Table 3. The SMILES notation of each compound was subsequently submitted to the DIA-DB web server that employs an inverse virtual screening of compounds with Autodock Vina against a given set of 18 protein targets associated with diabetes. The 18 protein targets can be classified into the three following groups: (i) regulation of insulin secretion and/or sensitivity (DPP4, FFAR1, HSD11B1, INSR, PTPN9, RBP4), (ii) regulation of glucose metabolism (AKR1B1, AMY2A, FBP1, GCK, MGAM, PDK2, PYGL), and (iii) regulation of lipid metabolism (NR5A2, PPARA, PPARD, PPARG, RXRA) [2,175]. More specifically, the targets were aldose reductase (AKR1B1), AMY2A, FBP1, free fatty acid receptor 1 (FFAR1), glucokinase (GCK), 11B-hydroxysteroid dehydrogenase type 1 (HSD11B1), insulin receptor (INSR), MGAM, NR5A2, pyruvate dehydrogenase kinase isoform 2 (PDK2), PPARA, PPARD, PPARG, PTPN9, liver glycogen phosphorylase (PYGL), RBP4, and retinoid X receptor alpha (RXRA). A cut-off docking score of -8 kcal mol^{-1} was set to distinguish between potential active and inactive compounds.

3.3. Evaluation of Molecular Similarity of Predicted Active Phytochemicals and Known/Experimental Antidiabetic Drugs

The similarities studies with known/experimental antidiabetic drugs were performed according to a previous work [176]. The molecular similarity was performed using the metric of the Tanimoto similarity on the calculated ECFP4 molecular fingerprints of the compounds. The molecular similarity network was generated with Cytoscape and the ChemViz2 Application version 1.1.0.

4. Conclusions

The present study was undertaken to shed light on the antidiabetic properties of edible Mediterranean plants. Our results showed that the most active compounds within examined Mediterranean plants are flavonoids that are widely distributed in herbs, vegetables, medicinal plants, and fruits. Our findings also show that the glycosylation of flavonoids potentially improves its antidiabetic activity. Both the evaluation of antidiabetic activity and molecular similarity of antidiabetic dugs highlighted the antidiabetic potential of hesperidin and rutin. Finally, the present study showed that the employment of the DIA-DB inverse virtual screening web server allows us to explain the antidiabetic effects of natural products and to pinpoint the most active compounds.

Author Contributions: V.G. conceived the project and designed the experiment. A.C. performed the literature review. A.J.B.-L. and H.P.-S. carried out in silico experiments. V.G., A.J.B.-L., H.P.-S., and A.B. interpreted the results and prepared the manuscript. All authors have read and agreed to the published version of the manuscript.

Funding: This research was funded by Cyprus University Of Technology, grant number EX200146. This work was also funded by grants from the Fundación Séneca de la Región de Murcia under Project 20988/PI/18. This research was partially supported by the supercomputing infrastructure of Poznan Supercomputing Center, the e-infrastructure program of the Research Council of Norway via the supercomputer center of UiT—the Arctic University of Norway, and by the supercomputing infrastructure of the NLHPC (ECM-02), Powered@NLHPC. Finally, publication fees were covered by the Open Access Fund of Cyprus University of Technology.

Institutional Review Board Statement: Not applicable.

Informed Consent Statement: Not applicable.

Conflicts of Interest: The authors declare no conflict of interest.

References

1. World Health Organization. Estimates for the year 2000 and projections for 2030. *World Health* **2004**, *27*, 1047–1053.
2. Pereira, A.S.P.; Banegas-Luna, A.J.; Peña-García, J.; Pérez-Sánchez, H.; Apostolides, Z. Evaluation of the anti-diabetic activity of some common herbs and spices: Providing new insights with inverse virtual screening. *Molecules* **2019**, *24*, 4030. [CrossRef] [PubMed]
3. Li, W.; Yuan, G.; Pan, Y.; Wang, C.; Chen, H. Network pharmacology studies on the bioactive compounds and action mechanisms of natural products for the treatment of diabetes mellitus: A review. *Front. Pharmacol.* **2017**, *8*, 1–10. [CrossRef] [PubMed]
4. Newman, D.J.; Cragg, G.M. Natural products as sources of new drugs over the nearly four decades from 01/1981 to 09/2019. *J. Nat. Prod.* **2020**, *83*, 770–803. [CrossRef]
5. Yeung, A.W.K.; Tzvetkov, N.T.; Durazzo, A.; Lucarini, M.; Souto, E.B.; Santini, A.; Gan, R.Y.; Jozwik, A.; Grzybek, W.; Horbańczuk, J.O.; et al. Natural products in diabetes research: Quantitative literature analysis. *Nat. Prod. Res.* **2021**, *35*, 5813–5827. [CrossRef]
6. Xu, L.; Li, Y.; Dai, Y.; Peng, J. Natural products for the treatment of type 2 diabetes mellitus: Pharmacology and mechanisms. *Pharmacol. Res.* **2018**, *130*, 451–465. [CrossRef]
7. Prabhakar, P.K.; Doble, M. Mechanism of action of natural products used in the treatment of diabetes mellitus. *Chin. J. Integr. Med.* **2011**, *17*, 563–574. [CrossRef]
8. Georgoulis, M.; Kontogianni, M.D.; Yiannakouris, N. Mediterranean diet and diabetes: Prevention and treatment. *Nutrients* **2014**, *6*, 1406–1423. [CrossRef]
9. Martín-Peláez, S.; Fito, M.; Castaner, O. Mediterranean diet effects on type 2 diabetes prevention, disease progression, and related mechanisms. A review. *Nutrients* **2020**, *12*, 2236. [CrossRef]
10. Schröder, H. Protective mechanisms of the Mediterranean diet in obesity and type 2 diabetes. *J. Nutr. Biochem.* **2007**, *18*, 149–160. [CrossRef]
11. Wang, H.; Du, Y.J.; Song, H.C. α-Glucosidase and α-amylase inhibitory activities of guava leaves. *Food Chem.* **2010**, *123*, 6–13. [CrossRef]
12. Gray, A.M.; Flatt, P.R. Insulin-releasing and insulin-like activity of the traditional anti-diabetic plant *Coriandrum sativum* (coriander). *Br. J. Nutr.* **1999**, *81*, 203–209. [CrossRef] [PubMed]
13. Benhaddou-Andaloussi, A.; Martineau, L.C.; Spoor, D.; Vuong, T.; Leduc, C.; Joly, E.; Burt, A.; Meddah, B.; Settaf, A.; Arnason, J.T.; et al. Antidiabetic activity of *Nigella sativa* seed extract in cultured pancreatic β-cells, skeletal muscle cells, and adipocytes. *Pharm. Biol.* **2008**, *46*, 96–104. [CrossRef]
14. El-Houri, R.B.; Kotowska, D.; Olsen, L.C.B.; Bhattacharya, S.; Christensen, L.P.; Grevsen, K.; Oksbjerg, N.; Færgeman, N.; Kristiansen, K.; Christensen, K.B. Screening for bioactive metabolites in plant extracts modulating glucose uptake and fat accumulation. *Evid.-Based Complement. Altern. Med.* **2014**, *2014*, 156398. [CrossRef]
15. Kalhotra, P.; Chittepu, V.C.S.R.; Osorio-Revilla, G.; Gallardo-Velazquez, T. Phytochemicals in garlic extract inhibit therapeutic enzyme DPP-4 and induce skeletal muscle cell proliferation: A possible mechanism of action to benefit the treatment of diabetes mellitus. *Biomolecules* **2020**, *10*, 305. [CrossRef]
16. Perez Gutierrez, R.M. Inhibition of advanced glycation end-product formation by *Origanum majorana* L. in vitro and in streptozotocin-induced diabetic rats. *Evid.-Based Complement. Altern. Med.* **2012**, *2012*, 598638. [CrossRef]
17. Heshmati, J.; Namazi, N. Effects of black seed (*Nigella sativa*) on metabolic parameters in diabetes mellitus: A systematic review. *Complement. Ther. Med.* **2015**, *23*, 275–282. [CrossRef]
18. Benhaddou-Andaloussi, A.; Martineau, L.C.; Vallerand, D.; Haddad, Y.; Afshar, A.; Settaf, A.; Haddad, P.S. Multiple molecular targets underlie the antidiabetic effect of *Nigella sativa* seed extract in skeletal muscle, adipocyte and liver cells. *Diabetes Obes. Metab.* **2010**, *12*, 148–157. [CrossRef]
19. Kaleem, M.; Kirmani, D.; Asif, M.; Ahmed, Q.; Bano, B. Biochemical effects of *Nigella sativa* L seeds in diabetic rats. *Indian J. Exp. Biol.* **2006**, *44*, 745–748.
20. Agrawal, S.; Yallatikar, T.; Gurjar, P. Brassica Nigra: Ethopharmacological review of a routinely Used condiment. *Curr. Drug Discov. Technol.* **2018**, *16*, 40–47. [CrossRef]
21. El-Manawaty, M.A.; Gohar, L. In vitro alpha-glucosidase inhibitory activity of egyptian plant extracts as an indication for their antidiabetic activity. *Asian J. Pharm. Clin. Res* **2018**, *11*, 360–367. [CrossRef]
22. Anand, P.; Murali, K.Y.; Tandon, V.; Chandra, R.; Murthy, P.S. Preliminary studies on antihyperglycemic effect of aqueous extract of *Brassica nigra* (L.) Koch in streptozotocin induced diabetic rats. *Indian J. Exp. Biol.* **2007**, *45*, 696–701. [PubMed]
23. Hawash, M.; Jaradat, N.; Elaraj, J.; Hamdan, A.; Lebdeh, S.A.; Halawa, T. Evaluation of the hypoglycemic effect of seven wild folkloric edible plants from Palestine (Antidiabetic effect of seven plants from Palestine). *J. Complement. Integr. Med.* **2020**, *17*, 1–10. [CrossRef]
24. Thi Viet Huong, D.; Minh Giang, P.; Hoang Yen, N.; Nguyen, S.T. *Plantago major* L. extracts reduce blood glucose in streptozotocin-induced diabetic mice. *J. Chem.* **2021**, *2021*, 6688731. [CrossRef]
25. Angelica Abud, M.; Nardello, A.L.; Torti, J.F. Hypoglycemic effect due to insulin stimulation with *Plantago major* in wistar rats. *Med. Aromat. Plants* **2017**, *6*, 292. [CrossRef] [PubMed]

26. Lv, J.; Cao, L.; Li, M.; Zhang, R.; Bai, F.; Wei, P. Effect of hydroalcohol extract of lemon (*Citrus limon*) peel on a rat model of type 2 diabetes. *Trop J. Pharm. Res.* **2018**, *17*, 1367–1372. [CrossRef]
27. Sathiyabama, R.G.; Rajiv Gandhi, G.; Denadai, M.; Sridharan, G.; Jothi, G.; Sasikumar, P.; de Souza Siqueira Quintans, J.; Narain, N.; Cuevas, L.E.; Coutinho, H.D.M.; et al. Evidence of insulin-dependent signalling mechanisms produced by *Citrus sinensis* (L.) Osbeck fruit peel in an insulin resistant diabetic animal model. *Food Chem. Toxicol.* **2018**, *116*, 86–99. [CrossRef]
28. Benayad, O.; Bouhrim, M.; Tiji, S.; Kharchoufa, L.; Addi, M.; Drouet, S.; Hano, C.; Lorenzo, J.M.; Bendaha, H.; Bnouham, M.; et al. Phytochemical profile, α-glucosidase, and α-amylase inhibition potential and toxicity evaluation of extracts from *Citrus aurantium* (L) peel, a valuable by-product from Northeastern Morocco. *Biomolecules* **2021**, *11*, 1555. [CrossRef]
29. Obih, P.; Obih, J.-C.; Arome, O. Is alpha-glucosidase inhibition a nechanism of the antidiabetic action of garlic (*Allium sativum*)? *J. Biosci. Med.* **2019**, *7*, 42–49. [CrossRef]
30. Gandhi, G.R.; Vasconcelos, A.B.S.; Wu, D.T.; Li, H.B.; Antony, P.J.; Li, H.; Geng, F.; Gurgel, R.Q.; Narain, N.; Gan, R.Y. Citrus flavonoids as promising phytochemicals targeting diabetes and related complications: A systematic review of in vitro and in vivo studies. *Nutrients* **2020**, *12*, 2907. [CrossRef]
31. Andallu, B.; Ramya, V. Anti-hyperglycemic, cholesterol-lowering and HDL-raising effects of cumin (*Cuminum cyminum*) seeds in type 2 diabetes. *J. Nat. Remedies* **2007**, *7*, 142–149.
32. Srivsatava, R.; Srivastava, S.P.; Jaiswal, N.; Mishra, A.; Maurya, R.; Srivastava, A.K. Antidiabetic and antidyslipidemic activities of *Cuminum cyminum* L. in validated animal models. *Med. Chem. Res.* **2011**, *20*, 1656–1666. [CrossRef]
33. Mnif, S.; Aifa, S. Cumin (*Cuminum cyminum* L.) from traditional uses to potential biomedical applications. *Chem. Biodivers* **2015**, *12*, 733–742. [CrossRef]
34. Siow, H.L.; Gan, C.Y. Optimization study in extracting anti-oxidative and a-amylase inhibitor peptides from cumin seeds (*Cuminum Cyminum*). *J. Food Biochem.* **2017**, *41*, e12280. [CrossRef]
35. Saleem, F.; Sarkar, D.; Ankolekar, C.; Shetty, K. Phenolic bioactives and associated antioxidant and anti-hyperglycemic functions of select species of Apiaceae family targeting for type 2 diabetes relevant nutraceuticals. *Ind. Crops Prod.* **2017**, *107*, 518–525. [CrossRef]
36. Mishra, N. Haematological and hypoglycemic potential *Anethum graveolens* seeds extract in normal and diabetic Swiss albino mice. *Vet. World* **2013**, *6*, 502–507. [CrossRef]
37. Haidari, F.; Zakerkish, M.; Borazjani, F.; Ahmadi Angali, K.; Amoochi Foroushani, G. The effects of *Anethum graveolens* (dill) powder supplementation on clinical and metabolic status in patients with type 2 diabetes. *Trials* **2020**, *21*, 483. [CrossRef]
38. Eidi, A.; Eidi, M.; Esmaeili, E. Antidiabetic effect of garlic (*Allium sativum* L.) in normal and streptozotocin-induced diabetic rats. *Phytomedicine* **2006**, *13*, 624–629. [CrossRef]
39. Shabani, E.; Sayemiri, K.; Mohammadpour, M. The effect of garlic on lipid profile and glucose parameters in diabetic patients: A systematic review and meta-analysis. *Prim. Care Diabetes* **2019**, *13*, 28–42. [CrossRef]
40. Wang, J.; Zhang, X.; Lan, H.; Wang, W. Effect of garlic supplement in the management of type 2 diabetes mellitus (T2DM): A meta-analysis of randomized controlled trials. *Food Nutr. Res.* **2017**, *61*, 1377571. [CrossRef]
41. Pinent, M.; Blay, M.; Bladé, M.C.; Salvadó, M.J.; Arola, L.; Ardévol, A. Grape seed-derived procyanidins have an antihyperglycemic effect in streptozotocin-induced diabetic rats and insulinomimetic activity in insulin-sensitive cell lines. *Endocrinology* **2004**, *145*, 4985–4990. [CrossRef] [PubMed]
42. Asbaghi, O.; Nazarian, B.; Reiner, Ž.; Amirani, E.; Kolahdooz, F.; Chamani, M.; Asemi, Z. The effects of grape seed extract on glycemic control, serum lipoproteins, inflammation, and body weight: A systematic review and meta-analysis of randomized controlled trials. *Phyther. Res.* **2020**, *34*, 239–253. [CrossRef] [PubMed]
43. Bouyahya, A.; Chamkhi, I.; Benali, T.; Guaouguaou, F.E.; Balahbib, A.; El Omari, N.; Taha, D.; Belmehdi, O.; Ghokhan, Z.; El Menyiy, N. Traditional use, phytochemistry, toxicology, and pharmacology of *Origanum majorana* L. *J. Ethnopharmacol.* **2021**, *265*, 113318. [CrossRef] [PubMed]
44. Zhang, Y.; Wen, M.; Zhou, P.; Tian, M.; Zhou, J.; Zhang, L. Analysis of chemical composition in Chinese olive leaf tea by UHPLC-DAD-Q-TOF-MS/MS and GC–MS and its lipid-lowering effects on the obese mice induced by high-fat diet. *Food Res. Int.* **2020**, *128*, 108785. [CrossRef]
45. Al-Hayaly, L.; Al-Sultan, A.; Sultan, S.M.S. Effect of olive leaves extract on alloxan induced diabetes in male albino mice. In Proceedings of the 1st International Multi-Disciplinary Conference Theme: Sustainable Development and Smart Planning (IMDC-SDSP 2020), Cyberspace, 28–30 June 2020. [CrossRef]
46. Abd El-Moneim, M.R.A.; El-Beltagi, H.S.; Fayed, S.A.; El-Ansary, A.E. Enhancing effect of olive leaves extract on lipid profile and enzymes activity in streptozotocin induced diabetic rats. *Fresenius Environ. Bull.* **2018**, *27*, 1875–1883.
47. de Bock, M.; Derraik, J.G.B.; Brennan, C.M.; Biggs, J.B.; Morgan, P.E.; Hodgkinson, S.C.; Hofman, P.L.; Cutfield, W.S. Olive (*Olea europaea* L.) leaf polyphenols improve insulin sensitivity in middle-aged overweight men: A randomized, placebo-controlled, crossover trial. *PLoS ONE* **2013**, *8*, e57622. [CrossRef]
48. Abunab, H.; Dator, W.L.; Hawamdeh, S. Effect of olive leaf extract on glucose levels in diabetes-induced rats: A systematic review and meta-analysis. *J. Diabetes* **2017**, *9*, 947–957. [CrossRef]
49. Abouzed, T.K.; del Mar Contreras, M.; Sadek, K.M.; Shukry, M.; Abdelhady, D.H.; Gouda, W.M.; Abdo, W.; Nasr, N.E.; Mekky, R.H.; Segura-Carretero, A.; et al. Red onion scales ameliorated streptozotocin-induced diabetes and diabetic nephropathy in Wistar rats in relation to their metabolite fingerprint. *Diabetes Res. Clin. Pract.* **2018**, *140*, 253–264. [CrossRef]

50. Kang, M.J.; Kim, J.H.; Choi, H.N.; Kim, M.J.; Han, J.H.; Lee, J.H.; Kim, J.I. Hypoglycemic effects of Welsh onion in an animal model of diabetes mellitus. *Nutr. Res. Pract.* **2010**, *4*, 486–491. [CrossRef]
51. Lee, S.K.; Hwang, J.Y.; Kang, M.J.; Kim, Y.M.; Jung, S.H.; Lee, J.H.; Kim, J.I. Hypoglycemic effect of onion skin extract in animal models of diabetes mellitus. *Food Sci. Biotechnol.* **2008**, *17*, 130–134.
52. Islam, M.S.; Choi, H.; Loots, D.T. Effects of dietary onion (*Allium cepa* L.) in a high-fat diet streptozotocin-induced diabetes rodent model. *Ann. Nutr. Metab.* **2008**, *53*, 6–12. [CrossRef] [PubMed]
53. Yanardağ, R.; Bolkent, Ş.; Tabakoğlu-Oğuz, A.; Özsoy-Saçan, Ö. Effects of *Petroselinum crispum* extract on pancreatic B cells and blood glucose of streptozotocin-induced diabetic rats. *Biol. Pharm. Bull.* **2003**, *26*, 1206–1210. [CrossRef] [PubMed]
54. Abou Khalil, N.S.; Abou-Elhamd, A.S.; Wasfy, S.I.A.; El Mileegy, I.M.H.; Hamed, M.Y.; Ageely, H.M. Antidiabetic and antioxidant impacts of desert date (*Balanites aegyptiaca*) and Parsley (*Petroselinum sativum*) aqueous extracts: Lessons from experimental rats. *J. Diabetes Res.* **2016**, *2016*, 8408326. [CrossRef] [PubMed]
55. Cazzola, R.; Camerotto, C.; Cestaro, B. Anti-oxidant, anti-glycant, and inhibitory activity against α-amylase and α-glucosidase of selected spices and culinary herbs. *Int. J. Food Sci. Nutr.* **2011**, *62*, 175–184. [CrossRef]
56. Agatonovic-Kustrin, S.; Kustrin, E.; Gegechkori, V.; Morton, D.W. Bioassay-guided identification of α-amylase inhibitors in herbal extracts. *J. Chromatogr. A* **2020**, *1620*, 460970. [CrossRef]
57. Eidi, A.; Eidi, M. Antidiabetic effects of sage (*Salvia officinalis* L.) leaves in normal and streptozotocin-induced diabetic rats. *Diabetes Metab. Syndr Clin. Res. Rev.* **2009**, *3*, 40–44. [CrossRef]
58. Mahdi, S.; Azzi, R.; Lahfa, F.B. Evaluation of in vitro α-amylase and α-glucosidase inhibitory potential and hemolytic effect of phenolic enriched fractions of the aerial part of *Salvia officinalis* L. *Diabetes Metab. Syndr Clin. Res. Rev.* **2020**, *14*, 689–694. [CrossRef]
59. Behradmanesh, S.; Derees, F.; Kopaei, M.R. Effect of *Salvia officinalis* on diabetic patients. *J. Ren. Inj. Prev.* **2013**, *2*, 51–54. [CrossRef]
60. Chen, L.; Lin, X.; Fan, X.; Qian, Y.; Lv, Q.; Teng, H. *Sonchus oleraceus* Linn extract enhanced glucose homeostasis through the AMPK/Akt/ GSK-3β signaling pathway in diabetic liver and HepG2 cell culture. *Food Chem. Toxicol.* **2020**, *136*, 111072. [CrossRef] [PubMed]
61. Chen, L.; Fan, X.; Lin, X.; Qian, L.; Zengin, G.; Delmas, D.; Paoli, P.; Teng, H.; Xiao, J. Phenolic extract from *Sonchus oleraceus* L. protects diabetes-related liver injury in rats through TLR4/NF-κB signaling pathway. *eFood* **2019**, *1*, 77. [CrossRef]
62. Kim, H.Y.; Lim, S.H.; Park, Y.H.; Ham, H.J.; Lee, K.J.; Park, D.S.; Kim, K.H.; Kim, S. Screening of α-amylase, α-glucosidase and lipase inhibitory activity with Gangwon-do wild plants extracts. *J. Korean Soc Food Sci. Nutr.* **2011**, *40*, 308–315. [CrossRef]
63. Teugwa, C.M.; Mejiato, P.C.; Zofou, D.; Tchinda, B.T.; Boyom, F.F. Antioxidant and antidiabetic profiles of two African medicinal plants: *Picralima nitida* (Apocynaceae) and *Sonchus oleraceus* (Asteraceae). *BMC Complement. Altern Med.* **2013**, *13*, 175. [CrossRef] [PubMed]
64. El-Hawary, S.S.; Mohammed, R.; El-Din, M.E.; Hassan, H.M.; Ali, Z.Y.; Rateb, M.E.; Bellah El Naggar, E.M.; Othman, E.M.; Abdelmohsen, U.R. Comparative phytochemical analysis of five Egyptian strawberry cultivars (*Fragaria* × *ananassa* Duch.) and antidiabetic potential of Festival and Red Merlin cultivars. *RSC Adv.* **2021**, *11*, 16755–16767. [CrossRef] [PubMed]
65. Mandave, P.; Khadke, S.; Karandikar, M.; Pandit, V.; Ranjekar, P.; Kuvalekar, A.; Mantri, N. Antidiabetic, lipid normalizing, and nephroprotective actions of the strawberry: A potent supplementary fruit. *Int. J. Mol. Sci.* **2017**, *18*, 124. [CrossRef] [PubMed]
66. Da Silva Pinto, M.; de Carvalho, J.E.; Lajolo, F.M.; Genovese, M.I.; Shetty, K. Evaluation of antiproliferative, anti-type 2 diabetes, and antihypertension potentials of ellagitannins from strawberries (*Fragaria* × *ananassa* Duch.) Using In Vitro Models. *J. Med. Food* **2010**, *13*, 1027–1035. [CrossRef] [PubMed]
67. Takács, I.; Szekeres, A.; Takács, Á.; Rakk, D.; Mézes, M.; Polyák, Á.; Lakatos, L.; Gyémánt, G.; Csupor, D.; Kovács, K.J.; et al. Wild strawberry, blackberry, and blueberry leaf extracts alleviate starch-induced hyperglycemia in prediabetic and diabetic mice. *Planta Med.* **2020**, *86*, 790–799. [CrossRef]
68. Kemertelidze, É.P.; Sagareishvili, T.G.; Syrov, V.N.; Khushbaktova, Z.A. Chemical composition and pharmacological activity of garden savory (*Satureja hortensis* L.) occurring in Georgia. *Pharm. Chem. J.* **2004**, *38*, 319–322. [CrossRef]
69. Saravanan, S.; Pari, L. Protective effect of thymol on high fat diet induced diabetic nephropathy in C57BL/6J mice. *Chem. Biol. Interact.* **2016**, *245*, 1–11. [CrossRef]
70. Deligiannidou, G.-E.; Philippou, E.; Vidakovic, M.; Berghe, W.V.; Heraclides, A.; Grdovic, N.; Mihailovic, M.; Kontogiorgis, C. Natural products derived from the Mediterranean diet with antidiabetic Activity: From insulin mimetic hypoglycemic to nutriepigenetic modulator compounds. *Curr. Pharm. Des.* **2019**, *25*, 1760–1782. [CrossRef]
71. Alu'datt, M.H.; Rababah, T.; Johargy, A.; Gammoh, S.; Ereifej, K.; Alhamad, M.N.; Brewer, M.S.; Saati, A.A.; Kubow, S.; Rawshdeh, M. Extraction, optimisation and characterisation of phenolics from *Thymus vulgaris* L.: Phenolic content and profiles in relation to antioxidant, antidiabetic and antihypertensive properties. *Int. J. Food Sci. Technol.* **2016**, *51*, 720–730. [CrossRef]
72. Yazdanparast, R.; Ardestani, A.; Jamshidi, S. Experimental diabetes treated with *Achillea santolina*: Effect on pancreatic oxidative parameters. *J. Ethnopharmacol.* **2007**, *112*, 13–18. [CrossRef] [PubMed]
73. Hamdan, I.I.; Afifi, F.U. Capillary electrophoresis as a screening tool for alpha amylase inhibitors in plant extracts. *Saudi Pharm. J.* **2010**, *18*, 91–95. [CrossRef] [PubMed]
74. Al-Snafi, A.E. Chemical constituents and pharmacological activities of milfoil (*Achillea santolina*). A review. *Int. J. PharmTech Res.* **2013**, *5*, 1373–1377.

75. Đorđević, B.S.; Todorović, Z.B.; Troter, D.Z.; Stanojević, L.P.; Stojanović, G.S.; Đalović, I.G.; Mitrović, P.M.; Veljković, V.B. Extraction of phenolic compounds from black mustard (*Brassica nigra* L.) seed by deep eutectic solvents. *J. Food Meas. Charact.* **2021**, *15*, 1931–1938. [CrossRef]
76. Petrović, M.; Jovanović, M.; Lević, S.; Nedović, V.; Mitić-Ćulafić, D.; Semren, T.Ž.; Veljović, S. Valorization potential of *Plantago major* L. solid waste remaining after industrial tincture production: Insight into the chemical composition and bioactive properties. *Waste Biomass Valoriz.* **2022**, *13*, 1639–1651. [CrossRef]
77. Singh, B.; Singh, J.P.; Kaur, A.; Singh, N. Phenolic composition, antioxidant potential and health benefits of citrus peel. *Food Res. Int.* **2020**, *132*, 109114. [CrossRef]
78. Demir, S.; Korukluoglu, M. A comparative study about antioxidant activity and phenolic composition of cumin (*Cuminum cyminum* L.) and coriander (*Coriandrum sativum* L.). *Indian J. Tradit. Knowl.* **2020**, *19*, 383–393.
79. Wasli, H.; Jelali, N.; Silva, A.M.S.; Ksouri, R.; Cardoso, S.M. Variation of polyphenolic composition, antioxidants and physiological characteristics of dill (*Anethum graveolens* L.) as affected by bicarbonate-induced iron deficiency conditions. *Ind. Crops Prod.* **2018**, *126*, 466–476. [CrossRef]
80. Beato, V.M.; Orgaz, F.; Mansilla, F.; Montaño, A. Changes in phenolic compounds in garlic (*Allium sativum* L.) owing to the cultivar and location of growth. *Plant. Foods Hum. Nutr.* **2011**, *66*, 218–223. [CrossRef]
81. Nile, S.H.; Kim, S.H.; Ko, E.Y.; Park, S.W. Polyphenolic contents and antioxidant properties of different grape (*V. vinifera*, *V. labrusca*, and *V. hybrid*) cultivars. *Biomed. Res. Int.* **2013**, *2013*, 718065. [CrossRef]
82. Talhaoui, N.; Taamalli, A.; Gómez-Caravaca, A.M.; Fernández-Gutiérrez, A.; Segura-Carretero, A. Phenolic compounds in olive leaves: Analytical determination, biotic and abiotic influence, and health benefits. *Food Res. Int.* **2015**, *77*, 92–108. [CrossRef]
83. Wang, H.; Provan, G.J.; Helliwell, K. Determination of rosmarinic acid and caffeic acid in aromatic herbs by HPLC. *Food Chem.* **2004**, *87*, 307–311. [CrossRef]
84. Ferreira, F.S.; de Oliveira, V.S.; Chávez, D.W.H.; Chaves, D.S.; Riger, C.J.; Sawaya, A.C.H.F.; Guizellini, G.M.; Sampaio, G.R.; da Silva Torres, E.A.F.; Saldanha, T. Bioactive compounds of parsley (*Petroselinum crispum*), chives (*Allium schoenoprasum* L) and their mixture (Brazilian cheiro-verde) as promising antioxidant and anti-cholesterol oxidation agents in a food system. *Food Res. Int.* **2022**, *151*, 110864. [CrossRef] [PubMed]
85. Afshari, M.; Rahimmalek, M.; Miroliaei, M. Variation in polyphenolic profiles, antioxidant and antimicrobial activity of different Achillea species as natural sources of antiglycative compounds. *Chem. Biodivers.* **2018**, *15*, e1800075. [CrossRef] [PubMed]
86. Nour, V.; Trandafir, I.; Cosmulescu, S. Bioactive compounds, antioxidant activity and nutritional quality of different culinary aromatic herbs. *Not. Bot Horti Agrobot Cluj-Napoca* **2017**, *45*, 179–184. [CrossRef]
87. Roby, M.H.H.; Sarhan, M.A.; Selim, K.A.H.; Khalel, K.I. Evaluation of antioxidant activity, total phenols and phenolic compounds in thyme (*Thymus vulgaris* L.), sage (*Salvia officinalis* L.), and marjoram (*Origanum majorana* L.) extracts. *Ind. Crops Prod.* **2013**, *43*, 827–831. [CrossRef]
88. Kobeasy, I.; Abdel-Fatah, M.; Abd El-Salam, S.M.; El-Ola Mohamed, Z.M. Biochemical studies on *Plantago major* L. and *Cyamopsis tetragonoloba* L. *Int. J. Biodivers. Conserv.* **2011**, *3*, 83–91.
89. Mohanty, S.; Maurya, A.K.; Jyotshna; Saxena, A.; Shanker, K.; Pal, A.; Bawankule, D.U. Flavonoids rich fraction of *Citrus limetta* fruit peels reduces proinflammatory cytokine production and attenuates Malaria pathogenesis. *Curr. Pharm. Biotechnol.* **2015**, *16*, 544–552. [CrossRef]
90. Sandhu, A.K.; Gu, L. Antioxidant capacity, phenolic content, and profiling of phenolic compounds in the seeds, skin, and pulp of *Vitis rotundifolia* (Muscadine Grapes) as determined by HPLC-DAD-ESI-MSn. *J. Agric. Food Chem.* **2010**, *58*, 4681–4692. [CrossRef]
91. Saleh, A.M.; Selim, S.; Al Jaouni, S.; AbdElgawad, H. CO_2 enrichment can enhance the nutritional and health benefits of parsley (*Petroselinum crispum* L.) and dill (*Anethum graveolens* L.). *Food Chem.* **2018**, *269*, 519–526. [CrossRef]
92. Williner, M.R.; Pirovani, M.E.; Güemes, D.R. Ellagic acid content in strawberries of different cultivars and ripening stages. *J. Sci. Food Agric.* **2003**, *83*, 842–845. [CrossRef]
93. Bourgou, S.; Ksouri, R.; Bellila, A.; Skandrani, I.; Falleh, H.; Marzouk, B. Phenolic composition and biological activities of Tunisian *Nigella sativa* L. shoots and roots. *Comptes Rendus Biol.* **2008**, *331*, 48–55. [CrossRef]
94. Prakash, D.; Singh, B.N.; Upadhyay, G. Antioxidant and free radical scavenging activities of phenols from onion (*Allium cepa*). *Food Chem.* **2007**, *102*, 1389–1393. [CrossRef]
95. Barbieri, J.B.; Goltz, C.; Batistão Cavalheiro, F.; Theodoro Toci, A.; Igarashi-Mafra, L.; Mafra, M.R. Deep eutectic solvents applied in the extraction and stabilization of rosemary (*Rosmarinus officinalis* L.) phenolic compounds. *Ind. Crops Prod.* **2020**, *144*, 112049. [CrossRef]
96. Russell, W.R.; Scobbie, L.; Labat, A.; Duthie, G.G. Selective bio-availability of phenolic acids from Scottish strawberries. *Mol. Nutr. Food Res.* **2009**, *53*, 85–91. [CrossRef]
97. Rebey, I.B.; Zakhama, N.; Karoui, I.J.; Marzouk, B. Polyphenol composition and antioxidant activity of cumin (*Cuminum Cyminum* L.) seed extract under drought. *J. Food Sci.* **2012**, *77*, C734–C739. [CrossRef]
98. Fratianni, F.; Ombra, M.N.; Cozzolino, A.; Riccardi, R.; Spigno, P.; Tremonte, P.; Coppola, R.; Nazzaro, F. Phenolic constituents, antioxidant, antimicrobial and anti-proliferative activities of different endemic Italian varieties of garlic (*Allium sativum* L.). *J. Funct. Foods* **2016**, *21*, 240–248. [CrossRef]

99. Obreque-Slier, E.; Herrera-Bustamante, B.; López-Solís, R. Ripening-associated flattening out of inter-varietal differences in some groups of phenolic compounds in the skins of six emblematic grape wine varieties. *J. Food Compos. Anal.* **2021**, *99*, 103858. [CrossRef]
100. Yilmaz, Y.; Toledo, R.T. Major Flavonoids in grape seeds and skins: Antioxidant capacity of catechin, epicatechin, and gallic acid. *J. Agric. Food Chem.* **2004**, *52*, 255–260. [CrossRef]
101. Mahmood, T.; Anwar, F.; Abbas, M.; Saari, N. Effect of maturity on phenolics (Phenolic acids and flavonoids) profile of strawberry cultivars and mulberry species from Pakistan. *Int. J. Mol. Sci.* **2012**, *13*, 4591–4607. [CrossRef]
102. Erdogan Orhan, I.; Senol, F.S.; Ozturk, N.; Celik, S.A.; Pulur, A.; Kan, Y. Phytochemical contents and enzyme inhibitory and antioxidant properties of *Anethum graveolens* L. (dill) samples cultivated under organic and conventional agricultural conditions. *Food Chem. Toxicol.* **2013**, *59*, 96–103. [CrossRef] [PubMed]
103. Tepe, B.; Sokmen, A. Production and optimisation of rosmarinic acid by *Satureja hortensis* L. callus cultures. *Nat. Prod. Res.* **2007**, *21*, 1133–1144. [CrossRef] [PubMed]
104. Poureini, F.; Mohammadi, M.; Najafpour, G.D.; Nikzad, M. Comparative study on the extraction of apigenin from parsley leaves (*Petroselinum crispum* L.) by ultrasonic and microwave methods. *Chem. Pap.* **2020**, *74*, 3857–3871. [CrossRef]
105. Singh, B.; Singh, J.P.; Kaur, A.; Yadav, M.P. Insights into the chemical composition and bioactivities of citrus peel essential oils. *Food Res. Int.* **2021**, *143*, 110231. [CrossRef] [PubMed]
106. Alrekabi, D.G.; Hamad, M.N. Phytochemical investigation of *Sonchus oleraceus* (Family:Asteraceae) cultivated in Iraq, isolation and identification of quercetin and apigenin. *J. Pharm. Sci. Res.* **2018**, *10*, 2242–2248.
107. Boroja, T.; Katanić, J.; Rosić, G.; Selaković, D.; Joksimović, J.; Mišić, D.; Stanković, V.; Jovičić, N.; Mihailović, V. Summer savory (Satureja hortensis L.) extract: Phytochemical profile and modulation of cisplatin-induced liver, renal and testicular toxicity. *Food Chem. Toxicol.* **2018**, *118*, 252–263. [CrossRef]
108. Lee, Y.H.; Choo, C.; Waisundara, V.Y. Determination of the total antioxidant capacity and quantification of phenolic compounds of different solvent extracts of black mustard seeds (*Brassica nigra*). *Int. J. Food Prop.* **2015**, *18*, 2500–2507. [CrossRef]
109. Indah, K.P.; Menara, B.; Mpaj, P.; Utama, J.P.; Lumpur, K. GC-MS Analysis of various extracts from leaf of *Plantago major* used as traditional medicine. *World Appl. Sci. J.* **2012**, *17*, 67–70.
110. Insani, E.M.; Cavagnaro, P.F.; Salomón, V.M.; Langman, L.; Sance, M.; Pazos, A.A.; Carrari, F.O.; Filippini, O.; Vignera, L.; Galmarini, C.R. Variation for health-enhancing compounds and traits in onion (*Allium cepa* L.) germplasm. *Food Nutr. Sci.* **2016**, *7*, 577–591. [CrossRef]
111. Kelebek, H.; Selli, S. Characterization of phenolic compounds in strawberry fruits by RP-HPLC-DAD and investigation of their antioxidant capacity. *J. Liq. Chromatogr. Relat. Technol.* **2011**, *34*, 2495–2504. [CrossRef]
112. Soliman, M.A.; Galal, T.M.; Naim, M.A.; Khalafallah, A.A. Seasonal variation in the secondary metabolites and antimicrobial activity of *Plantago major* L. from Egyptian heterogenic habitats. *Egypt J. Bot.* **2022**, *62*, 255–273. [CrossRef]
113. El-Sayed, M.A.; Al-Gendy, A.A.; Hamdan, D.I.; El-Shazly, A.M. Phytoconstituents, LC-ESI-MS profile, antioxidant and antimicrobial activities of *Citrus* x *limon* L. Burm. F. cultivar variegated pink lemon. *J. Pharm. Sci. Res.* **2017**, *9*, 375–391.
114. Oganesyan, E.T.; Nersesyan, Z.M.; Parkhomenko, A.Y. Chemical composition of the above-ground part of *Coriandrum sativum*. *Pharm. Chem. J.* **2007**, *41*, 149–153. [CrossRef]
115. Pieroni, A.; Heimler, D.; Pieters, L.; van Poel, B.; Vlietinck, A.J. In vitro anti-complementary activity of flavonoids from olive (*Olea europaea* L.) leaves. *Pharmazie* **1996**, *51*, 765–768.
116. Zeng, X.; Shi, J.; Zhao, M.; Chen, Q.; Wang, L.; Jiang, H.; Luo, F.; Zhu, L.; Lu, L.; Wang, X.; et al. Regioselective glucuronidation of diosmetin and chrysoeriol by the interplay of glucuronidation and transport in UGT1A9-Overexpressing hela cells. *PLoS ONE* **2016**, *11*, e0166239. [CrossRef]
117. Nie, J.Y.; Li, R.; Wang, Y.; Tan, J.; Tang, S.H.; Jiang, Z.T. Antioxidant activity evaluation of rosemary ethanol extract and their cellular antioxidant activity toward HeLa cells. *J. Food Biochem.* **2019**, *43*, e12851. [CrossRef]
118. Mustafa, A.M.; Angeloni, S.; Abouelenein, D.; Acquaticci, L.; Xiao, J.; Sagratini, G.; Maggi, F.; Vittori, S.; Caprioli, G. A new HPLC-MS/MS method for the simultaneous determination of 36 polyphenols in blueberry, strawberry and their commercial products and determination of antioxidant activity. *Food Chem.* **2022**, *367*, 130743. [CrossRef]
119. Chkhikvishvili, I.; Sanikidze, T.; Gogia, N.; McHedlishvili, T.; Enukidze, M.; Machavariani, M.; Vinokur, Y.; Rodov, V. Rosmarinic acid-rich extracts of summer savory (*Satureja hortensis* L.) protect jurkat T cells against oxidative stress. *Oxid. Med. Cell Longev.* **2013**, *2013*, 456253. [CrossRef]
120. Chen, Y.; Wen, J.; Deng, Z.; Pan, X.; Xie, X.; Peng, C. Effective utilization of food wastes: Bioactivity of grape seed extraction and its application in food industry. *J. Funct. Foods* **2020**, *73*, 104113. [CrossRef]
121. Stan, M.; Soran, M.L.; Varodi, C.; Lung, I. Extraction and identification of flavonoids from parsley extracts by HPLC analysis. *AIP Conf. Proc.* **2012**, *1425*, 50–52. [CrossRef]
122. Vallverdú-Queralt, A.; Regueiro, J.; Martínez-Huélamo, M.; Rinaldi Alvarenga, J.F.; Leal, L.N.; Lamuela-Raventos, R.M. A comprehensive study on the phenolic profile of widely used culinary herbs and spices: Rosemary, thyme, oregano, cinnamon, cumin and bay. *Food Chem.* **2014**, *154*, 299–307. [CrossRef]
123. Mašković, P.; Veličković, V.; Mitić, M.; Đurović, S.; Zeković, Z.; Radojković, M.; Cvetanović, A.; Švarc-Gajić, J.; Vujić, J. Summer savory extracts prepared by novel extraction methods resulted in enhanced biological activity. *Ind. Crops Prod.* **2017**, *109*, 875–881. [CrossRef]

124. Wojdyło, A.; Oszmiański, J.; Czemerys, R. Antioxidant activity and phenolic compounds in 32 selected herbs. *Food Chem.* **2007**, *105*, 940–949. [CrossRef]
125. Hasan, F.; Aziz, S.A.; Melati, M. Leaf and flavonoid production of perennial sow-thistle (*Sonchus arvensis* L.) at different growth stages. *Int. J. Biosci.* **2017**, *10*, 147–155. [CrossRef]
126. Świąder, K.; Hallmann, E.; Piotrowska, A. The effect of organic practices on the bioactive compounds content in strawberry fruits. *J. Res. Appl. Agric. Eng.* **2016**, *61*, 176–179.
127. Kulišić, T.; Kriško, A.; Dragovic-Uzelać, V.; Miloš, M.; Pifat, G. The effects of essential oils and aqueous tea infusions of oregano (*Origanum vulgare* L. spp. hirtum), thyme (*Thymus vulgaris* L.) and wild thyme (*Thymus serpyllum* L.) on the copper-induced oxidation of human low-density lipoproteins. *Int. J. Food Sci. Nutr.* **2007**, *58*, 87–93. [CrossRef] [PubMed]
128. Bianchin, M.; Pereira, D.; de Florio Almeida, J.; de Moura, C.; Pinheiro, R.S.; Heldt, L.F.S.; Haminiuk, C.W.I.; Carpes, S.T. Antioxidant properties of lyophilized rosemary and sage extracts and its effect to prevent lipid oxidation in poultry pátê. *Molecules* **2020**, *25*, 5160. [CrossRef]
129. Herrera, M.C.; Luque De Castro, M.D. Ultrasound-assisted extraction for the analysis of phenolic compounds in strawberries. *Anal. Bioanal. Chem.* **2004**, *379*, 1106–1112. [CrossRef]
130. Mocan, A.; Babotă, M.; Pop, A.; Fizeșan, I.; Diuzheva, A.; Locatelli, M.; Carradori, S.; Campestre, C.; Menghini, L.; Sisea, C.R.; et al. Chemical constituents and biologic activities of sage species: A comparison between *Salvia officinalis* L., *S. glutinosa* L and *S. transsylvanica* (schur ex griseb. & schenk) schur. *Antioxidants* **2020**, *9*, 480. [CrossRef]
131. Al-Maqtari, M.A.A.; Alghalibi, S.M.; Alhamzy, E.H. Chemical composition and antimicrobial activity of essential oil of *Thymus vulgaris* from yemen. *Turkish J. Biochem.* **2011**, *36*, 342–349.
132. Lakwani, M.A.S.; Kenanoğlu, O.N.; Taştan, Y.; Bilen, S. Effects of black mustard (*Brassica nigra*) seed oil on growth performance, digestive enzyme activities and immune responses in rainbow trout (*Oncorhynchus mykiss*). *Aquac. Res.* **2022**, *53*, 300–313. [CrossRef]
133. Lasram, S.; Zemni, H.; Hamdi, Z.; Chenenaoui, S.; Houissa, H.; Saidani Tounsi, M.; Ghorbel, A. Antifungal and antiaflatoxinogenic activities of *Carum carvi* L., *Coriandrum sativum* L. seed essential oils and their major terpene component against *Aspergillus flavus*. *Ind. Crops Prod.* **2019**, *134*, 11–18. [CrossRef]
134. Ben Miri, Y.; Djenane, D. Evaluation of protective impact of algerian *Cuminum cyminum* L. and *Coriandrum sativum* L. essential oils on *Aspergillus flavus* growth and aflatoxin B1 production. *Pakistan J. Biol. Sci.* **2018**, *21*, 67–77. [CrossRef] [PubMed]
135. Kaur, V.; Kaur, R.; Bhardwaj, U. A review on dill essential oil and its chief compounds as natural biocide. *Flavour Fragr. J.* **2021**, *36*, 412–431. [CrossRef]
136. Wu, Y.; Duan, S.; Zhao, L.; Gao, Z.; Luo, M.; Song, S.; Xu, W.; Zhang, C.; Ma, C.; Wang, S. Aroma characterization based on aromatic series analysis in table grapes. *Sci. Rep.* **2016**, *6*, 31116. [CrossRef]
137. Konoz, E.; Abbasi, A.; Moazeni, R.S.; Parastar, H.; Jalali-Heravi, M. Chemometrics-assisted gas chromatographic-mass spectrometric analysis of volatile components of olive leaf oil. *J. Iran. Chem. Soc.* **2013**, *10*, 169–179. [CrossRef]
138. Woo, J.-H.; Mok, M.-G.; Han, K.-W.; Lee, S.-Y.; Park, K.-W. Aroma components and Aatioxidant activities of pure rosemary essential oil goods produced in different countries. *Korean J. Hortic Sci. Technol.* **2010**, *28*, 696–700.
139. Skubij, N.; Dzida, K. Essential oil composition of summer savory (*Satureja hortensis* L.) cv. Saturn depending on nitrogen nutrition and plant development phases in raw material cultivated for industrial use. *Ind. Crops Prod.* **2019**, *135*, 260–270. [CrossRef]
140. Dubey, P.N.; Saxena, S.N.; Mishra, B.K.; Solanki, R.K.; Vishal, M.K.; Singh, B.; Sharma, L.K.; John, S.; Agarwal, D.; Yogi, A. Preponderance of cumin (*Cuminum cyminum* L.) essential oil constituents across cumin growing agro-ecological sub regions, India. *Ind. Crops Prod.* **2017**, *95*, 50–59. [CrossRef]
141. Craft, J.D.; Setzer, W.N. The volatile components of parsley, *Petroselinum crispum* (Mill.) Fuss. *Am. J. Essent. Oils Nat. Prod.* **2017**, *5*, 27–32.
142. Jiang, Y.; Wu, N.; Fu, Y.J.; Wang, W.; Luo, M.; Zhao, C.J.; Zu, Y.G.; Liu, X.L. Chemical composition and antimicrobial activity of the essential oil of Rosemary. *Environ. Toxicol. Pharmacol.* **2011**, *32*, 63–68. [CrossRef] [PubMed]
143. Çalişkan, T.; Maral, H.; Pala, C.; Kafkas, E.; Kirici, S. Morphogenetic variation for essential oil content and composition of sage (*Salvia officinalis* L.) In Çukurova condition introduction. *Arab. J. Med. Aromat. Plants* **2019**, *5*, 32–38
144. Mohtashami, S.; Rowshan, V.; Tabrizi, L.; Babalar, M.; Ghani, A. Summer savory (*Satureja hortensis* L.) essential oil constituent oscillation at different storage conditions. *Ind. Crops Prod.* **2018**, *111*, 226–231. [CrossRef]
145. Hudaib, M.; Speroni, E.; Di Pietra, A.M.; Cavrini, V. GC/MS evaluation of thyme (*Thymus vulgaris* L.) oil composition and variations during the vegetative cycle. *J. Pharm. Biomed. Anal.* **2002**, *29*, 691–700. [CrossRef]
146. El-Shazly, A.M.; Hafez, S.S.; Wink, M. Comparative study of the essential oils and extracts of *Achillea fragrantissima* (Forssk.) Sch. Bip. and *Achillea santolina* L. (Asteraceae) from Egypt. *Pharmazie* **2004**, *59*, 226–230.
147. Palmieri, S.; Pellegrini, M.; Ricci, A.; Compagnone, D.; lo Sterzo, C. Chemical composition and antioxidant activity of thyme, hemp and coriander extracts: A comparison study of maceration, soxhlet, UAE and RSLDE techniques. *Foods* **2020**, *9*, 1221. [CrossRef]
148. Patel, B.D.; Bhadada, S.V.; Ghate, M.D. Design, synthesis and anti-diabetic activity of triazolotriazine derivatives as dipeptidyl peptidase-4 (DPP-4) inhibitors. *Bioorg. Chem.* **2017**, *72*, 345–358. [CrossRef]

149. Governa, P.; Caroleo, M.C.; Carullo, G.; Aiello, F.; Cione, E.; Manetti, F. FFAR1/GPR40: One target, different binding sites, many agonists, no drugs, but a continuous and unprofitable tug-of-war between ligand lipophilicity, activity, and toxicity. *Bioorg. Med. Chem. Lett.* **2021**, *41*, 127969. [CrossRef]
150. Nguyen Vo, T.H.; Tran, N.; Nguyen, D.; Le, L. An in silico study on antidiabetic activity of bioactive compounds in *Euphorbia thymifolia* Linn. *Springerplus* **2016**, *5*, 1359. [CrossRef]
151. Kong, W.J.; Zhang, H.; Song, D.Q.; Xue, R.; Zhao, W.; Wei, J.; Wang, Y.M.; Shan, N.; Zhou, Z.X.; Yang, P.; et al. Berberine reduces insulin resistance through protein kinase C-dependent up-regulation of insulin receptor expression. *Metabolism* **2009**, *58*, 109–119. [CrossRef]
152. Sharma, C.; Kim, Y.; Ahn, D.; Chung, S.J. Protein tyrosine phosphatases (PTPs) in diabetes: Causes and therapeutic opportunities. *Arch. Pharm. Res.* **2021**, *44*, 310–321. [CrossRef] [PubMed]
153. Liu, J.; Song, C.; Nie, C.; Sun, Y.; Wang, Y.; Xue, L.; Fan, M.; Qian, H.; Wang, L.; Li, Y. A novel regulatory mechanism of geniposide for improving glucose homeostasis mediated by circulating RBP4. *Phytomedicine* **2022**, *95*, 153862. [CrossRef]
154. Reddy, A.B.M.; Tammali, R.; Mishra, R.; Srivastava, S.; Srivastava, S.K.; Ramana, K.V. Aldose reductase deficiency protects sugar-induced lens opacification in rats. *Chem. Biol. Interact.* **2011**, *191*, 346–350. [CrossRef] [PubMed]
155. Usher, C.L.; Handsaker, R.E.; Esko, T.; Tuke, M.A.; Weedon, M.N.; Hastie, A.R.; Cao, H.; Moon, J.E.; Kashin, S.; Fuchsberger, C.; et al. Structural forms of the human amylase locus and their relationships to SNPs, haplotypes and obesity. *Nat. Genet.* **2015**, *47*, 921–925. [CrossRef] [PubMed]
156. Petersen, M.C.; Vatner, D.F.; Shulman, G.I. Regulation of hepatic glucose metabolism in health and disease. *Nat. Rev. Endocrinol.* **2017**, *13*, 572–587. [CrossRef] [PubMed]
157. Lee, B.H.; Eskandari, R.; Jones, K.; Reddy, K.R.; Quezada-Calvillo, R.; Nichols, B.L.; Rose, D.R.; Hamaker, B.R.; Pinto, B.M. Modulation of starch digestion for slow glucose release through "Toggling" of activities of mucosal α-glucosidases. *J. Biol. Chem.* **2012**, *287*, 31929–31938. [CrossRef]
158. Sugden, M.C.; Holness, M.J. Recent advances in mechanisms regulating glucose oxidation at the level of the pyruvate dehydrogenase complex by PDKs. *Am. J. Physiol. Endocrinol. Metab.* **2003**, *284*, E855–E862. [CrossRef]
159. Zois, C.E.; Favaro, E.; Harris, A.L. Glycogen metabolism in cancer. *Biochem. Pharmacol.* **2014**, *92*, 3–11. [CrossRef]
160. Fayard, E.; Auwerx, J.; Schoonjans, K. LRH-1: An orphan nuclear receptor involved in development, metabolism and steroidogenesis. *Trends Cell Biol.* **2004**, *14*, 250–260. [CrossRef]
161. Bionaz, M.; Thering, B.J.; Loor, J.J. Fine metabolic regulation in ruminants via nutrient-gene interactions: Saturated long-chain fatty acids increase expression of genes involved in lipid metabolism and immune response partly through PPAR-α activation. *Br. J. Nutr.* **2012**, *107*, 179–191. [CrossRef]
162. Luquet, S.; Lopez-Soriano, J.; Holst, D.; Gaudel, C.; Jehl-Pietri, C.; Fredenrich, A.; Grimaldi, P.A. Roles of peroxisome proliferator-activated receptor delta (PPARδ) in the control of fatty acid catabolism. A new target for the treatment of metabolic syndrome. *Biochimie* **2004**, *86*, 833–837. [CrossRef] [PubMed]
163. Janani, C.; Ranjitha Kumari, B.D. PPAR gamma gene - A review. *Diabetes Metab. Syndr Clin. Res. Rev.* **2015**, *9*, 46–50. [CrossRef] [PubMed]
164. Grzegorzewska, A.E.; Niepolski, L.; Mostowska, A.; Warchoł, W.; Jagodziński, P.P. Involvement of adropin and adropin-associated genes in metabolic abnormalities of hemodialysis patients. *Life Sci.* **2016**, *160*, 41–46. [CrossRef] [PubMed]
165. Harakeh, S.; Almuhayawi, M.; Al Jaouni, S.; Almasaudi, S.; Hassan, S.; Al Amri, T.; Azhar, N.; Abd-Allah, E.; Ali, S.; El-Shitany, N.; et al. Antidiabetic effects of novel ellagic acid nanoformulation: Insulin-secreting and anti-apoptosis effects. *Saudi J. Biol. Sci.* **2020**, *27*, 3474–3480. [CrossRef]
166. Fatima, N.; Hafizur, R.M.; Hameed, A.; Ahmed, S.; Nisar, M.; Kabir, N. Ellagic acid in *Emblica officinalis* exerts anti-diabetic activity through the action on β-cells of pancreas. *Eur. J. Nutr.* **2017**, *56*, 591–601. [CrossRef]
167. Huang, D.A.W.; Shen, S.C.; Swi-Bea Wu, J. Effects of caffeic acid and cinnamic acid on glucose uptake in insulin-resistant mouse hepatocytes. *J. Agric. Food Chem.* **2009**, *57*, 7687–7692. [CrossRef]
168. Narasimhan, A.; Chinnaiyan, M.; Karundevi, B. Insulin-signalling molecules in the liver of high-fat diet and fructose-induced type-2 diabetic adult male rat. *Appl. Physiol. Nutr. Metab.* **2015**, *40*, 769–781. [CrossRef]
169. Jadhav, R.; Puchchakayala, G. Hypoglycemic and antidiabetic activity of flavonoids: Boswellic acid, ellagic acid, quercetin, rutin on streptozotocin-nicotinamide iduced type 2 diabetic rats. *Int. J. Pharm. Pharm. Sci* **2012**, *4*, 2–7.
170. Xiong, H.; Wang, J.; Ran, Q.; Lou, G.; Peng, C.; Gan, Q.; Hu, J.; Sun, J.; Yao, R.; Huang, Q. Hesperidin: A therapeutic agent for obesity. *Drug Des. Devel Ther.* **2019**, *13*, 3855–3866. [CrossRef]
171. Yoshida, H.; Tsuhako, R.; Sugita, C.; Kurokawa, M. Glucosyl hesperidin has an anti-diabetic effect in high-fat diet-induced obese mice. *Biol. Pharm. Bull.* **2021**, *44*, 422–430. [CrossRef]
172. Available online: https://www.scopus.com/search/form.uri?display=basic#basic (accessed on 20 April 2022).
173. Available online: https://scholar.google.com/ (accessed on 20 April 2022).
174. Available online: https://pubchem.ncbi.nlm.nih.gov/ (accessed on 10 March 2021).
175. Pérez-Sánchez, H.; den-Haan, H.; Peña-García, J.; Lozano-Sánchez, J.; Martínez Moreno, M.E.; Sánchez-Pérez, A.; Muñoz, A.; Ruiz-Espinosa, P.; Pereira, A.S.P.; Katsikoudi, A.; et al. DIA-DB: A database and web server for the prediction of diabetes drugs. *J. Chem. Inf. Model.* **2020**, *60*, 4124–4130. [CrossRef] [PubMed]

176. Pereira, A.S.P.; den Haan, H.; Peña-García, J.; Moreno, M.M.; Pérez-Sánchez, H.; Apostolides, Z. Exploring african medicinal plants for potential anti-diabetic compounds with the DIA-DB inverse virtual screening web server. *Molecules* **2019**, *24*, 2002. [CrossRef] [PubMed]

plants

2, 11, 1388

MDPI

Article

Isolation and Characterization of Werneria Chromene and Dihydroxyacidissimol from *Burkillanthus malaccensis* (Ridl.) Swingle

Masyitah Zulkipli [1], Nuzum Mahbub [1], Ayesha Fatima [2], Stefanie Lim Wan-Lin [1], Teng-Jin Khoo [1], Tooba Mahboob [3], Mogana Rajagopal [4], Chandramathi Samudi [3], Gheetanjali Kathirvalu [3], Nor Hayati Abdullah [5], Ana Rita Pinho [6,7], Sonia M. R. Oliveira [8,9], Maria de Lourdes Pereira [6,8], Mohammed Rahmatullah [10], Anamul Hasan [10], Alok K. Paul [11], Mark S. Butler [12], Muhammad Nawaz [13], Polrat Wilairatana [14,*], Veeranoot Nissapatorn [15,*] and Christophe Wiart [16,*]

1 School of Pharmacy, University of Nottingham Malaysia Campus, Semenyih 43500, Malaysia; khyy6mzz@nottingham.edu.my (M.Z.); hyxnm1@nottingham.edu.my (N.M.); hyysl1@nottingham.edu.my (S.L.W.-L.); tengjin.khoo@nottingham.edu.my (T.-J.K.)

2 Beykoz Institute of Life Sciences and Biotechnology, Bezmialem Vakif University, Istanbul 34093, Turkey; afatima@bezmialem.edu.tr

3 Department of Medical Microbiology, University of Malaya, Kuala Lumpur 50603, Malaysia; tooba666@hotmail.com (T.M.); chandramathi@um.edu.my (C.S.); geetha.medmicrob@gmail.com (G.K.)

4 Faculty of Pharmaceutical Sciences, UCSI University, Kuala Lumpur 56000, Malaysia; mogana@ucsiuniversity.edu.my

5 Natural Product Division, Forest Research Institute Malaysia (FRIM), Kepong 52109, Malaysia; norhayatiab@frim.gov.my

6 Department of Medical Sciences, University of Aveiro, 3810-193 Aveiro, Portugal; arapinho@ua.pt (A.R.P.); mlourdespereira@ua.pt (M.d.L.P.)

7 Neuroscience and Signaling Laboratory, Institute of Biomedicine-IBIMED, University of Aveiro, 3810-193 Aveiro, Portugal

8 CICECO-Aveiro Institute of Materials, University of Aveiro, 3810-193 Aveiro, Portugal; sonia.oliveira@ua.pt

9 Hunter Medical Research Institute (HMRI), New Lambton Heights, NSW 2305, Australia

10 Department of Biotechnology & Genetic Engineering, University of Development Alternative, Lalmatia, Dhaka 1207, Bangladesh; rahamatm@hotmail.com (M.R.); anamulhasanoris@gmail.com (A.H.)

11 School of Pharmacy and Pharmacology, University of Tasmania, Hobart, TAS 7001, Australia; alok.paul@utas.edu.au

12 Institute for Molecular Bioscience, University of Queensland, Brisbane, QLD 4072, Australia; m.butler2@imb.uq.edu.au

13 Department of Nano-Medicine, Institute for Research and Medical Consultations (IRM), Imam Abdulrahman Bin Faisal University, Dammam 34212, Saudi Arabia; nawwaz@gmail.com

14 Department of Clinical Tropical Medicine, Faculty of Tropical Medicine, Mahidol University, Bangkok 10400, Thailand

15 School of Allied Health Sciences, World Union for Herbal Drug Discovery (WUHeDD), Research Excellence Center for Innovation and Health Products (RECIHP), Walailak University, Nakhon Si Thammarat 80160, Thailand

16 Institute for Tropical Biology and Conservation, Universiti Malaysia Sabah, Jalan UMS, Kota Kinabalu 88400, Malaysia

* Correspondence: polrat.wil@mahidol.ac.th (P.W.); veeranoot.ni@wu.ac.th (V.N.); asianpharmcog@gmail.com (C.W.)

Citation: Zulkipli, M.; Mahbub, N.; Fatima, A.; Wan-Lin, S.L.; Khoo, T.-J.; Mahboob, T.; Rajagopal, M.; Samudi, C.; Kathirvalu, G.; Abdullah, N.H.; et al. Isolation and Characterization of Werneria Chromene and Dihydroxyacidissimol from *Burkillanthus malaccensis* (Ridl.) Swingle. *Plants* **2022**, *11*, 1388. https://doi.org/10.3390/plants11111388

Academic Editors: Bistra Galunska, Ilian Badjakov and Ivayla Dincheva

Received: 30 March 2022
Accepted: 17 May 2022
Published: 24 May 2022

Abstract: The secondary metabolites of endemic plants from the Rutaceae family, such as *Burkillanthus malaccensis* (Ridl.) Swingle from the rainforest of Malaysia, has not been studied. *Burkillanthus malaccensis* (Ridl.) Swingle may produce antibacterial and antibiotic-potentiating secondary metabolites. Hexane, chloroform, and methanol extracts of leaves, bark, wood, pericarps, and endocarps were tested against bacteria by broth microdilution assay and their antibiotic-potentiating activities. Chromatographic separations of hexane extracts of seeds were conducted to investigate effective phytochemicals and their antibacterial activities. Molecular docking studies of werneria chromene and dihydroxyacidissiminol against SARS-CoV-2 virus infection were conducted using AutoDock Vina. The methanol extract of bark inhibited the growth of *Staphylococcus aureus*, *Escherichia coli*, and *Pseudomonas aeruginosa* with the minimum inhibitory concentration of 250, 500, and 250 μg/mL, respectively. The chloroform extract of endocarps potentiated the activity of imipenem against imipenem-resistant *Acinetobacter baumannii*. The hexane extract of seeds increased the sensitivity of *P. aeruginosa* against ciprofloxacin and levofloxacin. The hexane extract of seeds and chloroform extract of endocarps were chromatographed, yielding werneria chromene and dihydroxyacidissiminol. Werneria chromene was bacteriostatic for *P. aeruginosa* and *P. putida*, with MIC/MBC values of 1000 > 1000 μg/mL. Dihydroxyacidissiminol showed the predicted binding energies of −8.1, −7.6, −7.0, and −7.5 kcal/mol with cathepsin L, nsp13 helicase, SARS-CoV-2 main protease, and SARS-CoV-2 spike protein receptor-binding domain S-RBD. *Burkillanthus malaccensis* (Ridl.) Swingle can be a potential source of natural products with antibiotic-potentiating activity and that are anti-SARS-CoV-2.

Keywords: *Burkillanthus malaccensis*; antibiotic potentiator; werneria chromene; dihydroxyacidissiminol; SARS-CoV-2; cathepsin L; nsp13 helicase; spike protein

---

1. Introduction

Nowadays, clinicians are confronted with the huge burden of treating patients infected by multidrug-resistant nosocomial bacteria [1,2]. Nosocomial bacteria, such as *Acinetobacter baumannii* and *Pseudomonas aeruginosa*, in intensive care units resist the main classes of antibiotics, such as oxazolidinooxazolidinones, lipopeptides, macrolides, fluoroquinolones, tetracyclines, β-lactams, β-lactam-β-lactamase inhibitor combinations, carbapenems, and glycopeptides antibiotics [3–6]. Besides bacterial species, drug-resistant parasites have emerged, such as *Plasmodium* species [7] and Brugia species [8].

Another challenge is the treatment of zoonotic viruses, such as the severe acute respiratory syndrome-associated coronavirus (SARS-CoV), the Middle-East respiratory syndrome-associated coronavirus (MERS-CoV), and the severe acute respiratory syndrome-associated coronavirus 2 (SARS-CoV-2) [1,2]. Since December 2019, SARS-CoV-2, the causative agent of the coronavirus disease 2019 (COVID-19) pandemic, has been infecting millions of people globally [9–11]. SARS-CoV-2 is an enveloped, monopartite, linear, single-stranded (+)-RNA zoonotic virus in the Coronaviridae that replicates in the lower respiratory tract of humans, leading in some cases to lethal pneumonia [3,12]. Vaccines have been developed, affording good protection rates, but there is still the need to develop leads to be taken orally for the prevention and/or treatment of COVID-19, as complementary to vaccination. Further, if the human pressure on the environment is not halted, other new zoonotic viruses are bound to emerge, for which an armamentarium of anti-viral molecules is required to attempt to limit casualties.

SARS-CoV-2 infects cells expressing the surface receptors angiotensin-converting enzyme 2 (ACE2) and cellular serine protease TMPRSS2 (transmembrane protease serine 2). SARS-CoV-2 enters the host cells by attaching its surface spike proteins to the surface ACE2 of human host cells [9,11,13]. TMPRSS2 and cysteine protease cathepsin L are also required to facilitate host cell entry. After fusion of the viral membrane with the host cell membrane, the viral genomic material is released in the cytoplasm and replicated, via several enzymes, including Nsp13 helicase [10,11,14–16]. Targeting these three proteins (ACE2, cathepsin L, Nsp13 helicase) can assist in the development of anti-COVID-19 drugs [11]. COVID-19 infection can progress to acute respiratory distress syndrome (ARDS) that favors secondary bacterial infection, leading to sepsis and, ultimately, death [11,17–19].

The alternative for the discovery of new antibiotics is to develop antibiotic-potentiators, also known as antibiotic adjuvants. Therefore, antibiotics, antibiotic-potentiators, and anti-viral compounds (more so against COVID-19 are urgently needed. Furthermore, since promising leads for new drugs have usually come from the plant kingdom [20,21], it becomes more important to study plants to help discover novel drugs that can play important therapeutic/prophylactic roles against newly emerging viruses and antibiotic-resistant microbial species. Unfortunately, various plants are becoming extinct for many reasons, including irresponsible deforestation, an extension of human habitat, conversion of forests and water bodies into agricultural land, and global warming [22].

The primary rainforests of Malaysia used to be one of the hotspots of plant biodiversity. A study carried out in 2008 found that a randomly chosen 5-hectare area in Ayer Hitam Forest, Puchong, contained 6621 trees belonging to 319 species in 148 genera and 51 families [23]. These plants can be a source of diverse secondary metabolites, which are yet to be explored [24]. As various therapeutic medications originate from the secondary metabolites (phytochemicals) of plants, any extinction or even rarity of plant species may make it impossible to ever know of the phytochemical properties of any given plant. Since these rainforests of Malaysia are steadily declining because of various commercial interests [25–27], the chances of discovering new drugs from plant species, any of which may be novel antibiotics, antibiotic-potentiators, or anti-viral drugs, are also decreasing rapidly. Thus, there is an urgent need to study these rainforest plants.

Burkillanthus malaccensis (Ridl.) Swingle. (synonym: Citrus malaccensis Ridl.), in the family Rutaceae, is a low-land primary rainforest tree known to the Malays as "*limau hantu*" (meaning the lemon tree of ghosts) and commonly known as Burkill's lime tree [28–34]. The current study investigated for the development of safe, effective, and inexpensive plant-based antibiotic potentiators that can improve current treatment strategies for treating bacteria-resistant infections and simultaneously can contribute to the development of anti-SARS-CoV-2 agents. The aims of this study were to examine the antibacterial properties of *B. malaccensis* leaves, bark, wood, and fruits (exocarp, endocarp, and seeds) against *Enterococcus faecium, Staphylococcus aureus, Klebsiella pneumoniae, Acinetobacter baumannii, Pseudomonas aeruginosa, Pseudomonas putida,* and *Enterobacter* spp.) [35]; to examine the antibiotic-potentiating properties of extracts with antibiotics, to isolate the major constituents in the active extracts, and to test their antibacterial effects in vitro and anti-SARS-CoV-2 effects in silico against (a) a complex of SARS-CoV-2 spike protein (S) receptor-binding domain (RBD) bound with human receptor angiotensin-converting enzyme 2 (hACE2) or S-RBD-hACE2; (b) cathepsin L; (c) SARS-CoV-2 Nsp13 helicase; (d) SARS-CoV-2 main protease Mpro (plays an integral part in viral replication); and (e) S-RBD. While ACE-2 is the receptor for the viral spike protein S, cathepsin L cleaves the spike protein and enhances the entry of virus [36], and SARS-CoV-2 Nsp13 possesses NTPase (nucleotide triphosphatase) and RNA helicase activities [37]. In this study, a plant extract that has not been investigated for its antiviral activities yielded two isolated compounds. Thus, it was of interest to study the antiviral activities against some SARS-CoV-2 proteins and their human targets so as not to miss out on anything of importance, but instead get a composite picture of the anti-SARS-CoV-2 potential of the isolated novel compounds.

2. Material and Methods

2.1. Plant Collection

Chemotaxonomic collection of *B. malaccensis* from Manong village close to the Kuala Kangsar Forest, State of Perak, Malaysia (4.7746° N, 100.9520° E), was performed in February 2017. It is a beautiful plant with thorny stems, trifoliate leaves, large white flowers, and large grapefruit-like fruits, containing numerous seeds covered with a yellow resin (personal observation). Samples of leaves, bark, wood, seeds, and fruits were collected. A voucher herbarium specimen with vernacular names, collection localities, and dates were deposited (Voucher number NB0541) for identification by taxonomists at the Forest Research Institute of Malaysia (FRIM). The use of plants in the present study complies with international guidelines. The plant was selected in fieldwork according to the following three criteria: (i) chemotaxonomic features; (ii) the presence of fruits or flowers to allow an accurate botanical identification; and (iii) a sufficient amount available.

2.2. Preparation of Plant Extracts

The collected leaves, bark, wood, seeds, endocarps (sac juice), and fruit pericarps from *B. malaccensis* were separated and air-dried at room temperature for two weeks. The dry materials were then finely pulverized by grinding using an aluminum collection blender (Philips, Guangdong, China); the powders obtained were weighed with a top-loading balance (Sartorius AG, Göttingen, Germany). Dried plant powders (200 g) were successively soaked at room temperature with hexane, chloroform, and methanol. Each extraction was performed using the maceration technique with a plant powder-to-solvent ratio of 1:5 (w/v) for 3 days at room temperature with 3 successive repetitions. The liquid extracts were subsequently filtered through qualitative filter papers, No. 1 (Whatman International Ltd., Maidstone, UK), using an aspirator pump (EW-35031-00, 18 L/min, 9.5 L Bath, 115 VAC), and the filtrates were concentrated to dryness under reduced pressure at 40 °C using a rotary evaporator (Buchi Labortechnik AG, Flawil, Switzerland). The dry extracts obtained were weighed and stored in tightly closed glass scintillation vials (Kimble, NY, USA) at −20 °C until further use.

The yields of extracts were calculated using the following formula:

$$\%\ \text{yield} = ((\text{Mass of dried extract})/(\text{Mass of dried plant part})) \times 100\%$$

2.3. Tested Bacterial Strains

Stock cultures of bacteria used for this study were kindly provided by the Department of Medical Microbiology, Faculty of Medicine, University of Malaya, Malaysia. The following bacteria were used as test organisms from the American Type Culture Collection (ATCC, Manassas, VA, USA): *S. aureus* (ATCC11632), *B. subtilis* (ATCC 6633), *E. coli* (ATCC 25218), *P. aeruginosa* (ATCC 10145), and *A. baumannii* (clinical stain, imipenem-resistant) were sub-cultured in nutrient agar. *Pseudomonas putida* (ATCC 49128) was used for the antibacterial testing of isolated compounds. All sub-cultured bacterial specimens were aseptically transferred using an inoculating loop and prepared in 10 mL suspensions of Mueller-Hinton Broth (Oxoid, Hampshire, UK) and were used 15 min after inoculation. A fraction equivalent to 1 mL of the bacterial suspensions was transferred to a cuvette and subjected to a spectrophotometer (Biochrom, Cambridge, UK), where the UV absorbance value was monitored to be in the range of 0.008 to 0.10 at 625 nm, in order to be adjusted to 0.5 McFarland turbidity standards (Healthlink, Orlando, FL, USA), which correspond to a bacterial cell count of 1.5×10^8 CFU/mL [38]. The extracts and phytoconstituents were prepared by dissolving in 10% DMSO and in a minimum essential medium to make a stock solution with a final concentration of 1% DMSO. This stock solution was diluted in liquid broth to further to obtain concentrations ranging from 500 to 5000 μg/mL.

2.4. Broth Microdilution Assay

Minimum inhibitory concentration (MIC) values were determined according to the guidelines of the Clinical and Laboratory Standards of the Institute. Briefly, bacterial strains were grown for 18–24 h at 37 °C. Colonies were directly suspended in cautiously adjusted Müller–Hinton broth (CAMHB) and adjusted to OD625 0.08–0.1, which corresponds to $1{\sim}2 \times 10^8$ CFU/mL, followed by 10-fold serial dilutions to give 1×10^6 CFU/mL. Bacterial suspensions (1000 μL) were added to the 96-well round-bottom microtiter plates. Each well was then filled with 100 μL of liquid broth containing extracts of phytoconstituents to yield final concentrations of extracts or phytoconstituents of 250, 625, 1000, 1500, and 2500 μg/mL, respectively. The 96-well plates were incubated for 24 h at 37 °C. The MIC was defined as the lowest concentration of extract or phytoconstituents that completely inhibited the growth of bacteria. Negative controls consisted of bacterial suspensions (100 μL) added to 96-well round-bottom microtiter plates containing 100 μL of liquid broth with a minimum amount of DMSO, as in extracts of phytoconstituent stock solution preparations. Minimum bactericidal concentration (MBC) was determined (for extracts with MIC values equal to or less than 250 μg/mL) by sub-culturing the test dilutions onto a sterile agar plate and further incubated for 18–24 h. The highest dilution that yielded 0% bacterial growth on agar plates was taken as the MBC, MIC and MBC values were calculated as the mean of triplicate experiments. Chloramphenicol, tetracycline, and imiperem were used as positive control antibiotics.

2.5. Antibiotic-Potentiating Assay

The ability of *B. malaccensis* extracts to increase the sensitivity of bacteria to antibiotics was measured by the technique described by Boonyanugomol et al., with slight modifications [39]. Standard antibiotic discs of ampicillin (10 μg/disc), gentamicin (10 μg/disc), imipenem (10 μg/disc), levofloxacin (5 μg/disc), penicillin G (10 μg/disc), and ciprofloxacin (5 μg/disc) (Sigma-Aldrich, St. Louis, MO, USA) were loaded with 10 μL of a 100 μg/μL solution of extracts of leaves, bark, wood, seeds, endocarps, or pericarps. The zones of inhibition were measured after overnight incubation (12 h) and estimated as follows:

1. Zone of combined extract and antibiotic > zone of extract + zone of antibiotic: synergy.
2. Zone of combined extract and antibiotic = zone of extract + zone of antibiotic: additive.
3. Zone of combined extract and antibiotic < zone of extract + zone of antibiotic: antagonism.

2.6. Cytotoxicity Assay

To study in vitro the cytotoxicity of extracts against human fibroblast cells (MRC-5 cell line, MRC-5 ATCC CCL-171 Homo sapiens lung normal), the reduction of (3-(4,5-dimethylthiazol-2-yl)-2,5-diphenyltetrazolium bromide) by a colorimetric assay (MTT) was used [40]. The extracts were prepared by dissolving in DMSO and in a minimum essential medium (MEM) to make a stock solution with a final concentration of 1% DMSO. This stock solution was diluted further to obtain concentrations ranging from 0 to 200 μg/mL. Cells were cultured in Rosewell Park Memorial Institute media (RPMI) supplemented with 10% Fetal Bovine Serum (FBS). Cells were incubated with the diluted plant extracts for 48 h at 37 °C in 5% CO_2. After cells were washed twice with saline, a solution of MTT (0.5 mg/mL) in phosphate buffer saline (PBS) was added to the wells. After 4 h of incubation, the wells were washed, and the formazan residue dissolved in DMSO (0.1 mL per well). The absorbance was then measured in a spectrophotometer (SpectraMax M3, Multi-Mode Microplate reader, Molecular Devices, San Jose, CA, USA) at a wavelength of 570 nm and plotted against the concentration of the extracts. Cells with no added test reagents were taken as untreated cells with 100% viability, and cells with RPMI 1640 medium were used as blanks. The cells viability percentage was plotted against the extract's concentration. All experiments were performed in triplicate, and results were expressed as the concentration by reducing the number of live MRC-5 cells by 50% (CC_{50}). The percentage of viability was calculated using the following formula:

$$\% \text{ Viability} = [(\text{Abs sample} - \text{Abs blank})/(\text{Abs untreated} - \text{Abs blank})] \times 100\%$$

The effects of extracts on the % viability of cells (y-axis) was plotted against log concentrations (g/L) (x-axis) and interpolated sigmoidal curves (4-parameter logistic curve, 4 PL) using GraphPad Prism and the CC_{50} was automatically determined using GraphPad Prism v6 software (GraphPad Software Inc., La Jolla, CA, USA).

2.7. Isolation and Identification of Compounds

Hexane extract from seeds (8.5 g) was loaded onto preparative thin-layer chromatography (TLC) plates in a mobile phase of hexane: ethyl acetate: dichloromethane (40:30:30). TLC analysis revealed the presence of a major layer under UV light (254 nm) with a retention factor (Rf) value of 0.25. TLC layers were stained with vanillin sulfuric acid. The layer with the lowest purple Rf stained with vanillic sulfuric acid was collected, redissolved, filtered, and isolated to yield the compound 1 (5 mg) that was identified as werneria chromene ($C_{15}H_{16}O_3$, molecular mass = 244.28 g/mol) by comparing its ^1H-NMR (proton nuclear magnetic resonance) and EIMS (electron impact mass spectrometry) data with those from the literature [31]. This compound spontaneously forms a translucent crystal analyzed by X-ray diffraction, which confirmed the interpretation of NMR data.

Chloroform extracts of the fruit endocarp exhibited on TLC the presence of nine layers under UV light (254 nm), with Rf values of 0.187, 0.242, 0.286, 0.341, 0.396, 0.462, 0.659, 0.769, and 0.846, respectively, when a mobile phase of chloroform: ethyl acetate: diethyl ether (40:40:20) was used. After spraying with vanillin sulfuric acid, the stains observed here were pink and dark pink for most of them, which indicates the presence of phenols or steroids. The main compound (3 mg) with a Rf value of 0.187 was collected, redissolved, filtered, and isolated to yield dihydroxyacidissimol ($C_{25}H_{34}NO_5$, molecular mass 427.2359 g/mol) by comparing its ^{1}H-NMR and EIMS data with those from the literature [41,42].

2.8. In Silico Studies—Auto Dock Vina (Blind Docking Methodology)

2.8.1. Protein Preparation

Main Protease

We took Mpro (Pdb: 6LU7 with 2.16 Å resolution), which has 306 amino acid residues, as our target protein, which contained a bound inhibitor known as N3 in its crystal structure [43]. For docking preparation, we removed the water molecules from the crystallographic structure of Mpro and removed the N3 molecule as well. Thus, ligands can be docked within every pocket of the protein. Next, we added a polar hydrogen atom because crystallographic structures usually lack hydrogen atoms. The addition of polar hydrogen atoms and removals of the water molecules and N3 were done with Pymol software [44]. Then, the protein molecule was saved in pdb format.

Spike Protein Receptor-Binding Domain (S-RBD) Bound with the ACE2 Complex

The SARS-CoV-2 S-RBD bound with the ACE2 complex X-ray diffraction structure (PDB: 6LZG with 2.50 Å resolution) was downloaded from the protein data bank (6LRG. Available online: https://www.rcsb.org/structure/6LZG (accessed on 1 January 2022)). This complex structure has two protein chains (Chain A is ACE2 and Chain B is S-RBD)

Spike Protein Receptor-Binding Domain (S-RBD)

The S-RBD structure was extracted from a SARS-CoV-2 S-RBD bound with the ACE2 complex X-ray diffraction structure (PDB: 6M0J with 2.45 Å resolution) (6M0J. Available online: https://www.rcsb.org/structure/6M0J (accessed on 1 January 2022)). Here, we took only the S-RBD as our target receptor and removed the ACE2 protein chain.

Cathepsin L

The X-ray crystal structure of cathepsin L was found in the PDB (PDB Id: 3HHA with 1.27 Å resolution) [45]. The crystal structure has four identical protein chains. We took the monomeric form of cathepsin L.

Nsp13 Helicase

Crystal structure of SARS-CoV-2 helicase at 1.94 Å (PDB:6ZSL) was downloaded from pdb (6ZSL. Available online: https://www.rcsb.org/structure/6ZSL (accessed on 1 January 2022)) [46].

2.8.2. Ligand Preparation

Ligand molecules were downloaded from PubChem [47] in sdf format. They were optimized with the force field type MMFF94 using Openbable software and saved in pdbqt format.

Docking

We have used here the blind docking method for screening phytochemicals. So, the grid box in Autodock Vina was generated, aiming to cover up the whole protein molecule. In which region the ligand binds effectively with protein molecule can be found in blind docking. We have used exhaustively "16" for better ligand and protein binding. AutoDock Vina tool [48] provides a total of nine docked poses for each ligand; among them, pose1 is the best pose with the highest binding affinity. We have saved pose1 in pdb format by using Pymol for further analysis. 2D diagrams and the interactions between the ligand and amino acids of the protein were obtained in Discovery Studio Software [49].

2.9. Statistical Analysis

All values presented in the results section are the mean or mean ± standard deviation of the mean of three independent analyses, calculated using GraphPad Prism v6 Software (GraphPad Software Inc., La Jolla, CA, USA). Interpolated sigmoidal curves (4-parameter logistic curve, 4 PL) were determined automatically using GraphPad Prism v6 software (GraphPad Software Inc., La Jolla, CA, USA).

3. Results

3.1. Plant Extraction

The air-dried parts of the plant were extracted successively with hexane, chloroform, and methanol, respectively, to obtain lipophilic (non-polar), amphiphilic (mid-polar), and hydrophilic (polar) extracts from *B. malaccensis*. The average yield values ranged from 1.0 to 9.5% (Table 1). The average yields calculated for the hexane, chloroform, and methanol extracts were 2.7, 4.9, and 2.3%, respectively.

Table 1. Percentage extraction yields of 18 organic extracts from *B. malaccensis*.

Part Extracted	Plant Extracts Yield (%)		
	Hexane	Chloroform	Methanol
Leaves	1.8	3.3	1.9
Bark	3.8	5.4	1.2
Wood	1.0	3.4	4.2
Fruit pericarp	5.1	2.9	2.1
Fruit endocarp	3.4	9.5	1.0
Seeds	1.5	5.1	3.6
Average yields	2.7	4.9	2.3

Chloroform fruit endocarps extract gave the highest extraction yield (9.5%), while the same plant part extracted with methanol yielded the lowest (1%).

3.2. Broth Microdilution

The broth microdilution method [38] was used to determine the minimum inhibitory concentration (MIC) of extracts against a panel of five bacteria (Table 2). The broth microdilution assay results confirmed that Gram-positive bacteria were more susceptible than Gram-negative bacteria to *B. malaccensis* extracts. The chloroform extract of leaves exhibited the lowest MIC against *S. aureus* and *B. subtilis*, with values of 250 µg/mL and a minimum bactericidal concentration (MBC) above 1000 µg/mL. The lowest MICs against *E. coli* and *P. aeruginosa* were observed with the methanol extract of bark (500 and 250 µg/mL, respectively) and MBC value above 1000 µg/mL. This extract inhibited the growth of *S. aureus* with MIC/MBC values of 250/1000 µg/mL, and, as such, it had the broadest spectrum of activity out of the 18 extracts tested. None of the extracts was active against *A. baumannii* (imipenem-resistant).

Table 2. Minimum inhibitory concentration (MIC) by broth microdilution (μg/mL).

Plant Part	Solvent	*S. aureus* (ATCC 11632)	*B. subtilis* (ATCC 6633)	*E. coli* (ATCC 8379)	*P. aeruginosa* (ATCC 10145)	*A. baumannii* (Imipenem-Resistant)
Leaves	Hexane	1000	-	-	-	-
Leaves	Chloroform	**250** (>1000)	**250** (>1000)	-	-	-
Leaves	Methanol	500	1000	-	-	-
Bark	Hexane	-	-	-	1000	-
Bark	Chloroform	1000	-	-	-	-
Bark	Methanol	**250** (>1000)	-	500	**250** (>1000)	-
Wood	Hexane	625	1250	5000	2500	-
Wood	Chloroform	2500	2500	2500	625	-
Wood	Methanol	-	2500	2500	2500	-
Endocarp	Chloroform	-	-	-	1000	-
Endocarp	Methanol	-	-	-	1000	-
Seeds	Hexane	-	-	-	1000	-
Chloramphenicol		0.03	0.02	Nt	Nt	Nt
Tetracycline		Nt	Nt	0.02	0.01	-
Imipenem		Nt	Nt	Nt	Nt	12.0
Negative control		Fg	Fg	Fg	Fg	Fg

Abbreviations: Nt, Not tested; Fg, Full bacterial growth; '-', No activity. Extracts with no activity against all the bacteria tested are not included here in this table. Bold data indicate the lower MIC values. Values are given as the mean of triplicates. Second values in parentheses represent corresponding minimum bactericidal concentrations (MBC).

3.3. Antibiotic-Potentiating Activities

Of the 18 extracts tested, the most effective antibiotic-potentiator for Gram-positive bacteria was the methanol extract of wood with the β-lactam amoxicillin against *S. aureus* (Table 3). Regarding Gram-negative bacteria, the hexane and chloroform extracts of wood potentiated the aminoglycoside gentamicin against *E. coli*, with increments of inhibition zones of about 8 and 10 mm, respectively. We observed that the methanol extract of endocarps was able to rend penicillin G active against *E. coli* and the chloroform extract of endocarps relinquished the resistance of clinical isolates of *A. baumannii* to imipenem. In addition, the hexane extract of seeds acted potentiator for the quinolones (ciprofloxacin and levofloxacin) action against *P. aeruginosa*.

Table 3. Antibiotic-potentiating activities (mm).

Treatment with	*S. aureus* (ATCC 11632)	*B. subtilis* (ATCC 6633)	*E. coli* (ATCC 8379)	*P. aeruginosa* (ATCC 10145)	*A. baumannii* (Imipenem-Resistant)
Extracts			-	-	-
I	-	7.0 ± 1.4	-	-	-
II	12 ± 0.0	-	-	-	-
III	-	-	-	-	-
IV	-	-	-	-	-
V	-	-	-	-	-
VI	-	-	-	-	-
VII	-	-	-	-	-
Amoxicillin	16 ± 0.0	14.3 ± 0.5	-	-	-
Ampicillin	41 ± 0.3	20 ± 0.1	-	-	-
Ciprofloxacin	-	38 ± 0.0	-	35 ± 0.02	-
Gentamicin	-	-	25 ± 0.01	-	-
Levofloxacin	-	35 ± 1.0	38 ± 0.2	28 ± 0.3	-
Penicillin G	-	-	-	-	-
Imipenem	-	-	-	-	10 ± 0.04

Table 3. *Cont.*

Treatment with	*S. aureus* (ATCC 11632)	*B. subtilis* (ATCC 6633)	*E. coli* (ATCC 8379)	*P. aeruginosa* (ATCC 10145)	*A. baumannii* (Imipenem-Resistant)
Amoxicillin + I	-	26.7 ± 0.0	-	-	-
Amoxicillin + II	22.5 ± 0.5	-	-	-	-
Amoxicillin + III	**23.7 ± 0.5**	-	-	-	-
Ampicillin + II	40.7 ± 0.8	-	-	-	-
Ampicillin + III	**42.0 ± 0.5**	-	-	-	-
Ampicillin + IV	-	22 ± 0.3	-	-	-
Ciprofloxacin + IV	-	39 ± 0.1	-	**38 ± 1.1**	-
Ciprofloxacin + V	-	38.5 ± 0.0	-	36 ± 0.1	-
Gentamicin + I	-	-	**34.3 ± 1.7**	-	-
Gentamicin + II	-	-	**35.3 ± 1.3**	-	-
Levofloxacin + IV	-	-	-	**30.7 ± 0.6**	-
Levofloxacin + V	-	38.5 ± 0.3	-	30 ± 1.0	-
Penicillin G + VI	-	-	**6.5 ± 0.02**	-	-
Imipenem + V	-	-	-	-	**11 ± 1.2**

Abbreviations: I = Wood hexane (1 mg/disc); II = Wood chloroform (1 mg/disc); III = Wood methanol (1 mg/disc); IV = Seeds hexane; V = Endocarp chloroform; VI = Endocarp methanol; VII: Leaves chloroform. Amoxicillin (10 µg/disc); Ampicillin (10 µg/disc); Ciprofloxacin (5 µg/disc); Gentamicin (10 µg/disc0; Imipenem (10 µg/disc); Levofloxacin (5 µg/disc); '-', No activity. Extracts without any synergy are not included. Synergies are indicated in bold. The values are expressed as the mean ± standard deviation.

3.4. Cytotoxic Activities

The toxicity of the 18 extracts against MRC-5 (Medical Research Council cell strain 5) human fibroblast cells was evaluated. The lowest 50% cytotoxic concentration (CC_{50}) was obtained with the chloroform extract of pericarps, with a value of 0.36 g/L (Figure 1A). The methanolic extract pericarps exhibited a CC_{50} value of 0.09 g/L (Figure 1B).

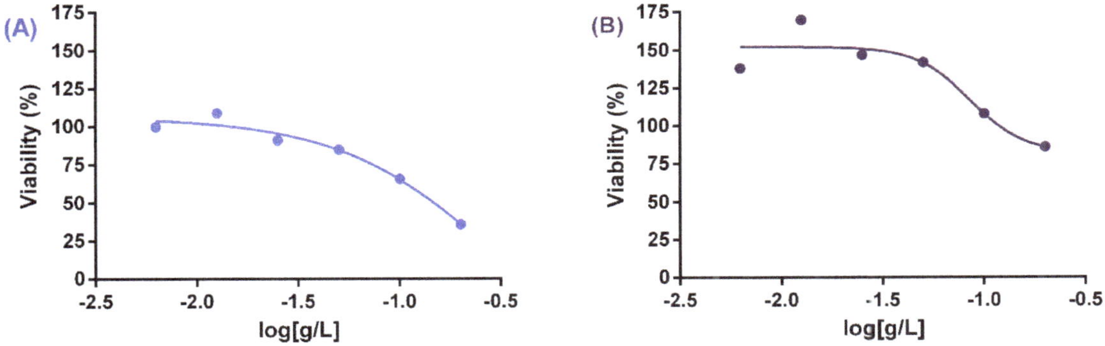

Figure 1. Cytotoxicity of chloroform (**A**) and methanolic (**B**) extracts of pericarps of *B. malaccensis* using Sigmoidal 4PL dose-response curves.

3.5. Isolation of the Main Constituents from Active Extracts and Antibacterial Effects

The hexane extract of seeds that increased the potencies of quinolone antibiotics against *P. aeruginosa* was subjected to preparative TLC, yielding werneria chromene from the extract (Figure 2a, Table 4). From the chloroform extract of endocarps, the tyramine alkaloid dihydroxyacidissiminol was isolated (Figure 2b; Table 5). The absolute structure of werneria chromene was further confirmed using X-ray diffraction (Figure 2c,d). Werneria chromene was inactive against all bacteria tested except *P. aeruginosa*, with the MIC value of 1000 µg/mL and the MBC value of 1000 µg/mL, whereby both compounds repressed the growth of *P. putida* the MIC value of 1000 µg/mL and the MBC of 1000 µg/mL.

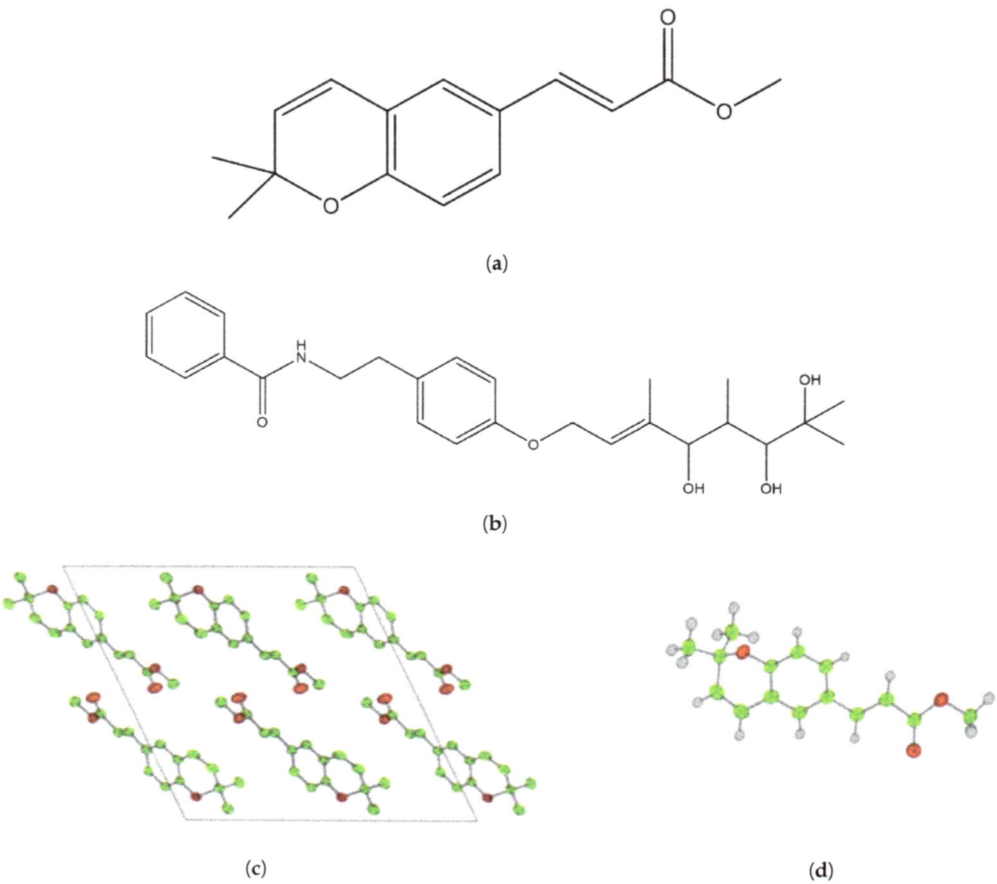

Figure 2. (**a**) Werneria chromene; (**b**) Dihydroxyacidissiminol; (**c**,**d**) Crystal structure of werneria chromene; (**c**) Cell packing in monoclinic state; (**d**) Crystal structure of isolated methyl (Z)-3-(2,2-dimethyl-2H-chromen-6-yl) acrylate from the ORTEP diagram at 50% ellipsoid probability.

Table 4. NMR data of werneria chromene (p.p.m.).

Position	δ-H [31]	δ-H Werneria Chromene	Integration	Position	δ-C [31]	δ-C Werneria Chromene
3	5.62 d	5,65 d	1	2	77.1	78
4	6.28 d	6.30 d	1	3	131.3	132
5	7.12 d	7.18 d	1	4	121.7	122
7	7.26 dd	7.25 dd	1	5	134.3	134
8	6.74 d	6.78 d	1	6	127.1	128
9	7.58 d	7.60 d	1	7	129.4	129
10	6.26 d	6.20 d	1	8	116.7	116
12,13	7.58 d	7.57 d	2,2	10	121.3	122
O-CH_3	3.76 d	3.80 s	3	11	115	115
				12	144.6	145

Table 5. NMR data (p.p.m.) of dihydroxyacidissiminol.

Position	δ-H [42]	δ-H Dihydroxyacidissiminol	Integration	δ-C [42]	δ-C Dihydroxyacidissiminol
1′	-	-	-	134.60	135.00
2′,6′	7.69 d	7.70 d	1,1	126.80	126.50
3′,5′	7.41 t	7.40 t	1,1	128.60	128.90
4′	7.49 t	7.49 t	1	131.40	130.00
CO-NH	6.10 m	6.15 m	1	167.60	167.50
N-CH_2	3.70 q	3.70 m	2	41.30	41.50
Ar-CH_2	2.88 t	2.80 t	2	34.80	35.00
1″	-	-	-	157.30	157.00
2″, 6″	6.87 d	6.87 d	1,1	114.90	115.00
3″, 5″	7.18 d	7.16 d	1,1	129.80	129.90
4″	-	-	-	131.00	131.50
1	4.58 d	4.60 d	2	64.50	64.50
2	5.80 d	5.78 d	1	121.10	121.50
3	-	-	-	142.00	140.50
3-Me	1.76 s	1.75 s	3	12.40	12.90
4	4.35 dd	4.35 dd	1	77.40	79.00
5	-	-	-	-	-
6	3.64 m	3.65 m	1	78.70	77.50
7	-	-	-	72.60	74.00
7-Me	1.18 s	1.17 s	3	23.70	24.00
4-OH	1.55 s	1.75 s	-	-	-
6,7 OH	1.55 s	1.65 s	-	-	-

Abbreviation: '-', No peak.

3.6. Crystal Structure of Isolated Methyl (Z)-3-(2,2-dimethyl-2H-chromen-6-yl) Acrylate Werneria Chromene

During the isolation process of werneria chromene, translucent crystals were obtained, and its chemical structure was confirmed by X-ray diffraction as methyl (E)-3-(2,2-dimethyl-2H-chromen-6-yl) acrylate. The crystal structure (Figure 2c,d) of werneria chromene indicated that the core structure is based on 2H-chromene, known as benzopyran, whereas the cyclic pyran ring takes the half boat conformation shape. The bond length of C8-O3 at 1.367 Å is shorter than C14-O3 at 1.465 Å, which forms the fused pyran. A planar geometry is exerted throughout the chemical structure while being extended by the side group in the form of methyl ester. Configuration E allows the structure to take a highly conjugated property, as the 10-atom chain packing of werneria chromene is monoclinic in space group P 1 21/c 1, as shown in Figure 2c,d. The core structure of benzopyran exists in cyclin-dependent kinase (CDK) inhibitors and can be found in drugs such as flavopiridol (also known as alvocidib) [50]. Werneria chromene does not have a chiral center and thus does not have conformers.

3.7. In Silico Studies with Werneria Chromene and Dihydroxyacidimissinol

The reported structure of the SARS-CoV-2 spike protein receptor-binding domain complex with human ACE2 (S-RBD-hACE2) (PDB ID: 6LZG) was used for docking studies with werneria chromene and dihydroxyacidimissinol (2D interactions shown in Figure 3a,b, respectively). For cathepsin L, the reported structure PDB ID: 3HHA was used. PDB ID: 6ZSL was used for NSP13 helicase; PDB ID: 6LU7 was used for Mpro; and PDB ID: 6M0J was used for the spike protein receptor-binding domain (S-RBD). The binding energies of the two compounds (ΔG = kcal/mol) to the target proteins are shown in Table 6.

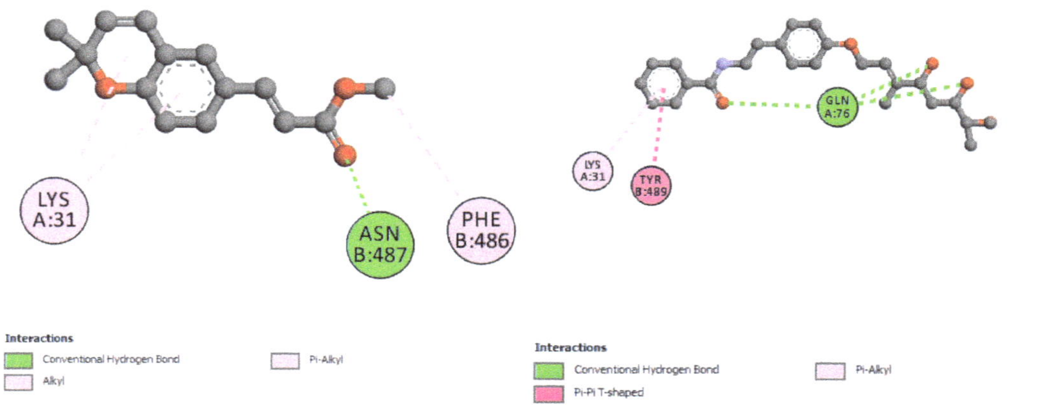

(**a**) (**b**)

Figure 3. 2D-interactions between (**a**) werneria chromene and S-RBD-hACE2; (**b**) dihydroxy acidissiminol and S-RBD-hACE2.

Table 6. Predicted binding energies (ΔG = kcal/mol) of werneria chromene and dihydroxyacidissiminol with various SARS-CoV-2 and human target proteins.

Phytochemical	Spike Protein RBD Bound with ACE2 PDB: 6LZG	Cathepsin L PDB: 3HHA	Nsp13 Helicase PDB: 6ZSL	Mpro PDB: 6LU7	Spike Protein RBD PDB: 6M0J *
Dihydroxyacidissiminol	−5.8	−8.1	−7.6	−7.0	−7.5
Werneria chromene	−6.6	−6.4	−6.4	−5.9	−6.0

* 6M0J is also spike protein RBD bound with ACE2, but unlike 6LZG, ACE2 was removed from spike protein RBD and molecular docking studies conducted with spike protein RBD only.

Werneria chromene did not display a good binding affinity to any of the five target proteins. The least predicted binding energy was observed with 6LZG (S-RBD-hACE2) and showed the predicted binding energy (ΔG) of −6.6 kcal/mol. On the other hand, dihydroxyacidissiminol showed the predicted binding energies of −8.1, −7.6, −7.0, and −7.5 kcal/mol with cathepsin L, nsp13 helicase, Mpro, and S-RBD, respectively (Tables 6 and 7). Cathepsin L (PDB ID: 3HHA) with 220 amino acid residues has a two-chain form, R and L. The L domain contains 3 α-helices, while the R domain is a β-barrel closed at the bottom by a α-helix. The reactive site comprises His163 located at the top of the β-barrel and Cys25, which is located at the N-terminus of the central helix in the L domain [51]. The interacting amino acid residues of cathepsin L, forming hydrogen, hydrophobic, or electrostatic bonds with dihydroxyacidissiminol, include Gly23, Cys25, Ser24, Trp26, Met70, Ala135, His163, Gly164, and Trp189. The involvement of dihydroxyacidissiminol in interacting with both reactive site amino acids Cys25 and His163 makes this compound a potential potent inhibitor for cathepsin L. The 2D interactions of cathepsin L with werneria chromene and dihydroxyacidissiminol are shown in Figure 4a,b, respectively. PDB ID: 6ZSL represents the Nsp13 helicase of SARS-CoV-2. The structure of Nsp13 of SARS-CoV-2, like that of SARS-CoV, shows five domains, namely, RecA-like domains 1A and 2A, the 2B domain, the zinc-binding domain (ZBD), and the stalk domain.

Table 7. Interaction of dihydroxyacidissiminol with amino acid residues of cathepsin L, Nsp13 helicase, and spike protein receptor-binding domain (S-RBD).

Residue (Cathepsin L)	Distance	Category	Type
		Dihydroxyacidissiminol	
TRP26	2.68	Hydrogen Bond	Conventional Hydrogen Bond
GLY164	2.06	Hydrogen Bond	Conventional Hydrogen Bond
TRP189	2.10	Hydrogen Bond	Conventional Hydrogen Bond
GLY23	3.60	Hydrogen Bond	Carbon Hydrogen Bond
HIS163	3.64	Electrostatic	Pi-Cation
TRP189	3.67	Hydrophobic	Pi-Sigma
TRP189	3.99	Hydrophobic	Pi-Sigma
CYS25	4.81	Other	Pi-Sulfur
MET70	5.04	Other	Pi-Sulfur
GLY23, SER24	4.38	Hydrophobic	Amide-Pi Stacked
ALA135	4.15	Hydrophobic	Pi-Alkyl
Residue (Nsp13 helicase)	**Distance**	**Category**	**Type**
		Dihydroxyacidissiminol	
PRO514	2.29	Hydrogen Bond	Conventional Hydrogen Bond
TYR515	4.64	Hydrophobic	Pi-Alkyl
HIS554	4.06	Hydrophobic	Pi-Alkyl
PRO406	4.13	Hydrophobic	Pi-Alkyl
Residue (S-RBD)	**Distance**	**Category**	**Type**
		Dihydroxyacidissiminol	
ASP364	1.89	Hydrogen Bond	Conventional Hydrogen Bond
B:TRP436	2.29	Hydrogen Bond	Conventional Hydrogen Bond
CYS336	2.38	Hydrogen Bond	Conventional Hydrogen Bond
PHE342	2.70	Hydrogen Bond	Conventional Hydrogen Bond
ASN343	1.94	Hydrogen Bond	Conventional Hydrogen Bond
LEU441	3.58	Hydrophobic	Pi-Sigma
TRP436	3.98	Hydrophobic	Pi-Sigma
PHE374	4.75	Hydrophobic	Pi-Pi T-shaped
VAL367	4.58	Hydrophobic	Alkyl

The key amino acid residues of the ATP-binding site are six in number and are Lys288, Ser289, Asp374, Glu375, Gln404, and Arg567 [52]. Dihydroxyacidissiminol showed the predicted binding energy of −7.6 kcal/mol with Nsp13 helicase. The interacting amino acids of Nsp13 helicase, forming hydrophobic and hydrogen bonds with dihydroxyacidissiminol, were found to be Pro406, Pro514, Tyr515, and His554. Interestingly, none of the interacting amino acid residues of Nsp13 helicase with dihydroxyacidissiminol were from the ATP-binding site. Further studies are therefore needed to determine whether the binding of dihydroxyacidissiminol to Nsp13 helicase will lead to inhibition of helicase activities or not. What is noteworthy is that dihydroxyacidissiminol interacts with the C-terminus domain of the Nsp13 helicase, which is necessary for its helicase activities. The 2D interactions of Nsp13 helicase with werneria chromene and dihydroxyacidissiminol are shown in Figure 5a,b, respectively. The spike protein (S) of SARS-CoV-2 comprises two subunits S1 and S2, and the first subunit is responsible for binding to its receptor hACE2.

Human ACE2 has two hotspots for the receptor-binding domain (RBD) of S, hotspot 31 and hotspot 353. SARS-CoV-2 recognizes hACE2 hotspot 31 through two amino acid residues on its RBD, Gln493 and Leu 455.

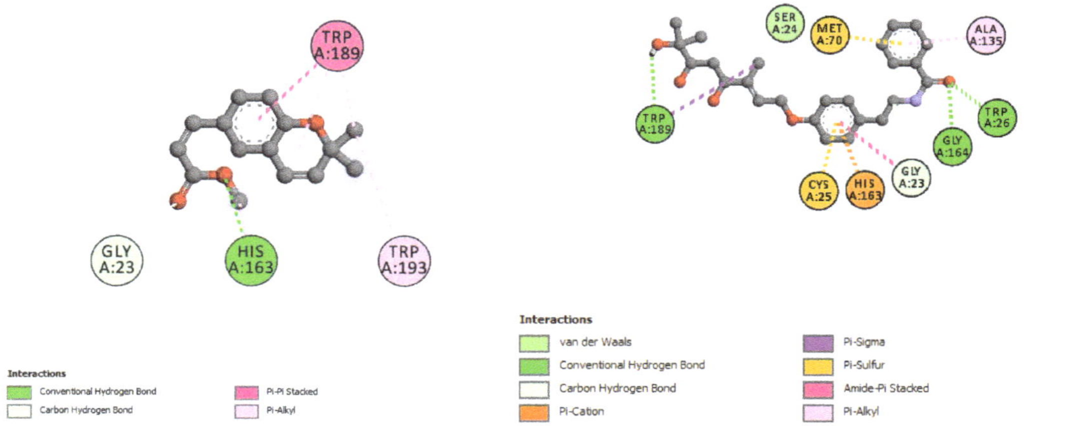

(**a**) (**b**)

Figure 4. (**a**) 2D-interaction between werneria chromene and cathepsin L and S-RBD-hACE2; (**b**) 2D-interaction between dihydroxyacidissiminol and cathepsin L.

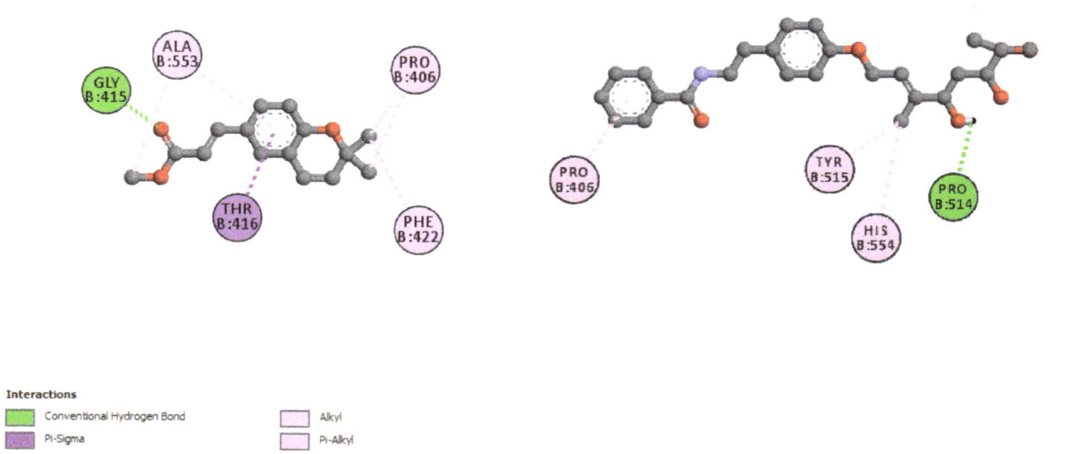

(**a**) (**b**)

Figure 5. (**a**) 2D-interaction between werneria chromene and Nsp13 helicase; (**b**) 2D-interaction between dihydroxyacidissiminol and Nsp13 helicase.

Other amino acid residues playing significant roles in RBD interactions with hotspots 31 and 353 include Phe486, and Ser494 [53]. Dihydroxyacidissiminol interacts through hydrogen and hydrophobic bonding with amino acid residues Cys336, Phe342, Asn343, Asp364, Val367, Phe374, Trp436, and Leu441 of S-RBD. The S-RBD of SARS-CoV-2 comprises amino acid residues 387-516 [54]. Apart from the last two amino acids of S-RBD, dihydroxyacidissiminol shows interactions with other amino acid residues outside S-RBD. The conclusion formed is that despite showing a high binding affinity for S-RBD, dihydroxyacidissiminol possibly will have little or no inhibitory influences on S-RBD binding to hACE2. Both phytochemicals werneria chromene and dihyroxyacidissiminol did not show predicted good binding energies to Mpro. The 2D interactions of werneria chromene and dihydroxyacidissiminol with Mpro are shown in Figure 6a,b, respectively; the 2D interactions of werneria chromene and dihydroxyacidissiminol with S-RBD are shown, respectively, in Figure 7a,b. Taken cumulatively, dihydroxyacidissiminol shows predicted low-binding energies for cathepsin L. Since it interacts with both reactive site amino acids Cys25 and His163, it makes this compound a potential potent inhibitor for cathepsin L. The protease (cathepsin L or CTSL) plays a major role in SARS-CoV-2 infectivity. The circulating level of CTSL increases after SARS-CoV-2 infection and is positively correlated with disease course and severity. Scientists have postulated that the enzyme can make a good therapeutic target [36].

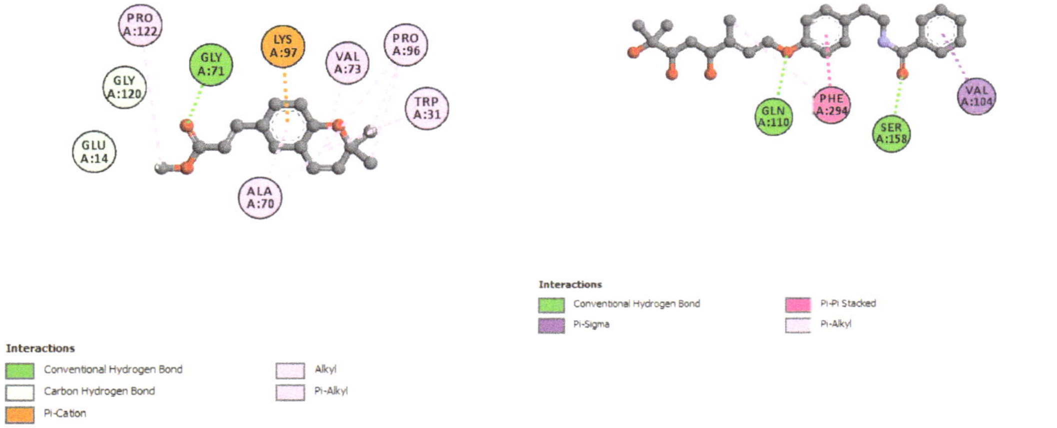

(a) (b)

Figure 6. (**a**) 2D-interaction between werneria chromene and Mpro; (**b**) 2D-interaction between dihydroxy acidissiminol and Mpro.

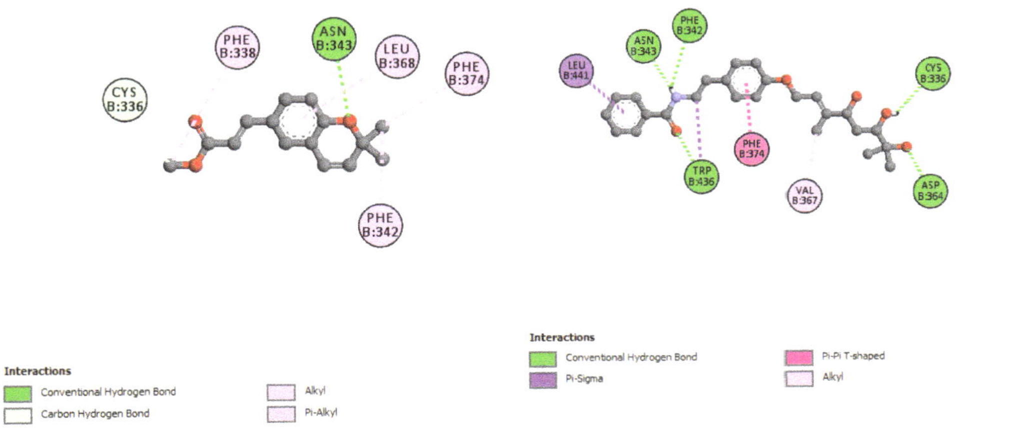

(a) (b)

Figure 7. (**a**) 2D-interaction between werneria chromene and S-RBD; (**b**) 2D-interaction between dihydroxyacidissiminol and S-RBD.

4. Discussion

COVID-19 caused by SARS-CoV-2 is the first but possibly not the last zoonotic virus that can paralyze global human activities. This pandemic reminded researchers to be vigilantly prepared to protect human lives from any potential viral or bacterial infections that may lead to cause a pandemic-like situation. Various phytochemicals from rainforests can be investigated in search for potential drugs to prevent these infections, although these are yet to validate the speculation. However, human activities, such as the increased need of food for the global growing population and industrialization, lead to deforestation. The Malaysian primary rainforest may disappear in the face of intense logging and palm-oil plantations, which has claimed around 1.1 million hectares of rainforest between 1990 and 2005 [55]. It must be grasped that the disappearance of primary rainforest trees signifies the disappearance of potential drugs [56]. *B. malaccensis* was collected in the primary rainforest (one of the few remaining pockets of the primary rainforest) of Manong, located in the north of Peninsular Malaysia, on the banks of the Perak River. This plant belongs to the subfamily Aurantioideae and the tribe Citrinae [57]. We selected that plant because (i) it was having fruits and flowers, allowing botanical identificationl (ii) it belongs to the family Rutaceae, which is a rich source of antimicrobial compounds; and (iii) and it is traditionally used by the Malays of Perak as medicine. Members of the Rutaceae family have been reported to have antibacterial, antibiotic-potentiating, and anti-viral properties [29–32]. Since there has been no study thus far to investigate the antimicrobial activities of *B. malaccensis*, its antibacterial and synergistic antibacterial, as well as its potent anti-viral properties against SARS-CoV-2 were explored. Some plants from the Rutaceae family, including *Citrus sinensis* (L.) Osbeck, have been used for the treatment of flu in traditional medicine and have antimicrobial effects [33]. *Citrus limon* (L.) Osbeck, *C. sinensis*, and *Citrus pardisi* Macfad. from Rutaceae reportedly showed significant activity against Hepatitis A virus (HAV) [28]. It is important to note that naringenin, a flavanone almost exclusively found in members of the genus *Citrus* L. fruits, has been reported to be a potent inhibitor of SARS-CoV-2 [34]. The average yield values ranged from 1.0 to 9.5%, indicating a fair extraction process [58,59].

First, the antibacterial activities of 18 extracts were assessed by a microbroth dilution assay against a panel of Gram-positive and Gram-negative bacteria and most of these extracts displayed levels of inhibition, especially the chloroform extract of leaves and methanol extract of bark, as these had the highest potencies; this result is reported for the first time. The β-lactam antibiotic imipenem is one of the last-resort treatments for *A. baumannii* infections [60]. Outbreaks of nosocomial infection caused by *A. baumannii* resistant to imipenem have reached the proportions of a global health emergency [52]. According to the International Network for the Study and Emergency Prevention of Antimicrobial Resistance [61], multi-resistant *A. baumannii* infection is a "sentinel event that warrants a coordinated response to control this multi-resistant pathogen" [62]. Gram-positive bacteria were more sensitive to the extracts than Gram-negative bacteria, in line with the literature [63]. Gram-positive bacteria are also more susceptible to xenobiotics since they only have a peptidoglycan wall, which is not an effective permeability barrier compared to Gram-negative bacteria, and are equipped with an outer lipopolysaccharide layer, porins, and an arsenal of efflux pumps. *P. aeruginosa* resists most antibiotics via β-lactamases, efflux pumps, loss, or alteration of the outer membrane porin [64]. *P. aeruginosa* is also resistant to fluoroquinolone exposure through mutations in their DNA gyrase and topoisomerase IV enzymes, as well as in efflux pumps [65].

The MIC of *B. malaccensis* extracts was determined by the microdilution broth assay. Rios and Recio suggested that a crude extract with MIC greater than 1000 μg/mL is inactive and proposed interesting antibacterial activity for MICs of 100 μg/mL or lower [66]. Earlier, Fabry and colleagues have defined crude active extracts as having MIC values below 8000 μg/mL [67]. While, more recently, Kuete used stricter endpoint criteria, in which crude extracts with MIC values less than 100 μg/mL are considered active [68]. Further, Kuete classified MICs above 625 μg/mL as weakly active extracts [68]. Following Kuete (2010) reports, it can be said that chloroform extract from *B. malaccensis* leaves exhibited mild antibacterial effects against the two Gram-positive bacteria tested (*S. aureus* and *B. Subtilis*) [69]. Moreover, extracts of antibacterial compounds can be categorized into two classes: bacteriostatic (MBC/MIC ratio greater than 4) and bactericidal (MBC/MIC ratio less than or equal to 4), according to Krishnan et al. [27]. Following this classification, the chloroform extract of leaves was bacteriostatic, while the methanol extract of bark was mildly bacteriostatic against *S. aureus* and *P. aeruginosa*. None of the plant extracts were active against *A. baumannii* (imipenem-resistant).

Regarding the antibiotic-potentiating activity of the *B. malaccensis* extracts tested, the strongest potentiation for Gram-positive bacteria was observed with the methanol extract from wood (0 mm) with amoxicillin (16 ± 0.0 mm) against *S. aureus* (23.7 ± 0.5 mm). Of note, it was observed that the chloroform extract of endocarps was able to render penicillin G active against *E. coli* and potentiated imipenem activity against imipenem-resistant *A. baumannii*. The hexane extract of *B. malaccensis* seeds also potentiated the antibiotic activities of ciprofloxacin and levofloxacin against the multidrug-resistant bacteria *P. aeruginosa*. These antibiotic-potentiating properties are reported for the first time. The bacterial strains we tested are, except for imipenem-resistant *A. baumannii*, antibiotic sensitive, and the extract tested further increased their vulnerability to antibiotics.

In this study, the strongest potentiation was observed with the methanol extract of wood with the β-lactam antibiotic amoxicillin towards *S. aureus*. It is an important finding because amoxicillin resistance represents a severe clinical burden worldwide in hospitals [70]. *S. aureus* resists amoxicillin via β-lactamases and changes in penicillin-binding protein 2a [71]. The low standard deviation obtained indicate that data are close to the mean. The hexane extract of the seeds potentiated ciprofloxacin and levofloxacin against *P. aeruginosa*. The nosocomial bacterium *A. baumannii* resists imipenem via intrinsic and acquired metallo-β-lactamases and oxacillinases, as well as porin loss [43]. In China, for instance, more than 50% of isolates were found to be imipenem-resistant in 2009, and in Thailand, the rate of resistance to imipenem increased significantly from 2% in 2000 to 67% in 2011 [72]. *E. coli* is known to resist penicillin G via penicillin acylase and β-lactamases [73], as well as efflux pumps [74].

Cell line cytotoxicity (at concentrations of 0-200 μg/mL of extracts against human MRC-5 cells) was assessed following the confirmed antibacterial activities of some *B. malaccensis* extracts. The American National Cancer Institute defines a plant extract as toxic to human cells when the CC_{50} values are below 30 μg/mL after an exposure time of 72 h [75]. Accordingly, none of the extracts studied were toxic in vitro for the cell line tested in this study. Among all tested extracts, the chloroform extract of *B. malaccensis* demonstrated the lowest cytotoxicity. Additionally, the identification of the main constituents of the hexane extract of seeds and the chloroform extract of fruits was carried out, which elicited interesting antibiotic-potentiating effects against problematic Gram-negative bacteria, resulting in the isolation and characterization of werneria chromene and dihydroxyacidissiminol.

Since werneria chromene was selectively active against *P. aeruginosa*, its activity against *P. putida* and observed and activity was observed. Likewise, hydroxyacidissimol was specifically active against *P. putida*. These activities, although weak, are specific and are reported for the first time, although the reasons for this specificity against *Pseudomonas* spp. remain unknown. We did not attempt to isolate compounds from methanol extracts as we looked for mid-polar to non-polar compounds that may have better ADME.

The occurrence of these constituents in *B. malaccensis* was not known previously. Both compounds were weakly but specifically bacteriostatic against *P. putida* and inactive for all other bacteria tested. Further, the crystal structure of werneria chromene is reported here for the first time. However, their synergistic activities were not examined due to insufficient amounts of available extracts, which requires further investigation. We are currently examining minor constituents in these extracts and found a series of prenylated flavonols (unpublished data) that may work synergistically to bring about antibiotic-potentiating effects and we are looking into this matter.

In silico molecular docking of werneria chromene and dihydroxyacidissimol was examined with the reported structure of the SARS-CoV-2 spike protein receptor-binding domain complexed with human ACE2 (S-RBD-hACE2) (PDB ID: 6LZG), human cathepsin L (PDB ID: 3HHA); PDB ID: 6ZSL was used for NSP13 helicase; PDB ID: 6LU7 was used for Mpro; and PDB ID: 6M0J was used for the spike protein receptor-binding domain (S-RBD) [50,76–79]. The 6M0J is also a spike protein RBD bound with ACE2, but unlike 6LZG ACE2 was removed from the spike protein RBD and molecular docking studies conducted with the spike protein RBD only. Dihydroxyacidissiminol showed a good affinity for essential target proteins (i.e., its binding energy values were −8.1, −7.6, and −7.5 kcal/mol for cathepsin L, nsp13 helicase, and spike protein receptor-binding domain, respectively) associated with SARS-CoV-2 entry and replication in human cells, such as spike protein receptor-binding domain (S-RBD), cathepsin L, and Nsp13 helicase.

5. Conclusions

The decrease in the development of new and effective antibiotics by pharmaceutical industries and the concomitant and steady increase in bacterial resistance leaves clinicians with the increasing difficulty to save the life of patients infected by nosocomial bacteria globally. As the world is going through a pandemic caused by SARS-CoV-2, a part of the mortality is derived from SARS-CoV-2 infection associated with bacterial infections. The development of resistance-modifying agents can be an additional strategy to overcome bacteria multidrug resistance. To our knowledge, the antimicrobial effect potentiating properties of *B. malaccensis* have not been previously reported. Most of the extracts demonstrated inhibiting the growth of Gram-positive bacteria in particular. Two compounds, werneria chromene and dihydroxylacidissiminol, isolated from this plant's extracts, inhibited the growth of *P. putida*, and dihydroxyacidissiminol demonstrated a good affinity for cathepsin L. The amount of available plant was a limitation to our study, and we plan to obtain larger collections. The principle involved in antibiotic-potentiation, the in vitro activity of dihydroxyacidissiminol against SARS-CoV-2 and other coronaviruses, and the therapeutic potential of the compound isolated as specific inhibitors of *Pseudomonas* spp. need to be examined.

Author Contributions: Conceptualization, M.Z., V.N., C.W., P.W. and M.R. (Mohammed Rahmatullah); methodology, M.Z., N.M., A.F., S.L.W.-L., T.-J.K., T.M., M.R. (Mogana Rajagopal), C.S., G.K., N.H.A., A.R.P., A.H., A.K.P., M.R. (Mohammed Rahmatullah), V.N. and C.W.; validation, S.M.R.O., M.d.L.P., M.R. (Mohammed Rahmatullah), P.W., A.K.P., C.W. and V.N.; investigation, M.Z., N.M., A.F., S.L.W.-L., T.-J.K., T.M., M.R. (Mogana Rajagopal), C.S., G.K., N.H.A., A.R.P., A.H., A.K.P., M.S.B., M.N. and C.W.; writing—original draft preparation, M.Z., N.M., A.F., S.L.W.-L., T.-J.K., T.M., M.R. (Mohammed Rahmatullah), C.S., G.K., N.H.A., A.R.P., A.K.P., S.M.R.O., V.N. and C.W.; writing—review and editing, S.M.R.O., M.d.L.P., M.R. (Mogana Rajagopal), A.H., A.K.P., M.S.B., M.N., P.W., V.N. and C.W.; funding, V.N., P.W. and C.W.; supervision, C.W., V.N., P.W. and M.R. (Mohammed Rahmatullah). All authors have read and agreed to the published version of the manuscript.

Funding: This project was funded by a grant from the Malaysian Ministry of Education (FRGS/1/2018/WAB07/UNIM/02/1, Malaysian Ministry of Higher Education, Malaysia). Project CICECO-Aveiro Institute of Materials, UIDB/50011/2020, UIDP/50011/2020 & LA/P/0006/2020, financed by national funds through the FCT/MEC (PIDDAC) is acknowledged. We also would like to acknowledge the project entitled "Medicinal plants as anti-viral activity against important emerging viruses" The Plant Genetics Conservation Project under the Royal Initiation of Her Royal Highness Princess Maha Chakri Sirindhorn, Walailak University, Thailand (Grant No. 032-2565). Dr. Muhammad Nawaz is thankful to the Deanship of Scientific Research (DSR) at Imam Abdulrahman Bin Faisal University (IAU) for financial support through project no. COVID-19-2020-003-IRMC. In addition, we would like to thank Mr. Hassan and Mr. Pidji for guiding us in the rainforest of Perak.

Conflicts of Interest: The authors declare no conflict of interest.

References

1. World Health Organization. WHO Publishes List of Bacteria for Which New Antibiotics Are Urgently Needed. 2017. Available online: https://www.who.int/news/item/27-02-2017-who-publishes-list-of-bacteria-for-which-new-antibiotics-are-urgently-needed (accessed on 1 January 2022).
2. Rice, L.B. Federal funding for the study of antimicrobial resistance in nosocomial pathogens: No eskape. *J. Infect. Dis.* **2008**, *197*, 1079–1081. [CrossRef] [PubMed]
3. De Oliveira, D.M.P.; Forde, B.M.; Kidd, T.J.; Harris, P.N.A.; Schembri, M.A.; Beatson, S.A.; Paterson, D.L.; Walker, M.J. Antimicrobial resistance in eskape pathogens. *Clin. Microbiol. Rev.* **2020**, *33*, e00181-19. [CrossRef] [PubMed]
4. Herc, E.S.; Kauffman, C.A.; Marini, B.L.; Perissinotti, A.J.; Miceli, M.H. Daptomycin nonsusceptible vancomycin resistant *Enterococcus* bloodstream infections in patients with hematological malignancies: Risk factors and outcomes. *Leuk. Lymphoma* **2017**, *58*, 2852–2858. [CrossRef] [PubMed]
5. Iguchi, S.; Mizutani, T.; Hiramatsu, K.; Kikuchi, K. Rapid acquisition of linezolid resistance in methicillin-resistant *Staphylococcus aureus*: Role of hypermutation and homologous recombination. *PLoS ONE* **2016**, *11*, e0155512. [CrossRef] [PubMed]
6. Naylor, N.R.; Atun, R.; Zhu, N.; Kulasabanathan, K.K.; Silva, S.; Chatterjee, A.; Knight, G.M.; Robotham, J.V. Estimating the burden of antimicrobial resistance: A systematic literature review. *Antimicrob. Resist. Infect. Control* **2018**, *7*, 58. [CrossRef]
7. Wongsrichanalai, C.; Pickard, A.L.; Wernsdorfer, W.H.; Meshnick, S.R. Epidemiology of drug-resistant malaria. *Lancet Infect. Dis.* **2002**, *2*, 209–218. [CrossRef]
8. Cobo, F. Determinants of parasite drug resistance in human lymphatic filariasis. *Rev. Esp. Quimioter.* **2016**, *29*, 288–295.
9. Li, Q.; Guan, X.; Wu, P.; Wang, X.; Zhou, L.; Tong, Y.; Ren, R.; Leung, K.S.M.; Lau, E.H.Y.; Wong, J.Y.; et al. Early transmission dynamics in Wuhan, China, of novel Coronavirus-infected pneumonia. *N. Engl. J. Med.* **2020**, *382*, 1199–1207. [CrossRef]
10. Shereen, M.A.; Khan, S.; Kazmi, A.; Bashir, N.; Siddique, R. COVID-19 infection: Origin, transmission, and characteristics of human Coronaviruses. *J. Adv. Res.* **2020**, *24*, 91–98. [CrossRef]
11. Tay, M.Z.; Poh, C.M.; Rénia, L.; MacAry, P.A.; Ng, L.F.P. The trinity of COVID-19: Immunity, inflammation and intervention. *Nat. Rev. Immunol.* **2020**, *20*, 363–374. [CrossRef]
12. Mousavizadeh, L.; Ghasemi, S. Genotype and phenotype of COVID-19: Their roles in pathogenesis. *J. Microbiol. Immunol. Infect.* **2021**, *54*, 159–163. [CrossRef] [PubMed]
13. Jannat, K.; Paul, A.K.; Bondhon, T.A.; Hasan, A.; Nawaz, M.; Jahan, R.; Mahboob, T.; Nissapatorn, V.; Wilairatana, P.; Pereira, M.L.; et al. Nanotechnology applications of flavonoids for viral diseases. *Pharmaceutics* **2021**, *13*, 1895. [CrossRef] [PubMed]
14. Fehr, A.R.; Perlman, S. Coronaviruses: An overview of their replication and pathogenesis. *Methods Mol. Biol.* **2015**, *1282*, 1–23. [CrossRef] [PubMed]
15. Huang, I.C.; Bosch, B.J.; Li, F.; Li, W.; Lee, K.H.; Ghiran, S.; Vasilieva, N.; Dermody, T.S.; Harrison, S.C.; Dormitzer, P.R.; et al. SARS Coronavirus, but not human Coronavirus nl63, utilizes cathepsin l to infect ACE2-expressing cells. *J. Biol. Chem.* **2006**, *281*, 3198–3203. [CrossRef] [PubMed]
16. Astuti, I.; Ysrafil. Severe Acute Respiratory Syndrome Coronavirus 2 (SARS-CoV-2): An overview of viral structure and host response. *Diabetes Metab. Syndr.* **2020**, *14*, 407–412. [CrossRef]
17. Hendaus, M.A.; Jomha, F.A. COVID-19 induced superimposed bacterial infection. *J. Biomol. Struct. Dyn.* **2021**, *39*, 4185–4191. [CrossRef]
18. Hendaus, M.A.; Jomha, F.A.; Alhammadi, A.H. Virus-induced secondary bacterial infection: A concise review. *Ther. Clin. Risk Manag.* **2015**, *11*, 1265–1271. [CrossRef]
19. Rusic, D.; Vilovic, M.; Bukic, J.; Leskur, D.; Perisin, A.S.; Kumric, M.; Martinovic, D.; Petric, A.; Modun, D.; Bozic, J. Implications of COVID-19 pandemic on the emergence of antimicrobial resistance: Adjusting the response to future outbreaks. *Life* **2021**, *11*, 220. [CrossRef]
20. Rates, S.M. Plants as source of drugs. *Toxicon* **2001**, *39*, 603–613. [CrossRef]
21. Veeresham, C. Natural products derived from plants as a source of drugs. *J. Adv. Pharm. Technol. Res.* **2012**, *3*, 200–201. [CrossRef]
22. Bennet, L. *Deforestation and Climate Change*; A Publication of the Climate Institute: Washington, DC, USA, 2017.
23. Faridah-Hanum, I.; Philip, L.; Noor, A.A. Sampling species diversity in a Malaysian rain forest: The case of a logged-over forest. *Pak. J. Bot.* **2008**, *40*, 1729–1733.
24. Koh, L.P.; Wilcove, D.S. Is oil palm agriculture really destroying tropical biodiversity? *Conserv. Lett.* **2008**, *1*, 60–64. [CrossRef]
25. Bryan, J.E.; Shearman, P.L.; Asner, G.P.; Knapp, D.E.; Aoro, G.; Lokes, B. Extreme differences in forest degradation in Borneo: Comparing practices in Sarawak, Sabah, and Brunei. *PLoS ONE* **2013**, *8*, e69679. [CrossRef] [PubMed]
26. Cushman, S.A.; Macdonald, E.A.; Landguth, E.L.; Malhi, Y.; Macdonald, D.W. Multiple-scale prediction of forest loss risk across borneo. *Landsc. Ecol.* **2017**, *32*, 1581–1598. [CrossRef]
27. Krishnan, N.; Ramanathan, S.; Sasidharan, S.; Murugaiyah, V.; Mansor, S. Antimicrobial activity evaluation of *Cassia spectabilis* leaf extracts. *Int. J. Pharmacol.* **2010**, *6*, 510–514. [CrossRef]
28. Salem, M.A.; Ezzat, S.M. The use of aromatic plants and their therapeutic potential as antiviral agents: A hope for finding anti-COVID-19 essential oils. *J. Essent. Oil Res.* **2021**, *33*, 105–113. [CrossRef]
29. Al-Majmaie, S.; Nahar, L.; Rahman, M.M.; Nath, S.; Saha, P.; Talukdar, A.D.; Sharples, G.P.; Sarker, S.D. Anti-MRSA constituents from *Ruta chalepensis* (rutaceae) grown in Iraq, and in silico studies on two of most active compounds, chalepensin and 6-hydroxy-rutin 3′,7-dimethyl ether. *Molecules* **2021**, *26*, 1114. [CrossRef]

30. Fratianni, F.; Cozzolino, A.; De Feo, V.; Coppola, R.; Ombra, M.N.; Nazzaro, F. Polyphenols, antioxidant, antibacterial, and biofilm inhibitory activities of peel and pulp of *Citrus medica* L., *Citrus bergamia*, and *Citrus medica* cv. Salò cultivated in southern italy. *Molecules* **2019**, *24*, 4577. [CrossRef]
31. Bohlmann, F.; Zdero, C.; King, R.M.; Robinson, H. Prenylated p-coumarates from *Werneria stuebelii*. *Phytochemistry* **1984**, *23*, 1135–1137. [CrossRef]
32. Santhi, V.P.; Masilamani, P.; Sriramavaratharajan, V.; Murugan, R.; Gurav, S.S.; Sarasu, V.P.; Parthiban, S.; Ayyanar, M. Therapeutic potential of phytoconstituents of edible fruits in combating emerging viral infections. *J. Food Biochem.* **2021**, *45*, e13851. [CrossRef]
33. Ulasli, M.; Gurses, S.A.; Bayraktar, R.; Yumrutas, O.; Oztuzcu, S.; Igci, M.; Igci, Y.Z.; Cakmak, E.A.; Arslan, A. The effects of *Nigella sativa (ns), Anthemis hyalina (ah)* and *Citrus sinensis (cs)* extracts on the replication of Coronavirus and the expression of trp genes family. *Mol. Biol. Rep.* **2014**, *41*, 1703–1711. [CrossRef] [PubMed]
34. Clementi, N.; Scagnolari, C.; D'Amore, A.; Palombi, F.; Criscuolo, E.; Frasca, F.; Pierangeli, A.; Mancini, N.; Antonelli, G.; Clementi, M.; et al. Naringenin is a powerful inhibitor of SARS-CoV-2 infection in vitro. *Pharmacol. Res.* **2021**, *163*, 105255. [CrossRef] [PubMed]
35. Mukhopadhyay, S.; Prasad, A.S.B.; Mehta, C.H.; Nayak, U.Y. Antimicrobial peptide polymers: No escape to eskape pathogens-A review. *World J. Microbiol. Biotechnol.* **2020**, *36*, 131. [CrossRef] [PubMed]
36. Zhao, M.-M.; Yang, W.-L.; Yang, F.-Y.; Zhang, L.; Huang, W.-J.; Hou, W.; Fan, C.-F.; Jin, R.-H.; Feng, Y.-M.; Wang, Y.-C. Cathepsin l plays a key role in SARS-CoV-2 infection in humans and humanized mice and is a promising target for new drug development. *Signal Trans. Target. Ther.* **2021**, *6*, 134. [CrossRef] [PubMed]
37. Shu, T.; Huang, M.; Wu, D.; Ren, Y.; Zhang, X.; Han, Y.; Mu, J.; Wang, R.; Qiu, Y.; Zhang, D.Y.; et al. SARS-Coronavirus-2 nsp13 possesses NTPase and RNA helicase activities that can be inhibited by bismuth salts. *Virol. Sin.* **2020**, *35*, 321–329. [CrossRef] [PubMed]
38. Darwish, A.M.; Farmer, B.D.; Hawke, J.P. Improved method for determining antibiotic susceptibility of *Flavobacterium columnare* isolates by broth microdilution. *J. Aquat. Anim. Health* **2008**, *20*, 185–191. [CrossRef] [PubMed]
39. Boonyanugomol, W.; Kraisriwattana, K.; Rukseree, K.; Boonsam, K.; Narachai, P. In vitro synergistic antibacterial activity of the essential oil from *Zingiber cassumunar* Roxb against extensively drug-resistant *Acinetobacter baumannii* strains. *J. Infect. Public Health* **2017**, *10*, 586–592. [CrossRef]
40. Mosmann, T. Rapid colorimetric assay for cellular growth and survival: Application to proliferation and cytotoxicity assays. *J. Immunol. Methods* **1983**, *65*, 55–63. [CrossRef]
41. Ghosh, P.; Ghosh, M.K.; Thakur, S.; Akihisa, T.; Tamura, T.; Kimura, Y. Dihydroxy acidissiminol and acidissiminol epoxide, two tyramine derivatives from *Limonia acidissima*. *Phytochemistry* **1994**, *37*, 757–760. [CrossRef]
42. Ghosh, P.; Sil, P.; Das, S.; Thakur, S.; Kokke, W.; Akihisa, T.; Shimizu, N.; Tamura, T.; Matsumoto, T. Tyramine derivatives from the fruit of *Limonia acidissima*. *J. Nat. Prod.* **1991**, *54*, 1389–1393. [CrossRef]
43. Liu, X.; Zhang, B.; Jin, Z.; Yang, H.; Rao, Z. The crystal structure of COVID-19 main protease in complex with an inhibitor n3. *Nature* **2020**, *582*, 289–293.
44. Yuan, S.; Chan, H.S.; Hu, Z. Using pymol as a platform for computational drug design. *Wiley Interdiscip. Rev. Comput. Mol. Sci.* **2017**, *7*, e1298. [CrossRef]
45. Asaad, N.; Bethel, P.A.; Coulson, M.D.; Dawson, J.E.; Ford, S.J.; Gerhardt, S.; Grist, M.; Hamlin, G.A.; James, M.J.; Jones, E.V.; et al. Dipeptidyl nitrile inhibitors of cathepsin L. *Bioorg. Med. Chem. Lett.* **2009**, *19*, 4280–4283. [CrossRef] [PubMed]
46. RCSB pdb-6ZSL: Crystal Structure of the SARS-CoV-2 Helicase at 1.94 Angstrom Resolution. 2021. Available online: https://www.rcsb.org/structure/6ZSL (accessed on 1 January 2022).
47. Hähnke, V.D.; Kim, S.; Bolton, E.E. Pubchem chemical structure standardization. *J. Cheminform.* **2018**, *10*, 36. [CrossRef]
48. Trott, O.; Olson, A.J. Autodock vina: Improving the speed and accuracy of docking with a new scoring function, efficient optimization, and multithreading. *J. Comput. Chem.* **2010**, *31*, 455–461. [CrossRef]
49. Studio, D. *Dassault Systemes Biovia, Discovery Studio Modelling Environment, version 4.5*; Accelrys Softw Inc.: San Diego, CA, USA, 2015; pp. 98–104.
50. Baby, K.; Maity, S.; Mehta, C.H.; Suresh, A.; Nayak, U.Y.; Nayak, Y. SARS-CoV-2 entry inhibitors by dual targeting tmprss2 and ace2: An in silico drug repurposing study. *Eur. J. Pharmacol.* **2021**, *896*, 173922. [CrossRef]
51. Madadlou, A. Food proteins are a potential resource for mining cathepsin l inhibitory drugs to combat SARS-CoV-2. *Eur. J. Pharmacol.* **2020**, *885*, 173499. [CrossRef]
52. Vivek-Ananth, R.P.; Krishnaswamy, S.; Samal, A. Potential phytochemical inhibitors of SARS-CoV-2 helicase nsp13: A molecular docking and dynamic simulation study. *Mol. Divers.* **2022**, *26*, 429–442. [CrossRef]
53. Choudhary, S.; Malik, Y.S.; Tomar, S. Identification of SARS-CoV-2 cell entry inhibitors by drug repurposing using in silico structure-based virtual screening approach. *Front. Immunol.* **2020**, *11*, 1664. [CrossRef]
54. Lan, J.; Ge, J.; Yu, J.; Shan, S.; Zhou, H.; Fan, S.; Zhang, Q.; Shi, X.; Wang, Q.; Zhang, L.; et al. Structure of the SARS-CoV-2 spike receptor-binding domain bound to the ACE2 receptor. *Nature* **2020**, *581*, 215–220. [CrossRef]
55. Fitzherbert, E.B.; Struebig, M.J.; Morel, A.; Danielsen, F.; Brühl, C.A.; Donald, P.F.; Phalan, B. How will oil palm expansion affect biodiversity? *Trends Ecol. Evol.* **2008**, *23*, 538–545. [CrossRef] [PubMed]
56. Mishra, J.; Mishra, P.; Arora, N.K. Linkages between environmental issues and zoonotic diseases: With reference to COVID-19 pandemic. *Environ. Sustain.* **2021**, *4*, 455–467. [CrossRef]

57. Bayer, R.J.; Mabberley, D.J.; Morton, C.; Miller, C.H.; Sharma, I.K.; Pfeil, B.E.; Rich, S.; Hitchcock, R.; Sykes, S. A molecular phylogeny of the orange subfamily (Rutaceae: Aurantioideae) using nine cpDNA sequences. *Am. J. Bot.* **2009**, *96*, 668–685. [CrossRef] [PubMed]
58. Harborne, A. *Phytochemical Methods a Guide to Modern Techniques of Plant Analysis*; Springer Science & Business Media: Berlin, Germany, 1998.
59. Parthasarathy, V.; Chempakam, B.; Zachariah, T. *Chemistry of Spices*; CAB International: Wallingford, UK, 2008.
60. Garnacho-Montero, J.; Ortiz-Leyba, C.; Fernández-Hinojosa, E.; Aldabó-Pallás, T.; Cayuela, A.; Marquez-Vácaro, J.A.; Garcia-Curiel, A.; Jiménez-Jiménez, F.J. *Acinetobacter baumannii* ventilator-associated pneumonia: Epidemiological and clinical findings. *Intensiv. Care Med.* **2005**, *31*, 649–655. [CrossRef]
61. Richet, H.M.; Mohammed, J.; McDonald, L.C.; Jarvis, W.R. Building communication networks: International network for the study and prevention of emerging antimicrobial resistance. *Emerg. Infect. Dis.* **2001**, *7*, 319–322. [CrossRef]
62. Arias, M.E.; Gomez, J.D.; Cudmani, N.M.; Vattuone, M.A.; Isla, M.I. Antibacterial activity of ethanolic and aqueous extracts of *Acacia aroma* gill. Ex hook et arn. *Life Sci.* **2004**, *75*, 191–202. [CrossRef]
63. Scherrer, R.; Gerhardt, P. Molecular sieving by the *Bacillus megaterium* cell wall and protoplast. *J. Bacteriol.* **1971**, *107*, 718–735. [CrossRef]
64. Poole, K. *Pseudomonas aeruginosa*: Resistance to the max. *Front. Microbiol.* **2011**, *2*, 65. [CrossRef]
65. Jean, S.S.; Hsueh, P.R. High burden of antimicrobial resistance in asia. *Int. J. Antimicrob. Agents* **2011**, *37*, 291–295. [CrossRef]
66. Ríos, J.L.; Recio, M.C. Medicinal plants and antimicrobial activity. *J. Ethnopharmacol.* **2005**, *100*, 80–84. [CrossRef]
67. Fabry, W.; Okemo, P.O.; Ansorg, R. Antibacterial activity of east African medicinal plants. *J. Ethnopharmacol.* **1998**, *60*, 79–84. [CrossRef]
68. Kuete, V. Potential of Cameroonian plants and derived products against microbial infections: A review. *Planta Med.* **2010**, *76*, 1479–1491. [CrossRef] [PubMed]
69. Kuete, V.; Betrandteponno, R.; Mbaveng, A.T.; Tapondjou, L.A.; Meyer, J.J.; Barboni, L.; Lall, N. Antibacterial activities of the extracts, fractions and compounds from *Dioscorea bulbifera*. *BMC Complement. Altern. Med.* **2012**, *12*, 228. [CrossRef] [PubMed]
70. Pantosti, A.; Sanchini, A.; Monaco, M. Mechanisms of antibiotic resistance in *Staphylococcus aureus*. *Future Microbiol.* **2007**, *2*, 323–334. [CrossRef] [PubMed]
71. Poirel, L.; Nordmann, P. Carbapenem resistance in *Acinetobacter baumannii*: Mechanisms and epidemiology. *Clin. Microbiol. Infect.* **2006**, *12*, 826–836. [CrossRef]
72. Ohashi, H.; Katsuta, Y.; Nagashima, M.; Kamei, T.; Yano, M. Expression of the *Arthrobacter viscosus* penicillin G acylase gene in *Escherichia coli* and *Bacillus subtilis*. *Appl. Environ. Microbiol.* **1989**, *55*, 1351–1356. [CrossRef]
73. Lehtinen, J.; Lilius, E.M. Promethazine renders *Escherichia coli* susceptible to penicillin g: Real-time measurement of bacterial susceptibility by fluoro-luminometry. *Int. J. Antimicrob. Agents* **2007**, *30*, 44–51. [CrossRef]
74. Kuete, V.; Efferth, T. Cameroonian medicinal plants: Pharmacology and derived natural products. *Front. Pharmacol.* **2010**, *1*, 123. [CrossRef]
75. Ogbole, O.O.; Segun, P.A.; Adeniji, A.J. In vitro cytotoxic activity of medicinal plants from Nigeria ethnomedicine on rhabdomyosarcoma cancer cell line and hplc analysis of active extracts. *BMC Complement. Altern. Med.* **2017**, *17*, 494. [CrossRef]
76. Keum, Y.S.; Jeong, Y.J. Development of chemical inhibitors of the SARS Coronavirus: Viral helicase as a potential target. *Biochem. Pharmacol.* **2012**, *84*, 1351–1358. [CrossRef]
77. Simmons, G.; Gosalia, D.N.; Rennekamp, A.J.; Reeves, J.D.; Diamond, S.L.; Bates, P. Inhibitors of cathepsin l prevent Severe Acute Respiratory Syndrome Coronavirus entry. *Proc. Natl. Acad. Sci. USA* **2005**, *102*, 11876–11881. [CrossRef] [PubMed]
78. Benítez-Cardoza, C.G.; Vique-Sánchez, J.L. Potential inhibitors of the interaction between ACE2 and SARS-CoV-2 (RBD), to develop a drug. *Life Sci.* **2020**, *256*, 117970. [CrossRef] [PubMed]
79. Ton, A.T.; Gentile, F.; Hsing, M.; Ban, F.; Cherkasov, A. Rapid identification of potential inhibitors of SARS-CoV-2 main protease by deep docking of 1.3 billion compounds. *Mol. Inform.* **2020**, *39*, e2000028. [CrossRef] [PubMed]

MDPI
St. Alban-Anlage 66
4052 Basel
Switzerland
Tel. +41 61 683 77 34
Fax +41 61 302 89 18
www.mdpi.com

Plants Editorial Office
E-mail: plants@mdpi.com
www.mdpi.com/journal/plants

www.ingramcontent.com/pod-product-compliance
Lightning Source LLC
Chambersburg PA
CBHW061620220926
51850CB00005B/557
9783036565101